Lecture Notes in Computer Science 15983

The series Lecture Notes in Computer Science (LNCS), including its subseries Lecture Notes in Artificial Intelligence (LNAI) and Lecture Notes in Bioinformatics (LNBI), has established itself as a medium for the publication of new developments in computer science and information technology research, teaching, and education.

LNCS enjoys close cooperation with the computer science R & D community, the series counts many renowned academics among its volume editors and paper authors, and collaborates with prestigious societies. Its mission is to serve this international community by providing an invaluable service, mainly focused on the publication of conference and workshop proceedings and postproceedings. LNCS commenced publication in 1973.

Fedor V. Fomin · Mingyu Xiao
Editors

Computing and Combinatorics

31st International Computing and Combinatorics Conference
COCOON 2025, Chengdu, China, August 15–17, 2025
Proceedings, Part I

Editors
Fedor V. Fomin
Department of Informatics
University of Bergen
Bergen, Norway

Mingyu Xiao
University of Electronic Science
and Technology of China
Chengdu, China

ISSN 0302-9743 ISSN 1611-3349 (electronic)
Lecture Notes in Computer Science
ISBN 978-981-95-0214-1 ISBN 978-981-95-0215-8 (eBook)
https://doi.org/10.1007/978-981-95-0215-8

This Springer imprint is published by the registered company Springer Nature Singapore Pte Ltd.
The registered company address is: 152 Beach Road, #21-01/04 Gateway East, Singapore 189721, Singapore

Preface

It is our great pleasure to present the proceedings of the 31st International Computing and Combinatorics Conference (COCOON 2025), held in the vibrant city of Chengdu, China, from August 15 to 17, 2025. As a premier international forum, COCOON 2025 brought together leading researchers and practitioners in algorithms, theory of computation, computational complexity, and combinatorics to share cutting-edge discoveries, exchange ideas, and cultivate collaborative opportunities.

The conference encompassed a broad spectrum of topics, reflecting the dynamic and interdisciplinary nature of the field. Key areas included: Approximation Algorithms, Combinatorial Optimization, Computational Complexity, Computational Geometry, Economics and Computation, Graph Algorithms and Graph Theory, Learning and Data-Related Theory, Parameterized Algorithms, and String Algorithms and Discrete Structures.

This year, COCOON received 191 valid submissions from 24 countries and regions, demonstrating its global reach and academic significance. Each non-withdrawn submission underwent a rigorous peer-review process by the Program Committee, assisted by external referees, with at least three independent double-blind reviews. After careful evaluation, 54 papers were selected for presentation, yielding an acceptance rate of 28.3%.

We were privileged to host four distinguished invited speakers, whose thought-provoking talks enriched the conference:

Andrew Chi-Chih Yao (Tsinghua University, China)
Andrei Bulatov (Simon Fraser University, Canada)
Saket Saurabh (Institute of Mathematical Sciences, India & University of Bergen, Norway)
Shang-Hua Teng (University of Southern California, USA)

Their contributions provided deep insights and stimulated engaging discussions among attendees.

The success of COCOON 2025 would not have been possible without the collective efforts of many individuals and organizations. We extend our deepest gratitude to: The authors for their high-quality submissions; The Program Committee members and external reviewers for their diligent and constructive evaluations; The invited speakers for their enlightening presentations; The local organizers for their meticulous planning and execution; The Steering Committee for their invaluable guidance and support. We also acknowledge the generous sponsorship of: The Algorithms and Logic Group, University of Electronic Science and Technology of China, and The Theoretical Computer Science Committee of the China Computer Federation (CCF). Their support was pivotal in ensuring the conference's success.

Finally, we sincerely thank Springer for publishing the COCOON 2025 proceedings in their prestigious Lecture Notes in Computer Science (LNCS) series, enabling the dissemination of these research contributions to a global audience.

We hope this volume serves as a valuable resource for researchers and inspires further advancements in the fields of computing and combinatorics.

August 2025

Fedor V. Fomin
Mingyu Xiao

Organization

General Chair

Bakh Khoussainov	University of Electronic Science and Technology of China, China

PC Co-chairs

Fedor V. Fomin	University of Bergen, Norway
Mingyu Xiao	University of Electronic Science and Technology of China, China

Program Committee

Faisal Abu-Khzam	Lebanese American University, Lebanon
Xiaohui Bei	Nanyang Technological University, Singapore
René van Bevern	Huawei Technologies Co, Russia
Davide Bilò	University of L'Aquila, Italy
Ivan Bliznets	University of Groningen, Netherlands
Zhipeng Cai	Georgia State University, USA
Karthekeyan Chandrasekaran	University of Illinois, Urbana-Champaign, USA
Xue Chen	University of Science and Technology of China, China
Yong Chen	Hangzhou Dianzi University, China
Pål Grønås Drange	University of Bergen, Norway
Fedor V. Fomin	University of Bergen, Norway
Zhiguo Fu	Northeast Normal University, China
Takuro Fukunaga	Chuo University, Japan
Robert Ganian	Technische Universität Wien, Austria
Serge Gaspers	UNSW Sydney, Australia
Archontia Giannopoulou	National and Kapodistrian University of Athens, Greece
Alexander Grigoriev	Maastricht University, Netherlands
Gregory Gutin	Royal Holloway, University of London, UK
Tanmay Inamdar	IIT Jodhpur, India
Haitao Jiang	Shandong University, China

Bakh Khoussainov	University of Electronic Science and Technology of China, China
Ralf Klasing	CNRS and University of Bordeaux, France
Arie Koster	RWTH Aachen University, Germany
Jan Kratochvil	Charles University, Czechia
Alexander Kulikov	JetBrains, Cyprus
Van Bang Le	University of Rostock, Germany
Yi Li	Nanyang Technological University, Singapore
Chung-Shou Liao	National Tsing Hua University, Taiwan
Guohui Lin	University of Alberta, Canada
Jiamou Liu	University of Auckland, New Zealand
Jingcheng Liu	Nanjing University, China
Shengxin Liu	Harbin Institute of Technology, China
Markus Lohrey	University of Siegen, Germany
Kazuhisa Makino	Kyoto University, Japan
Matthias Mnich	Hamburg University of Technology, Germany
Nicolas Nisse	Inria, France
Yoshio Okamoto	University of Electro-Communications, Japan
Alexander Okhotin	St. Petersburg State University, Russia
Yota Otachi	Nagoya University, Japan
Qi Qi	Renmin University of China, China
Ignasi Sau	Université de Montpellier, France
Dominik Scheder	Chemnitz University of Technology, Germany
Alexander Shen	University of Montpellier, France
Kirill Simonov	University of Bergen, Norway
Frank Stephan	National University of Singapore, Singapore
Toru Takisaka	University of Electronic Science and Technology of China, China
Zhihao Gavin Tang	Shanghai University of Finance and Economics, China
Biaoshuai Tao	Shanghai Jiao Tong University, China
Ioan Todinca	University of Orléans, France
Ryuhei Uehara	Japan Advanced Institute of Science and Technology, Japan
Erik Jan van Leeuwen	Utrecht University, Netherlands
Xiaowei Wu	University of Macau, China
Mingyu Xiao	University of Electronic Science and Technology of China, China
Chao Xu	University of Electronic Science and Technology of China, China
Boting Yang	University of Regina, Canada
Meirav Zehavi	Ben-Gurion University of the Negev, Israel

Jialin Zhang	Chinese Academy of Sciences, China
Louxin Zhang	National University of Singapore, Singapore
Yong Zhang	SIAT, Chinese Academy of Sciences, China
Zhao Zhang	Zhejiang Normal University, China
Binhai Zhu	Montana State University, USA

Local Organizing Committee

Mingyu Xiao	University of Electronic Science and Technology of China, China
Chao Xu	University of Electronic Science and Technology of China, China
Yi Zhou	University of Electronic Science and Technology of China, China
Dong Hao	University of Electronic Science and Technology of China, China
Toru Takisaka	University of Electronic Science and Technology of China, China
Yuting Fang	University of Electronic Science and Technology of China, China
Ting Gou	University of Electronic Science and Technology of China, China
Yiping Liu	University of Electronic Science and Technology of China, China
Junqiang Peng	University of Electronic Science and Technology of China, China
Kangyi Tian	University of Electronic Science and Technology of China, China
Jingyang Zhao	University of Electronic Science and Technology of China, China
Yuxi Liu	University of Electronic Science and Technology of China, China
Binglin Tao	Sichuan Agricultural University, China

Additional Reviewers

Juliette Achdou
Isolde Adler
Tian Bai
Sayan Bandyapadhyay
Aritra Banik
Nikhil Bansal
Sebastian Bielfeldt
Vittorio Bilò
Arthur Braida
Robert Brijder

Anton A. Bukov
Cristian S. Calude
Dipayan Chakraborty
Jou-Ming Chang
Jiejiang Chen
Shengminjie Chen
Xiaoyu Chen
Xiuyang Chen
Yike Chen
Zishang Chen
Eddie Cheng
Kyungjin Cho
Nikolai Chukhin
Yu Cong
Gennaro Cordasco
Rajni Dabas
Renu Dalal
Sebastian Debus
Francesco Diana
Yuejia Dou
Maël Dumas
Michal Dvořák
Ajaykrishnan E. S.
Leah Epstein
Qilong Feng
Yi Feng
Kaito Fujii
Ameet Gadekar
Jiacheng Gao
Claudio Gentile
Valentin Gledel
Petr Golovach
Luciano Grippo
Luciano Gualà
Dong Hao
David Hartman
Klaus Heeger
Milan Hladík
Stepan Holub
Haoqiang Huang
Zengfeng Huang
Lucas Isenmann
Satyabrata Jana
Vít Jelínek
Hua Jiang
Zhile Jiang
Kaspar Kasche
Akinori Kawachi
Marco Kemmerling
Asif Khan
Kei Kimura
Naoki Kitamura
Masashi Kiyomi
Dimitris Kolonelos
Alexander Kononov
Sotiris Kotsiantis
Shubhang Kulkarni
Greg Kuperberg
Jan Matyáš Křišťan
Manuel Lafond
Anissa Lamani
Michael Lampis
Alexandra Lassota
Patrick Lederer
Chia-Wei Lee
Amit Levi
Fei Li
Shuai Li
Weidong Li
Wei Liang
Ya-Chun Liang
Mathieu Liedloff
Bingkai Lin
Cheng-Kuan Lin
Honghao Lin
Hanbin Liu
Maël Luce
Kelin Luo
Zihan Luo
Chensheng Ma
Mengfan Ma
Yuchao Ma
Raghunath Reddy Madireddy
Diego Marcos
Mathurin Massias
Filippos Mavropoulos
Sally Mcclean
Lili Mei
Alexander Melnikov
Ivan Mikhailin

Pedro Montealegre
Junya Nakamura
Anurag Murty Naredla
Daniel Neuen
Meike Neuwohner
André Nichterlein
Maksim Nikolaev
Fedor Noskov
Nacim Oijid
Shmuel Onn
Ioannis Panagiotas
Artem Panin
Anton Paramonov
Bo Peng
Junqiang Peng
Pan Peng
Anthony Perez
Matthias Pfretzschner
Romila Pradhan
Ioannis Psarros
Lianrong Pu
Clemens Puppe
C. Ramya
Adele Rescigno
Andreas Rosowski
Benjamin Rossman
Danil Sagunov
Parikshit Saikia
Alice Sayutina
Christiane Schmidt
Timo Schneider
Sanjay Seetharaman
François t'Serstevens
Carlos Eduardo Silva de Oliveira
M. P. Singh
Alexander Skopalik
Siani Smith
Anastasia Sofronova
Farehe Soheil
Tasuku Soma
Matej Stehlik
Yuichi Sudo
Akira Suzuki
Arman Tadevosian
Arman Tadevosian
Ayman Tajeddine
Kenjiro Takazawa
Karolina Tammemaa
Johannes Tantow
Binglin Tao
Gabor Tardos
İstenç Tarhan
Sergio Thoumi
Alexander Thumm
Kangyi Tian
Ankit Titoriya
Ron Triepels
Dimitra Tsigkari
Freija van Lent
Rob van Stee
Vsevolod Vaskin
Changjun Wang
Chenhao Wang
Dingyu Wang
Hongjie Wang
Qi Wang
Rongquan Wang
Xao Wang
Yanheng Wang
Ye Wang
Zhiqi Wang
Xuan Wu
Kelin Xia
Chenyang Xu
Renzhe Xu
Kuan Yang
Ran Yingli
Peter Zeman
Mengxiao Zhang
Jingyang Zhao
Lei Zhao
Muyang Zhao
Yajie Zhao
Da Wei Zheng
Mingchao Zhou
Yi Zhou
Jiadong Zhu
Weihao Zhu

Contents – Part I

Approximation Algorithms

Combinatorial Optimization

Contents – Part II

Graph Algorithms and Graph Theory

Learning and Data-Related Theory

Parameterized Algorithms

String Algorithms and Discrete Structures

Approximation Algorithms

Improved Approximation Algorithms for Combinatorial Contracts with Type Constraints

Qinqin Gong[1], Chunlin Hao[1], Donglei Du[2], and Ruiqi Yang[1(✉)]

[1] Institute of Operations Research and Information Engineering, Beijing University of Technology, Beijing 100124, People's Republic of China
gongqinqin@emails.bjut.edu.cn, {haochl,yangruiqi}@bjut.edu.cn
[2] Faculty of Business Administration, University of New Brunswick, Fredericton, NB E3B 5A3, Canada
ddu@unb.ca

Abstract. We introduce a multi-agent combinatorial contract design problem with type constraints. A principal assigns a task to agents divided into k types. Agents of each type decide whether to exert costly effort. The principal's reward is a non-negative, monotone, and submodular function of the agents exerting effort across all types. Extending prior work by Dütting et al. (2023) on the single-agent-per-type ($k = n$) case, we allow multiple agents per type, where n denotes the number of agents. We formulate the contract design as a bi-level optimization problem, which we transform into a single-level subset selection problem using backward induction. We provide a parameterized approximation algorithm using value and demand query oracles, achieving approximation ratios approximately 5 times better than Dütting et al. (2023) under suitable parameter settings.

Keywords: Combinatorial contracts · Principal-agent model · Approximation algorithms · Submodularity

1 Introduction

Algorithmic contract theory [15] is an emerging interdisciplinary research area at the intersection of computer science and economics. Contract theory [1] provides a systematic framework for designing incentive mechanisms. Computer science contributes by offering tools to analyze computational complexity, facilitating the exploration of the trade-offs between simple and optimal solutions. As an increasing number of classic contract applications rapidly transition to online platforms and the scale of data continues to grow, algorithmic contract theory has garnered increasing attention from researchers [2,16]. Applications range from crowdsourcing platforms [20] to online labor markets [23] and blockchain-based smart contracts [9].

F. V. Fomin and M. Xiao (Eds.): COCOON 2025, LNCS 15983, pp. 3–17, 2026.
https://doi.org/10.1007/978-981-95-0215-8_1

In the classic hidden-action principal-agent model [19,21], a principal hires an agent to perform a task. The agent takes a costly action, which is private information and cannot be directly observed by the principal. Instead, the principal can only observe the incurred outcome resulting from the agent's action and obtain a reward based on the outcome. Information asymmetry may lead to *moral hazard* [21,22], as the principal wants the agent to act in her best interest, but the agent may exploit his information advantage to pursue his own interests. To address the problem, the principal needs to design a contract to incentivize the agent, thereby maximizing her own utility. The contract specifies the payment obtained by the agent. The agent's goal is to choose an action that maximizes his own utility after receiving the contract. In the case of one principal and one agent, the contract design problem can be efficiently solved in polynomial time by solving a linear programming for each action [19].

A central topic in algorithmic contract theory is *combinatorial contracts* [12], which is a natural focus of research due to its inherent computational complexity and mechanism design challenges. Combinatorial contracts primarily involves two settings: multi-action and multi-agent. In a multi-action model [10], a principal assigns a task to an agent who can take more than one action. The possible outcomes are binary: success or failure. The reward function is defined on the subset of actions taken by the agent. In a multi-agent model [11], a principal interacts with multiple agents, each agent has two actions to choose from: effort or no effort. The reward function is defined on the subset of agents that exert effort. Although previous studies on combinatorial contract design have considered agents heterogeneity, they have been limited to the case where each type contains a single agent. We extend the problem of multi-agent contract design to a setting where each type consists of multiple agents, meaning that agents may belong to different types due to differences in competence or risk preferences. The scenario has many applications in real life. For example, open source projects are split into groups based on the skills of the developers, with front-end developers responsible for interface design, back-end developers responsible for server logic, and testers responsible for code validation.

In this paper, we focus on the multi-agent combinatorial contract design problem with k distinct types. The set of agents of type $t \in [k]$ is denoted as N_t, and the ground set of all agents is $N = \bigcup_{t\in[k]} N_t$, where $N_t \cap N_t' = \emptyset$ for any $t, t' \in [k] = \{1, ..., k\}$. The agents of type t who choose to exert effort form a subset $X_t \subseteq N_t$, and the total set of agents exerting effort is $X = \bigcup_{t\in[k]} X_t$. Each agent $e \in X$ incurs a cost $\mathcal{C}_e \in \mathbb{R}_{\geq 0}$. The set X produces a stochastic outcome $o \in O$, and the principal receives a reward $r(o)$. The principal's expected reward function $\mathcal{F}$ is defined on the set X. We focus on linear contracts. Given type $t \in [k]$, a linear contract is defined as a vector $\boldsymbol{\alpha}^t = (\alpha_1^t, ..., \alpha_{n_t}^t) \in [0,1]^{n_t}$, where n_t is the number of agents in type t. Then, the payment obtained by agent $e \in N_t$ for outcome o is $\alpha_e^t r(o)$, which specifies the proportion of the reward allocated to agent e. Assuming that $\mathcal{F}$ is non-negative monotone and submodular, we show that the optimal linear contract design problem can be reformulated as an optimal subset selection problem. When we restrict the number of agents to one for each type, the problem degenerates into the principal-agent model studied by Dütting et al. [11]. For submodular reward functions, they proposed an approximation algorithm with an approximation ratio of $1/(256(1+\varepsilon)+2)$ (≈ 0.00387). Our work gives a parameterized approximation algorithm, which can achieve an approximation ratio nearly 5 times better than that of Dütting et al. [11] when appropriate parameters are chosen.

1.1 Our Contributions and Techniques

We introduce the k-type constrained multi-agent contract design problem, which extends the single-type setting originally introduced by Dütting et al. [11]. In their model, each type contains a single agent, and the principal incentivizes a subset of agents to exert effort by designing a contract. They proposed a first constant approximation algorithm for the linear contract. In our work, we extend this scenario by allowing each type to contain multiple agents, each of whom decides whether to exert effort. The union of these agent subsets across all types forms the set of agents who exert effort, and the reward function is defined on the set. We assume that the reward function is non-negative monotone and submodular.

We model the problem as a bi-level optimization problem, with the goal is to design a contract maximizing the principal's expected utility, while ensuring that the action choices of each agent within each type satisfy the incentive compatibility constraint. Using backward induction, we transform the problem into a single-level subset selection maximization problem. Thus, we reformulate the optimization problem as follows:

$$\max_{X\subseteq N} \mathcal{G}(X) = \left(1 - \sum_{t\in[k]}\sum_{e\in X_t} \frac{\mathcal{C}_e}{\mathcal{F}(e|X\setminus\{e\})}\right)\mathcal{F}(X). \tag{1}$$

By utilizing value and demand query oracles, we develop a parameterized approximation algorithm. The main result is summarized below.

Main Result (Formally, Theorem 1): *Assuming that the expected reward function $\mathcal{F}$ is non-negative monotone and submodular, there exists a parameterized approximation algorithm for the multi-agent combinatorial contract problem with k types of agents.*

The key elements in designing an approximation algorithm when the reward function is submodular are outlined as follows.

Although the reward function $\mathcal{F}$ exhibits desirable structural properties, such as monotonicity and submodularity, the objective function $\mathcal{G}$ generally fails to preserve these properties. This poses significant challenges for directly applying standard optimization techniques. However, by leveraging the structural properties of $\mathcal{F}$ and analyzing the relationship between $\mathcal{G}$ and $\mathcal{F}$, we are still able to provide certain bounds and estimates for the optimal value. Specifically, we show that the optimal value $\mathcal{G}(X^*)$ can be upper bounded by the sum of

$\sum_{t\in[k]} \mathcal{F}(X_t^* \cap Y_t)$ and $\max_{e\in N} \mathcal{G}(\{e\})$. This serves as a foundation for our subsequent analysis. By introducing the curvature κ, we can quantify the nonlinearity of the submodular reward function $\mathcal{F}$, ensuring bounds on marginal gains. In particular, we utilize the following observation: $\mathcal{F}(\bigcup_{t\in[k]} E_t) \geq (1-\kappa)\sum_{t\in[k]} \mathcal{F}(E_t)$ and $\mathcal{F}(e|E\setminus\{e\}) \geq (1-\kappa)\,[\mathcal{F}(E_t) - \mathcal{F}(E_t \setminus \{e\})]$, where $E_t \subseteq E$ for any $t \in [k]$ and κ denotes the curvature of $\mathcal{F}$. Furthermore, by integrating geometric scaling technique with the scaling property of submodular functions proposed by Dütting et al. [11], we develop the approximation algorithm which achieves an improved approximation guarantee.

1.2 Related Work

The contract theory was initially proposed by [26], and its core principal-agent problem has been extensively examined in the economic literature [21,27]. In the basic version of single principal and single agent, the agent can choose an action to perform a task. By solving the linear programming for each action, the optimal contract can be solved efficiently [19]. Linear contracts are a type of contract that has been widely studied in the existing literature. For each possible outcome, a linear contract specifies the proportion of the reward that is allocated to the agent. Carroll [7] and Dütting et al. [16] demonstrated that linear contracts are robustly optimal.

Combinatorial Contracts. Many studies have explored the contract design problems in various combinatorial settings from an algorithmic perspective. Combinatorial contracts mainly involve two settings: multi-action and multi-agent, where the reward function is defined as a set function over a subset of actions or agents. In the multi-action setting, Dütting et al. [10] demonstrated that the optimal contract can be computed in polynomial time when the reward function is gross substitute, while the computation of the optimal contract becomes an NP-hard problem when the reward function is submodular, with its difficulty stemming from the need for an exponential number of demand query oracles [14]. Ezra et al. [17] further showed that no constant-factor approximation algorithm exists for the problem. Vuong et al. [29] proposed a polynomial-time algorithm for computing the optimal contract when the reward function is supermodular and the cost function is modular. Dütting et al. [13] further demonstrated that the optimal contract can still be computed in polynomial time when the cost function is submodular.

Another research direction focuses on multi-agent setting, which is closely related to our work. Babaioff et al. [2] first introduced the combinatorial agent model, along with its extended version [5] and subsequent research [3,4]. Dütting et al. [11] studied a combinatorial model where each agent chooses whether to exert effort, with the reward function defined on the set of agents that exert effort. They proposed an approximation algorithm with an approximation ratio of $1/(256(1+\varepsilon)+2)$ for XOS or submodular reward functions in the complement-free hierarchy [24]. Ezra et al. [17] proved that the upper bound of the approximation ratio is 0.7 when the reward function is submodular. However, for the case where the reward function is XOS, no constant-factor approximation can be achieved using only value query oracle. Vuong et al. [29] showed that there is no constant factor approximation algorithm when the reward function is supermodular. Dütting et al. [12] combined the multi-agent and multi-action settings. They proposed a constant-factor approximation algorithm for the multi-agent combinatorial contract problem with submodular reward functions.

Submodular Maximization Algorithms. From resource allocation and feature selection to sensor deployment, discrete situations are ubiquitous in real life. Submodularity is a highly attractive property when designing discrete objective functions. It naturally exhibits the characteristic of diminishing marginal returns. Since the introduction of the greedy algorithm by Nemhauser et al. [25], a series of theoretical methods and algorithmic tools have been gradually developed. As early as 1978, Fisher et al. [18] provided a deterministic 0.5-approximation greedy algorithm under the matroid constraint setting. Calinescu et al. [6] proposed a continuous greedy algorithm, combined with the pipage rounding method to give a $(1 - 1/e)$-approximation.

To more accurately characterize the behavior of submodular functions, various assumptions can be made about the inputs. One important assumption is the concept of the total curvature, introduced by Conforti and Cornuéjols [8]. The total curvature $\kappa \in [0, 1]$ of a non-decreasing submodular function $f : 2^N \rightarrow R_{\geq 0}$ is defined as $\kappa = 1 - \min_{j \in N} \frac{f(N) - f(N \setminus \{j\})}{f(\{j\})}$, which measures the degree to which the function f deviates from linearity. When $\kappa = 0$, the submodular function reduces to modular case. The definition intituvely helps us better understand the characteristics of submodular functions and supports the design of more efficient algorithms. Conforti and Cornuéjols [8] proved that the greedy algorithm can provide an approximation ratio of $(1 - e^{-\kappa})/\kappa$ for non-decreasing submodular functions. When $\kappa \rightarrow 1$, the approximation ratio converges to the classical $(1 - 1/e)$. Vondrák [28] further extended the $(1 - e^{-\kappa})/\kappa$-approximation of [8] to any matroid constraints. Additionally, Vondrák [28] showed that for instances with curvature κ, this factor is indeed optimal.

1.3 Organization

Section 2 formally introduces the model. Section 3 analyzes the properties of the optimal contract and presents auxiliary lemmas for submodular reward functions. Section 4 provides and analyzes an approximation algorithm. Finally, Sect. 5 concludes our work.

2 Preliminaries

In this paper, we study a scenario involving a principal and a set of agents, denoted by $N = \{1, ..., n\}$. The agents are divided into k distinct types, where each type t corresponds to a subset $N_t \subseteq N$ of agents with similar characteristics or capabilities. We assume that any two type set of agents are disjoint, i.e.,

$N_t \cap N_t' = \emptyset$ for any pair of types $t \neq t'$ with $t, t' \in [k]$. Formally, we denote $N = \bigcup_{t=1}^{k} N_t$ as the set of agents and refer to the total number of agents as $n = \sum_{t=1}^{k} n_t$, where n_t is the number of agents in type t. For each type $t \in [k]$, the subset of agents that exert effort is $X_t \subseteq N_t$. Let $X = \bigcup_{t=1}^{k} X_t \subseteq N$ denote the subset of agents who exert effort across all types. Each agent $e \in X$ incurs a cost $\mathcal{C}_e \in \mathbb{R}_{\geq 0}$. There is an outcome space O, and the subset of agents X that exert effort is associated with a probability distribution $p_X : O \to [0,1]$ over the outcomes. A stochastic outcome o occurs with probability $p_X(o)$, and the principal receives a reward $r(o) \in \mathbb{R}_{\geq 0}$ associated with the outcome. We define the expected reward function of the principal as $\mathcal{F} : 2^N \to \mathbb{R}_{\geq 0}$. Specifically, $\mathcal{F}(X) = \sum_{o \in O} p_X(o) r(o)$. Denote the marginal gain of agent e by $\mathcal{F}(e|X) = \mathcal{F}(X \cup \{e\}) - \mathcal{F}(X)$. We consider $\mathcal{F}(X)$ is assumed as monotone and submodular that is $\mathcal{F}(e|X) \geq 0$ and $\mathcal{F}(e|X) \geq \mathcal{F}(e|X')$ for any $X \subseteq X' \subseteq N$. Additionally, we define $\mathcal{F}(\emptyset) = 0$.

Since the action taken by each agent is unobservable, this may lead to a moral hazard problem. The agent may choose not to exert effort to avoid personal cost, thereby harming the principal's interests. To address this problem, the principal needs to design a contract to incentivize the agent to exert effort. We focus on linear contracts. Given type $t \in [k]$, a linear contract is defined as a vector $\boldsymbol{\alpha}^t = (\alpha_1^t, ..., \alpha_{n_t}^t) \in [0,1]^{n_t}$. Then, the payment obtained by agent $e \in N_t$ for outcome o is $\alpha_e^t r(o)$, which specifies the proportion of the reward allocated to agent e. Therefore, the expected payment of agent e is $\alpha_e^t \mathcal{F}(X)$. Given the contract $\boldsymbol{\alpha}^t$ for each type t, the expected utility of the principal is given by $\left(1 - \sum_{t \in [k]} \sum_{e \in N_t} \alpha_e^t\right) \mathcal{F}(X)$ and for each type $t \in [k]$, the expected utility of agent $e \in N_t$ is $\alpha_e^t \mathcal{F}(X) - \mathbf{1}_{e \in X_t} \cdot \mathcal{C}_e$.

Consider all agents within each type participating in an induced game and gradually converge to a locally optimal solution. Each agent observes the contract provided by the principal and chooses whether to exert effort in response. Under the constraint that all agents adopt the best response strategy, the principal's goal is to design a contract that maximizes her own utility. Therefore, the optimization problem is formulated as

$$\begin{aligned} &\max_{\boldsymbol{\alpha}^t \in [0,1]^{n_t}, t \in [k]} \left(1 - \sum_{t \in [k]} \sum_{e \in N_t} \alpha_e^t\right) \mathcal{F}(X) \\ &\text{s.t.} \quad \alpha_e^t \mathcal{F}(X) - \mathcal{C}_e \geq \alpha_e^t \mathcal{F}(X \setminus \{e\}), \forall t \in [k], e \in X_t, \\ &\qquad\;\; \alpha_e^t \mathcal{F}(X) \geq \alpha_e^t \mathcal{F}(X \cup \{e\}) - \mathcal{C}_e, \forall t \in [k], e \notin X_t. \end{aligned}$$

For a fixed set of agents X, the best way for the principal to incentivize the set is by setting $\alpha_e^t = \frac{\mathcal{C}_e}{\mathcal{F}(e|X \setminus \{e\})}$ for any $e \in X_t, t \in [k]$, and $\alpha_e^t = 0$ for any $e \notin X_t, t \in [k]$. Therefore, our model above can be reformulated as a single level optimization Problem (1). We denote by $X^* = \arg\max_{X \subseteq N} \mathcal{G}(X)$ the optimal solution of the problem.

In this paper, we consider the reward function $\mathcal{F} : 2^N \to \mathbb{R}_{\geq 0}$ that belongs to the following classes of complement-free hierarchy set functions [11,24]:

• *Modular:* The reward function $\mathcal{F}$ is modular if $\mathcal{F}(S) = \sum_{e \in S} w_e$ for some $w_e \in \mathbb{R}_{\geq 0}$ and any subset $S \in 2^N$, and is also called additive.

• *Submodular:* The reward function $\mathcal{F}$ is submodular if $\mathcal{F}(e|S) \geq \mathcal{F}(e|T)$ for any set $S \subseteq T$ and $e \in N \backslash T$.

• *Subadditive:* The reward function $\mathcal{F}$ is subadditive if $\mathcal{F}(S \cup T) \leq \mathcal{F}(S) + \mathcal{F}(T)$ for any $S, T \in 2^N$.

As is typical in the submodular optimization literature related to set functions, we assume two basic primitives for accessing $\mathcal{F}$.

• A *value query oracle* for $\mathcal{F}$ return $\mathcal{F}(S)$ in $O(1)$ time for any given set S.

• A *demand query oracle* for $\mathcal{F}$ takes a price vector $p \in \mathbb{R}^n_{\geq 0}$ as input and outputs the optimal set $S \in \arg\max_{S \in 2^N} \mathcal{F}(S) - \sum_{e \in S} p_e$ in $O(1)$ time.

3 Warm-Up: Structural Insights

We firstly show that there exists an upper bound on the optimal value $\mathcal{G}(X^*)$ that is associated with $\sum_{t \in [k]} \mathcal{F}(X_t^* \cap Y_t)$ and $\max_{e \in N} \mathcal{G}(\{e\})$, where we denote by $Y_t = \{e \in N_t : \frac{\mathcal{C}_e}{\mathcal{F}(\{e\})} \leq \frac{1}{2}\}$ as the candidate high cost-effective agents of type t. The set Y_t is constructed to screen out elements that satisfy a particular cost-benefit ratio, thereby providing a basis for the upper bound analysis. The proof appears in the full version of this paper.

Lemma 1. *If $\mathcal{F}$ is submodular, then*

$$\mathcal{G}(X^*) \leq \sum_{t \in [k]} \mathcal{F}(X_t^* \cap Y_t) + \max\{0, \max_{e \in N} \mathcal{G}(\{e\})\}.$$

The next lemma presents a novel property of set scaling, which approximately preserves the marginal gains of a submodular function. This property effectively controls the value of the reward function $\mathcal{F}$ and its marginal gains.

Lemma 2. *[11] Consider a submodular function $\mathcal{F}$. For any $E = \bigcup_{t=1}^k E_t \subseteq N$, along with parameters $\theta \in (0,1)$ and $\rho \in (0, \mathcal{F}(E))$, there exists a polynomial time algorithm called Threshold-Scaling Submodular Selection $(\mathcal{F}, E, \rho, \theta)$, which queries the value oracle to access $\mathcal{F}$ and outputs a set $H \subseteq E$ satisfying*

$$(1-\theta)\rho \leq \mathcal{F}(H) \leq \rho + \max_{e \in E} \mathcal{F}(\{e\}),$$

and

$$\mathcal{F}(e|H \setminus \{e\}) \geq \theta \mathcal{F}(e|E \setminus \{e\})$$

for any $e \in H$.

Note that the Threshold-Scaling Submodular Selection $(\mathcal{F}, E, \rho, \theta)$ is originally introduced by [11]. For the sake of completeness, we restate it here. The details of the algorithm are summarized in Algorithm 1.

Algorithm 1. Threshold-Scaling Submodular Selection $(\mathcal{F}, E, \rho, \theta)$ [11]

Input: A submodular function $\mathcal{F}$, parameters $\rho \in (0, \mathcal{F}(E))$ and $\theta \in (0, 1)$
Output: A set $H \subseteq E$

Define E_0 as the smallest subset of E for which $\mathcal{F}(E_0) = \mathcal{F}(E)$
for $\ell = 1, ..., |E_0|$ **do**
    Set $e_\ell \in \arg\min_{e \in E_{\ell-1}} \frac{\mathcal{F}(e|E_{\ell-1} \setminus \{e\})}{\mathcal{F}(e|E_0 \setminus \{e\})}$
    Set $E_\ell \leftarrow E_{\ell-1} \setminus \{e_\ell\}$
    Set $\theta_\ell = \frac{\mathcal{F}(e_\ell|E_{\ell-1} \setminus \{e_\ell\})}{\mathcal{F}(e_\ell|E_0 \setminus \{e_\ell\})}$
end for
Set $j^* = \min\{j | \mathcal{F}(E_j) \leq \rho\}$
Set $k^* = \min\left\{k | \frac{\mathcal{F}(E_k)}{\mathcal{F}(E_{j^*-1})} \leq 1 - \theta\right\}$
Set $\ell^* = \arg\max_{\ell \in \{j^*, \dots, k^*\}} \theta_\ell$
return $H \leftarrow E_{\ell^*-1}$

4 Algorithms and Theoretical Analysis

Before proceeding with the design of the algorithm and the analysis of the approximation ratio, we first introduce some useful auxiliary lemmas. These lemmas establish an important connection between the principal utility, expected rewards and costs. They provide the theoretical foundation for our analytical approach, which uses value and demand query oracle to identify suitable agents and incentivize them to join a contract. The proofs of these lemmas appear in the full version of this paper.

Lemma 3. *[11] If $\mathcal{F}$ is submodular, then for each subset $X \subseteq X^*$, we have*

$$\sum_{e \in X} \sqrt{\mathcal{C}_e} \leq \sqrt{\mathcal{F}(X)}.$$

Lemma 4. *If $\mathcal{F}$ is submodular, then for any agent subset $X \subseteq N$ with a reward $\mathcal{F}(X) > 0$ and parameter $\lambda > 1$, the following inequality holds:*

$$\mathcal{G}(X) \geq \left(1 - \frac{1}{\lambda}\right) \mathcal{F}(X)$$

suppose that $\mathcal{F}(e|X \setminus \{e\}) \geq \sqrt{\lambda \mathcal{C}_e \mathcal{F}(X)}$ for any $e \in X$.

Lemma 5. *If $\mathcal{F}$ is submodular with curvature of $\kappa = 1 - \min_{e \in N} \frac{\mathcal{F}(N) - \mathcal{F}(N \setminus \{e\})}{\mathcal{F}(\{e\})}$, then for any set $X = \bigcup_{t=1}^{k} X_t$, the following inequalities hold:*

$$\mathcal{F}(\bigcup_{t \in [k]} X_t) \geq (1 - \kappa) \sum_{t \in [k]} \mathcal{F}(X_t)$$

and

$$\mathcal{F}(e|X \setminus \{e\}) \geq (1 - \kappa) \left[\mathcal{F}(X_t) - \mathcal{F}(X_t \setminus \{e\})\right].$$

Algorithm 2. Approximating Optimal Contract

Input: A submodular function $\mathcal{F}$ with curvature $\kappa \in [0, 0.48]$, costs $\{\mathcal{C}_e\}_{e\in N}$, a set Y_t for any $t \in [k]$, parameters $\omega > 1$, $\eta > 1$, $\varepsilon \in (0,1)$, $\theta \in (0,1)$

Output: A set I

$e^* \leftarrow \arg\max_{e\in N} \mathcal{G}(\{e\})$
$\mathcal{A} \leftarrow \{e^*\}$
Let $z = \frac{1}{2}\max_{e\in Y_t, t\in[k]} \mathcal{F}(\{e\})$
for $i = 0, ..., \lceil \log_{1+\varepsilon} \frac{2n}{\omega} \rceil$ **do**
  Let $z_i = \omega z(1+\varepsilon)^i$
  Let $\rho_i = (1-\kappa)\left[\left(1-\frac{1}{\eta}\right) z_i - \max_{e\in Y=\bigcup_{t\in[k]} Y_t} \mathcal{F}(\{e\})\right]$
  For every $e \in Y_t$, let $q_e^i = \frac{1}{\eta}\sqrt{\mathcal{C}_e z_i}$
  for $t = 1, ..., k$ **do**
    $E_t \leftarrow \arg\max_{E_t' \subseteq Y_t} \left\{\mathcal{F}(E_t') - \sum_{e\in E_t'} q_e^i\right\}$
  end for
  $E = \bigcup_{t\in[k]} E_t$
  if $\rho_i \in (0, \mathcal{F}(E))$ **then**
    $H^i \leftarrow$the output of Algorithm 1 $(\mathcal{F}, E, \rho_i, \theta)$
  else
    $H^i \leftarrow \emptyset$
  end if
  $\mathcal{A} \leftarrow \mathcal{A} \cup H^i$
end for
return $I \leftarrow \arg\max_{I'\in\mathcal{A}} \mathcal{G}(I')$

We next formally analyze Problem (1). In the case where the reward function is submodular, we develop an algorithm that combines the geometric scaling technique with the subroutine developed in Sect. 3. We prove that the algorithm runs in polynomial time and achieves the desired performance bound. The details are presented in Algorithm 2, and our main result is summarized in Theorem 1.

Theorem 1. *Consider the multi-agent combinatorial contract problem with k types. Assuming that the expected reward function $\mathcal{F}$ is non-negative monotone and submodular, we propose a parameterized approximation algorithm that runs in polynomial time. The performance guarantees are determined by the values of the input parameters η, θ, κ and ε.*

Proof. With the fixed parameter $\omega > 1$, we divide the argument into two cases. **Case 1.** Consider the case where $\sum_{t\in[k]} \mathcal{F}(X_t^* \cap Y_t) \leq \omega \max\{0, \max_{e\in N} \mathcal{G}(\{e\})\}$, then by Lemma 1, we obtain

$$\mathcal{G}(X^*) \leq (1+\omega)\max\{0, \max_{e\in N} \mathcal{G}(\{e\})\}.$$

Case 2. Assume that $\sum_{t\in[k]} \mathcal{F}(X_t^* \cap Y_t) > \omega \max\{0, \max_{e\in N} \mathcal{G}(\{e\})\}$. Utilizing the subadditivity of $\mathcal{F}$ and the expression for z, we have

$$\sum_{t\in[k]} \mathcal{F}(X_t^* \cap Y_t) \leq \sum_{t\in[k]} \sum_{e\in X_t^*\cap Y_t} \mathcal{F}(\{e\}) \leq \sum_{t\in[k]} n_t \max_{e\in Y_t, t\in[k]} \mathcal{F}(\{e\}) = 2nz.$$

Moreover, we obtain

$$\sum_{t\in[k]} \mathcal{F}(X_t^* \cap Y_t) > \omega \max_{e\in Y_t, t\in[k]} \mathcal{G}(\{e\}) \geq \frac{\omega}{2} \max_{e\in Y_t, t\in[k]} \mathcal{F}(\{e\}) = \omega z.$$

Therefore, we derive $\omega z \leq \sum_{t\in[k]} \mathcal{F}(X_t^* \cap Y_t) \leq 2nz$, which implies that there exist a unique $i^* \in \{0, ..., \lceil \log_{1+\varepsilon} \frac{2n}{\omega} \rceil\}$ such that $z_{i^*} \leq \sum_{t\in[k]} \mathcal{F}(X_t^* \cap Y_t) \leq (1+\varepsilon) z_{i^*}$. We focus on the case where $\rho_{i^*} > 0$, since the case $\rho_{i^*} \leq 0$ is trivial. In this case, we have

$$\begin{aligned}
\mathcal{F}(E) = \mathcal{F}(\bigcup_{t\in[k]} E_t) &\geq (1-\kappa) \sum_{t\in[k]} \mathcal{F}(E_t) \geq (1-\kappa) \sum_{t\in[k]} \left(\mathcal{F}(E_t) - \sum_{e\in E_t} q_e^{i^*} \right) \\
&\geq (1-\kappa) \sum_{t\in[k]} \left(\mathcal{F}(X_t^* \cap Y_t) - \sum_{e\in X_t^* \cap Y_t} q_e^{i^*} \right) \\
&= (1-\kappa) \left(\sum_{t\in[k]} \mathcal{F}(X_t^* \cap Y_t) - \sum_{e\in \bigcup_{t\in[k]} (X_t^* \cap Y_t)} \frac{1}{\eta} \sqrt{\mathcal{C}_e z_{i^*}} \right) \\
&\geq (1-\kappa) \left(\sum_{t\in[k]} \mathcal{F}(X_t^* \cap Y_t) - \frac{\sqrt{z_{i^*}}}{\eta} \sqrt{\mathcal{F}(\bigcup_{t\in[k]} (X_t^* \cap Y_t))} \right) \\
&\geq (1-\kappa) \left(\sum_{t\in[k]} \mathcal{F}(X_t^* \cap Y_t) - \frac{\sqrt{\sum_{t\in[k]} \mathcal{F}(X_t^* \cap Y_t)}}{\eta} \sqrt{\mathcal{F}(\bigcup_{t\in[k]} (X_t^* \cap Y_t))} \right) \\
&\geq (1-\kappa) \left(\sum_{t\in[k]} \mathcal{F}(X_t^* \cap Y_t) - \frac{\sqrt{\sum_{t\in[k]} \mathcal{F}(X_t^* \cap Y_t)}}{\eta} \sqrt{\sum_{t\in[k]} \mathcal{F}(X_t^* \cap Y_t)} \right) \\
&\geq (1-\kappa) \left(1 - \frac{1}{\eta} \right) z_{i^*} \\
&> \rho_{i^*},
\end{aligned}$$

where the first inequality holds since Lemma 5, the second inequality follows by the non-negativity of q_e^i, the third inequality derives from E_t is a demand set over the set Y_t, the fourth inequality uses that Lemma 3, the fifth inequality and the penultimate inequality result from $\sum_{t\in[k]} \mathcal{F}(X_t^* \cap Y_t) \geq z_{i^*}$, the sixth inequality follows from the subadditive of $\mathcal{F}$, the last inequality holds by the definition of ρ_{i^*}.

According to Lemma 2, there exist a set H^{i^*} such that

$$\begin{aligned}\mathcal{F}(H^{i^*}) &\leq \rho_{i^*} + \max_{e\in E}\mathcal{F}(\{e\})\\ &= (1-\kappa)\left[\left(1-\frac{1}{\eta}\right)z_{i^*} - \max_{e\in Y}\mathcal{F}(\{e\})\right] + \max_{e\in E}\mathcal{F}(\{e\})\\ &\leq (1-\kappa)\left(1-\frac{1}{\eta}\right)z_{i^*} + \kappa\max_{e\in Y}\mathcal{F}(\{e\})\\ &\leq \left(1-\frac{1}{\eta}\right)z_{i^*}, \end{aligned} \tag{2}$$

where the first inequality derives from Lemma 2, the second inequality follows from the non-negativity of $\mathcal{F}$, the third inequality holds by $\rho_{i^*} > 0$. For any $e \in H^{i^*}$, we have

$$\begin{aligned}\mathcal{F}(e|H^{i^*}\setminus\{e\}) &\geq \theta\mathcal{F}(e|E\setminus\{e\}) \geq \theta(1-\kappa)\left[\mathcal{F}(E_t) - \mathcal{F}(E_t\setminus\{e\})\right]\\ &\geq \theta(1-\kappa)q_e^{i^*} = \frac{\theta(1-\kappa)}{\eta}\sqrt{\mathcal{C}_e z_{i^*}}\\ &\geq \sqrt{\frac{\theta^2(1-\kappa)^2}{\eta(\eta-1)}\mathcal{C}_e\mathcal{F}(H^{i^*})},\end{aligned}$$

where the first inequality follows from Lemma 2, the second inequality holds since Lemma 5, the third inequality derives from E_t is a demand set, the last inequality holds by Inequality (2). If $\frac{\theta^2(1-\kappa)^2}{\eta(\eta-1)} > 1$, we derive

$$\begin{aligned}\mathcal{G}(H^{i^*}) &\geq \left(1-\frac{\eta(\eta-1)}{\theta^2(1-\kappa)^2}\right)\mathcal{F}(H^{i^*}) \geq \left(1-\frac{\eta(\eta-1)}{\theta^2(1-\kappa)^2}\right)(1-\theta)\rho_{i^*}\\ &= \left(1-\frac{\eta(\eta-1)}{\theta^2(1-\kappa)^2}\right)(1-\theta)(1-\kappa)\left[\left(1-\frac{1}{\eta}\right)z_{i^*} - \max_{e\in Y}\mathcal{F}(\{e\})\right],\end{aligned}$$

where the first inequality follows by Lemma 4, the second inequality holds since Lemma 2.

For the sake of simplicity, let $\Delta_1 = \left(1 - \frac{\eta(\eta-1)}{\theta^2(1-\kappa)^2}\right)(1-\theta)(1-\kappa)$, $\Delta_2 = 1 - \frac{1}{\eta}$. We now analyze the principal's utility on the set H^{i^*}.

$$\begin{aligned}
\mathcal{G}(H^{i^*}) &\geq \Delta_1 \left[\Delta_2 z_{i^*} - \max_{e\in Y} \mathcal{F}(\{e\})\right] \\
&\geq \Delta_1 \left[\frac{\Delta_2}{1+\varepsilon} \sum_{t\in[k]} \mathcal{F}(X_t^* \cap Y_t) - \max_{e\in Y} \mathcal{F}(\{e\})\right] \\
&\geq \Delta_1 \left[\frac{\Delta_2}{1+\varepsilon} \sum_{t\in[k]} \mathcal{F}(X_t^* \cap Y_t) - 2\max_{e\in N} \mathcal{G}(\{e\})\right] \\
&\geq \Delta_1 \left(\frac{\Delta_2}{1+\varepsilon} - \frac{2}{\omega}\right) \sum_{t\in[k]} \mathcal{F}(X_t^* \cap Y_t) \\
&\geq \Delta_1 \left(\frac{\Delta_2}{1+\varepsilon} - \frac{2}{\omega}\right) \frac{\omega}{\omega+1} \mathcal{G}(X^*),
\end{aligned}$$

where the third inequality is by $\mathcal{G}(\{e\}) = \left(1 - \frac{c_e}{\mathcal{F}(\{e\})}\right)\mathcal{F}(\{e\}) \geq \frac{1}{2}\mathcal{F}(\{e\})$ for any $e \in Y$, the penultimate and last inequalities both hold since $\sum_{t\in[k]} \mathcal{F}(X_t^* \cap Y_t) > \omega \max\{0, \max_{e\in N} \mathcal{G}(\{e\})\}$. In addition, the last inequality also depends on Lemma 1.

Let the approximations of the two cases be equal, i.e.,

$$\frac{1}{\omega+1} = \Delta_1 \left(\frac{\Delta_2}{1+\varepsilon} - \frac{2}{\omega}\right) \frac{\omega}{\omega+1},$$

then we obtain

$$\omega = \frac{\eta(1+\varepsilon)}{(1-\kappa-\frac{\eta(\eta-1)}{\theta^2(1-\kappa)})(1-\theta)(\eta-1)} + \frac{2\eta(1+\varepsilon)}{\eta-1}.$$

Therefore, we obtain the following parametric approximation ratio:

$$\frac{1}{\omega+1} = \frac{1}{\frac{\eta(1+\varepsilon)}{(1-\kappa-\frac{\eta(\eta-1)}{\theta^2(1-\kappa)})(1-\theta)(\eta-1)} + \frac{2\eta(1+\varepsilon)}{\eta-1} + 1}.$$

The rest is to maximize the approximation ratio $1/(\omega+1)$ by adjusting the parameters η, θ, κ and ε. Specifically, we summarized the computational results for different values of κ, η and θ under $\varepsilon = 0.1, 0.01$, and 0.001 in Table 1. Furthermore, we compared these results with those obtained by Dütting et al. (2023).

Corollary 1. *When $\kappa = 0.01$ and $\varepsilon = 0.001$ is fixed, Algorithm 2 achieves a 0.01890-approximation for Problem (1) in polynomial time by setting $\eta = 1.1811$ and $\theta = 0.66316$. Meanwhile, when the success probability reduces to linear (i.e., $\kappa = 0.00$), our algorithm achieves a 0.01931-approximation which is nealy 5 times better than the result obtained by Dütting et al. [11].*

Table 1. The approximation ratios calculated using the parameters η and θ with κ from 0.00 to 0.48 and fix $\varepsilon = 0.1, 0.01$, and 0.001, comparing them with the results from Dütting et al. (2023).

ε	κ	η	θ	Approx. Ratios	
				Ours	Dütting et al. [11]
	0.00	1.1831	0.66143	0.01761	
	0.01	1.1811	0.66316	0.01723	
	0.10	1.1551	0.66530	0.01394	
0.1	0.20	1.1271	0.66811	0.01062	0.00352
	0.30	1.1011	0.67064	0.00769	
	0.40	1.0771	0.67429	0.00523	
	0.48	1.0591	0.67413	0.00360	
	0.00	1.1831	0.66143	0.01914	
	0.01	1.1811	0.66316	0.01873	
	0.10	1.1551	0.66530	0.01516	
0.01	0.20	1.1271	0.66811	0.01155	0.00383
	0.30	1.1011	0.67064	0.00837	
	0.40	1.0771	0.67429	0.00569	
	0.48	1.0591	0.67413	0.00392	
	0.00	1.1831	0.66143	0.01931	
	0.01	1.1811	0.66316	0.01890	
	0.10	1.1551	0.66530	0.01530	
0.001	0.20	1.1271	0.66811	0.01165	0.00387
	0.30	1.1011	0.67064	0.00845	
	0.40	1.0771	0.67429	0.00574	
	0.48	1.0591	0.67413	0.00396	

5 Conclusion

In this paper, we studied k-type constrained combinatorial contract design for a principal dealing with multiple agents. We reformulated a bi-level optimization problem into a single-level subset selection problem using backward induction. Leveraging curvature and combining geometric scaling techniques with submodular function scaling, we developed a parameterized approximation algorithm that significantly improves upon Dütting et al. [11]. Several open questions warrant further exploration. For example, analyzing approximation ratios under non-monotonic reward functions, which may be more realistic, is an important direction. Additionally, extending the reward function beyond joint effort to k-submodular functions could lead to improved algorithms.

Acknowledgments. The first and third authors are supported by Beijing Natural Science Foundation (Nos. IS24001, Z220004) and Natural Science Foundation of China (No. 12201619). The fourth author is supported by Natural Science Foundation of China (No. 12101587) and Natural Science Foundation of Shandong Province (No. ZR2023MA031).

References

1. Scientific background on the 2016 nobel price in economic sciences. Royal Swedish Academy of Sciences (2016)
2. Babaioff, M., Feldman, M., Nisan, N.: Combinatorial agency. In: Proceedings of ACM EC, pp. 18-28 (2006)
3. Babaioff, M., Feldman, M., Nisan, N.: Free-riding and free-labor in combinatorial agency. In: Proceedings of SAG, pp. 109-121 (2009)
4. Babaioff, M., Feldman, M., Nisan, N.: Mixed strategies in combinatorial agency. J. Artifi. Intell. Res. **38**, 339–369 (2010)
5. Babaioff, M., Feldman, M., Nisan, N., Winter, E.: Combinatorial agency. J. Econ. Theory **147**(3), 999–1034 (2012)
6. Calinescu, G., Chekuri, C., Pal, M., Vondrák, J.: Maximizing a monotone submodular function subject to a matroid constraint. SIAM J. Comput. **40**(6), 1740–1766 (2011)
7. Carroll, G.: Robustness and linear contracts. Am. Econ. Rev. **105**(2), 536–563 (2015)
8. Conforti, M., Cornuejols, G.: Submodular set functions, matroids and the greedy algorithm: tight worst-case bounds and some generalizations of the rado-edmonds theorem. Discret. Appl. Math. **7**(3), 251–274 (1984)
9. Cong, L., He, Z.: Blockchain disruption and smart contracts. Rev. Finan. Stud. **32**(5), 1754–1797 (2019)
10. Dütting, P., Ezra, T., Feldman, M., Kesselheim, T.: Combinatorial contracts. In: Proceedings of FOCS, pp. 815-826 (2022)
11. Dütting, P., Ezra, T., Feldman, M., Kesselheim, T.: Multi-agent contracts. In: Proceedings of STOC, pp. 1311-1324 (2023)
12. Dütting, P., Ezra, T., Feldman, M., Kesselheim, T.: Multi-agent combinatorial contracts. In: Proceedings of SODA, pp. 1857-1891 (2025)
13. Dütting, P., Feldman, M., Gal Tzur, Y.: Combinatorial contracts beyond gross substitutes. In: Proceedings of SODA, pp. 92-108 (2024)
14. Dütting, P., Feldman, M., Gal-Tzur, Y., Rubinstein, A.: The query complexity of contracts. arXiv **2403**, 09794 (2024)
15. Dütting, P., Feldman, M., Talgam-Cohen, I.: Algorithmic contract theory: a survey. Foundations Trends Theoret. Comput. Sci. **16**(3–4), 211–412 (2024)
16. Dütting, P., Roughgarden, T., Talgam-Cohen, I.: Simple versus optimal contracts. In: Proceedings of ACM EC, pp. 369-387 (2019)
17. Ezra, T., Feldman, M., Schlesinger, M.: On the (In) approximability of Combinatorial Contracts. In: Proceedings of ITCS (2024)
18. Fisher, L., Nemhauser, L., Wolsey, A.: An analysis of approximations for maximizing submodular set functions-II, pp. 73–87. Springer, Berlin Heidelberg (1978)
19. Grossman, S., Hart, O.: An analysis of the principal-agent problem. Econometrica **51**(1), 7–45 (1983)

20. Ho, C.J., Slivkins, A., Vaughan, J.W.: Adaptive contract design for crowdsourcing markets: bandit algorithms for repeated principal-agent problems. J. Artifi. Intell. Res. **55**, 317–359 (2016)
21. Holmström, B.: Moral hazard and observability. Bell J. Econ. **10**(1), 74–91 (1979)
22. Holmström, B.: Moral hazard in teams. Bell J. Econ. **13**, 324–340 (1982)
23. Kaynar, N., Siddiq, A.: Estimating effects of incentive contracts in online labor platforms. Manage. Sci. **69**(4), 2106–2126 (2023)
24. Lehmann, B., Lehmann, D., Nisan, N.: Combinatorial auctions with decreasing marginal utilities. Games Econom. Behav. **55**(2), 270–296 (2006)
25. Nemhauser, L., Wolsey, A., Fisher, L.: An analysis of approximations for maximizing submodular set functions-I. Math. Program. **14**, 265–294 (1978)
26. Ross, A.: The economic theory of agency: the principal's problem. Am. Econ. Rev. **63**(2), 134–139 (1973)
27. Shavell, S.: Risk sharing and incentives in the principal and agent relationship. Bell J. Econ. 55-73 (1979)
28. Vondrák, J.: Submodularity and curvature: the optimal algorithm. RIMS Kokyuroku Bessatsu **B23**, 253–266 (2010)
29. R. Vuong, S. Dughmi, N. Patel, and A. Prasad. On supermodular contracts and dense subgraphs. *In: Proceedings of SODA*, 2024, pp. 109-132

Approximation Algorithms for the Maximum Connected Submodular Functions

Qinqin Gong, Zixuan Wang, Yang Lv, and Ruiqi Yang(✉)

Institute of Operations Research and Information Engineering, Beijing University of Technology, Beijing 100124, People's Republic of China
{gongqinqin,wzx1009,lvyang}@emails.bjut.edu.cn, yangruiqi@bjut.edu.cn

Abstract. Motivated by the challenge of maximizing connected coverage with limited UAVs in communication networks, we address the problem within a graph network framework $G = (V, E)$, where V represents potential UAV deployment positions and E denotes communication links between nodes. A utility function $f : 2^V \to \mathbb{R}_+$ is defined to characterize coverage efficiency. Under the constraint of limited field-of-view (FoV) UAVs, the objective is to identify a subset $S \subseteq V$ with $|S| \leq K$ that maximizes $f(S)$ while ensuring the induced subgraph $G[S]$ remains connected. We formulate this as the Maximum Connected Submodular function with Cardinality constraint (MCSC) problem and propose a $\frac{1-e^{-1}}{2\sqrt{K-1}+5}$-approximation algorithm, leveraging a novel tree decomposition technique. Additionally, we present a bicriteria $\left(\frac{(1-e^{-1})\alpha}{2\sqrt{K}+3\alpha}, \alpha^2\right)$-approximation algorithm for the problem, where $\alpha > 1$ is a constant. For a special case of the MCSC problem, where the submodular utility exhibits partial additivity when subsets are sufficiently far apart, we define the Maximum Connected h-Hop Submodular function with a Cardinality constraint (MCHSC) problem. We provide an approximation algorithm with a ratio of $(1-2\varepsilon)\left(\frac{1-e^{-1}}{5(h+1)+1} - \delta\right)$ when $K > 25h(h+1)-5$, where ε, δ are small positive constants and h captures the partial additivity property.

Keywords: Submodular optimization · Connectivity · Cardinality · Approximation algorithms

1 Introduction

In recent years, the deployment of Unmanned Aerial Vehicle (UAV) technology has facilitated its gradual integration into various fields. UAVs can function as aerial base stations, delivering wireless coverage to targeted areas or users. They play a pivotal role in disaster management by supporting emergency communication [6,15,17,23,27]. However, in many real-world scenarios, the number of available UAVs is limited. Consequently, the challenge of effectively deploying

F. V. Fomin and M. Xiao (Eds.): COCOON 2025, LNCS 15983, pp. 18–31, 2026.
https://doi.org/10.1007/978-981-95-0215-8_2

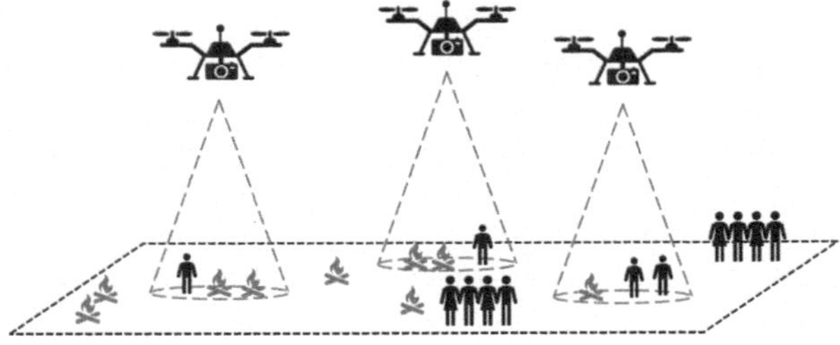

Fig. 1. A fire-affected area with trapped individuals is marked by black dashed lines. Three UAVs monitor the disaster, with their coverage zones shown by blue circular dashed lines. (Color figure online)

a communication network with a constrained number of UAVs has garnered significant attention from researchers.

Research on UAV network deployment can be broadly categorized into two main areas: (i) minimizing the number of UAVs required for deployment, and (ii) maximizing connected coverage with a limited number of UAVs. In the first category, Zhao et al. [29] proposed a centralized algorithm for optimal UAV placement. Sawalmeh et al. [21] introduced an iterative clustering and 3D placement algorithm to maximize coverage with minimal UAVs and transmit power. Zhang et al. [28] minimized UAVs for delay-bounded data collection, offering approximation algorithms with guarantees. Sabzehali et al. [22] optimized UAV counts for backhaul connectivity and full ground coverage, proposing a low-complexity algorithm. In the second category, Zhao et al. [29] developed a distributed motion control algorithm for bi-connected UAV networks. Gupta et al. [9] maximized throughput with a single UAV. Danilchenko et al. [4] addressed NP-hard connectivity constraints with fixed UAVs, proposing approximation algorithms with performance guarantees.

Our work falls into the second category of UAV deployment research, focusing on optimizing network deployment with a limited number of UAVs. The goal is to maximize coverage area and the number of rescued individuals while ensuring network connectivity (see Fig. 1). Connectivity is crucial for collaboration, conflict prevention, real-time adjustments, and data sharing, as highlighted in [14,26]. We model this problem using an undirected graph, where UAVs are nodes, and edges represent communication links. The coverage area and served users are modeled via a submodular prize function on the node set, while connectivity is reflected in the graph's structure. Since the problem is NP-hard, we focus on designing approximation algorithms.

1.1 Related Work

Research on UAV deployment problem primarily falls into two categories: one focuses on deploying networks with the minimum number of UAVs [21,22,28,29], while the other aims to optimize the deployment of a limited number of UAVs to maximize connected coverage.

Zhao et al. [29] proposed a distributed motion control algorithm for optimal UAV deployment. Gupta et al. [9] addressed 3D deployment for a single UAV using alternating optimization. Danilchenko et al. [4] developed approximation algorithms with performance guarantees for various scenarios. When UAV connectivity is ignored, the problem reduces to cardinality-constrained monotone submodular maximization. A $(1 - e^{-1})$-approximation greedy algorithm was proposed by [18], and [14] provided a $(1 - e^{-1})$-approximation when a root node must be included. With connectivity, Kuo et al. [14] proved the MCSC problem is NP-hard and proposed an $\Omega\left(\frac{1}{\sqrt{K}}\right)$-approximation algorithm for real-valued functions, specifically $\frac{1-e^{-1}}{2\sqrt{K}+11}$. They modeled the UAV network as an undirected graph. D'Angelo et al. [3] studied directed information dissemination, proposing a bicriteria algorithm with ratio $\left(\Omega\left(\frac{\varepsilon^3}{\sqrt{B}}\right), 1+\varepsilon\right)$, $\varepsilon \in (0, 1]$ Khuller et al. [12,13] studied the Budgeted Connected Dominating Set (BCDS) problem, aiming to find a connected subset S of at most K nodes to maximize dominated nodes. They proposed a $\frac{1-e^{-1}}{12}$-approximation algorithm. Xu et al. [25] developed an algorithm for maximizing throughput in UAV networks with an approximation ratio of $\frac{1-e^{-1}}{\lfloor\sqrt{K}\rfloor}$. Additionally, [26] introduced the h-hop independent submodular function, proposing a $\frac{1-e^{-1}}{2h+3}$-approximation algorithm for $\sqrt{K} \geq 2h + 3$, where h is a positive integer.

For the Maximum Connected Submodular set function with Budget constraint (MCSB) problem, ignoring connectivity reduces it to the classic knapsack problem, while considering connectivity leads to the maximum connected coverage problem with a budget, widely studied [1,11,16,19]. The MCSC algorithm builds on [14], which focuses on high-marginal-profit nodes within a smaller range using a $(1 - e^{-1})$-approximate algorithm for cardinality constraints. In contrast, [25] considers high-marginal-profit nodes with additional constraints, employing a greedy algorithm for knapsack constraints. While [25] improves the approximation ratio, [14] offers a faster running time. The MCHSC algorithm builds on [26], which maximizes users served by a connected network of K UAVs, mapping objectives to non-negative integers. This approach may limit coverage precision, as coverage sizes are not always integers. To address this, we propose an approximation algorithm for the real-valued MCHSC problem.

1.2 Organization

In Sect. 2, we define submodular functions and related concepts. In Sect. 3, we present the MCSC problem and its approximation algorithm. In Sect. 4, we introduce the MCHSC problem and its approximation algorithm. In Sect. 5, we conduct numerical experiments to validate the effectiveness of the algorithms. Finally, we conclude our paper in Sect. 6.

2 Preliminaries

We are given an undirected connected graph $G = (V, E)$, where V is the set of nodes and E is the set of edges. A path in the graph G is a sequence of distinct vertices $(v_1, v_2, \cdots, v_k)$ with a sequence of undirected edges (v_i, v_{i+1}) for $i \in \{1, 2, \cdots, k-1\}$. For any two nodes u and v in V, a path between u and v with the minimum cost of edges is called the shortest path between u and v. The cost of this shortest path, denoted by $d(u, v)$, represents the distance between u and v in the graph G. In our text, each edge has a cost of one, meaning that the shortest path between u and v is the path with minimum number of edges. For any two non-empty subsets A and B of V, let $d(A, B) = \min\{d(u, v) \mid u \in A, v \in B\}$. For a subset $S \subseteq V$, let $G[S] = (S, E[S])$, where $E[S] = \{(u, v) \in E \mid u, v \in S\}$. Given a tree T in G, let $w(T)$ denote the number of edges in T.

Definition 1. *Given a finite set V, a real-valued function $f : 2^V \to \mathbb{R}_+$ is defined as a normalized submodular function if it satisfies the following:*

(1) Normalization: $f(\emptyset) = 0$;
(2) Submodularity: $f(A) + f(B) \geq f(A \cup B) + f(A \cap B)$, $\forall A, B \subseteq V$.

The submodularity property can also be equivalently expressed as: $f(A \cup \{v\}) - f(A) \geq f(B \cup \{v\}) - f(B)$ for any subsets $A \subseteq B \subseteq V$ and $v \notin B$. Additionally, f is called a monotone submodular function if it satisfies $f(A) \leq f(B)$ for any subsets $A \subseteq B \subseteq V$. Next we restate the h-hop submodular function on a graph $G = (V, E)$ as follows.

Definition 2. *[26] Given a graph $G = (V, E)$, a fnction $f : 2^V \to \mathbb{R}_+$ is an h-hop submodular function if it satisfies:*

(1) f is a monotone submodular function, and
(2) $f(A \cup B) = f(A) + f(B)$ if $d(A, B) \geq h, \forall A, B \subseteq V$.

3 Approximating MCSC Problem

We consider the MCSC problem and give an improved approximation algorithm for the problem. Given an undirected and connected graph $G = (V, E)$ and a monotone submodular function $f : 2^V \to \mathbb{R}_+$, a positive integer K, the MCSC problem aims to find a set of nodes S such that $G[S]$ is a connected subgraph of G, maximizing the value of $f(S)$ while ensuring the cardinality of S does not exceed K. Let S_K^* represent the optimal solution of this instance and $OPT = f(S_K^*)$. Additionally, let T_K^* denote the spanning tree of $G[S_K^*]$.

3.1 Algorithm and Performance Guarantees

We propose an approximation algorithm for the MCSC problem in Algorithm 1 inspired by [14]. We introduce a new parameter m, which plays an important role in analyzing our algorithm. The main results are summarized in Theorem 1 and Theorem 2. Before analyzing these two theorems, we firstly establish the following three lemmas. The proofs appear in the full version of this paper.

Algorithm 1. Algorithm for the MCSC problem

Input: Graph $G = (V, E)$, a monotone submodular function $f : 2^V \to \mathbb{R}^{\geq 0}$, positive integer K, m
Output: S.
$S = \emptyset$;
for $r \in V$ **do**
  $V_r = \{v \mid d(r, v) \leq \max\{m-1, \lfloor\sqrt{K}\rfloor\}\}$;
  S_r is the solution of the cardinality constrained rooted monotone submodular maximization problem with root r must be included [14];
  if $f(S_r) > f(S)$ **then**
    $r^* = r$;
    $S = S_{r^*}$;
  end if
end for
for $v \in S_{r^*}$ **do**
  Find the shortest path from v to r^*, and add the new nodes in the path to S;
end for
return S.

Lemma 1. *Assume $m \leq \lfloor\sqrt{K-1}\rfloor + 1$, Algorithm 1 returns a feasible solution for the MCSC problem.*

Lemma 2. *For any tree T, there always exist $n_t \leq \frac{2w(T)}{m-1} + 1$ subtrees $T^i = (V^i, E^i)$ of T, where $|V^i| \leq m$, $\forall 1 \leq i \leq n_t$, $m \geq 2$, such that $\bigcup_{i=1}^{n_t} V^i = V(T)$ and $\bigcup_{i=1}^{n_t} E^i = E(T)$.*

Lemma 3. *Assume $m = \lfloor\sqrt{K}\rfloor$, Algorithm 1 achieves $\frac{1-e^{-1}}{2\sqrt{K}+11}$-approximation which reduces to the result stated by [14].*

Theorem 1. *Assume that $m = \lfloor\sqrt{K-1}\rfloor + 1$, Algorithm 1 achieves $\frac{1-e^{-1}}{2\sqrt{K-1}+5}$-approximation for the MCSC problem.*

Proof. We first assume that $K \geq 5$. Then, we find that $m = \lfloor\sqrt{K-1}\rfloor + 1 \geq 3$. In this case, we decompose the tree T_K^* into n_t subtrees, ensuring that the number of nodes in each subtree does not exceed m. Therefore, we derive

$$\begin{aligned} n_t &\leq 2\,\frac{K-1}{m-1} + 1 = 2\,\frac{K-1}{\lfloor\sqrt{K-1}\rfloor} + 1 \leq 2\,\frac{K-1}{\sqrt{K-1}-1} + 1 \\ &= 2\left(\sqrt{K-1} + 1 + \frac{1}{\sqrt{K-1}-1}\right) + 1 \leq 2\sqrt{K-1} + 5. \end{aligned}$$

For the case of $1 \leq K \leq 4$, the tree T_K^* contains K nodes, which can be covered by $n_t \leq K \leq 4 < 2\sqrt{K-1} + 5$ subtrees. Let $V' = \arg\max_{i \in \{1, \cdots, n_t\}} f(V^i)$. Since $|V'| \leq m$, and the distance between any two nodes is at most $m-1$. Therefore, $f(S_m^*) \geq f(V')$. By the submodularity of f, we have $OPT = f(V(T_K^*)) = f\left(\bigcup_{i=1}^{n_t} V^i\right) \leq \sum_{i=1}^{n_t} f(V^i) \leq n_t f(V')$. Select any node $r \in S_m^*$, where $S_m^* \subseteq V_r$

in Algorithm 1 and $|S_m^*| = m$. Then, we have $f(S_r) \geq (1-e^{-1})OPT' \geq (1-e^{-1})f(S_m^*)$, where OPT' denotes the optimal value of the maximum rooted submodular function f with constraints $|S| \leq m$, and $r \in S$ such that $f(S)$ is maximized. Therefore, we have $f(S) \geq f(S_{r^*}) \geq f(S_r) \geq (1-e^{-1})f(S_m^*) \geq (1-e^{-1})f(V') \geq (1-e^{-1})\frac{OPT}{n_t} \geq \frac{1-e^{-1}}{2\sqrt{K-1}+5}OPT$. □

Assume parameter m are chosen by $m = \alpha\sqrt{K} > \lfloor\sqrt{K-1}\rfloor + 1$, there exists a bicriteria approximation as follows:

Theorem 2. *For the MCSC problem, Algorithm 1 obtains a bicriteria ratio of* $\left(\frac{(1-e^{-1})\alpha}{2\sqrt{K}+3\alpha}, \alpha^2\right)$ *when* $m = \alpha\sqrt{K} > \lfloor\sqrt{K-1}\rfloor + 1$.

Proof. Let $m = \alpha\sqrt{K}$, where $m > \lfloor\sqrt{K-1}\rfloor + 1$. It follows that $m - 1 \geq \lfloor\sqrt{K-1}\rfloor + 1 \geq \lfloor\sqrt{K}\rfloor$. Consequently, $V_r = \{v \mid d(v,r) \leq m-1\}$ in line 4 of Algorithm 1. Additionally, we have $\alpha = \frac{m}{\sqrt{K}} \geq \frac{\lfloor\sqrt{K-1}\rfloor+2}{\sqrt{K}} \geq \frac{\sqrt{K-1}+1}{\sqrt{K}} \geq 1$. If $\alpha = 1$, then $\sqrt{K}$ must be an integer. However, this would imply $\sqrt{K} = \lfloor\sqrt{K}\rfloor \leq \lfloor\sqrt{K-1}\rfloor + 1$, which is impossible. Therefore, $\alpha > 1$ and $m = \alpha\sqrt{K} \geq 2$ since m is an integer. We obtain

$$\begin{aligned} n_t &= 2\,\frac{K-1}{m-1} + 1 = 2\,\frac{K-1}{\alpha\sqrt{K}-1} + 1 \\ &= 2\left(\frac{\sqrt{K}}{\alpha} + \frac{1}{\alpha^2}\right) + 2\frac{\frac{1}{\alpha^2}-1}{\alpha\sqrt{K}-1} + 1 \\ &\leq 2\frac{\sqrt{K}}{\alpha} + 3. \end{aligned}$$

Similar to the proof of Theorem 1, we derive $f(S) \geq f(S_{r^*}) \geq f(S_r) \geq (1-e^{-1})f(S^*_{\alpha\sqrt{K}}) \geq (1-e^{-1})\frac{OPT}{n_t} \geq \frac{(1-e^{-1})\alpha}{2\sqrt{K}+3\alpha}OPT$. The number of nodes in the solution S satisfies $|S| \leq m + (m-1)(m-2) = \alpha\sqrt{K} + (\alpha\sqrt{K}-1)(\alpha\sqrt{K}-2) \leq \alpha^2K - 2\alpha\sqrt{K} + 2 \leq \alpha^2K$. Therefore, Algorithm 1 derives a bicriteria ratio $\left(\frac{(1-e^{-1})\alpha}{2\sqrt{K}+3\alpha}, \alpha^2\right)$ when $m = \alpha\sqrt{K} > \lfloor\sqrt{K-1}\rfloor + 1$. □

4 Approximating the MCHSC Problem

In this section, we consider the MCHSC problem and propose an approximation algorithm for this problem.

Algorithm 2. Generalized Prize Assignment Algorithm

Input: An undirected, connected graph $G = (V, E)$, an h-hop submodular function $f : 2^V \to \mathbb{R}_+$ on G, a starting node $v \in V$.
Output: A prize mapping function, i.e., $p : V \to \mathbb{R}_+$.

$p(v) = f(\{v\})$;
$A = \{v\}$;
$U = V \setminus A$;
while $U \neq \emptyset$ **do**
  $u = \arg\max\{f(A \cup \{v\}) - f(A) \mid v \in U\}$;
  $p(u) = f(A \cup \{u\}) - f(A)$;
  $A = A \cup \{u\}$;
  $U = U \setminus \{u\}$;
end while
return $p : V \to \mathbb{R}^{\geq 0}$.

4.1 Node Number Minimization Quota Problem

Inspired by the classic quota problem [1], we firstly state a Node number Minimization Quota (NMQ) problem as follows. Given an undirected and connected graph $G = (V, E)$, an additive function $p : V \to \mathbb{R}_+$, and a positive number $Q > 0$, the goal is to find a tree in G with the minimum number of nodes such that the total prize satisfies $p(V(T)) \geq Q$. For convenience, we use the shortcut $p(T) = p(V(T)) = \sum_{v \in V(T)} p(v)$.

4.2 Warm-Up Algorithms

In this section, we introduce two algorithms that are utilized in the algorithm for the MCHSC problem. Given an instance of the MCHSC problem with $G = (V, E)$, a function f, and a positive integer K, let $n = |V|$ and $m = |E|$.

• *Generalized Prize Assignment Algorithm:* We can use Algorithm 2 inspired by [26] to assign prizes to the nodes in V. We extend the initial prize assignment algorithm, no longer requiring the prize allocation function to be an integer function.

• *Bicriteria Approximation for the NMQ problem:* An algorithm is a bicriteria (α, β)-approximation for the NMQ problem if for any instance of the problem, it returns a solution T such that $|T| \leq \alpha|T^*|$ and $p(T) \geq \beta Q$, where T^* is the optimal tree.

Our algorithm mainly follows by the idea proposed by Bateni et al. in [1]. The pseudo-codes are summarized in Algorithm 3 and the approximation ratio is $(5, 1 - 2\varepsilon)$ for any $\varepsilon > 0$.

Lemma 4. *[1] Given any α-approximation algorithm for the k-ST problem, there exists an $(\alpha, 1 - 2\varepsilon)$-approximation algorithm of the NMQ problem.*

Corollary 1. *Algorithm 3 is a $(5, 1 - 2\varepsilon)$-approximation algorithm for the NMQ problem.*

Lemma 5. *The tree* $T = T^* \bigcup(\cup_{i=1}^{K} P_i)$ *with fewer than* $(K-1)h+1$ *nodes can be decomposed into* $t \le \alpha(h+1)+1$ *subtrees* $T^j = (V^j, E^j)$*. Each subtree satisfies* $|V^j| \le \lfloor \frac{K}{\alpha} \rfloor$ *for all* $j \in \{1, 2, \cdots, t\}$ *when* $K > \alpha^2 h^2 + \alpha^2 h - \alpha$ *and* $\alpha > 1$*, where* $\alpha(h+1)$ *is a positive integer.*

Algorithm 3. Approximation Algorithm for the NMQ problem [1,2]

Input: An undirected, connected graph $G = (V, E)$, an additive function $p : V \to \mathbb{R}_+$, a positive number $Q > 0$, and $\varepsilon > 0$.
Output: A tree T in G, such that $|V(T)|$ is minimized and $p(T) \ge Q$.

$V(T) \leftarrow \emptyset, E(T) = \emptyset, R = \emptyset, E_0 = \emptyset, n = |V|$;
for $e \in E$ **do**
  $c(e) = 1$;
end for
for $v \in V$ **do**
  $n_v = \left\lceil \dfrac{np(v)}{\varepsilon Q} \right\rceil$;
  Let $R_v = \{r_1, r_2, \cdots, r_{n_v}\}$;
  for $r \in R_v$ **do**
    Connect the node v and node r, and let $c((v, r)) = 0$;
    $E_v = E_v \cup (v, r)$;
  end for
  $R = R \cup R_v$;
  $E_0 = E_0 \cup E_v$;
end for
Let $G' = (V', E')$, $V' = V \cup R, E' = E \cup E_0$;
Let $k = \left\lfloor \dfrac{n}{\varepsilon} \right\rfloor$, V is the set of steiner nodes, R is the set of terminals, $c : E' \to \{0, 1\}$;
Obtain the tree T' by using 5-approximation algorithm of k-ST problem[2] to the instance $I' = < G' = (V', E'), R, c, k >$;
$V(T) = V(T') \cap V$, $E(T) = E(T') \cap E$;
Return $T = (V(T), E(T))$.

4.3 Algorithms and Performance Guarantees

We propose our algorithm for the MCHSC problem in Algorithm 4. The main results are summarized as follows.

Theorem 3. *When* $K > 25h(h+1) - 5$ *and fix* $\varepsilon > 0, \delta > 0$*, Algorithm 4 achieves* $(1-2\varepsilon)\left(\frac{1-e^{-1}}{5(h+1)+1} - \delta\right)$*-approximation for the MCHSC problem.*

Proof. Assume that K and h satisfy $K > 25h(h+1) - 5$. We focus on the iteration in which z is the starting node firstly. According to Lemma 5, there exists a tree T' satisfying $|V(T')| \le \lfloor \frac{K}{5} \rfloor$ and $p(T') \ge \frac{1-e^{-1}}{5(h+1)+1} OPT$.

In the case of $f(\{z\}) \leq q \leq \frac{1-e^{-1}}{5(h+1)+1}OPT$, the optimal value of the NMQ problem satisfies $OPT_Q \leq \lfloor \frac{K}{5} \rfloor$. After applying Algorithm 3, we obtain a tree T_q such that $|V(T_q)| \leq 5OPT_Q \leq K$. It implies that $|V(T_q)| > K$ if and only if $q > \frac{1-e^{-1}}{5(h+1)+1}OPT$. In Algorithm 4, the upper bound *right* of search interval $[left, right]$ is a value of q such that $|V(T_q)| > K$, and $right > \frac{1-e^{-1}}{5(h+1)+1}OPT$ always holds. At the end of any iteration in Algorithm 4, the lower bound *left* and the upper bound *right* satisfy $left + \delta \geq right$. Therefore, $left \geq right - \delta > \frac{1-e^{-1}}{5(h+1)+1}OPT - \delta > \left(\frac{1-e^{-1}}{5(h+1)+1} - \delta\right)OPT$, where we assume $OPT \geq 1$. As a result, the iteration with the starting node z will return a tree T_q satisfying $p(T_q) \geq (1-2\varepsilon)left > (1-2\varepsilon)\left(\frac{1-e^{-1}}{5(h+1)+1} - \delta\right)OPT$.

In any iteration i, assume the tree searched in this iteration by the Algorithm 4 is T_i. At the end of our analysis, we will establish the relationship between $f(T_i)$ and $p(T_i)$. For any node $v \in V$, let D_v denote the set of nodes that have been assigned a prize before v in this iteration. Let $D'_v = D_v \cap V(T_i)$. Therefore, we have

$$\begin{aligned} p(T_i) &= \sum_{v \in V(T_i)} p(v) = \sum_{v \in V(T_i)} f(D_v \cup \{v\}) - f(D_v) \\ &\leq \sum_{v \in V(T_i)} f(D'_v \cup \{v\}) - f(D'_v) = f(T_i). \end{aligned}$$

At the end of Algorithm 4, the solution satisfies $f(S) = \max_i\{f(T_i)\} \geq (1-2\varepsilon)\left(\frac{1-e^{-1}}{5(h+1)+1} - \delta\right)OPT$. □

A feasible solution of the MCHSC problem: Let T^* denote the spanning tree of $G[S^*]$, where $S^* = \{s_1, s_2, \cdots, s_K\}$. Define $z = \arg\max\{f(\{v\}) \mid v \in S^*\}$, $S_{h-1} = \{v \mid d(v, S^*) \leq h-1\}$, and $S_h = \{v \mid d(v, S^*) > h\}$. It can be observed that if a node is in S^*, then it must also be in S_{h-1} since $d(v, S^*) = 0 \leq h-1$. Algorithm 2 assigns prizes to nodes in V with the starting node z, and label the nodes as $\{v_1, v_2, \cdots, v_n\}$ according to the order of prize assignment, where $\{v_1\} = \{z\}$. Search for the first K nodes in the set S_{h-1}, and denote their subscripts as $l_1, \cdots, l_K$. Let $Z_K = \{v_{l_1}, v_{l_2}, \cdots, v_{l_K}\}$, and represent these nodes as $\{z_1, z_2, \cdots, z_K\}$. This ensures $p(Z_K) \geq (1-e^{-1})OPT$. Each node $z_i \in Z_K$ has a path P_i to any node in T^*. Let $T = T^* \bigcup (\cup_{i=1}^K P_i)$. Therefore, the number of the edges in T is bounded by

$$\begin{aligned} |E(T)| &\leq |E(T^*)| + \sum_{i=1}^{K} |P_i| = |E(T^*)| + \sum_{i=2}^{K} |P_i| \\ &\leq (K-1) + (K-1)(h-1) = (K-1)h. \end{aligned}$$

The first inequality accounts for possible intersections between the edges in $E(T^*)$ and P_i, the second equality is due to $E(P_1) = 0$ since $\{z_1\} = \{z\} \subseteq S^*$, the third inequality reflects that P_i is the shortest path between some node in S_{h-1} and any node in T^*, ensuring $|P_i| \leq h-1$. Simultaneously, we have

Algorithm 4. Approximation Algorithm for the MCHSC problem

Input: Graph $G = (V, E)$, an h-hop submodular function $f : 2^V \to \mathbb{R}_{\geq 0}$, a positive integer K, small positive numbers ε, δ.
Output: A feasible subset $S \subseteq V$.

```
S ← ∅;
for v ∈ V do
    if f({v}) > 0 then
        Apply Algorithm 2 to get prizes for each node with the starting node v;
    else
        continue; /* consider the next node being the starting node.*/
    end if
    Let left ← f({v}) and right ← f(V)
    while left + δ < right do
        q ← (left + right) / 2;
        Apply Algorithm 3 to find a tree T_q in G, where p(v) ← p(v), Q ← q;
        if |V(T_q)| ≤ K then
            left ← q;
        else
            right ← q;
        end if
    end while
    q ← left;
    Apply Algorithm 3 to find a tree T_q in G, where p(v) ← p(v), Q ← q;
    if f(V(T_q)) > f(S) then
        Let S ← V(T_q);
    end if
end for
return S
```

$|V(T)| = |E(T)| + 1 \leq (K-1)h + 1$. Since the nodes in Z_K are included in $V(T)$, we have $p(T) \geq p(Z_K) \geq (1 - e^{-1})OPT$. The tree T may not necessarily be a feasible solution, but we can decompose it to obtain a feasible solution of the MCHSC problem.

According to Lemma 5, let $T' = (V', E')$ with $V' = \arg\max\{p(V^j) \mid j \in \{1, 2, \cdots, t\}\}$ where the starting node is z. The additivity of p implies $p(T) = \sum_{v \in V(T)} p(v) \leq \sum_{j=1}^{t} p(T^j) \leq tp(V')$. Additionally, $|V(T')| \leq \lfloor \frac{K}{\alpha} \rfloor$ when $K > \alpha^2 h^2 + \alpha^2 h - \alpha$. Let $\alpha = 5$. When $K > 25h(h+1) - 5$, there exists a tree T' satisfing $p(T') \geq \frac{1-e^{-1}}{5(h+1)+1} OPT$ and $|V(T')| \leq \lfloor \frac{K}{5} \rfloor$.

In Algorithm 4, each node with $f(\{v\}) > 0$ is used as the starting node once. Since $z = \arg\max\{f(\{v\}) \mid v \in S^*\}$, it follows that $f(S^*) \leq Kf(\{z\})$. If $f(\{z\}) = 0$, then $OPT = f(S^*) = 0$, and this case can be ignored. Therefore, the node z will be used as the starting node once. As a result, there exists a tree T' such that $p(T') \geq \frac{1-e^{-1}}{5(h+1)+1} OPT$ and $|V(T')| \leq \lfloor \frac{K}{5} \rfloor$.

5 Numerical Experiments

Consider the MCSC and MCHSC problems with K UAVs in a $4000 \times 4000\,\mathrm{m}^2$ area serving 200 users. The users have real-valued weights and follow a fat-tailed, nonuniform distribution. Each UAV hovers at $H = 300$ m, with inter-UAV communication range $R = 800$ m and user-UAV communication range r'. The coverage range of each UAV is $r = \sqrt{r'^2 - H^2}$. We divide the $4000 \times 4000\,\mathrm{m}^2$ area into 64 small squares of 500 m side length, as shown in Fig. 2. UAVs can be positioned at 49 candidate points at intersections of dashed lines, represented by coordinates (x_i, y_i) for $i \in \{1, 2, \cdots, 49\}$, ignoring altitude. Users are represented as points (x, y) in the same plane. A user is served if $(x - x_i)^2 + (y - y_i) \leq r^2$. The submodular function f is then defined. We solve this problem using Algorithm 1 and a simplified version of Algorithm 4 (running steps 2–23 from a single node, following HopAlg [26]).

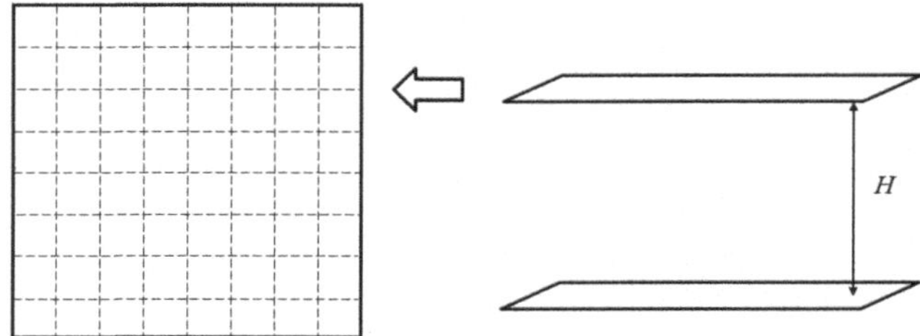

Fig. 2. A square with the size $4000 \times 4000\,\mathrm{m}^2$ and the divided method

In Algorithm 1, we set $m = \lfloor \sqrt{K-1} \rfloor + 1$, ensuring a feasible solution with an approximation ratio of $\frac{1-e^{-1}}{2\sqrt{K-1}+5}$. In Algorithm 4, real values are rounded down to integers, as the algorithm operates on integer fields. Additionally, we set $\varepsilon = 0.1$, $\delta = 0.02$. Initially, we set $K = 8$ UAVs and vary the coverage range from 150 to 350. We compare the coverage weight of Algorithm 1 and Algorithm 4 with GreedyAlg, RandAlg, HopAlg [26], and MCSAlg [14], as shown in Fig. 3. The corresponding running times are illustrated in Fig. 4. Algorithm 1 consistently outperforms MCSAlg, achieving up to 85.1% higher coverage weight at $r = 200$ m, with similar running times. Thus, Algorithm 1 is superior to MCSAlg. Algorithm 4 outperforms HopAlg in some ranges, with up to 30.3% higher weight at $r = 300$ m. Although its running times are higher, Algorithm 4 directly handles real-valued scenarios. Next, we set the UAV coverage range to 150 m and vary the number of UAVs K from 10 to 18 in steps of 2. We compare the coverage weight of Algorithm 1 and Algorithm 4 with GreedyAlg, RandAlg, HopAlg [26], and MCSAlg, as shown in Fig. 5. Algorithm 1 consistently outperforms MCSAlg, achieving up to 15.8% higher coverage weight with similar running times, demonstrating its superiority. Algorithm 4 outperforms HopAlg in some cases, with up to 8% higher weight at $K = 10$. Although its running times are longer, Algorithm 4 directly handles real-valued scenarios, unlike HopAlg, which requires adjustments. For Algorithm 4, with $h = 2$ as derived in [26], it provides a $\left(\frac{1-e^{-1}}{20} - 0.016\right)$-approximation solution when K and h meet the specified conditions.

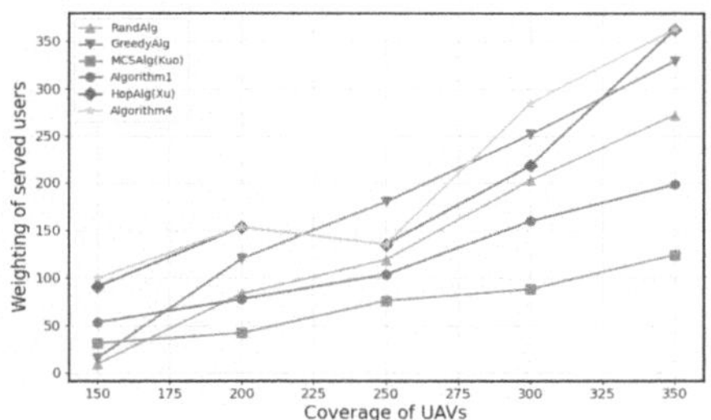

Fig. 3. The performance of different algorithms by varying r from 150 m to 350 m, R = 800 m, 200 users and K = 8.

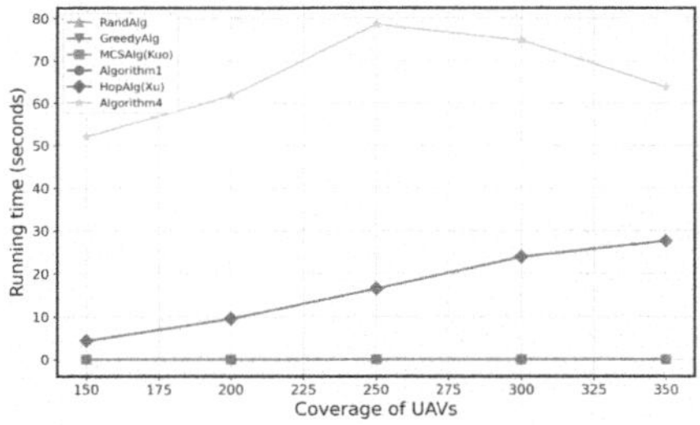

Fig. 4. The performance of different algorithms by varying the coverage range r of a UAV from 150 m to 350 m while fixing R = 800 m, when there are 200 users. K = 8.

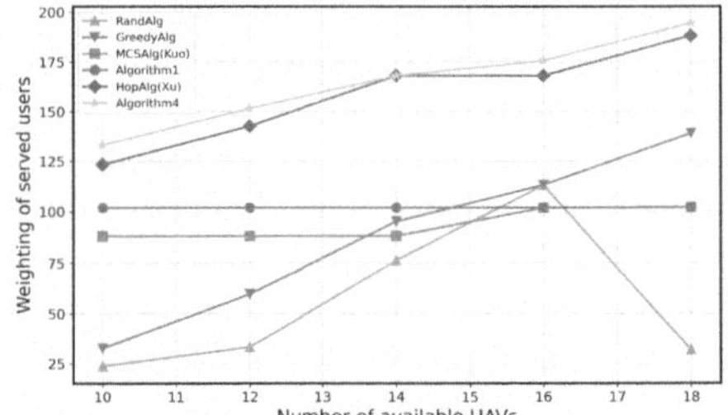

Fig. 5. The performance of different algorithms by varying the number of available UAVs K from 10 to 18 while fixing $R = 800$ m, when there are 200 users and the coverage range of UAVs is 150 m.

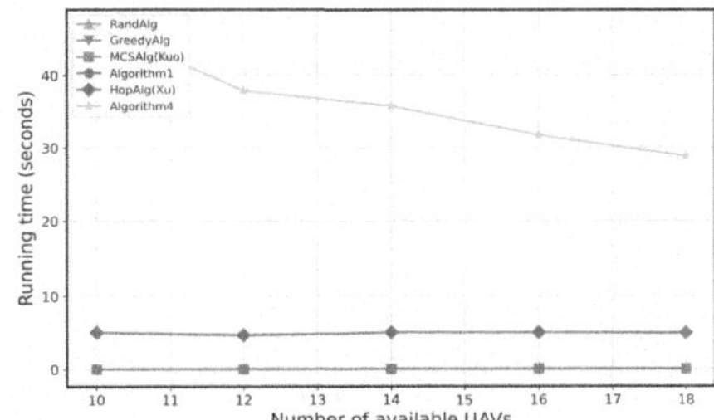

Fig. 6. The performance of different algorithms by varying the number of available UAVs K from 10 to 18 while fixing $R = 800$ m, when there are 200 users and the coverage range of UAVs is 150 m.

6 Conclusion

We addressed the maximum connected coverage problem with limited UAVs by formulating as the MCSC and the MCHSC problems, respectively. For the MCSC problem, we proposed an improved approximation algorithm by introducing a new tree-decomposition technique. For the special setting of MCHSC problem, we provided an improved algorithm with partial additive parameter. Future work could explore extending these algorithms to dynamic environments, where user positions or network conditions change over time. Additionally, incorporating constraints like energy consumption or UAV mobility patterns could enhance the proposed solutions. This would further improve their real-world applicability.

Acknowledgments. The first three authors are supported by Beijing Natural Science Foundation (Nos. Z220004, IS24001). The fourth author is supported by Natural Science Foundation of China (Nos. 12101587, 12201619).

References

1. Bateni, M.H., Hajiaghayi, M.T., Liaghat, V.: Improved approximation algorithms for (budgeted) node-weighted Steiner problems. SIAM J. Comput. **47**(4), 1275–1293 (2018)
2. Chudak, F.A., Roughgarden, T., Williamson, D.P.: Approximate k-MSTs and k-Steiner trees via the primal-dual method and Lagrangean relaxation. Math. Program. **100**, 411–421 (2004)
3. D'Angelo, G., Delfaraz, E., Gilbert, H.: Budgeted out-tree maximization with submodular prizes. arXiv preprint arXiv:2204.12162 (2022)
4. Danilchenko, K., Nutov, Z., Segal, M.: Covering users with QoS by a connected swarm of drones: graph theoretical approach and experiments. IEEE/ACM Trans. Netw. (2022)
5. Demaine, E.D., Hajiaghayi, M.T., Klein, P.N.: Node-weighted steiner tree and group steiner tree in planar graphs. In: International Colloquium on Automata, Languages, and Programming, pp. 328–340 (2009)
6. Erdelj, M., Natalizio, E.: UAV-assisted disaster management: applications and open issues. In: International Conference on Computing, Networking and Communications, pp. 1–5 (2016)
7. Garg, N.: Saving an epsilon: a 2-approximation for the k-MST problem in graphs. In: Proceedings of ACM Symposium on Theory of Computing, pp. 396-402 (2005)
8. Goemans, M.X., Williamson, D.P.: A general approximation technique for constrained forest problems. SIAM J. Comput. **24**(2), 296–317 (1995)
9. Gupta, N., Agarwal, S., Mishra, D.: UAV deployment for throughput maximization in a UAV-assisted cellular communications. In: Proceedings of International Symposium on Personal, Indoor and Mobile Radio Communications, pp. 1055-1060 (2021)
10. Huang, L., Li, J., Shi, Q.: Approximation algorithms for the connected sensor cover problem. Theoret. Comput. Sci. **809**, 563–574 (2020)
11. Johnson, D.S., Minkoff, M., Phillips, S.: The prize collecting steiner tree problem: theory and practice. In: Proceedings of ACM - SIAM Symposium on Discrete Algorithms, vol. 1(0.6), p. 4 (2000)
12. Khuller, S., Purohit, M., Sarpatwar, K.K.: Analyzing the optimal neighborhood: algorithms for budgeted and partial connected dominating set problems. In: Proceedings of ACM-SIAM Symposium on Discrete Algorithms, pp. 1702-1713 (2014)
13. Khuller, S., Purohit, M., Sarpatwar, K.K.: Analyzing the optimal neighborhood: Algorithms for partial and budgeted connected dominating set problems. SIAM J. Discret. Math. **34**(1), 251–270 (2020)
14. Kuo, T.W., Lin, K., Tsai, M.J.: Maximizing submodular set function with connectivity constraint: theory and application to networks. IEEE/ACM Trans. Netw. **23**(2), 533–546 (2014)

15. Liu, M., Yang, J., Gui, G.: DSF-NOMA: UAV-assisted emergency communication technology in a heterogeneous Internet of Things. IEEE Internet Things J. **6**(3), 5508–5519 (2019)
16. Moss, A., Rabani, Y.: Approximation algorithms for constrained node weighted steiner tree problems. SIAM J. Comput. **37**(2), 460–481 (2007)
17. Nguyen, L.D., Nguyen, K.K., Kortun, A., Duong, T.Q.: Real-time deployment and resource allocation for distributed UAV systems in disaster relief. In: Proceedings of International Workshop on Signal Processing Advances in Wireless Communications, pp. 1-5 (2019)
18. Nemhauser, G.L., Wolsey, L.A., Fisher, M.L.: An analysis of approximations for maximizing submodular set functions-I. Math. Program. **14**, 265–294 (1978)
19. Rabani, Y., Scalosub, G.: Bicriteria approximation tradeoff for the node-cost budget problem. ACM Trans. Algorithms **5**(2), 1–14 (2009)
20. Sviridenko, M.: A note on maximizing a submodular set function subject to a knapsack constraint. Oper. Res. Lett. **32**(1), 41–43 (2004)
21. Sawalmeh, A., Othman, N.S., Liu, G., Khreishah, A., Alenezi, A., Alanazi, A.: Power-efficient wireless coverage using minimum number of UAVs. Sensors **22**(1), 223 (2021)
22. Sabzehali, J., Shah, V.K., Fan, Q., Choudhury, B., Liu, L., Reed, J.H.: Optimizing number, placement, and backhaul connectivity of multi-UAV networks. IEEE Internet Things J. **9**(21), 21548–21560 (2022)
23. Wang, B., Sun, Y., Sun, Z., Nguyen, L.D., Duong, T.Q.: UAV-assisted emergency communications in social IoT: a dynamic hypergraph coloring approach. IEEE Internet Things J. **7**(8), 7663–7677 (2020)
24. Williamson, D.P., Shmoys, D.B.: The design of approximation algorithms. Cambridge university press (2011)
25. Xu, W., et al.: Throughput maximization of UAV networks. IEEE/ACM Trans. Netw. **30**(2), 881–895 (2021)
26. Xu, W., et al.: An approximation algorithm for the h-Hop independently submodular maximization problem and its applications. IEEE/ACM Trans. Netw. **31**(3), 1216–1229 (2022)
27. Zhao, N., et al.: UAV-assisted emergency networks in disasters. IEEE Wirel. Commun. **26**(1), 45–51 (2019)
28. Zhang, J., et al.: Minimizing the number of deployed UAVs for delay-bounded data collection of IoT devices. In: IEEE INFOCOM 2021-IEEE Conference on Computer Communications, pp. 1-10 (2021)
29. Zhao, H., Wang, H., Wu, W., Wei, J.: Deployment algorithms for UAV airborne networks toward on-demand coverage. IEEE J. Sel. Areas Commun. **36**(9), 2015–2031 (2018)

Approximating Per-Scenario Bound for the Two-Stage Stochastic Facility Location Problem

Mengzhen Li[1], Chenchen Wu[2](✉), Dachuan Xu[1], and Yicheng Xu[3]

[1] Institute of Operations Research and Information Engineering, Beijing University of Technology, Beijing 100124, People's Republic of China
lmz9710@emails.bjut.edu.cn, xudc@bjut.edu.cn

[2] Institute of Operations Research and Systems Engineering, College of Science, Tianjin University of Technology, Tianjin 300384, People's Republic of China
wu_chenchen_tjut@163.com

[3] Shenzhen Insitute of Advanced Technology, Chinese Academy of Sciences, Beijing, People's Republic of China
yc.xu@siat.ac.cn

Abstract. The two-stage stochastic facility location problem (2-SFLP) involves selecting initial facility locations under uncertainty, given known probabilities for each demand scenario. After the actual demand scenario is realized, additional facilities in the second stage, which incur higher costs, may be added to reduce the overall expected cost, including both opening and connection expenses. We present an improved per-scenario bound of 2.322 for 2-SFLP using the LP-rounding algorithm from prior work. By introducing the integrated distance estimation technique, we offer a more refined analysis. This technique, which involves a detailed estimation through a non-negative linear combination of the maximum and average distances within the neighborhood of an arbitrary client-scenario pair, has potential applications for the analysis of other facility location problems.

Keywords: Facility location · Approximation algorithm · Two-stage stochastic · Per-scenario bound

1 Introduction

The facility location problem arises from the determination of locations for factories, stations, schools, proxy servers, and other facilities, and it has a strong practical background. As a result, many researchers are interested in this problem. The un-capacitated facility location problem (UFLP) is the most fundamental type of facility location problem. In the UFLP, we have a set of facilities F and a set of clients C. Each facility $i \in F$ has an associated opening cost f_i, and the cost to connect facility i to client $j \in C$ is c_{ij}. The objective is to select a subset of facilities $\hat{F}$ from F to open and serve all clients while minimizing the total costs of opening and connecting.

F. V. Fomin and M. Xiao (Eds.): COCOON 2025, LNCS 15983, pp. 32–43, 2026.
https://doi.org/10.1007/978-981-95-0215-8_3

The facility location problem has several variants, with 2-SFLP as a key variant particularly relevant to practical applications. In the 2-SFLP, the facility set F, the client set C, and information about K scenarios are given. Each scenario k corresponds to a client subset $C_k \subseteq C$, with the probability of scenario k occurring given by $p_k \geq 0$ for $k = 1, 2, \ldots, K$, where $\sum_{k=1}^{K} p_k = 1$ and K is a polynomial in $|C|$. The problem consists of two stages: In the first stage (also called scenario 0), facilities are selected to serve all potential clients. In the second stage, for each scenario and its associated probability, determine which facilities to open. Facilities opened in a given scenario can only serve clients in that same scenario. The facility opening cost during the first stage is f_i^0, and during the second stage is f_i^k, with $f_i^0 \leq f_i^k$. The objective of the 2-SFLP is to decide which facilities to open at each stage and connect clients to the appropriate facilities in each scenario, aiming to minimize the expected total cost of opening facilities and serving clients.

The standard approximation ratio for 2-SFLP fails to reflect the relationships among distinct scenarios. Gupta et al. [5] from the perspective of risk aversion, propose the local algorithm, which inspires many scholars to study per-scenario bounds. Therefore, the per-scenario bound is introduced, providing an upper limit on the cost ratio between a feasible solution and the optimal LP fraction for each scenario. Specifically, an algorithm is said to have a per-scenario bound of γ if, for any scenario k, we have

$$\frac{\sum_{i\in\mathcal{F}_0} f_i^0 + \sum_{i\in\mathcal{F}_k} f_i^k + \sum_{j\in C_k} \min_{i\in\mathcal{F}_0\cup\mathcal{F}_k} c_{ij}}{\sum_{i\in F} f_i^0 \tilde{y}_i^0 + \sum_{i\in F} f_i^k \tilde{y}_i^k + \sum_{i\in F}\sum_{j\in C_k} c_{ij}^{0k}\tilde{x}_{ij}^{0k} + \sum_{i\in F}\sum_{j\in C_k} c_{ij}^{kk}\tilde{x}_{ij}^{kk}} \leq \gamma,$$

where $\tilde{y}_i^0$ and $\tilde{y}_i^k$ denote the fractional variables for facility opening in scenarios 0 and k, respectively; $\tilde{x}_{ij}^{0k}$ and $\tilde{x}_{ij}^{kk}$ denote the fractional variables for connecting client-scenario pairs in different scenarios; $\mathcal{F}_0$ and $\mathcal{F}_k$ represent the sets of facilities opened in the first stage and in scenario k of the second stage, respectively. Thus, while the per-scenario bound can be used as an approximation ratio, conventional approximation ratios typically do not serve as per-scenario bounds.

1.1 Literature Review

For the UFLP, Shmoys et al. [12] are the first to propose an approximation algorithm with a constant factor of 3.16, introduced in 1997. Jain and Vazirani [6] later improve the approximation to 3 with the primal-dual algorithm. Since then, advancements have led to improvements in the approximation ratio. Chudak and Shmoys [3] advance it to $(1 + 2/e)$. Jain et al. [7] improve it to 1.61. Mahdian et al. [10] further improve this to 1.52. Lately, the approximation ratio is improved to 1.50 by Byrka and Aardal [1]. Li [8] propose the best 1.488-approximation algorithm.

Ravi and Sinha [11] introduce the 2-SFLP and propose an 8-approximation algorithm using linear programming rounding. Mahdian [9] improves the approximation ratio to 3, and Srinivasan [13] tightens it to 2.369 with the dual-fitting algorithm. Byrka and Srinivasan [2] obtain a 2.2975-approximation algorithm.

Swamy [14] introduces a per-scenario bound of 3.225 for the problem. Srinivasan [13] groups clients based on the fractional optimal solution and applies a dual-fitting algorithm to solve the linear programming problem, achieving a per-scenario bound of 3.095. Byrka and Srinivasan [2] enhance this further, applying the linear programming rounding technique to reduce the per-scenario bound to 2.4957. Wu et al. [15] improve the per-scenario bound to 2.3613 by an LP rounding approximation algorithm.

1.2 Contributions

In algorithm analysis, given a client-scenario pair and its neighborhood, we need to estimate a non-negative linear combination involving both the maximum and average distances within the neighborhood. In the analysis of [15], these two distances were analyzed separately for upper bounds. This led to a relatively large estimation of the upper bound for the non-negative linear combination of these two distances.

For the 2-SFLP, we improve the per-scenario bound to 2.322 by using integrated distance estimation. To our knowledge, this is the first time to adopt the integrated distance estimation technique in the two-stage stochastic facility location area. We believe this unified approach to combined distance estimation has the potential to extend to the analysis of other similar problems.

1.3 Organization

The rest of the paper is structured as follows. Section 2 introduces the model of the 2-SFLP and provides a detailed description of the LP-rounding algorithm. Section 3 presents the algorithm analysis using new techniques. Section 4 provides a summary of the paper. We defer the proofs of Lemmas 1, 3-4 to the journal version.

2 Algorithm

To clarify the association between clients and facilities across different stages or scenarios, we introduce the following notation. Client-scenario pair (j, k) refers to client j under scenario k. Facility-scenario pair (i, v) represents facility i corresponding to scenario v. Facility-scenario pair $(i, 0)$ denotes facility i in the first-stage setting with $v = 0$. Furthermore, we introduce the following notations.

- $\mathcal{F} := \{(i, v) \mid i \in F, v = 0, 1, 2, \ldots, K\}$, the set of facility-scenario pairs.
- $\mathcal{F}_0 := \{(i, 0) \mid i \in F\}$, the set established in first-stage setting with $v = 0$.
- $\mathcal{F}_k := \{(i, k) \mid i \in F, k = 1, 2, \ldots, K\}$, the set established in scenario k.
- $\mathcal{C} := \{(j, k) \mid j \in C_k, k = 1, 2, \ldots, K\}$, the set of client-scenario pairs.
- $c_{ij}^{vk} := c_{ij}$ (if $v = 0$ and k) or $+\infty$ (if $v \neq 0$ and k), the cost of serving (j, k) by (i, v).

Let $p_0 := 1$. We obtain the integer programming formulation for 2-SFLP (see Fig. 1):

$$
\begin{aligned}
\min \quad & \sum_{(i,v)\in\mathcal{F}} p_v f_i^v y_i^v + \sum_{(i,v)\in\mathcal{F}} \sum_{(j,k)\in\mathcal{C}} p_k c_{ij}^{vk} x_{ij}^{vk} \\
\text{s. t.} \quad & \sum_{(i,v)\in\mathcal{F}} x_{ij}^{vk} \geq 1, \quad \forall (j,k)\in\mathcal{C}, \\
& x_{ij}^{vk} \leq y_i^v, \quad \forall (i,v)\in\mathcal{F}, (j,k)\in\mathcal{C}, \\
& x_{ij}^{vk}, y_i^v \in \{0,1\}, \quad \forall (i,v)\in\mathcal{F}, (j,k)\in\mathcal{C}.
\end{aligned}
\tag{1}
$$

We provide the interpretation of the variables and constraints in the integer programming model.

- $y_i^v = 1$ if (i,v) is open, and $y_i^v = 0$ otherwise.
- $x_{ij}^{vk} = 1$ if (i,v) serves (j,k), and $x_{ij}^{vk} = 0$ otherwise.
- The first set of constraints denotes at least one facility-scenario pair (i,v) serve the client-scenario pair (j,k).
- The second set of constraints indicates that the client-scenario pair (j,k) is served by the opened facility-scenario pair (i,v).

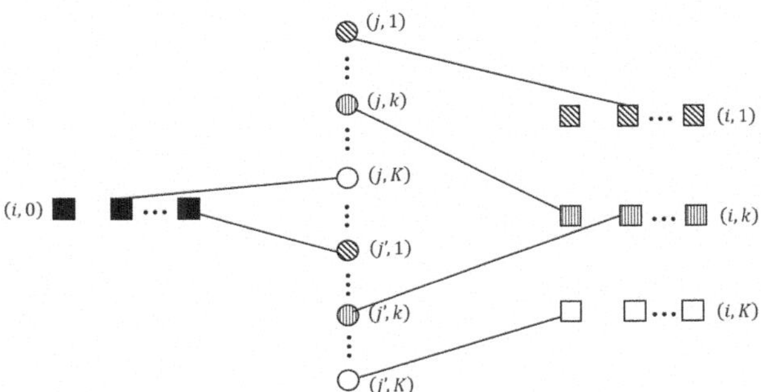

Fig. 1. 2-SFLP Diagram. Different styles of squares represent facilities under different scenarios, and different styles of circles represent clients under different scenarios.

By relaxing the integer constraints in (1), we get the linear programming relaxation of (1).

$$
\begin{aligned}
\min \quad & \sum_{(i,v)\in\mathcal{F}} p_v f_i^v y_i^v + \sum_{(i,v)\in\mathcal{F}} \sum_{(j,k)\in\mathcal{C}} p_k c_{ij}^{vk} x_{ij}^{vk} \\
\text{s. t.} \quad & \sum_{(i,v)\in\mathcal{F}} x_{ij}^{vk} \geq 1, \quad \forall (j,k)\in\mathcal{C}, \\
& x_{ij}^{vk} \leq y_i^v, \quad \forall (i,v)\in\mathcal{F}, (j,k)\in\mathcal{C}, \\
& x_{ij}^{vk}, y_i^v \geq 0, \quad \forall (i,v)\in\mathcal{F}, (j,k)\in\mathcal{C}.
\end{aligned}
\tag{2}
$$

Let $(\tilde{x}, \tilde{y})$ denote the fractional optimal solution of (2). We derive a linear programming rounding algorithm to establish the per-scenario bound based on (2). In fact, the splitting technique from [1] allows transforming the fractional optimal solution to the complete solution. Without loss of generality, assume $(\tilde{x}, \tilde{y})$ constitutes a complete solution. This is defined as a feasible solution to (2) where for all facilities $(i, v) \in \mathcal{F}$, clients $(j, k) \in \mathcal{C}$, the condition $x_{ij}^{vk} > 0$ implies $x_{ij}^{vk} = y_i^v$.

From the definition of $\{c_{ij}^{vk}\}$, it is evident that

$$\sum_{(i,v)\in\mathcal{F}} \tilde{x}_{ij}^{vk} = \sum_{i\in F} \tilde{x}_{ij}^{0k} + \sum_{i\in F} \tilde{x}_{ij}^{kk} = 1.$$

Scaling the complete solution $(\tilde{x}, \tilde{y})$ by a factor $\alpha > 2$, we obtain $(\bar{x}, \bar{y}) := (\alpha\tilde{x}, \alpha\tilde{y})$. By utilizing the splitting technique, we can assume that for any $(i, v) \in \mathcal{F}$ and $(j, k) \in \mathcal{C}$, the inequalities $\bar{x}_{ij}^{vk} \leq 1$ and $\bar{y}_i^v \leq 1$ hold.

To simplify the discussion, we define the following notation.

- $\mathcal{F}_{jk}^0 := \{(i, 0) : \tilde{x}_{ij}^{0k} > 0\}$, i.e., the set formed by $(i, 0)$ that are fractionally connected to (j, k) in the solution $(\tilde{x}, \tilde{y})$;
- $\mathcal{F}_{jk}^k := \{(i, k) : \tilde{x}_{ij}^{kk} > 0\}$, i.e., the set formed by (i, k) that are fractionally connected to (j, k) in the solution $(\tilde{x}, \tilde{y})$;
- $\mathcal{F}_{jk} := \mathcal{F}_{jk}^0 \cup \mathcal{F}_{jk}^k$, i.e., the set consisting of $(i, 0)$ and (i, k) that are fractionally connected to (j, k) in the solution $(\tilde{x}, \tilde{y})$;
- $\tilde{C}_{jk} := \sum_{(i,v)\in\mathcal{F}_{jk}} c_{ij}^{vk}\tilde{x}_{ij}^{vk}$, i.e., the average connection cost (j, k) in the solution $(\tilde{x}, \tilde{y})$.

For any (j, k), provide permutation π which is obtained in increasing order of facility-scenario pair connection costs such that

$$c_{\pi(1)j}^{0k} \leq \cdots \leq c_{\pi(i)j}^{0k} \leq \cdots \leq c_{\pi(n)j}^{0k},$$

and

$$c_{\pi(1)j}^{kk} \leq \cdots \leq c_{\pi(i)j}^{kk} \leq \cdots \leq c_{\pi(n)j}^{kk}.$$

Since $\sum_{(i,v)} \bar{x}_{ij}^{vk} = \alpha > 2$, we have $\sum_{i=1}^n \bar{x}_{\pi(i)j}^{0k} \geq 1$ or $\sum_{i=1}^n \bar{x}_{\pi(i)j}^{kk} \geq 1$, where at least one of these inequalities holds. The definitions of N_{jk}^0, N_{jk}^k, d_{jk}^0 and d_{jk}^k are given below.

- If $\sum_{i=1}^n \bar{x}_{\pi(i)j}^{0k} \geq 1$, then define $i_0^* := \min\{i' \mid \sum_{i=1}^{i'} \bar{x}_{\pi(i)j}^{0k} \geq 1\}$, set d_{jk}^0 to $c_{\pi(i_0^*)j}^{0k}$, and define $N_{jk}^0 := \{(\pi(i), 0) \mid i \in \{1, \ldots, i_0^*\}\}$. If $\sum_{i=1}^n \bar{x}_{\pi(i)j}^{0k} < 1$, then set $d_{jk}^0 := +\infty$.
- If $\sum_{i=1}^n \bar{x}_{\pi(i)j}^{kk} \geq 1$, then define $i_k^* := \min\{i' \mid \sum_{i=1}^{i'} \bar{x}_{\pi(i)j}^{kk} \geq 1\}$, set d_{jk}^k to $c_{\pi(i_k^*)j}^{kk}$, and define $N_{jk}^k := \{(\pi(i), k) \mid i \in \{1, \ldots, i_k^*\}\}$. If $\sum_{i=1}^n \bar{x}_{\pi(i)j}^{kk} < 1$, then set $d_{jk}^k := +\infty$.

To facilitate the subsequent analysis, we apply the splitting technique to ensure that $\sum_{i=1}^{i_0^*} \bar{x}_{\pi(i)j}^{0k} = 1$ and $\sum_{i=1}^{i_k^*} \bar{x}_{\pi(i)j}^{kk} = 1$ hold.

By comparing d_{jk}^0 and d_{jk}^k, we define $N(j,k)$ as the neighbor set associated with (j,k).

- If $d_{jk}^0 \leq d_{jk}^k$, set $d_{jk} := d_{jk}^0$ and $N(j,k) := N_{jk}^0$.
- If $d_{jk}^0 > d_{jk}^k$, set $d_{jk} := d_{jk}^k$ and $N(j,k) := N_{jk}^k$.

Let $\Delta_{jk} := \sum_{(i,v)\in N(j,k)} c_{ij}^{vk} \bar{x}_{ij}^{vk}$, representing the average connection cost of (j,k) in the neighborhood set $N(j,k)$. Let $\mathcal{C}' \subseteq \mathcal{C}$ represent the set of centers. Based on each center, we cluster the customer-scenario pairs and facility-scenario pairs. Let $\hat{\mathcal{F}}$ and $\mathcal{U}$ denote the sets of opened facility-scenario pairs and unconnected client-scenario pairs, respectively.

The high level of our algorithm is provided as follows. By scaling the optimal solution of the linear program to obtain a modified solution, we aim to balance connection costs and facility costs. Based on this modified solution, we construct clusters according to specific criteria. Specifically, we select the client-scenario pair (j,k) with the smallest combined distance as the cluster center, and identify both the facility-scenario neighbor set and the client-scenario neighbor set associated with this center to form a cluster centered at (j,k). For each cluster, we open exactly one facility-scenario pair from the center's facility-scenario neighbor set with probability x_{ij}^{vk}. For facility-scenario pairs not belonging to any cluster, we open them independently with probability y_i^v. Finally, each client-scenario pair is connected to the nearest opened facility-scenario pair.

Building upon the aforementioned preparations, we now present our detailed algorithm with a predetermined non-negative parameter α. Although the algorithm was originally proposed by Wu et al. [15], we restate it here for the self-containment of our paper.

Algorithm 1: LP(α)

Step 1 (Solving LP) Solve (2) to obtain the optimal solution $(\tilde{x}, \tilde{y})$.

Step 2 (Scaling) Scale the optimal solution using parameter α to derive the modified solution $(\bar{x}, \bar{y})$.

Step 3 (LP Rounding)

Step 3.1 (Initialization) Set $\mathcal{C}' := \emptyset$, $\hat{\mathcal{F}} := \emptyset$, and $\mathcal{U} := \mathcal{C}$.

Step 3.2 Select the client-scenario pair with the smallest $d_{jk} + \Delta_{jk}$ in $\mathcal{U}$ as the center, i.e., $(j',k') := \arg\min_{(j,k)\in\mathcal{U}}\{d_{jk} + \Delta_{jk}\}$. Update $\mathcal{C}' := \mathcal{C}' \cup \{(j',k')\}$. If $N(j',k') = N_{j'k'}^0$, set

$$B(j',k') := \{(j,k) : N(j,k) = N_{jk}^0, N_{jk}^0 \cap N_{j'k'}^0 \neq \emptyset\};$$

otherwise, set

$$B(j',k') := \{(j,k') : N(j,k') = N_{jk'}^{k'}, N_{jk'}^{k'} \cap N_{j'k'}^{k'} \neq \emptyset\}.$$

Update $\mathcal{U} := \mathcal{U} \setminus B(j',k')$. Repeat this step until $\mathcal{U} = \emptyset$.

Step 3.3 For any $(j',k') \in \mathcal{C}'$, open exactly one (i,v) in $N(j',k')$ with probability $\bar{x}_{ij'}^{vk'}$. Define $\mathcal{L} := \bigcup_{(j',k')\in\mathcal{C}'} N(j',k')$, and $\mathcal{R} := \mathcal{F} \setminus \mathcal{L}$. For any facility-scenario pair $(i,v) \in \mathcal{R}$, open (i,v) independently with probability $\bar{y}_i^v$.

Step 3.4 For any (j,k) in $\mathcal{C}$, the nearest open facility-scenario pairs from both scenario 0 and scenario k provide the service.

In Step 3.2 of Algorithm 1, we identify all client-scenario pairs that intersect with the neighbor set of (j', k'), and denote them by the set $B(j', k')$. The cluster centered at (j', k') is $\{(j', k'), N(j', k'), B(j', k')\}$ (see Fig. 2).

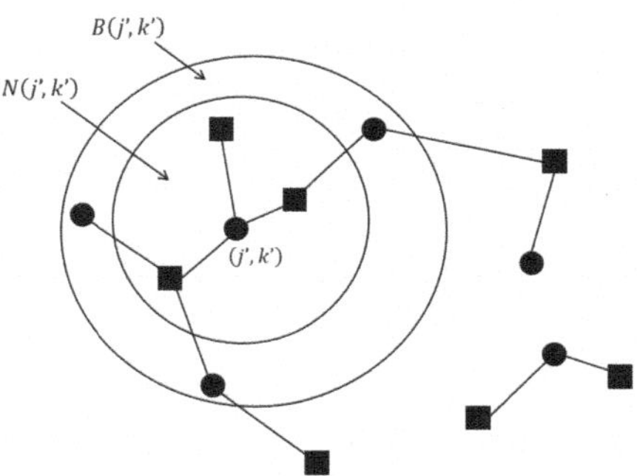

Fig. 2. Clustering centered on (j', k').

3 Analysis

To estimate the connection cost in each scenario k, we need to analyze d_{jk} and Δ_{jk}. Here, we provide a unified upper bound estimate for $2d_{jk}+\Delta_{jk}$. This unified estimate results in a tighter bound, which enhances the final approximation ratio. Note that this analysis only relies on the construction of the neighborhood set $N(j, k)$ for the client-scenario pair (j, k), we do not require this point to be a center.

Lemma 1. *For every client-scenario pair (j, k), it holds that*

$$a_1 d_{jk} + a_2 \Delta_{jk} \leq \max\left\{\frac{a_1\alpha}{\alpha-2}, a_2\alpha\right\} \cdot \tilde{C}_{jk},$$

where a_1 and a_2 are positive constants.

Let $a_1 = 2$, $a_2 = 1$. For $2 < \alpha \leq 4$, the following inequality holds

$$2d_{jk} + \Delta_{jk} \leq \frac{2\alpha}{\alpha-2} \cdot \tilde{C}_{jk}.$$

Remark 1. In [15], the upper bound for $2d_{jk} + \Delta_{jk}$ is given as $(\alpha^2/(\alpha-2)) \cdot \tilde{C}_{jk}$, while the upper bound obtained in this paper is $(2\alpha/(\alpha-2)) \cdot \tilde{C}_{jk}$. Clearly, we have

$$\frac{2\alpha}{\alpha-2} \leq \frac{\alpha^2}{\alpha-2}.$$

Thus, this inequality immediately implies an improvement in the approximation ratio.

We use the following lemmas to tighten the upper bound on the average connection cost.

Lemma 2. *[4] Let $\{u(q)\}_{q=1}^{m}$, $\{v(q)\}_{q=1}^{m}$, and $\{w(q)\}_{q=1}^{m}$ be non-negative sequences, with $\{u(q)\}_{q=1}^{m}$ decreasing and $\{v(q)\}_{q=1}^{m}$ increasing. We have*

$$\frac{\sum_{q=1}^{m} u(q)w(q)v(q)}{\sum_{q=1}^{m} u(q)w(q)} \leq \frac{\sum_{q=1}^{m} w(q)v(q)}{\sum_{q=1}^{m} w(q)}.$$

For any facility-scenario pair set A, let event $\mathcal{A}$ indicate that one or more facility-scenario pairs in A are open. Before analyzing the upper bound on the average connection cost, we analyze the upper bound on the connection cost for a client-scenario pair to the closest open facility-scenario pair within A, assuming that event $\mathcal{A}$ has occurred. Let $\bar{C}_{jk}(A)$ represent the average cost of connecting the client-scenario pair (j,k) to the facility-scenario pairs within the set A, i.e.,

$$\bar{C}_{jk}\left(\mathrm{A}\right) := \frac{\sum_{(i,v)\in A} c_{ij}^{vk}\bar{x}_{ij}^{vk}}{\sum_{(i,v)\in A} \bar{x}_{ij}^{vk}} = \frac{\sum_{(i,v)\in A} c_{ij}^{vk}\bar{y}_{i}^{v}}{\sum_{(i,v)\in A} \bar{y}_{i}^{v}}.$$

Lemma 3. *For any client-scenario pair (j,k) and facility-scenario pair set A, we have*

$$E\left[\min_{(i,v)\in A \text{ is open}} c_{ij}^{vk} \mid \mathcal{A}\right] \leq \bar{C}_{jk}(A).$$

Lemma 4. *Let event D denote the occurrence of one or more open facility-scenario pairs in $\mathcal{F}_{jk}$, where $(j,k) \notin \mathcal{C}'$. Then the event D occurs with probability at least $1-e^{-\alpha}$.*

From the above lemmas, the upper bound on the expected connection cost for any (j,k) is established.

Lemma 5. *The expected connection cost for the client-scenario pair (j,k) is given by*

$$E\left[C_{jk}\right] \leq \max\left\{\alpha, 1+\frac{\alpha+2}{\alpha-2}e^{-\alpha}\right\}\tilde{C}_{jk},$$

where $\tilde{C}_{jk}$ denotes the connection cost for (j,k) in Algorithm 1.

Proof. We analyze the following two cases.

- If $(j,k) \in \mathcal{C}'$, then according to Algorithm 1, there exists exactly one $(i,v) \in N(j,k)$ is open. Thus,

$$E\left[C_{jk}\right] = \sum_{(i,v)\in N(j,k)} c_{ij}^{vk}\bar{x}_{ij}^{vk} = \Delta_{jk} \leq \alpha\tilde{C}_{jk}.$$

- If $(j,k) \notin \mathcal{C}'$, we first consider the case where one or more facility-scenario pairs in $\mathcal{F}_{jk}$ are open, followed by the case where no facility-scenario pairs in $\mathcal{F}_{jk}$ are open.

Case 1 Based on Lemmas 3-4, the probability of this event occurring is no less than $1 - e^{-\alpha}$. Consequently, the expected connection cost for (j, k) is bounded by

$$\frac{\sum_{(i,v)\in\mathcal{F}_{jk}} c_{ij}^{vk}\bar{x}_{ij}^{vk}}{\sum_{(i,v)\in\mathcal{F}_{jk}} \bar{x}_{ij}^{vk}} = \frac{\sum_{(i,v)\in\mathcal{F}_{jk}} c_{ij}^{vk}\tilde{x}_{ij}^{vk}}{\sum_{(i,v)\in\mathcal{F}_{jk}} \tilde{x}_{ij}^{vk}} = \tilde{C}_{jk}.$$

Case 2 If no facility-scenario pairs are open in $\mathcal{F}_{jk}$, Lemma 4 implies that the probability of this event occurring does not exceed $e^{-\alpha}$. Consider the cluster containing (j, k), with (j', k') as the cluster's center. Suppose that an open facility-scenario pair within $N(j', k')$ is (i', v). Using the triangle inequality, we analyze the following two cases.

Case 2.1 There exists $(i_0, v_0) \in N(j, k) \cap N(j', k')$ such that $c_{i_0 j'}^{v_0 k'} \leq \Delta_{j'k'}$. Then, the expected cost satisfies

$$\begin{aligned} E[C_{jk}] &\leq E\left[c_{i'j}^{vk}\right] \\ &\leq E\left[c_{i_0 j}^{v_0 k} + c_{i'j'}^{vk'} + c_{i_0 j'}^{v_0 k'}\right] \\ &\leq d_{jk} + d_{j'k'} + \Delta_{j'k'} \\ &\leq 2d_{jk} + \Delta_{jk}, \end{aligned}$$

where the final inequality follows from (j', k') being the center of the cluster.

Case 2.2 For all $(i, v) \in N(j, k) \cap N(j', k')$, it holds that $c_{ij'}^{vk'} > \Delta_{j'k'}$. Let $A := N(j', k') \setminus N(j, k)$. Recall that $\mathcal{A}$ denote the event where at less than one facility-scenario pair in A is open. According to the definition of $\Delta_{j'k'}$, we obtain that

$$\frac{\sum_{(i,v)\in A} c_{ij'}^{vk'}\bar{x}_{ij'}^{vk'}}{\sum_{(i,v)\in A} \bar{x}_{ij'}^{vk'}} \leq \Delta_{j'k'}.$$

Given a facility-scenario pair (i_0, v_0) in both $N(j, k)$ and $N(j', k')$. By applying Lemma 3, we have

$$\begin{aligned} E[C_{jk}] &\leq E\left[\min_{(i',v)\in A:(i',v) \text{ is open}} c_{i'j}^{vk} \mid \mathcal{A}\right] \\ &\leq E\left[c_{i_0 j}^{v_0 k} + c_{i_0 j'}^{v_0 k'} + c_{i'j'}^{vk'} \mid \mathcal{A}\right] \\ &\leq d_{jk} + d_{j'k'} + \frac{\sum_{(i',v)\in A} c_{i'j'}^{vk'}\tilde{x}_{i'j'}^{vk'}}{\sum_{(i',v)\in A} \tilde{x}_{i'j'}^{vk'}} \\ &= d_{jk} + d_{j'k'} + \frac{\sum_{(i',v)\in A} c_{i'j'}^{vk'}\bar{x}_{i'j'}^{vk'}}{\sum_{(i',v)\in A} \bar{x}_{i'j'}^{vk'}} \\ &\leq d_{jk} + d_{j'k'} + \Delta_{j'k'} \\ &\leq 2d_{jk} + \Delta_{jk}. \end{aligned}$$

The two cases are upper bounded by a unified upper bound of $2d_{jk} + \Delta_{jk}$. By setting $a_1 = 2$, $a_2 = 1$, applying Lemma 1, and noting that $\alpha > 2$, we further relax the upper bound to $(2\alpha/(\alpha-2)) \cdot \tilde{C}_{jk}$, which is always greater than $\tilde{C}_{jk}$. Consequently, we have

$$\mathbb{E}[C_{jk}] \leq \left(1 + \frac{\alpha+2}{\alpha-2}e^{-\alpha}\right)\tilde{C}_{jk}.$$

□

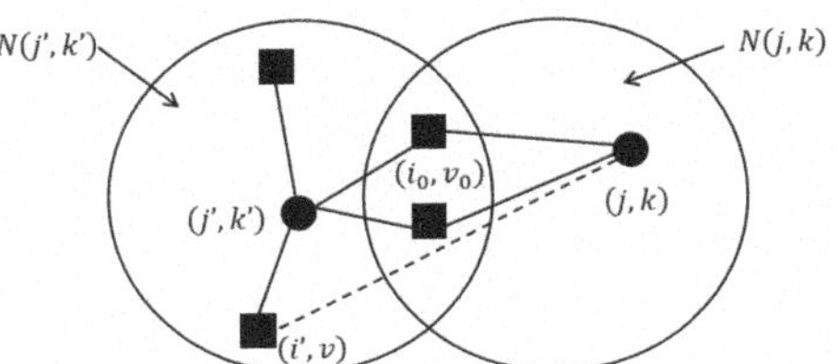

Fig. 3. Connecting (j,k) to (i',v).

Combined with Lemma 5, the theorem below gives an estimate for the per-scenario bound (Fig. 3).

Theorem 1. *For the solution obtained by Algorithm 1, the per-scenario bound for 2-SFLP is*

$$\max\left\{\alpha, 1 + \frac{\alpha+2}{\alpha-2}e^{-\alpha}\right\}.$$

Proof. For each scenario k, we analyse its expected facility cost. The facility-scenario pair (i,v) (where $v \in \{0,k\}$) open with two different probabilities.

1. If (i,v) belongs to the cluster focused on the client-scenario pair (j,k), the opening probability is $\bar{x}_{ij}^{vk}$ (since for the optimal solution we have $\bar{x}_{ij}^{vk} = \bar{y}_i^v$).
2. If (i,v) does not belong to any cluster, its opening probability is $\bar{y}_i^v$.

Thus, the expected opening cost for each facility-scenario pair (i,v) in scenario k is

$$\sum_{i\in F, v\in\{0,k\}} f_i^v \bar{y}_i^v = \alpha \sum_{i\in F, v\in\{0,k\}} f_i^v \tilde{y}_i^v.$$

By combining Lemma 5, we conclude that the expected cost for scenario k derived from Algorithm 1 is upper bounded by

$$\alpha \sum_{i\in F, v\in\{0,k\}} f_i^v \tilde{y}_i^v + \max\left\{\alpha,\ 1 + \frac{\alpha+2}{\alpha-2}e^{-\alpha}\right\} \sum_{j\in C_k} \tilde{C}_{jk}$$

$$= \max\left\{\alpha,\ 1 + \frac{\alpha+2}{\alpha-2}e^{-\alpha}\right\}\left(\sum_{i\in F}(f_i^0\tilde{y}_i^0 + f_i^k\tilde{y}_i^k) + \sum_{i\in F, s\in\{0,k\}}\sum_{j\in C_k} c_{ij}^{vk}\tilde{x}_{ij}^{vk}\right).$$

□

Choosing $\alpha = 2.322$ satisfies the equation

$$\alpha = 1 + \frac{\alpha + 2}{\alpha - 2} e^{-\alpha}.$$

This yields a per-scenario bound of 2.322, enhancing the previous approximation ratio of 2.3613.

4 Conclusion

s

We introduce the integrated distance estimation technique to refine the analysis of the per-scenario bound for the 2-SFLP. We establish an improved per-scenario bound of 2.322 by leveraging the LP-rounding algorithm from prior work. Building on our proposed integrated distance estimation technique, we identify three promising directions for future research:

1. Further improving the per-scenario bound for the 2-SFLP.
2. Enhancing the general 2.2975-approximation (cf. [2]) for the 2-SFLP beyond the per-scenario bound.
3. Exploring the applications of the integrated distance estimation technique in other facility location problems.

Acknowledgments. The first three authors are supported by National Natural Science Foundation of China (No. 12371320). The fourth author is supported by Guangdong Basic and Applied Basic Research Foundation (No. 2024A1515030197) and National Natural Science Foundation of China (No. 12371321).

References

1. Byrka, J.: Aardal, K: an optimal bifactor approximation algorithm for the metric uncapacitated facility location problem. SIAM J. Comput. **39**(6), 2212–2231 (2010)
2. Byrka, J., Srinivasan, A.: Approximation algorithms for stochastic and risk-averse optimization. SIAM J. Discret. Math. **32**(1), 44–63 (2018)
3. Chudak, F.A., Shmoys, D.B.: Improved approximation algorithms for the uncapacitated facility location problem. SIAM J. Comput. **33**(1), 1–25 (2003)
4. Fortuin, C.M., Kasteleyn, P.W., Ginibre, J.: Correlation inequalities on some partially ordered sets. Commun. Math. Phys. **22**, 89–103 (1971)
5. Gupta, A., Ravi, R., Sinha, A.: LP rounding approximation algorithms for stochastic network design. Math. Oper. Res. **32**(2), 345–364 (2007)
6. Jain, K., Vazirani, V.V.: Approximation algorithms for metric facility location and k-median problems using the primal-dual schema and Lagrangian relaxation. J. ACM **48**, 274–296 (2001)
7. Jain, K., Mahdian, M., Saberi, A.: A new greedy approach for facility location problems. In: Proceedings of the 34th annual ACM Symposium on Theory of Computing, pp. 731-740. (2002)

8. Li, S.: A 1.488 approximation algorithm for the uncapacitated facility location problem. Inform. Comput. **222**, 45-58 (2013)
9. Mahdian, M.: Facility Location and the Analysis of Algorithms through Factor-Revealing Programs. Ph.D. thesis, MIT, Cambridge, MA (2004)
10. Mahdian, M., Ye, Y., Zhang, J.: Approximation algorithms for metric facility location problems. SIAM J. Comput. **36**(2), 411–432 (2006)
11. Ravi, R., Sinha, A.: Hedging uncertainty: approximation algorithms for stochastic optimization problems. Math. Program. **108**, 97–114 (2006)
12. Shmoys, D.B., Tardos, É., Aardal, K.: Approximation algorithms for facility location problems. In: Proceedings of the 29th Annual ACM Symposium on Theory of Computing, pp. 265-274 (1997)
13. Srinivasan, A.: Approximation algorithms for stochastic and risk-averse optimization. In: Proceedings of the 18th Annual ACM-SIAM Symposium on Discrete Algorithms, pp. 1305-1313 (2007)
14. Swamy, C.: Approximation Algorithms for Clustering Problems. Ph.D. thesis. Cornell University, Ithaca, NY (2004)
15. Wu, C., Du, D., Xu, D.: An improved per-scenario bound for the two-stage stochastic facility location problem. SCIENCE CHINA Math. **58**, 213–220 (2015)

Bilevel Adversarial Scheduling Problem on Parallel Machines

Ruiqing Sun(✉)

School of Mathematics and Statistics, Yunnan University, Kunming, China
ruiqing2020@126.com

Abstract. In this paper, we consider a bilevel adversarial scheduling problem on parallel machines. Given a set of jobs that the leader has to select some jobs within the leader's budget. Then, the follower next schedules these jobs to minimize the makespan. The goal is to select the jobs so that the optimal (minimum) value of the makespan is maximum. We design a simple $(2+\varepsilon)$-approximation algorithm and a polynomial time approximation scheme (PTAS) for this problem. We also propose a simple efficient polynomial time approximation scheme (EPTAS) for this problem when the number of machines is fixed.

Keywords: Bilevel optimization · Scheduling · Approximation algorithm · PTAS · EPTAS

1 Introduction

Scheduling theory focuses on distributing a specific collection of jobs to resources across time intervals, with the aim of minimizing certain criteria and a large number of achievements have emerged [3,9–11]. Although the common assumption is that a sole decision-maker, known as an agent, is responsible for making all scheduling choices, a substantial portion of the academic literature explores scenarios where multiple agents engage in competition to schedule their respective sets of jobs. For instance, this is particularly evident in the domains of multiagent scheduling (as discussed by Agnetis et al. in [2]) and scheduling approaches grounded in game theory (such as those presented by Pascual et al. in [21]). In this paper, we turn our attention to a different kind of interaction among agents. This interaction is associated with the concept of bilevel optimization, as examined in the work of Dempe et al. [6], and it leading to bilevel scheduling problems.

Bilevel optimization models consider a leader and a follower playing a two-stage game. First, leader takes some decisions $x \in X(y)$ and wants to maximize its objective function f^L. Then, follower takes some decisions $y \in Y(x)$ and wants to minimize its objective function f^F. Here, leader's (follower's) feasible region may depends on the follower's (leader's) decisions, thus, it is denoted as

F. V. Fomin and M. Xiao (Eds.): COCOON 2025, LNCS 15983, pp. 44–53, 2026.
https://doi.org/10.1007/978-981-95-0215-8_4

$X(y)(Y(x))$. The bilevel model is following:

$\max f^L(x, y)$
$x \in X(y)$
y is an optimal solution to the follower's problem,
where the follower's problem is given as follows, where x is fixed:
$\min f^F(x, y)$
$y \in Y(x)$

Note that the choices of maximization and minimization for the leader and follower, respectively, are arbitrary.

To the best of our knowledge, very little results are known in the literature on bilevel scheduling problems. Karlof and Wang [12] first considered a flowshop scheduling problem with operators where the leader determines the schedule of operators to minimize the sum of job completion times while the follower determines the schedule of jobs to minimize the makespan. Guan et al. [8] presented a 3/2-approximation algorithm and a PTAS for the bilevel optimization models of parallel machine scheduling and path. Li and Sun [20] presented a 3/2-approximation algorithm and an asymptotic polynomial time approximation scheme for the bilevel optimization models of parallel machine scheduling and bin packing. More paper about bilevel scheduling problems please see the papers [1,14,16]. Recently, adversarial bilevel problems has received more and more attention.

In adversarial bilevel problems setting, the leader takes decision so that the optimal solution of the follower's problem is the worst possible. As an example, the bilevel knapsack problem introduced by DeNegre [7] is this setting. First, the leader selects items within the leader's budget. Then the follower selects items from the remaining items within the follower's budget. The goal of the leader is to minimize the maximum profit that the follower can obtain. They design an PTAS for this problem. Then, Chen et al. [5] considered a general case that bilevel problem with packing constraints. Sun and Li [23] presented a $(d+1)$-approximation algorithm for the bilevel optimization models of b-matching interdiction problem, where d is the dimension of the leader's budget. T'kindt et al. [24] consider single machine adversarial bilevel scheduling problems. In this problem, there are two decision-making agents: the leader and the follower. Both agents operate on the identical set of jobs. The leader takes the initiative in making decisions, with the intention of forcing the follower into a situation where the solution obtained is as detrimental as possible for the latter. After the leader's actions, the follower then schedules these jobs with the goal of optimizing a given criterion. The considered criteria are the total completion time, the total weighted completion time, the maximum lateness and the number of tardy jobs. The adversarial bilevel scheduling problems in this paper is also related to scheduling problems with cost or "knapsack" constraint. (see, e.g., [22]). Because when you select the jobs, some costs will be consumed and cannot exceed the given threshold. Li et al. [17] considered penalty cost constrained scheduling problem, they proposed a 2-approximation algorithm, followed by an PTAS. Li and Ou [18] considered bandwidth constraint scheduling problem, they proposed a 2-approximation algorithm, followed by an PTAS. Li and Ou [19] considered machine energy consumption constraint scheduling problem, they designed a $(\frac{4}{3} + \epsilon)$-approximation algorithm. Kones and Levin [15] presented a unified framework for designing EPTAS for machine activation cost constraint scheduling problem. Although we involve two types of decision-makers, the methods used in these issues can always offer us some inspiration for leader's problem.

A ρ-approximation algorithm for a minimization problem is a polynomial time algorithm that always finds a feasible solution with an objective value of at most ρ times the optimal value. The infimum value of ρ for which an algorithm is a ρ-approximation is called the approximation ratio or the performance guarantee of the algorithm. A PTAS for a given problem is a family of approximation algorithms such that the family has a $(1 + \varepsilon)$-approximation algorithm for any $\varepsilon > 0$. An EPTAS is a PTAS whose time complexity is upper bounded by a value of $f(\frac{1}{\varepsilon}) \cdot poly(n)$ where f is some computable (not necessarily polynomial) function and $poly(n)$ is a polynomial of the length of the (binary) encoding of the input. A fully polynomial time approximation scheme (FPTAS) is an EPTAS which satisfies that f must be a polynomial in $\frac{1}{\varepsilon}$.

In this paper, we consider a bilevel adversarial scheduling problem on parallel machines. Given a set of jobs that the leader has to select some jobs within the leader's budget. Then, the follower next schedules these jobs to minimize the makespan. Leader's goal is to select the jobs so that the optimal value of makespan is maximum. We first design a simple $(2+\varepsilon)$-approximation algorithm. Then, we propose an PTAS for this problem. Finally, we also propose a simple EPTAS for this problem when the number of machines is fixed. The remainder of this paper is structured as follows. In Sect. 2, we provide the problem formulation of our problem. In Sect. 3, we design a $(2+\varepsilon)$-approximation algorithm for this problem. In Sect. 4, we propose an PTAS for this problem. In Sect. 5, we also design a simple EPTAS for this problem when the number of machines is fixed. In the last section, we present conclusions and future work.

2 Preliminaries

Given a set of m machines $\mathcal{M} = \{M_1, \cdots, M_m\}$ and a set of $\mathcal{J} = \{J_1, \cdots, J_n\}$ jobs with processing $p_j > 0$ for all $j = 1, \cdots, n$ are available for the leader. First, the leader must select some jobs with a cost $c_j > 0$ for each job J_j such that the total cost can not exceed a given threshold B. Then the follower next schedules these jobs that the leader selected on machines to minimize the makespan, which is denoted as $P||C_{\max}$ problem. Let C_j be the completion time of job J_j in a schedule. Furthermore, we define $C_{\max} = \max\{C_j : J_j \in \mathcal{J}\}$ by the makespan. It is a classical scheduling problem which is to schedule n independent jobs on m parallel identical machines to minimize the makespan, i.e., the largest machine completion time. In sum, the leader's goal is to select the jobs so that the optimal value of the follower problem is maximized. Clearly, this problem is NP-hard, even the follower's $P||C_{\max}$ problem. Let $(\mathcal{J}, \mathcal{M}, p, c)$ be our problem instance. Let $P(S)$ and $c(S)$ be the total processing time and total cost of subset $S \subseteq \mathcal{J}$, respectively.

For a given instance for the problem, let $\mathcal{J}_{\max} \subseteq \mathcal{J}$ be the optimal set of items (jobs) in a maximum 0–1 knapsack problem with the profit (processing time) and weight (cost). There is also an FPTAS for the maximum 0–1 knapsack problem [13], let $\mathcal{J}'_{\max}$ be the job subset produced by an FPTAS. For an arbitrary desired accuracy $\varepsilon \in (0, 1/2]$, by scheduling the jobs in $\mathcal{J}_{\max}$ on m machines, we have

$$\begin{aligned} \frac{P(\mathcal{J}'_{\max})}{m} \leq \frac{P(\mathcal{J}_{\max})}{m} \leq OPT^* &\leq \frac{P(\mathcal{J}_{\max})}{m} + p_j \\ &\leq \frac{1}{1-\varepsilon} \cdot \frac{P(\mathcal{J}'_{\max})}{m} + p_j \\ &\leq (1+2\varepsilon)\frac{P(\mathcal{J}'_{\max})}{m} + p_j, \end{aligned} \tag{1}$$

where OPT^* is the optimal value of the makespan and p_j is the processing time of the latest processing job at the maximum load machine. Furthermore, given any feasible solution denoted as $\mathcal{J}'$ of the leader's problem, the follower's optimal value denoted as OPT' also satisfies

$$\max\{p_{\max}, \frac{P(\mathcal{J}')}{m}\} \leq OPT' \leq \frac{P(\mathcal{J}')}{m} + p_{j'}, \tag{2}$$

where $p_{\max}$ is the maximum processing time in $\mathcal{J}'$ and $p_{j'}$ is the processing time of the lastest processing job at the maximum load machine.

For the sake of analysis, we assume that $OPT^* = 1$. This can be down by dividing each processing time p_j by every possible value of $LB(1+\varepsilon)^k$ for $k = 0, 1, \cdots, \lceil \log_{1+\varepsilon}(1+2\varepsilon+m) \rceil - 1$ and $LB = \frac{P(\mathcal{J}'_{\max})}{m}$. Since OPT^* must be lie in one of the intervals $[LB(1+\varepsilon)^k, LB(1+\varepsilon)^{k+1})$. Among these $\lceil \log_{1+\varepsilon}(1+2\varepsilon+m) \rceil$ instances, there exists an instance such that the optimal value is close to 1 with error no more than ε. Moreover, we can also assume that $p_j \leq OPT^* = 1$ for every job $J_j \in \mathcal{J}$, since if there is an item such that $p_j > OPT^*$, we can delete the job J_j from $\mathcal{J}$. In sum, we assume that

$$OPT^* = 1 \text{ and } p_j \leq 1 \text{ for all } J_j \in \mathcal{J}.$$

3 A (2+ε)-Approximation Algorithm

In this section, we design a simple (2+ε)-approximation algorithm. The main idea of our algorithm is simply to choose as much total processing time as possible. Let $S \subseteq \mathcal{J}$ be any subset of $\mathcal{J}$, we solve the following problem:

$$\begin{cases} \max\limits_{S} P(S) \\ c(S) \leq B. \end{cases} \tag{3}$$

This is a 0–1 knapsack problem, and there is an FPTAS for this problem. Thus, we can find a subset S such that $P(\mathcal{J}_{\max}) \leq (1+\varepsilon)P(S)$. Let $p_{\max}$ be the job with the maximum processing time such that its cost does not exceed B. We choose a job set with $\max\{\frac{P(S)}{m}, p_{\max}\}$. Then, the follower find an optimal solution to the $P||C_{\max}$ problem. Let S_i be the set of jobs scheduled on machine M_i for the follower. Let the objective value OPT be the output value of the follower. It is sufficient to prove the following theorem.

Theorem 1. *$OPT^* \leq (2+\varepsilon)OPT$.*

Proof. Given an optimal solution of the leader's problem, let the optimal subset be $S^* \subseteq \mathcal{J}$, we know that the optimal value of the follower problem have

$$\begin{aligned} OPT^* \leq \frac{P(S^*)}{m} + p_j \leq \frac{P(\mathcal{J}_{\max})}{m} + p_j &\leq \frac{(1+\varepsilon)P(S)}{m} + p_j \\ &\leq (1+\varepsilon)\max\{p_{max}, \frac{P(S)}{m}\} + p_j \\ &\leq (2+\varepsilon)OPT. \end{aligned}$$

The second inequality follows from S^* is also the feasible solution of programming (3). The last inequality follows from (2) and $p_j \leq OPT$, since our algorithm considers the job with the maximum processing time such that its cost does not exceed B, that is, $p_j \leq p_{\max}$. □

4 A Polynomial Time Approximation Scheme

In this section, we design an PTAS for this problem. We first divide jobs into large jobs and small jobs and construct a generalized version of the programming (3), which can also be solved in polynomial time and possesses an PTAS. Then, we obtain an auxiliary instance for the problem $P||C_{\max}$ which follows an PTAS. Finally, we output a feasible subset with a maximum makespan that is calculated via an PTAS for the problem $P||C_{\max}$. The algorithm is described as follows.

Let $\varepsilon \in (0, 1/2]$ and $K = \lceil \log_{1+\varepsilon}(1/\varepsilon) \rceil$. For each job $J_j \in \mathcal{J}$, we construct a K-dimensional vector $p(j) = (p_1(j), p_2(j), \cdots, p_K(j))$, which is defined as follows. If $p_j < \varepsilon$, let $p_k(j) = 0$ for $k = 1, 2, \cdots, K$. Else if there is a positive integer τ such that $p_j \in [\varepsilon(1+\varepsilon)^{\tau-1}, \varepsilon(1+\varepsilon)^{\tau})$, let $p_\tau(j) = 1$ and $p_k(j) = 0$ for $k \neq \tau$. For convenience, if $p_j < \varepsilon$, the corresponding job J_j is called *small*. Otherwise, the corresponding job is called *large*.

For each vector $(n_1, n_2, \cdots, n_K)$ such that $\sum_{k=1}^{K} n_k \leq n$, we construct the following integer linear program $ILP(n_1, n_2, \cdots, n_K)$. It is easy to see that there are at most $\min\{(m \cdot 1/\epsilon)^K, n^K\}$ cases. The goal is to find a subset $S \subseteq \mathcal{J}$ such that it satisfies $p_k(S) = n_k$, for $k = 1, \cdots, K$ to maximize the total processing time, where $p_k(S) = \sum_{j \in S} p_k(j)$.

$$\begin{cases} \max\limits_{S} P(S) \\ c(S) \leq B, \\ p_k(S) = n_k, \text{ for } k = 1, \cdots, K. \end{cases} \tag{4}$$

Theorem 2. *There is an PTAS for $ILP(n_1, n_2, \cdots, n_K)$.*

Proof. $ILP(n_1, n_2, \cdots, n_K)$ can be solved as a $(K+1)$-dimensional knapsack problem by using an simple transformation, which possesses an PTAS [13], when $(K+1)$ is a constant. It also by just taking n_k minimum cost jobs for jobs in the interval $[\varepsilon(1+\varepsilon)^{\tau-1}, \varepsilon(1+\varepsilon)^{\tau})$. Then solve small jobs as a knapsack problem which possesses an FPTAS. However, it is sufficient that there is a result of PTAS for $ILP(n_1, n_2, \cdots, n_K)$.

Let $(S_1^*, S_2^*, \cdots, S_m^*)$ and S^* be the schedule and job subset in the optimal solution, respectively. Let $\mathcal{J}_0^* = \{J_j \in \mathcal{J} | J_j \in S^*, p_j < \varepsilon\}$ be the set of small jobs with processing time no more than ε. For $k = 1, 2, \cdots, K$, let

$$\mathcal{J}_k^* = \{J_j \in \mathcal{J} \mid J_j \in S^*, p_j \in [\varepsilon(1+\varepsilon)^{k-1}, \varepsilon(1+\varepsilon)^k)\}$$

be the set of jobs whose processing time lie in $[\varepsilon(1+\varepsilon)^{k-1}, \varepsilon(1+\varepsilon)^k)$ corresponding to the jobs in the subset S^*. For convenience, let $n_k^* = |\mathcal{J}_k^*|$. Let S be an $(1-\varepsilon)$-approximation job set for $ILP(n_1^*, n_2^*, \ldots, n_K^*)$. By the definitions of $p_k(j)$, S^* is a feasible solution for $ILP(n_1^*, n_2^*, \cdots, n_K^*)$. Thus,

$$P(S) \geq (1-\varepsilon)P^* \geq (1-\varepsilon)P(S^*), \tag{5}$$

where the P^* is the optimal value of $ILP(n_1^*, n_2^*, \cdots, n_K^*)$ and S^* is also the feasible solution of programming (4). Based on the $(1-\varepsilon)$ approximation solution S of $ILP(n_1^*, n_2^*, \cdots, n_K^*)$, we construct an instance $I^S = (\mathcal{J}^S, \mathcal{M}, p, c)$ for the problem $P||C_{\max}$, where $\mathcal{J}^S = \{J_j \in \mathcal{J} \mid J_j \in S\}$, $\mathcal{M} = \{M_1, M_2, \ldots, M_m\}$. For $k = 1, 2, \cdots, K$, let

$$\mathcal{J}_k^S = \{J_j \in \mathcal{J} \mid J_j \in S, p_j \in [\varepsilon(1+\varepsilon)^{k-1}, \varepsilon(1+\varepsilon)^k)\}$$

be the set of jobs in S whose processing time lie in $[\varepsilon(1+\varepsilon)^{k-1}, \varepsilon(1+\varepsilon)^k)$. By the definitions of $\mathcal{J}_k^*$ and $\mathcal{J}_k^S$, for any two jobs J_{j_1} and J_{j_2} in $\mathcal{J}_k^* \cup \mathcal{J}_k^S (k = 1, 2, \ldots, K)$, we have

$$p_{j_1} \leq p_{j_2}(1+\varepsilon), \text{ and } p_{j_2} \leq p_{j_1}(1+\varepsilon). \tag{6}$$

Let $OPT(I^S)$ be the optimal value of given S in instance I^S for the problem $P||C_{\max}$, which means that $\mathcal{J}^S$ can be partition into m disjoint subsets $S_1, S_2, \cdots, S_m$ such that $p(S_i) = \sum_{J_j \in S_i} p_j \leq OPT(I^S)$. Finally, our algorithm will output a maximum solution set for all given set S with instance I^S for the problem $P||C_{\max}$.

Theorem 3. *$OPT^* \leq (1+5\varepsilon)OPT(I^S) + \varepsilon$.*

Proof. Fixed a feasible solution S of the leader's problem with instance I^S as above, we can not optimally solve this problem of the follower in polynomial time. However, we can find a $(1+\varepsilon)$-approximation solution use the method in Hochbaum and Shmoys [9], Alon et al. [3], Jansen [10] and Jansen et al. [11] for the follower problem when the leader's solution is fixed. Finally, we have $OUT(I^S) \leq (1+\varepsilon)OPT(I^S)$ for the fixed S, where $OUT(I^S)$ is $(1+\varepsilon)$-approximation value.

Given an optimal solution S^* of leader's problem, let $(S_1^*, S_2^*, \cdots, S_m^*)$ be an optimal schedule for the jobs in $\mathcal{J}^* = \{J_j \in \mathcal{J} \mid p_j \in S^*\}$ in the optimal solution. By the definitions of $p(j)$ and n_k^*, we have

$$p_k(S) = p_k(S^*) = n_k^*, \text{ for } k = 1, 2, \cdots, K.$$

Next, we show that for the $(1+\varepsilon)$-approximation solution of instance I^S, there is a feasible solution of the follower for the optimal solution S^* of the leader's problem such that $OUT^* \leq (1+2\varepsilon)OUT(I^S) + \varepsilon$, where OUT^* is the objective value of the feasible solution of the follower for the optimal solution S^* of the leader's problem. Thus, we have $OPT^* \leq OUT^* \leq (1+2\varepsilon)OUT(I^S)+\varepsilon \leq (1+5\varepsilon)OPT(I^S)+\varepsilon$, since the follower's problem is minimized.

Let $(S_1, S_2, \cdots, S_m)$ be a schedule of an PTAS for the $P||C_{\max}$ problem and the jobs in $\mathcal{J}^S$. For $i = 1, 2, \cdots, m$, let $n_{ik} = \mathcal{J}_k^S \cap S_i$ be the number of jobs in $\mathcal{J}_k^S$ which are assigned to machine M_i. Clearly, $\sum_{i=1}^m n_{ik}^S = n_k^*$. For $i = 1, 2, \cdots, m$, assign n_{ik}^S large jobs in $\mathcal{J}_k^*$ to machine M_i for every k. For every small job in $\mathcal{J}_0^*$, assign it to the machine whose current load is least. Assume any feasible solution of any instance first processes large jobs and then processes small jobs in all machines. Let S_i^L be the set of large jobs assigned to machine M_i for $i = 1, 2, \cdots, m$.

If the last processing job is a large job in the constructed feasible solution, then we have $OUT^* \leq (1+\varepsilon)OUT(I^S)$. Since this "makespan" machine only processes large jobs. Assuming that the machine is M_i, by the (6) and construction rule as above, we have

$$P(S_i^L) = \sum_{J_j \in M_i} p_j \leq (1+\varepsilon)OUT(I^S).$$

If the last processing job is a small job in the constructed feasible solution. We claim that $OUT^* \leq (1+2\varepsilon)OUT(I^S)+\varepsilon$. This is because the last processed small job denoted as J_s is assigned to the machine with the least load and $p_s < \varepsilon$, then

$$\begin{aligned} OUT^* &= L_i^{current} + p_s \\ &\leq \frac{P(S')}{m} + p_s \\ &\leq \frac{P(S^*) - p_s}{m} + p_s \\ &\leq \frac{P(S^*)}{m} + \varepsilon, \end{aligned}$$

where $L_i^{current}$ is the load of the "makespan" machine denoted as M_i before the last job is processed and $P(S')$ is the total processing time before the last job is assigned. Thus, by the (2), (5) and $\varepsilon \in (0, 1/2]$, we have

$$\begin{aligned} OUT^* &\leq \frac{P(S^*)}{m} + \varepsilon \\ &\leq (1+2\varepsilon)\frac{P(S)}{m} + \varepsilon \\ &\leq (1+2\varepsilon)OPT(I^S) + \varepsilon \\ &\leq (1+2\varepsilon)OUT(I^S) + \varepsilon. \end{aligned}$$

Finally, we have $OPT^* \leq (1+2\varepsilon)OUT(I^S)+\varepsilon \leq (1+2\varepsilon)(1+\varepsilon)OUT(I^S)+\varepsilon \leq (1+5\varepsilon)OPT(I^S)+\varepsilon$. □

5 A Simple EPTAS When m is Fixed

In this section, we design a simple EPTAS when m is fixed for this problem. We first divide jobs into large jobs and small jobs. Let S^* be the optimal solution of our problem. The structure of the algorithm is as follows. We optimally "guess" the subset S^* of $m \cdot \lceil \frac{1}{\varepsilon} \rceil$ jobs of the largest jobs. Thus, we can enumerate all these subsets. For each guess, we only consider remaining small job sets. This can be find by solving a maximum 0–1 knapsack problem in remaining small jobs. That is, we want to find a small job set with the maximum total processing time. Finally, the algorithm is described as follows.

Let $\varepsilon \in (0, 1/2]$ and $K = \lceil \log_{1+\varepsilon}(1/\varepsilon) \rceil$. If $p_j \geq \varepsilon$, the job is called *large*. Otherwise, the job is called *small*. For each job $J_j \in \mathcal{J}$ and $p_j \geq \varepsilon$, there is a positive integer τ such that $p_j \in [\varepsilon(1+\varepsilon)^{\tau-1}, \varepsilon(1+\varepsilon)^{\tau})$ for $\tau = 1, \cdots, K$. Recall that $OPT^* = 1$, thus the number of large jobs in the optimal solution is at most $m \cdot \lceil \frac{1}{\varepsilon} \rceil$ and the number of large jobs is at most $\lceil \frac{1}{\varepsilon} \rceil$ with $p_j \in [\varepsilon(1+\varepsilon)^{\tau-1}, \varepsilon(1+\varepsilon)^{\tau})$ for each τ. For $m \cdot \lceil \frac{1}{\varepsilon} \rceil$ jobs, the choice of each job is at most $\lceil \frac{1}{\varepsilon} \rceil \cdot K$. Thus, via at most $O((\lceil \frac{1}{\varepsilon} \rceil \cdot K)^{m \cdot \lceil \frac{1}{\varepsilon} \rceil})$ enumerations, we can optimal "guess" the subset S^* of $m \cdot \lceil \frac{1}{\varepsilon} \rceil$ jobs of the largest jobs in constant time, since m and ε are constants.

Then, we consider small jobs in our algorithm. For each guess set S, if job $p_j \geq \varepsilon$ and have selected in our set S, then remove the job in the original job set and update $B = B - c_j$. If job $p_j \geq \varepsilon$ and not select in our set S, then let $c_j = +\infty$. We use $\mathcal{J}'$ and B' to denote the new remaining job set and remaining budget, respectively. Let $S' \subseteq \mathcal{J}'$ be any subset of $\mathcal{J}'$, we solve the following problem:

$$\begin{cases} \max\limits_{S'} P(S') \\ c(S') \leq B'. \end{cases} \tag{7}$$

This is a 0–1 knapsack problem and there is an FPTAS in [13] for the above problem. We can find a subset S' such that $P(\mathcal{J}'_{\max}) \leq (1+\varepsilon)P(S')$, where $\mathcal{J}'_{\max}$ is the optimal set of this problem. Finally, the follower will find an optimal solution to the $P||C_{\max}$ problem. Let S'^* be the optimal set of small jobs in the optimal solution. Let $S' \cup S$ be the final job set of each guess. There is an EPTAS by using the method in [3] to find a $(1+\varepsilon)$-approximation solution for the $P||C_{\max}$ problem. Finally, our algorithm will output a job set with the maximum makespan that is calculated via the above method. Let OPT be the optimal value of the output job set of our algorithm for the problem $P||C_{\max}$. We have the following theorem.

Theorem 4. $OPT^* \leq (1+3\varepsilon)OPT + \varepsilon$.

Proof. This can be proved by a similar argument as Theorem 3 but simple than it. Thus, we omit the detailed proof. □

6 Conclusion

In this paper, we consider a bilevel adversarial scheduling problem on parallel machines. We first design a simple $(2+\varepsilon)$-approximation algorithm for this problem. Then, we propose an PTAS for this problem. Finally, we also design a simple EPTAS for this problem when the number of machines is fixed. It is interesting to consider other criterion on parallel machines, such as the total completion time, the total weighted completion time, the maximum lateness and the number of tardy jobs. Finally, we will design a better approximation algorithm for this problem in the future.

Acknowledgments. The authors sincerely thank the editor and anonymous reviewers for their constructive comments that substantially improved the quality of this paper. The work was supported in part by Yunnan Fundamental Research Projects (Grants No. 202501AS070076, No. 202501AS070170).

References

1. Abass, S.: Bilevel programming approach applied to the flow shop scheduling problem under fuzsiness. CMS **2**, 279–293 (2005)
2. Agnetis, A., Billaut, J., Gawiejnowicz, S., Pacciarelli, D., Soukhal, A.: Multiagent scheduling. Springer, Berlin (2014)
3. Alon, N., Azar, Y., Woeginger, G.J., Yadid, T.: Approximation schemes for scheduling on parallel machines. J. Sched. **1**, 55–66 (1998)
4. Aissi, H., Bazgan, C., Vanderpooten, D.: Min-max and min-max regret versions of combinatorial optimization problems: a survey. Eur. J. Oper. Res. **197**, 427–438 (2009)
5. Chen, L., Wu, X., Zhang, G.: Approximation algorithms for interdiction problem with packing constraints. In: Proceedings of the 49th International Colloquium on Automata, Languages, and Programming (ICALP). LIPIcs, vol. 229, pp. 39:1–39:19 (2022)
6. Dempe, S., Kalashnikov, V., Pérez-Valdés, G., Kalashnikova, V.: Bilevel programming problems: Theory, Algorithms and Applications to Energy Networks. Springer, Berlin (2015)

7. DeNegre, S.: Interdiction and discrete bilevel linear programming (Ph.D. thesis), USA: Lehigh University (2011)
8. Guan, L., Li, J., Li, W., Lichen, J.: Improved approximation algorithms for the combination problem of parallel machine scheduling and path. J. Comb. Optim. **38**(3), 689–697 (2019). https://doi.org/10.1007/s10878-019-00406-0
9. Hochbaum, D.S., Shmoys, D.B.: Using dual approximation algorithms for scheduling problems theoretical and practical results. J. ACM **34**(1), 144–162 (1987)
10. Jansen, K.: An EPTAS for scheduling jobs on uniform processors: using an MILP relaxation with a constant number of integral variables. SIAM J. Discret. Math. **24**(2), 457–485 (2010)
11. Jansen, K., Klein, K.M., Verschae, J.: Closing the gap for makespan scheduling via sparsification techniques. Math. Oper. Res. **45**(4), 1371–1392 (2020)
12. Karlof, J., Wang, W.: Bilevel programming applied to the flow shop scheduling problem. Comput. Oper. Res. **23**, 443–451 (1996)
13. Kellerer, H., Pferschy, U., Pisinger, D.: Knapsack Problems. Springer, Heidelberg (2004)
14. Kis, T., Kovacs, A.: On bilevel machine scheduling problems. OR Spectrum **34**, 43–68 (2012)
15. Kones, I., Levin, A.: A unified framework for designing EPTAS for load balancing on parallel machines. Algorithmica **81**(7), 3025–3046 (2019)
16. Kovacs, A., Kis, T.: Constraint programming approach to a bilevel scheduling problem. Constraints **16**, 317–340 (2011)
17. Li, W., Li, J., Zhang, X., Chen, Z.: Penalty cost constrained identical parallel machine scheduling problem. Theoret. Comput. Sci. **607**, 181–192 (2015)
18. Li, W., Ou, J.: Machine scheduling with restricted rejection: an application to task offloading in cloud-edge collaborative computing. Eur. J. Oper. Res. **314**(3), 912–919 (2023)
19. Li, W., Ou, J.: Approximation algorithms for scheduling parallel machines with an energy constraint in green manufacturing. Eur. J. Oper. Res. **314**(3), 882–893 (2024)
20. Li, W., Sun, R.: Approximation algorithms for the integrated path and bin packing problem. RAIRO-Operat. Res. **59**(1), 325–333 (2025)
21. Pascual, F., Rzadca, K., Trystram, D.: Cooperation in multi-organization scheduling. Concurrency Comput. Pract. Exper. **21**, 905–921 (2009)
22. Shmoys, D.B., Tardos, É.: An approximation algorithm for the generalized assignment problem. Math. Program. **62**(1), 461–474 (1993)
23. Sun, R., Li, W. B-matching interdiction problem on bipartite graphs with unit weight and multi-dimensional budgets. In: Du, D., Han, L., Xu, D. (eds.) Combinatorial Optimization and Applications. COCOA 2024. LNCS, vol. 15435. Springer, Singapore (2025). https://doi.org/10.1007/978-981-96-4448-3_7
24. T'kindt, V., Della Croce, F., Agnetis, A.: Single machine adversarial bilevel scheduling problems. Euro. J. Operat. Res. **315**(1), 63-72 (2024)

A Randomized FPT Approximation Algorithm for Sorting Unsigned Genomes by Translocations: Breaking the 1.375 Approximation Barrier

Chengcheng Sun[1], Haitao Jiang[2](✉), Guojun Li[1], and Daming Zhu[2]

[1] Research Center for Mathematics and Interdisciplinary Sciences (Frontiers Science Center for Nonlinear Expectations), Shandong University, Qingdao, Shandong, China
{scc,gjli}@sdu.edu.cn

[2] School of Computer Science and Technology, Shandong University, Qingdao, Shandong, China
{htjiang,dmzhu}@sdu.edu.cn

Abstract. Comparing genomes based on gene order is a classical combinatorial optimization problem in computational biology, which seeks the minimum number of genome rearrangement operations required to transform one genome into another. The problem of sorting genomes by translocations has been extensively studied over the past few decades. Computing the translocation distance is NP-hard when the input genomes are unsigned, posing significant computational challenges. A widely adopted approach to approximating this problem involves decomposing the breakpoint graph into proper alternating cycles. However, this decomposition step becomes a bottleneck in calculating the corresponding rearrangement distances, hindering the ability to achieve approximation factors better than 1.375 in polynomial time. In this paper, we propose a novel FPT (fixed-parameter tractable) approximation algorithm for the problem of sorting genomes by translocations, improving the approximation factor to $4/3+\varepsilon$, thereby surpassing the long-standing best ratio of 1.375, which has held since 2016 [12]. Our algorithm employs a new randomized method for decomposing the breakpoint graph, which succeeds with high probability, $1 - \frac{1}{e^{O(n)}}$, as guaranteed by the Chernoff Bound. The time complexity of the algorithm is $O(2^{d^*} \cdot n^{O(\frac{1}{\epsilon})})$, where n represents the length of each genome and d^* denotes the optimal translocation distance.

Keyword: comparing genomes, approximation algorithm, translocation distance.

1 Introduction

Computing genomic distance on gene order is a fundamental area in computational biology as it reveals the evolutionary events between species, and attracts

F. V. Fomin and M. Xiao (Eds.): COCOON 2025, LNCS 15983, pp. 54–67, 2026.
https://doi.org/10.1007/978-981-95-0215-8_5

a lot of interest since it was formulated to be a group of combinatorial problems in the1990 s Given two genomes, which are composed of the same set of genes but different orders, an interesting combinatorial problem is to compute the rearrangement distance between two genomes, i.e., the minimum number of genome rearrangement operations to transform one genome into the other. Sankoff pioneered the proposal of three basic rearrangement operations: reversals, translocations and transpositions [14]. In this paper, we concentrate on the problem of sorting genomes by translocations.

The research on the signed genome rearrangement sorting problem dates back to 1995, when Kececioglu and Ravi first introduced the problem [8]. Later, Hannenhalli designed an algorithm for the problem with a time complexity of $O(n^3)$, where n represents the total number of genes in the genome [5]. However, Hannenhalli's algorithm contained an error in the computation of the translocation sequence. In response, Bergeron et al. revisited the problem and devised a new algorithm with a time complexity of $O(n^3)$, thereby correcting the flaw in Hannenhalli's approach [1]. Subsequently, researchers proposed several improved algorithms aiming to reduce the time complexity of solving this problem. Zhu and Ma first improved the algorithm to run in $O(n^2 \log n)$ time [16]. Wang et al. further reduced the time complexity to $O(n^2)$ [15]. Following this, Ozery-Flato and Shamir proposed an approximation algorithm with a time complexity of $O(n^{3/2}\sqrt{\log n})$ [10], which currently stands as the most efficient translocation sorting algorithm in the literature. The aforementioned translocation sorting algorithms are designed to compute both the translocation distance and the shortest translocation sequence. Specifically, Li et al. demonstrated that the translocation distance for signed genomes without duplicates can be computed in linear time [9], providing a significant optimization for scenarios where only the translocation distance is required.

Unfortunately, for unsigned genomes, the problem of sorting by translocations is NP-hard and APX-hard [17], and has a close relationship with the maximum cycle decomposition of the breakpoint graph, which is also NP-hard [2]. Consequently, the design of polynomial-time approximation algorithms has become a natural focus. Kececioglu and Ravi first designed a 2-approximation algorithm for the problem of sorting unsigned genomes by translocation [8]. Cui et al. improved the approximation ratio to 1.75 [3], and further to $1.5 + \varepsilon$ [4]. Jiang et al. proposed a $1.408 + \varepsilon$ approximation algorithm for this problem [7]. In 2016, Pu et al. designed a new polynomial-time approximation algorithm, improving the approximation ratio to 1.375 [12]. To date, this remains the best-known approximation ratio for the problem.

Currently, research on the problem of unsigned genomes by translocations primarily focuses on designing polynomial-time approximation algorithms, while algorithms based on parameterized time complexity remain underexplored. In fact, studies by Sankoff and Pevzner et al. have revealed that the number of translocation operations between homologous species is often subject to significant biological constraints [11,13]. This critical characteristic provides a natural advantage for the study of parameterized algorithms: by treating the translocation distance as a fixed parameter, more efficient fixed-parameter tractable algorithm can be designed, making them better suited for real biological data. Compared to the traditional approximation algorithm framework, the parameterized approach may demonstrate greater practical value.

In this paper, we propose an FPT (fixed-parameter tractable) approximation algorithm for the problem of sorting unsigned genomes by translocations. The core idea of the algorithm is to decompose the breakpoint graph of the unsigned genome based on minimum candidate sub-permutations. For most types of connected components, we perform cycle decomposition individually, decomposing as many cycles as possible within each component. For intractable components, we merge them with specific components to form combined components, which are then subjected to cycle decomposition. Finally, the cycle decompositions obtained from all connected components are combined to form the cycle decomposition of the breakpoint graph. The approximation factor of this algorithm can reach $4/3 + \varepsilon$, and its running time is $O(2^{d^*} \cdot n^{O(\frac{1}{\epsilon})})$, where n is the length of the genome and d^* represents the translocation distance. The algorithm is randomized and succeeds with high probability, specifically $1 - \frac{1}{e^{O(n)}}$.

2 Preliminaries

2.1 Gene, Chromosome and Genome

In the field of genome rearrangement, a set of integers $\Sigma = \{1, \ldots, n\}$ is typically employed, where each integer represents a long DNA sequence, such as a syntenic block or a gene. For simplicity, the term "gene" will be used hereafter. A *chromosome* is defined as a permutation of these integers, with each integer appearing at most once. A gene located at the terminus of a linear chromosome is referred to as a *telomere*. Genes g_i and g_j are said to form an *adjacency*, denoted by (g_i, g_j), if they are consecutive on a chromosome. An adjacency (g_i, g_j) is termed *trivial* if it satisfies $|g_j - g_i| = 1$. A linear chromosome is considered *trivial* if all its adjacencies are trivial. Similarly, a genome is *trivial* if all its chromosomes are trivial. Two genomes are deemed *related* if they share the same set of genes and the same set of telomeres. Throughout this paper, we assume that every linear chromosome contains two distinct telomeres.

In the context of genome sorting, the relative order of genes within the same chromosome is significant, whereas the orientation of the entire chromosome is not. This implies that each chromosome can be represented in either orientation. For signed genomes, a chromosome $\pi = [g_1, g_2, \ldots, g_k]$ is equivalent to $[-g_k, -g_{k-1}, \ldots, -g_1]$. For unsigned genomes, a chromosome $\pi = [g_1, g_2, \ldots, g_k]$ is equivalent to $[g_k, g_{k-1}, \ldots, g_1]$, where $g_i \in \Sigma$ for all $1 \leq i \leq k$.

2.2 The Problem of Sorting Unsigned Genomes by Translocations

For two signed chromosomes $\pi[1] = [g[1]_1, g[1]_2, \ldots, g[1]_n]$ and $\pi[2] = [g[2]_1, g[2]_2, \ldots, g[2]_m]$ in a genome, a translocation swaps the segments in the chromosomes and generates two new chromosomes. If we cut the adjacency $(g[1]_i, g[1]_{i+1})$ $(1 < i < n)$ in $\pi[1]$ and $(g[2]_j, g[2]_{j+1})$ $(1 < j < m)$ in $\pi[2]$ and reconnect them, then there are two types,

- prefix-prefix translocation: $\pi'[1] = [g[1]_1, \ldots, g[1]_i, g[2]_{j+1}, \ldots, g[2]_m]$, $\pi'[2] = [g[2]_1, \ldots, g[2]_j, g[1]_{i+1}, \ldots, g[1]_n]$,
- prefix-suffix translocation: $\pi''[1] = [g[1]_1, \ldots, g[1]_i, -g[2]_j, \ldots -g[2]_1]$, $\pi''[2] = [-g[2]_m, \ldots, -g[2]_{j+1}, g[1]_{i+1}, \ldots, g[1]_n]$.

If $\pi[1] = [g[1]_1, g[1]_2, \ldots, g[1]_n]$ and $\pi[2] = [g[2]_1, g[2]_2, \ldots, g[2]_m]$ are two unsigned genomes, the resulting chromosomes after the translocation are as follows,

- prefix-prefix translocation: $\pi'[1] = [g[1]_1, \ldots, g[1]_i, g[2]_{j+1}, \ldots, g[2]_m]$, $\pi'[2] = [g[2]_1, \ldots, g[2]_j, g[1]_{i+1}, \ldots, g[1]_n]$,
- prefix-suffix translocation: $\pi''[1] = [g[1]_1, \ldots, g[1]_i, g[2]_j, \ldots, g[2]_1]$, $\pi''[2] = [g[2]_m, \ldots, g[2]_{j+1}, g[1]_{i+1}, \ldots, g[1]_n]$.

Now, we formally put forward the problem we investigate in this paper.

Sorting unsigned genomes by translocations(abbreviated as SUT)

Input: Two related unsigned linear multichromosomal genomes A and B, where B is trivial.

Question: Transform A into B by a series of translocations $\rho_1, \rho_2, \ldots, \rho_k$, such that k is minimized. The minimum k is called the unsigned translocations distance between A and B, denoted by $d^u_{trl}(A, B)$.

3 Sorting Signed and Unsigned Genomes

First, let us review the computational method for signed genomes originally developed by Hannenhalli [5]. The breakpoint graph plays a crucial role in calculating the signed translocation distance.

3.1 The Breakpoint Graph of Signed Genomes

Given two related signed genomes A_s and B_s, the breakpoint graph $\overline{G}(A_s, B_s)$ is constructed as follows: For each chromosome $\pi = [g_1, g_2, \ldots, g_n]$ in A_s, replace each gene g_i with an ordered pair of vertices $(l(g_i), r(g_i))$. If g_i is positive, then $(l(g_i), r(g_i)) = (g_i^t, g_i^h)$; if g_i is negative, then $(l(g_i), r(g_i)) = (g_i^h, g_i^t)$. If the genes g_i and g_{i+1} are adjacent in A_s, connect $r(g_i)$ and $l(g_{i+1})$ with a black edge in $\overline{G}(A_s, B_s)$. Similarly, if g_i and g_{i+1} are adjacent in B_s, connect $r(g_i)$ and $l(g_{i+1})$ with a grey edge in $\overline{G}(A_s, B_s)$. Each vertex in $\overline{G}(A_s, B_s)$ (except those at the ends of chromosomes) is incident to one black edge and one grey edge. Thus, $\overline{G}(A_s, B_s)$ can be uniquely decomposed into cycles, with black and grey edges alternating. A cycle containing exactly i black (or grey) edges is called an i-cycle.

3.2 The Signed Translocation Distance Formula

Let $\overline{I} = [g[1]_1, g[1]_2, \ldots, g[1]_n]$ be a chromosome in A_s. A *sub-permutation*, denoted by SP, is a segment $[g[1]_i, g[1]_{i+1}, \ldots, g[1]_{i+l}]$ of at least four genes in S such that there is another segment $[g[2]_j, g[2]_{j+1}, \ldots, g[2]_{j+l}]$ of the same length in some chromosome $\overline{I'}$ of B_s satisfying that

- $g[1]_i = g[2]_j$, $g[1]_{i+l} = g[2]_{j+l}$ or $g[1]_i = -g[2]_{j+l}$, $g[1]_{i+l} = -g[2]_j$
- $\{|g[1]_i|, |g[1]_{i+1}|, \ldots, |g[1]_{i+l}|\} = \{|g[2]_j|, |g[2]_{j+1}|, \ldots, |g[2]_{j+l}|\}$
- I is not identical with I'

Here, $g[1]_i$ and $g[1]_{i+l}$ are the two ending genes of the SP, $(r(g[1]_i), l(g[1]_{i+1}))$ and $(r(g[1]_{i+l-1}), l(g[1]_{i+l}))$ denote the two boundary black edges of the SP. A *minSP* is defined as an SP that does not contain any other SP. If all minSPs in $\overline{G}(A_s, B_s)$ are part of an SP, and the total number of minSPs is even, then the SP is termed an *even isolation*.

Let b (resp. c, s) represent the number of black edges (resp. cycles, minSPs) in $\overline{G}(A_s, B_s)$. Let f represent an integer variable, which is defined as follows: (1) $f = 1$ if s is odd, (2) $f = 2$ if there exists an even isolation, and (3) $f = 0$ otherwise. Hannenhalli introduced the following formula to calculate the translocation distance $d_s(A_s, B_s)$ between A and B [5].

Lemma 1. $d_s(A_s, B_s) = b - c + s + f$.

Since we focus solely on designing an approximation algorithm for unsigned genomes (by converting them into signed ones), and f is at most 2, we disregard f throughout the paper. Our general approach involves identifying as many alternating cycles as possible from the breakpoint graph, while simultaneously minimizing the number of minSPs. Specifically, we aim to approximate $c - s$ and refer to $b - c + s$ as the translocation distance estimate.

3.3 The Breakpoint Graph of Unsigned Genomes

The breakpoint graph for two unsigned genomes differs slightly from that of signed genomes. In the breakpoint graph $G(A, B)$ of two unsigned genomes A and B, each vertex corresponds to a gene in A. Two genes are connected by a black edge if they are adjacent in a chromosome of A, and by a grey edge if they are adjacent in a chromosome of B. Consequently, the breakpoint graph $G(A, B)$ can be decomposed into a set of edge-disjoint cycles, denoted by **D**, where black and grey edges alternate within each cycle. Notably, each ending gene in A and B is incident to exactly one black edge and one grey edge, while each non-ending gene is incident to precisely two black edges and two grey edges. As a result, the decomposition of $G(A, B)$ into cycles may not be unique.

If each non-ending vertex g_i in $G(A, B)$ is split into two vertices $l(g_i)$ and $r(g_i)$, such that $l(g_i)$ is incident to one black edge and one grey edge connected to g_i in $G(A, B)$, and $r(g_i)$ is incident to the remaining black edge and grey edge connected to g_i in $G(A, B)$, a new graph **D** is obtained. This graph **D** is referred to as a cycle decomposition of $G(A, B)$. In the following subsection, we will show that each cycle decomposition of $G(A, B)$ is equivalent to the breakpoint graph $\overline{G}(A_s, B_s)$, where A_s and B_s are derived by assigning an appropriate sign to each gene in A and B.

In the context of unsigned genomes, a *candidate SP* (abbreviated as *CSP*) consists of a substring of at least four genes in A for which there exists a corresponding substring in B with identical gene content but a distinct gene order.

A *candidate minimal SP* (abbreviated as *minCSP*) is defined as a CSP that does not contain any other CSP within it.

3.4 Converting Unsigned Genomes Into Signed Ones

Once a cycle decomposition $\mathbf{D}$ of $G(A, B)$ is obtained, we can derive two signed genomes A_s and B_s by assigning a sign to each gene in A and B, such that $\overline{G}(A_s, B_s) = \mathbf{D}$. As previously mentioned, all genes in B are trivial. Consequently, all grey edges in $\overline{G}(A_s, B_s)$ take the form $((g_i)^h, (g_i + 1)^t)$.

Next, we describe the procedure for assigning appropriate signs to each gene in A to obtain A_s. An ending gene is assigned a positive sign if it occupies the same end (i.e., both left or both right) of a chromosome in A and a chromosome in B; otherwise, it is assigned a negative sign in A. For a non-ending gene g_i, based on the two grey edges $((g_i)^h, (g_i + 1)^t)$ and $((g_i - 1)^h, (g_i)^t)$ in the cycle decomposition, g_i is assigned a positive sign if $((g_i - 1)^h, l(g_i))$ is a grey edge in the given cycle decomposition. Conversely, if $((g_i - 1)^h, r(g_i))$ is a grey edge in the cycle decomposition, g_i is assigned a negative sign. An illustrative example is provided in Fig 1.

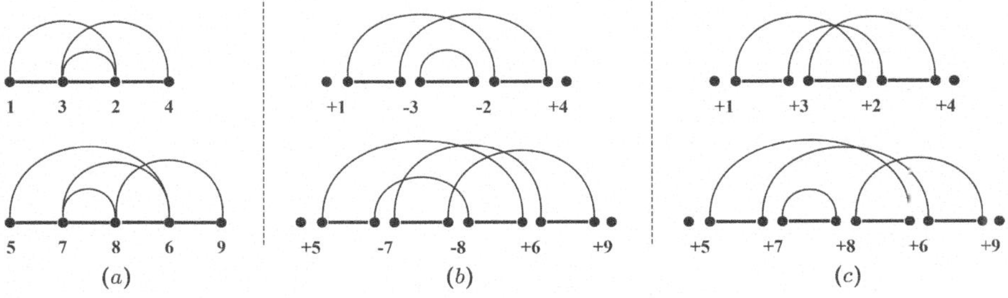

Fig. 1. (a) The breakpoint graph $G(A, B)$ for two unsigned genomes, where $A = \{[1, 3, 2, 4], [5, 7, 8, 6, 9]\}$ and $B = \{[1, 2, 3, 4], [5, 6, 7, 8, 9]\}$. (b) The breakpoint graph $\overline{G}(A_s, B_s)$ for two signed genomes, where $A_s = \{[+1, -3, -2, +4], [+5, -7, -8, +6, +9]\}$ and $B_s = \{[+1, +2, +3, +4], [+5, +6, +7, +8, +9]\}$. (c) The breakpoint graph $\overline{G}(\overline{A'}, \overline{B'})$ for another two signed genomes, where $\overline{A'} = \{[+1, +3, +2, +4], [+5, +7, +8, +6, +9]\}$ and $\overline{B'} = \{[+1, +2, +3, +4], [+5, +6, +7, +8, +9]\}$. Both (b) and (c) are cycle decompositions of (a).

Since the translocation distance between signed genomes can be computed in polynomial time, an efficient approach to solving the problem of sorting unsigned genomes by translocations is to assign appropriate signs to the unsigned genomes–equivalently, to identify a proper cycle decomposition of the breakpoint graph.

4 Sufficient Conditions for Solving the Problem of SUT

In this section, we will present sufficient conditions for a cycle decomposition of $G(A, B)$ to obtain an approximation translocation distance that is not greater than 1.375 times of the translocation distance. Previous studies have demonstrated that all potential 1-cycles can be preserved in the cycle decomposition, as supported by the following lemma.

Lemma 2. *There exists some optimal cycle decompositions containing all the existing 1-cycles [3].*

Owing to Lemma 2, we assume an arbitrary cycle decomposition of $G(A, B)$ always contain all 1-cycles in $G(A, B)$ in the following. Let $I = [g_i, \ldots, g_j]$ be a *CSP* in $G(A, B)$, $V(I) = \{g_i, \ldots, g_j\}$ and $N(I) = V(I) - \{g_i, g_j\}$. Let $G(A, B, I)$ denote the subgraph of $G(A, B)$ induced by $V(I)$, and $G(A, B, \overline{I})$ denote the subgraph of $G(A, B)$ induced by $V(G) - N(I)$. Since I is a *CSP*, each vertex in $G(A, B, I)$ or $G(A, B, \overline{I})$ must incident with one black edge and one grey edge or two black and two grey edges. Thus, both $G(A, B, I)$ and $G(A, B, \overline{I})$ can be decomposed into a set of edge disjoint cycles. Let $\overline{G}(A, B, I)$ and $\overline{G}(A, B, \overline{I})$ be the cycle decompositions of $G(A, B, I)$ and $G(A, B, \overline{I})$ respectively. Consequently, $\overline{G}(A, B, I)$ and $\overline{G}(A, B, \overline{I})$ can form a cycle decomposition of $G(A, B)$. For an arbitrary cycle decomposition $\overline{G}$ of $G(A, B)$, if grey edges incident with the two nodes $l(g_k)$ and $r(g_k)$ ($k \in \{i, j\}$) are crossing in $\overline{G}$, we call them as *cross edges* associated with I.

A cross edge in $\overline{G}$ either connects a vertex in $\{l(g_i), r(g_j)\}$ to a vertex in $IN(I) = \{r(g_i), l(g_{i+1}), r(g_{i+1}), \ldots, l(g_{j-1}), r(g_{j-1}), l(g_j)\}$, or connects a vertex in $\{r(g_i), l(g_j)\}$ to a vertex in $V(\overline{G}) - IN(I)$. If $\overline{G}$ has a cross edge connecting $l(g_i)$ (resp. $r(g_j)$) to a vertex in $IN(I)$, it must have another cross edge connecting $r(g_i)$ (resp. $l(g_j)$) to a vertex in $V(\overline{G}) - IN(I)$. Thus, $\overline{G}$ has either two or four cross edges. Let $\overline{G}^*(A, B)$ be an optimal cycle decomposition of $G(A, B)$ with the maximum number of *minSPs*.

Due to the page limit, the proofs of the following lemmas and theorems are omitted (Figs. 2, 3 and 4).

Lemma 3. *In an optimal cycle decomposition of $G(A, B)$ with the maximum number of minSPs, there exist no two cross edges associated with I.*

By Lemma 3, if $\overline{G}^*(A, B)$ has cross edges associated with I, it must have four cross edges. Let $(l(g_i), p_1), (r(g_i), q_1), (r(g_j), p_2), (l(g_j), q_2)$ be these four cross edges in $\overline{G}^*(A, B)$, where $\{r(g_i), l(g_j), p_1, p_2\} \subseteq IN(I)$ and $\{l(g_i), r(g_j), q_1, q_2, \} \subseteq V(\overline{G}^*(A, B)) - IN(I)$. The following lemma shows how these four cross edges can be connected with each other in $\overline{G}^*(A, B)$.

Lemma 4. *In an optimal cycle decomposition of $G(A, B)$ with the maximum number of minSPs, there exists no cycle containing both of the cross edges connecting to $l(g_i)$ and $r(g_i)$.*

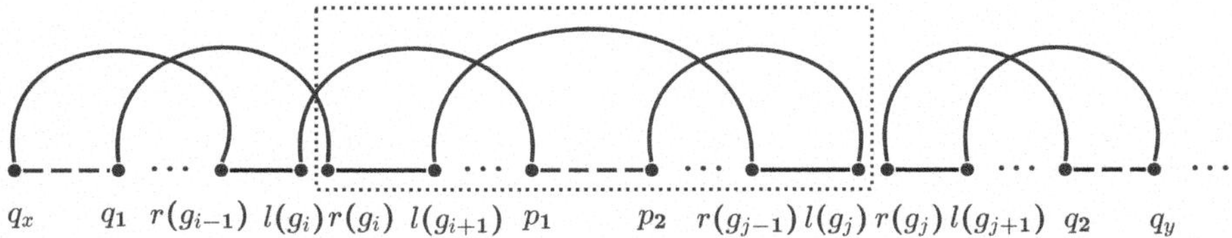

Fig. 2. A cycle decomposition of $G(A, B)$ with two cross edges $(l(g_i), p_1)$ and $(r(g_i), q_1)$. Replacing these two cross edges with $(l(g_i), q_1)$, $(r(g_i), p_1)$ can yield another cycle decomposition of $G(A, B)$ that includes one additional cycles compared to $\overline{G}^*(A, B)$.

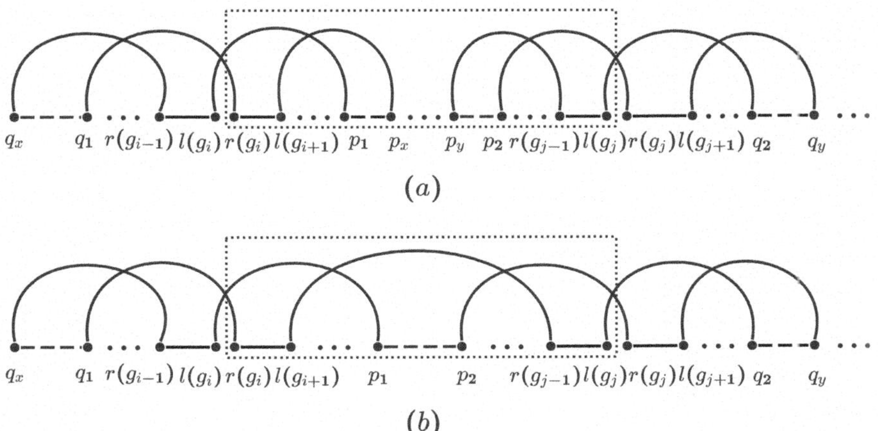

Fig. 3. (a) A cycle decomposition of $G(A, B)$ where four cross edges are contained within two cycles. (b) A cycle decomposition of $G(A, B)$ where four cross edges are contained within a single cycle. By replacing these four cross edges with $(l(g_i), q_1)$, $(r(g_i), p_1)$, $(l(g_j), p_2)$, and $(r(g_j), q_2)$, both (a) and (b) yield another cycle decomposition of $G(A, B)$ that includes two additional cycles compared to $\overline{G}^*(A, B)$.

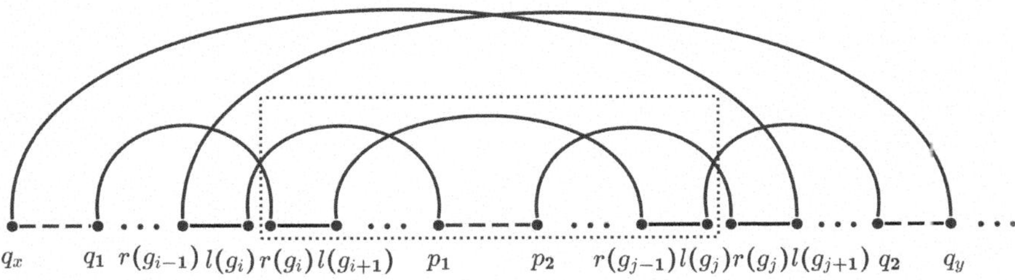

Fig. 4. A cycle decomposition of $G(A, B)$ where four cross edges are contained within a single cycle. Replacing these four cross edges with $(l(g_i), q_1)$, $(r(g_i), p_1)$, $(l(g_j), p_2)$, and $(r(g_j), q_2)$, can yield another cycle decomposition of $G(A, B)$ that includes one additional cycles compared to $\overline{G}^*(A, B)$.

5 Our Algorithm

To address the problem of sorting unsigned genomes by translocations, our primary objective is to identify an optimal cycle decomposition in the breakpoint graph. However, since the problem of finding the maximum cycle decomposition for two unsigned genomes is NP-hard [17], we employ approximation algorithms to tackle this challenge.

5.1 Cycle Decompositions Based on the CSPs

We say two *CSPs* in $G(A,B)$ are independent, if they share at most one gene. A black edge in $G(A,B)$ is *valid*, if it doesn't occur in any 1-cycle in $G(A,B)$. In what follows, all black edges we mentioned are valid.

A *CSP* that can be only decomposed into one 2-cycle (resp. 3-cycle) other than 1-cycles, is also called the 2-*knotty CSP* (resp. 3-*knotty CSP*) where the 2-cycle (resp. 3-cycle) are called as the 2-*knotty cycle* (resp. 3-*knotty cycle*). A *CSP* in $G(A,B)$ is called a *combined CSP* if, (1) it is a *CSP* and contains at least one *knotty minCSP* and at least one black edge not contained by any knotty *minCSP*; (2) all *minCSPs* contained are knotty *minCSPs*. A combined *minCSP* is minimal if it doesn't contain any other combined *CSP*.

Given a set of mutually independent *CSPs* in $G(A,B)$, denoted as $\mathfrak{Q}$. Then, based on the definitions provided in the previous sections, the subgraph of $G(A,B)$ induced by $V(I)$, is denoted by $G(A,B,I)$, and the subgraph of $G(A,B)$ induced by $\overline{\mathfrak{Q}} = V(G(A,B)) - \left(\bigcup_{I\in\mathfrak{Q}} N(I)\right)$, is denoted by $G(A,B,\overline{\mathfrak{Q}})$. For convenience, we refer to $\left(\bigcup_{I\in\mathfrak{Q}} G(A,B,I)\right) \cup G(A,B,\overline{\mathfrak{Q}})$ as a *CSP-partition* of $G(A,B)$ based on $\mathfrak{Q}$, such that $\left(\bigcup_{I\in\mathfrak{Q}} G(A,B,I)\right) \cup G(A,B,\overline{\mathfrak{Q}}) = G(A,B) = \bigcup_{\mathbb{J}\in\mathfrak{Q}\cup\overline{\mathfrak{Q}}} G(\mathbb{J})$.

Observe that any *CSP-partition* can be decomposed into a set of edge disjoint cycles, which form a cycle decomposition (such a cycle decomposition can be obtained from an arbitrary cycle decomposition by changing the signs of the ending gene of $I \in \mathfrak{Q}$, such that the number of cross edges is 0). In our algorithm, we firstly obtain a *CSP*-partition of $G(A,B)$ based on $\mathfrak{Q}$, then a cycle decomposition of $G(A,B)$ can be formed. Specially, while analyzing the performance ratio of our algorithm, we do not compare the cycle decomposition of our algorithm with the optimal cycle decomposition directly, but with one obtained by converting the optimal cycle decomposition into an union cycle decompositions of *CSP*-partition.

let $\overline{G}^*(A,B)$ be the optimal cycle decomposition of $G(A,B)$ with the maximum number of *minSPs*, and c^* and s^* be the number of cycles and *minSPs* in $\overline{G}^*(A,B)$ respectively.

Lemma 5. *There exists a cycle decomposition $\overline{G}^*_{\mathfrak{Q}}(A,B)$ of $G(A,B)$ based on $\mathfrak{Q}$, which contains $c^*_{\mathfrak{Q}}$ cycles, such that $c^*_{\mathfrak{Q}} = c^*$.*

Let $c^*_{\mathfrak{Q},i}$ be the number of i-cycles in $\overline{G}^*_{\mathfrak{Q}}(A,B)$. Let $\overline{G}_{\mathfrak{Q}}(A,B)$ be some cycle decomposition based on $\mathfrak{Q}$, and $c_{\mathfrak{Q}}$, $c_{\mathfrak{Q},1}$ and $s_{\mathfrak{Q}}$ be the number of cycles, 1-cycles and *minSPs* in $\overline{G}_{\mathfrak{Q}}(A,B)$, respectively. Since both $\overline{G}_{\mathfrak{Q}}(A,B)$ and $\overline{G}^*_{\mathfrak{Q}}(A,B)$ contain all 1-cycles in $G(A,B)$, $c_{\mathfrak{Q}^\circ} = c_{\mathfrak{Q}} - c_{\mathfrak{Q},1} = c_{\mathfrak{Q}} - c^*_{\mathfrak{Q},1}$. In order to guarantee an approximation ratio of 4/3, it is sufficient to show that $r = \frac{4}{3}(n - N - c^* + s^*) - (n - N - c_{\mathfrak{Q}} + s_{\mathfrak{Q}}) \geq 0$, where n and N represent the number of genes and chromosomes respectively.

$$
\begin{aligned}
r &= \frac{4}{3}(n - N - c^* + s^*) - (n - N - c_{\mathfrak{Q}} + s_{\mathfrak{Q}}) \\
&= \frac{1}{3}(n - N) - \frac{4}{3}c^* + c_{\mathfrak{Q}} + \frac{4}{3}s^* - s_{\mathfrak{Q}} \\
&= \frac{1}{3}(n - N) - \frac{4}{3}c^*_{\mathfrak{Q}} + c_{\mathfrak{Q}} + \frac{4}{3}s^* - s_{\mathfrak{Q}} \\
&= \frac{1}{3}(\sum_{1\leq i\leq n-1} i \cdot c^*_{\mathfrak{Q},i}) - \frac{4}{3}(\sum_{1\leq i\leq n-1} c^*_{\mathfrak{Q},i}) + c_{\mathfrak{Q}} + \frac{4}{3}s^* - s_{\mathfrak{Q}} \\
&= -c^*_{\mathfrak{Q},1} - \frac{2}{3}c^*_{\mathfrak{Q},2} - \frac{1}{3}c^*_{\mathfrak{Q},3} + \frac{1}{3}\sum_{4\leq i\leq n-1}(i-4)c^*_{\mathfrak{Q},i} + c_{\mathfrak{Q}} + \frac{4}{3}s^* - s_{\mathfrak{Q}} \\
&\geq -c^*_{\mathfrak{Q},1} - \frac{2}{3}c^*_{\mathfrak{Q},2} - \frac{1}{3}c^*_{\mathfrak{Q},3} + c_{\mathfrak{Q}} + \frac{4}{3}s^* - s_{\mathfrak{Q}} \\
&= -\frac{2}{3}c^*_{\mathfrak{Q},2} - \frac{1}{3}c^*_{\mathfrak{Q},3} + c_{\mathfrak{Q}^\circ} + \frac{4}{3}s^* - s_{\mathfrak{Q}}
\end{aligned}
$$

Let $\overline{G}^*_{\mathfrak{Q}}(A,B) = \bigcup_{\mathbb{J}\in\mathfrak{Q}\cup\overline{\mathfrak{Q}}} \overline{G}^*(\mathbb{J})$, where $\overline{G}^*(\mathbb{J})$ represent the cycle decomposition of $G(\mathbb{J})$. For each $\mathbb{J} \in \mathfrak{Q} \cup \overline{\mathfrak{Q}}$, let $c^*_{\mathfrak{Q}}(\mathbb{J})$ and $c^*_{\mathfrak{Q},i}(\mathbb{J})$ be the number of cycles and i-cycles in $\overline{G}^*(\mathbb{J})$ respectively, and let $s^*_{\mathfrak{Q}}(\mathbb{J})$ be the number of *minSPs* in $\overline{G}^*(\mathbb{J})$, which are *minSPs* in $\overline{G}^*_{\mathfrak{Q}}(A,B)$, then $c^*_{\mathfrak{Q}} = \sum_{\mathbb{J}\in\mathfrak{Q}\cup\overline{\mathfrak{Q}}} c^*_{\mathfrak{Q}}(\mathbb{J})$, $c^*_{\mathfrak{Q},i} = \sum_{\mathbb{J}\in\mathfrak{Q}\cup\overline{\mathfrak{Q}}} c^*_{\mathfrak{Q},i}(\mathbb{J})$, and $s^* \geq \sum_{\mathbb{J}\in\mathfrak{Q}\cup\overline{\mathfrak{Q}}} s^*_{\mathfrak{Q}}(\mathbb{J})$. Let $c_{\mathfrak{Q}^\circ}(\mathbb{J})$ and $s_{\mathfrak{Q}}(\mathbb{J})$ be the number of cycles other than 1-cycles and *minSPs* in the cycle decomposition of $G(\mathbb{J})$, thus, $c_{\mathfrak{Q}^\circ} = \sum_{\mathbb{J}\in\mathfrak{Q}\cup\overline{\mathfrak{Q}}} c_{\mathfrak{Q}^\circ}(\mathbb{J})$, and $s_{\mathfrak{Q}} = \sum_{\mathbb{J}\in\mathfrak{Q}\cup\overline{\mathfrak{Q}}} s_{\mathfrak{Q}}(\mathbb{J})$. Thus, it is sufficient to show that,

$$
r(\mathbb{J}) = -\frac{2}{3}c^*_{\mathfrak{Q},2}(\mathbb{J}) - \frac{1}{3}c^*_{\mathfrak{Q},3}(\mathbb{J}) + c_{\mathfrak{Q}^\circ}(\mathbb{J}) + \frac{4}{3}s^*_{\mathfrak{Q}}(\mathbb{J}) - s_{\mathfrak{Q}}(\mathbb{J}) \geq 0 \qquad (1)
$$

5.2 An Approximation Algorithm for SUT

In this subsection, we present an approximation for the problem of sorting unsigned genomes by translocations. For two related unsigned genomes A and B, the algorithm firstly computes a *CSP*-partition based on $\mathfrak{Q}$ in the breakpoint graph $G(A,B)$, and then finds a 'good' cycle decomposition for each $G(\mathbb{J})$, where $\mathbb{J} \in \mathfrak{Q} \cup \overline{\mathfrak{Q}}$.

Algorithm 1. The *CSP*-partition algorithm

Require: Two related unsigned genomes A and B, B is trivial.
Ensure: A *CSP*-partition of $G(A, B)$.
Construct the breakpoint graph $G(A, B)$.
$\mathfrak{Q} \leftarrow \{I \mid I \text{ is a } minCSP \text{ in } G(A, B)\}$
$\mathfrak{J} \leftarrow \{I' \mid I' \text{ is a combined } CSP \text{ in } G(A, B)\}$
for each $I' \in \mathfrak{J}$ **do**
  $\mathfrak{Q} \leftarrow \mathfrak{Q} - \{I'' \mid I'' \text{ is a knotty } minCSP \text{ contained by } I'\} + I'$
end for
return $\bigcup_{\mathbb{J} \in \mathfrak{Q} \cup \overline{\mathfrak{Q}}} G(\mathbb{J})$

Suppose that we have obtained a *CSP*-partition of $G(A, B)$, for each $G(\mathbb{J})$ with at most 27 black edges, an almost optimal cycle decomposition can be achieved by enumerating all possible combinations of gene signs within it. As for each $G(\mathbb{J})$ with more than 27 black edges, we employ Findcycles_Algorithm [6] to determine a cycle decomposition. However, during the construction of the conflict graph and the set packing system, a crucial optimization is applied: vertices and subsets corresponding to knotty cycles are intentionally omitted from the construction process.

Algorithm 2. Approximation Algorithm for SUT

Require: Two related unsigned genomes A and B, both of which contain n genes and N chromosomes, and B is trivial.
Ensure: $d_{trl}^{apr}(A, B)$ as the approximated unsigned translocation distance.
Call Algorithm 1 to obtain a *CSP*-partition $\bigcup_{\mathbb{J} \in \mathfrak{Q} \cup \overline{\mathfrak{Q}}} G(\mathbb{J})$.
for each $\mathbb{J} \in \mathfrak{Q} \cup \overline{\mathfrak{Q}}$ **do**
  if $G(\mathbb{J})$ has at most 27 black edge. **then**
    Enumerating all combinations of the signs of genes in $G(\mathbb{J})$, select a cycle decomposition $\overline{G}(\mathbb{J})$ of $G(\mathbb{J})$ with c cycles and s *minSPs* such that $c - s$ is maximized.
  else
    Preserve the maximum number of 1-cycles in $G(\mathbb{J})$.
    Invoke Findcycles_Algorithm with $G(\mathbb{J})$ excluding 1-cycles as the input.
    while There exists a knotty-*minCSPs* $\mathbb{J}_k$, whose edges are not used. **do**
      Assign proper signs to the nodes of $G(\mathbb{J}_k)$ to obtain a knotty-cycle.
    end while
    Arbitrarily assign signs to the rest of genes in $G(\mathbb{J})$, and obtain a cycle decomposition $\overline{G}(\mathbb{J})$ of $G(\mathbb{J})$.
  end if
end for
Let $\overline{G}_{\mathfrak{Q}}(A, B) = \bigcup_{\mathbb{J} \in \mathfrak{Q} \cup \overline{\mathfrak{Q}}} \overline{G}(\mathbb{J})$ be the cycle decomposition of $G(A, B)$, and $c_{\mathfrak{Q}}$ and $s_{\mathfrak{Q}}$ be the number of cycles and *minSPs* in $\overline{G}_{\mathfrak{Q}}(A, B)$ respectively.
return $d_{trl}^{apr}(A, B) = n - N - c_{\mathfrak{Q}} + s_{\mathfrak{Q}}$.

5.3 Analysis of the Approximation Ratio

In this subsection, we demonstrate that for each $G(\mathbb{J})$, where $\mathbb{J} \in \mathfrak{Q} \cup \overline{\mathfrak{Q}}$, the inequality $r(\mathbb{J}) \geq 0$ holds. This ensures that Algorithm 2 achieves an approximation ratio of 4/3.

Let $s^*_{\mathfrak{Q},k}(\mathbb{J})$, $s^*_{\mathfrak{Q},2}(\mathbb{J})$, $s^*_{\mathfrak{Q},3}(\mathbb{J})$ be the number of knotty-cycles, 2-knotty cycles, 3-knotty cycles in $\overline{G}^*(\mathbb{J})$, and $s^*_{\mathfrak{Q},k}(\mathbb{J}) = s^*_{\mathfrak{Q},2}(\mathbb{J}) + s^*_{\mathfrak{Q},3}(\mathbb{J})$. Because a knotty cycle generated by the cycle decomposition of combined *minCSPs* or $\overline{\mathfrak{Q}}$, must be a *minSPs* in $\overline{G}^*_{\mathfrak{Q}}(A, B)$, thus, for each $\overline{G}^*(\mathbb{J})$, $\mathbb{J} \in \mathfrak{Q} \cup \overline{\mathfrak{Q}}$, $s^*_{\mathfrak{Q}}(\mathbb{J}) \geq s^*_{\mathfrak{Q},k}(\mathbb{J})$.

Based on the cycle decomposition algorithm presented by Jiang et al., we have the following lemma and theorem.

Theorem 1. *With probability $1 - \frac{1}{e^{O(n)}}$, the number of cycles obtained by Findcycles_Algorithm is not less than $Max\{(\frac{5}{6} - \varepsilon') \times (1 - \delta)\frac{15}{16}c^*_2, (\frac{3}{5} - \varepsilon'') \times (1 - \delta)(c^*_2 + \frac{57}{64}c^*_3)\}$, where c^*_2(resp. c^*_3) be the number of 2-cycles(resp. 3-cycles) in the maximum cycle decomposition of $G(A, B)$, δ, ε' and ε'' are arbitrarily small constant [6].*

Corollary 1. *With probability $1 - \frac{1}{e^{O(n)}}$, the number of cycles obtained by Findcycles_Algorithm is not less than $(\frac{2}{3} - \varepsilon' - \delta)c^*_2 + (\frac{1}{3} - \varepsilon'' - \delta)c^*_3$, where c^*_2(resp. c^*_3) be the number of 2-cycles(resp. 3-cycles) in the maximum cycle decomposition of $G(A, B)$, δ, ε' and ε'' are arbitrarily small constant.*

Since Algorithm 2 do not construct vertices and subsets for these knotty *minCSPs*, then the number of cycles obtained by invoking Findcycles_Algorithm satisfies the following theorem.

Theorem 2. *If $G(\mathbb{J})$ has more than 27 black edges, and $c^1_{\mathfrak{Q}}(\mathbb{J})$ represents the number of cycles obtained by invoking Findcycles_Algorithm. With probability $1 - \frac{1}{e^{O(n)}}$, $c^1_{\mathfrak{Q}}(\mathbb{J}) \geq (\frac{2}{3} - \varepsilon' - \delta)c^*_{\mathfrak{Q},2}(\mathbb{J}) + (\frac{1}{3} - \varepsilon'' - \delta)c^*_{\mathfrak{Q},3}(\mathbb{J}) - s^*_{\mathfrak{Q},k}(\mathbb{J})$, δ, ε' and ε'' are arbitrarily small constant.*

Let $s_{\mathfrak{Q},k}(\mathbb{J})$ be the number of knotty cycles found in $G(\mathbb{J})$ by Algorithm 2.

Lemma 6. *For each $G(\mathbb{J})$ with more than 27 black edges, if $s_{\mathfrak{Q},k}(\mathbb{J}) = 0$, $s_{\mathfrak{Q}}(\mathbb{J}) \leq 1$; and if $s_{\mathfrak{Q},k}(\mathbb{J}) > 0$, $s_{\mathfrak{Q}}(\mathbb{J}) = s_{\mathfrak{Q},k}(\mathbb{J})$.*

For each $G(\mathbb{J})$ with more than 27 black edges, in addition to the cycles identified by Findcycles_Algorithm, some knotty cycle can also be found if none of their edges are used. Furthermore, if there are still black edges left, there exists at least one additional cycle obtained by step 11 in Algorithm 2.

Lemma 7. *For each $G(\mathbb{J})$ with more than 27 black edges, if $s_{\mathfrak{Q},k}(\mathbb{J}) \geq 1$, $c_{\mathfrak{Q}^\circ}(\mathbb{J}) \geq \frac{2}{3}c^*_{\mathfrak{Q},2}(\mathbb{J}) + \frac{1}{3}c^*_{\mathfrak{Q},3}(\mathbb{J}) - s^*_{\mathfrak{Q},k}(\mathbb{J}) + s_{\mathfrak{Q},k}(\mathbb{J})$, and if $s_{\mathfrak{Q},k}(\mathbb{J}) = 0$, $c_{\mathfrak{Q}^\circ}(\mathbb{J}) \geq \frac{2}{3}c^*_{\mathfrak{Q},2}(\mathbb{J}) + \frac{1}{3}c^*_{\mathfrak{Q},3}(\mathbb{J}) - s^*_{\mathfrak{Q},k}(\mathbb{J}) + 1$.*

Lemma 8. *For each $G(\mathbb{J})$ where $\mathbb{J} \in \mathfrak{Q} \cup \overline{\mathfrak{Q}}$, $r(\mathbb{J}) \geq 0$.*

Theorem 3. *With probability at least* $1 - e^{\frac{1}{O(n)}}$, *Algorithm 2 approximates the unsigned translocation distance within a factor of* $4/3 + \varepsilon$ *and runs in* $O(2^{d^*} \cdot n^{O(\frac{1}{\varepsilon})})$ *time, where* $\delta > 0$ *and* $\varepsilon > 0$ *are arbitrarily small constants,* n *denotes the length of each genome, and* d^* *represents the unsigned translocation distance.*

Proof. By Lemma 8, Algorithm 2 achieves an approximation factor of $4/3+\varepsilon$ for the unsigned translocation distance. The most computationally intensive step in Algorithm 2 is the invocation of Findcycles_Algorithm, whose time complexity is $O(2^{\frac{|V_{2\wedge3}|}{2}} \cdot n^{O(\frac{1}{\varepsilon})})$, where $V_{2\wedge3}$ denotes the set of nodes in $G(A,B)$ that are involved in at least two edge-disjoint 2-cycles/3-cycles. All other steps of the algorithm can be completed in $O(n^2)$ time.

Given that $d^* = n-N-c_1^*-\sum_{2\leq i\leq n-N} c_i^*$ and $\sum_{2\leq i\leq n-N} c_i^* \leq (n-N-c_1^*)/2$, where c_i^* represents the number of i-cycles ($1 \leq i \leq n-N$) in the optimal cycle decomposition, it follows that $d^* \geq (n-N-c_1^*)/2$. Each vertex in $V_{2\wedge3}$ is incident to exactly two black edges, and each black edge has at most two unsigned genes as its endpoints. Consequently, the number of unsigned genes in $V_{2\wedge3}$ is bounded by the number of black edges, i.e., $|V_{2\wedge3}| \leq n - N - c_1^* \leq 2d^*$.

In summary, the overall time complexity of Algorithm 2 is $O(2^{d^*} \cdot n^{O(\frac{1}{\varepsilon})})$.

6 Conclusion

In this paper, we propose an FPT approximation algorithm that transforms an unsigned genome A into another unsigned genome B using a sequence of at most $\frac{4}{3} \cdot d^u_{trl}(A,B) + 2$ translocation operations, where $d^u_{trl}(A,B)$ denotes the translocation distance between A and B, improving the previous approximation factor of 1.375. The algorithm is randomized and succeeds with high probability. The running time is bounded by $O(2^{d^*} \cdot n^{O(\frac{1}{\varepsilon})})$. When the algorithm is restricted to considering only 2-cycles and 3-cycles, the approximation factor of 4/3 is the best expected bound. Although the theoretical time complexity of the algorithm appears high, its practical performance is significantly influenced by the number of candidate 2-cycles and 3-cycles in the breakpoint graph. On real-world genome data, such cycles are rarely observed, which ensures the practical applicability of our algorithm. Nevertheless, the design of a polynomial-time approximation algorithm with a factor better than 1.375 for the problem of sorting by translocations remains an open and worthwhile research direction.

Acknowledgments. This research is supported by the National Natural Science Foundation of China (No. 62272279, 62272272, U24A20257).

References

1. Bergeron, A., Mixtacki, J., Stoye, J.: On sorting by translocations. J. Comput. Biol. **13**(2), 567–578 (2006)
2. Caprara, A.: Sorting permutations by reversals and eulerian cycle decompositions. SIAM J. Discret. Math. **12**(1), 91–110 (1999)

3. Cui, Y., Wang, L., Zhu, D.: A 1.75-approximation algorithm for unsigned translocation distance. J. Comput. Syst. Sci. **73**(7), 1045–1059 (2007)
4. Cui, Y., Wang, L., Zhu, D., Liu, X.: A $(1.5 + \epsilon)$-approximation algorithm for unsigned translocation distance. IEEE/ACM Trans. Comput. Biol. Bioinform. **5**(1), 56–66 (2008)
5. Hannenhalli, S.: Polynomial-time algorithm for computing translocation distance between genomes. Discret. Appl. Math. **71**(1–3), 137–151 (1996)
6. Jiang, H., Pu, L., Qingge, L., Sankoff, D., Zhu:, B.: A randomized fpt approximation algorithm for maximum alternating-cycle decomposition with applications. In: International Computing and Combinatorics Conference (COCOON), pp. 26–38 (2018)
7. Jiang, H., Wang, L., Zhu, B., Zhu, D.: A factor-$(1.408 + \varepsilon)$ approximation for sorting unsigned genomes by reciprocal translocations. Theoret. Comput. Sci. **607**, 166–180 (2015)
8. Kececioglu, J.D., Ravi, R.: Of mice and men: algorithms for evolutionary distances between genomes with translocation. In: ACM-SIAM Symposium on Discrete Algorithms (SODA), pp. 604–613 (1995)
9. Li, G., Qi, X., Wang, X., Zhu, B.: A linear-time algorithm for computing translocation distance between signed genomes. In: Sahinalp, S.C., Muthukrishnan, S., Dogrusoz, U. (eds.) CPM 2004. LNCS, vol. 3109, pp. 323–332. Springer, Heidelberg (2004). https://doi.org/10.1007/978-3-540-27801-6_24
10. Ozery-Flato, M., Shamir, R.: An $O(n^{\frac{3}{2}}\sqrt{logn})$ algorithm for sorting by reciprocal translocations. J. Dis. Algor. **9**(4), 344–357 (2011)
11. Pevzner, P., Tesler, G.: Genome rearrangements in mammalian evolution: lessons from human and mouse genomes. Genome Res. **13**(1), 37–45 (2003)
12. Pu, L., Zhu, D., Jiang, H.: A new approximation algorithm for unsigned translocation sorting. In: International Workshop on Algorithms in Bioinformatics (WABI), pp. 269–280 (2016)
13. Sankoff, D., Leduc, G., Antoine, N., Paquin, B., Lang, B.F., Cedergran, R.: Gene order comparisons for phylogenetic inference: evolution of the mitochondrial genome. Proc. Natl. Acad. Sci. **89**(14), 6575–6579 (1992)
14. Sankoff, D.: Edit distances for genome comparisons based on non-local operations. In: Annual Symposium on Combinatorial Pattern Matching (CPM), pp. 121–135 (1992)
15. Wang, L., Zhu, D., Liu, X., Ma, S.: An $O(n^2)$ algorithm for signed translocation. J. Comput. Syst. Sci. **70**(3), 284–299 (2005)
16. Zhu, D., Ma, S.: An improved algorithm for the translocation sorting problem of genomes. Chin. J. Comput. **25**(2), 189–196 (2002)
17. Zhu, D., Wang, L.: On the complexity of unsigned translocation distance. Theoret. Comput. Sci. **352**(1–3), 322–328 (2006)

On Online Approximation Algorithms for Two-Stage Bins

Guangwei Wu[1], Hongyun He[1], Guozhen Rong[2], Feng Shi[3,4(✉)], and Yongjie Yang[5]

[1] College of Computer Science and Mathematics, Central South University of Forestry and Technology, Changsha, People's Republic of China
[2] School of Computer, Changsha University of Science and Technology, Changsha, People's Republic of China
[3] School of Computer Science and Engineering, Central South University, Changsha, People's Republic of China
fengshi@csu.edu.cn
[4] Xiangjiang Laboratory, Changsha, People's Republic of China
[5] Chair of Economic Theory, Saarland University, Saarbrücken, Germany

Abstract. Motivated by applications in cloud computing, this paper studies a hybrid problem that combines the Parallel Two-stage Flowshop Scheduling problem with the Bin Packing problem, referred to as the Two-stage Bin Packing problem. Given a sequence of two-stage jobs, the problem aims to pack them into the minimum number of two-stage flowshops such that the completion time of each flowshop does not exceed a given time limit. To our best knowledge, the problem has not been studied before. Recognizing its NP-hardness, we investigate several approximation algorithms. First, we present two online algorithms based on the well-known FIRST-FIT and NEXT-FIT strategies, which achieve an absolute approximation ratio 4 with a lower bound 3.166 and a tight asymptotic approximation ratio 4, respectively. We then introduce an algorithm that applies Johnson's Order to each individual flowshop and assigns incoming jobs to flowshops using the FIRST-FIT strategy, which is shown to have an asymptotic approximation ratio 3.061 with a lower bound 2.66. Besides, we show that applying Johnson's Order in flowshops cannot improve the NEXT-FIT strategy in terms of approximation ratio.

Keywords: Parallel two-stage flowshops · Bin packing · Approximation algorithm · FIRST-FIT · NEXT-FIT · Cloud computing

1 Introduction

The Parallel Two-stage Flowshop problem has received increasing interest during the past several decades due to its wide applications in the areas such as

This work is supported in part by the National Natural Science Foundation of China under Grants 62072476 and 62302060, the Open Project of Xiangjiang Laboratory (No. 22XJ03005), and the Hunan Provincial Natural Science Foundation of China under Grant 2025JJ50395.

F. V. Fomin and M. Xiao (Eds.): COCOON 2025, LNCS 15983, pp. 68–80, 2026.
https://doi.org/10.1007/978-981-95-0215-8_6

industrial manufacturing, transportation and cloud computing [1–9]. Take an application of the problem in cloud computing for example. If a cloud server receives a user request, it needs to read the corresponding data from disk into memory and then send the data to the user via the network. Thus each user request can be treated as a two-stage job that contains the disk-read stage (R-operation) and network-transmit stage (T-operation), where the T-operation cannot start unless the corresponding R-operation is finished. Each cloud server can be treated as a two-stage flowshop, which has an R-processor and a T-processor that perform R-operations and T-operations of the two-stage jobs in parallel, respectively [7,10].

Given a collection of two-stage jobs and a set of two-stage flowshops, the original variant of the Parallel Two-stage Flowshop problem aims to schedule these jobs into the flowshops such that the *makespan* (i.e., the completion time of the last job) is minimized. The original variant is NP-hard if the number of flowshops is a constant, and becomes strongly NP-hard if the number of flowshops is a part of input [2,7]. Specifically, under the case where the number of flowshops is a constant, Kovalyov [5], Dong et al. [2], and Wu et al. [7] proposed FPTASs for the original variant independently. Under the case where the number of flowshops is part of the input, Wu et al. [8] proposed the first approximation algorithm with ratio 2.6. In the same year, Dong et al. gave a PTAS [3].

Besides the above works on approximation algorithms for the original variant, there are some works for the Parallel Two-stage Flowshop problem under makespan constraint, in which each job has a specific profit, and the aim is to select some jobs and pack them into multiple flowshops with the objective of maximizing the profit while subject to the makespan constraint. Chen et al. [1] showed that the problem is already strongly NP-hard if the number of flowshops is 2, and proposed several approximation algorithms for the problem. Tong et al. [6] gave a PTAS under the case where the number of flowshops is fixed.

It is known that cloud computing companies operate under the pay-per-use pricing model [11–13]. From the user's perspective, a primary concern is the cost of the cloud service used to complete their requests. For the Parallel Two-stage Flowshop problem, the cost is influenced by the number of flowshops and the makespan. Thus within the paper, we introduce a new variant called **Two-stage Bin Packing** problem, which can be treated as a dual problem to the original Parallel Two-stage Flowshop problem. The Two-stage Bin Packing problem can be roughly formulated as follows: Given a collection of two-stage jobs and a makespan constraint (i.e., time limit), the aim is to pack the jobs into the minimum number of two-stage flowshops such that the makespan does not exceed the constraint. The study on the Two-stage Bin Packing problem is obviously necessary and meaningful.

The name "Two-stage Bin Packing problem" stems from the observation that the classical Bin Packing problem is a special case of the Two-stage Bin Packing problem when all jobs consist solely of the disk-read stage (or the network-transmit stage). It is well-known that the Bin Packing problem is NP-hard, and no algorithm can achieve an absolute approximation ratio smaller than 3/2 unless

P = NP [14,15]. As a result, much research has focused on asymptotic approximation algorithms, which aim to find near-optimal solutions in polynomial time for most instances. Several AFPTASs have been developed [16–18].

The Bin Packing problem has also been extensively studied in the online setting, where each item must be packed into a bin as it arrives, without knowledge of future items. In contrast, the offline setting assumes that all items are known beforehand as a part of input. The study of the Bin Packing problem under the online setting was initiated by several key publications in the early1970 s[19–22]. Two widely used strategies for this problem are NEXT-FIT (NF) and FIRST-FIT (FF). Johnson analyzed several simple algorithms, including two online versions based on the two rules and proved that their tight asymptotic approximation ratios are 2 and 1.7, respectively [19,20]. The presorted offline versions of these algorithms, known as NFD and FFD (where items are sorted in non-increasing order by size before scheduling), were also studied. NFD has a tight asymptotic approximation ratio 1.6903 [23], while FFD has a tight ratio $\frac{11}{9}$ [19,24,25]. Then Lee and Lee [26] introduced a bounded-space online algorithm with an asymptotic approximation ratio 1.69103. Currently, for the online Bin Packing problem the best asymptotic approximation ratio is 1.57829 [27], and the optimal absolute approximation ratio is $\frac{5}{3}$ [28]. Additionally, the best known lower bound on the asymptotic approximation ratio of any online algorithm for the problem is 1.54278 [29].

Our research focuses on developing approximation algorithms for the Two-stage Bin Packing problem. We begin with two online algorithms that adopt the NF and FF rules independently. Through simple analysis, we show that the algorithm using FF achieves an absolute approximation ratio 4, while the one using NF has an asymptotic approximation ratio 4. By carefully constructing counterexamples, the two algorithms based on FF and NF are proved to have their asymptotic approximation ratio no smaller than 3.166 and 4, respectively. Unlike the classical Bin Packing problem, in the Two-stage Bin Packing problem, the job execution order affects the completion time of flowshop. Therefore, we propose an approximation algorithm, relaxing the limitation on repacking in the online setting. The algorithm still uses FF to assign jobs to flowshops, but assumes that all flowshops process (repack) their assigned jobs in Johnson's Order, which is crucial for scheduling on a single two-stage flowshop [30]. By a sophisticated amortized analysis, we successfully prove that applying Johnson's Order to each flowshop improves the asymptotic approximation ratio to 3.061 ($<$ 3.166). A thorough discussion shows a lower bound $\frac{8}{3} \approx 2.66$ on the asymptotic approximation ratio for the algorithm, and moreover, if an algorithm uses NF to assign jobs, the repacking of jobs on flowshop is of no help to improving its approximation ratio. To our best knowledge, these are the first approximation results for the two-stage generalization of the classical Bin Packing problem. Due to space constraints, the proofs of some lemmas and theorems are omitted in this version but will be provided in the full version of the paper.

2 Preliminaries

Consider a collection $G = \{J_1, J_2, \ldots, J_n\}$ of two-stage jobs, in which each two-stage job J_i is represented by a pair (r_i, t_i), where $r_i \geq 0$ (resp., $t_i \geq 0$) is the processing time of its R-operation (resp., T-operation), and the processing time of J_i is $r_i + t_i$. As all jobs and flowshops discussed are two-stage, we sometimes omit the term "two-stage" for brevity. Additionally, we simply referred to r_i and t_i as the R-time and T-time, respectively. W.l.o.g., we scale R-times and T-times of the jobs in G such that the time limit for each flowshop is 1.

Consider a flowshop and a collection C of jobs assigned to it. A *schedule* of C specifies not only the execution order of the R-operations of the jobs on the R-processor but also the order of the T-operations on the T-processor. In general, the execution order of the R-operations may differ from that of the T-operations. A *permutation schedule* is one, in which the execution order of the R-operations is identical to that of the T-operations. A permutation schedule specified by Johnson's Order is *optimal* as it minimizes the completion time of the flowshop (i.e., the time taken by the flowshop to finish all the assigned jobs). **Johnson's Order** [30]: *The jobs with their R-times no larger than their corresponding T-times are processed first in non-decreasing order of R-time, and are followed by the remaining jobs processed in non-increasing order of T-time.*

Consider a permutation schedule $S = \langle J_1, \ldots, J_s \rangle$ (which may not be optimal) of the jobs assigned to flowshop F_j. W.l.o.g., we assume that all R-operations and T-operations of the jobs are processed as early as possible. That is, the R-operations of the jobs in S are applied continuously without interruption. However, T-operation may not be applied continuously, and gaps can occur because the previous T-operation finishes while its corresponding R-operation has not yet completed. Under the above assumption, a job is called *critical* in S if its T-operation starts immediately after its R-operation finishes, and from that point onward, the T-operations are processed continuously (note that this job may be the last one in S). If job J_k $(1 \leq k \leq s)$ is critical, then the minimum completion time of F_j (or schedule S) is [8,30]

$$\tau_j = \sum_{i=1}^{k} r_i + \sum_{i=k}^{s} t_i. \tag{1}$$

For any other job J_l $(J_l \in S)$, we have

$$\sum_{i=1}^{l} r_i + \sum_{i=l}^{s} t_i \leq \tau_j = \sum_{i=1}^{k} r_i + \sum_{i=k}^{s} t_i, \tag{2}$$

since there may be a gap between the finish time of its R-operation and the start time of its T-operation in S. If multiple jobs satisfy the definition of a critical job, we designate the one with the lowest index in S as the critical job.

Let the *load* of the R-processor (resp., T-processor) of flowshop F_j be the sum of the R-times (resp., T-times) of the jobs assigned to F_j, and the *load* of flowshop F_j be the sum over the R-times and T-times of these assigned jobs, denoted by $E(j)$. By Equality (1), $\tau_j \leq E(j) \leq 2$.

Consider a collection C of jobs assigned to flowshop F, a new job $J \notin C$ and a schedule rule. If the completion time of the job sequence specified by the rule for the jobs of $C \cup \{J\}$ meets the time limit 1, then we say that with respect to the rule, J can *fit* into the flowshop F, or that F is *fit* for J.

Now we formally define the Two-stage Bin Packing problem.

Two-Stage Bin Packing

Input: A collection $G = \{J_1, J_2, \ldots, J_n\}$ of two-stage jobs, where $J_i = (r_i, t_i)$ with $0 < r_i + t_i \leq 1$ for each $1 \leq i \leq n$.
Task: Schedule the jobs in G into the minimum number of two-stage flowshops such that the completion times of these flowshops do not exceed 1.

Note that the schedule mentioned in the formulation not only considers the assignment of the jobs in G to the flowshops but also determines the execution order of the jobs assigned to each flowshop. As mentioned above, if the T-time of each job equals 0, then the Two-stage Bin Packing problem can be regarded as the classical Bin Packing problem, which is strongly NP-hard [15]. The following corollary follows.

Corollary 1. *The Two-stage Bin Packing problem is strongly NP-hard.*

Consider an algorithm A for the Two-stage Bin Packing problem. Denote by $A(G)$ the number of flowshops used by A to pack the jobs in G; and by $\text{OPT}(G)$ the number of flowshops used by an optimal schedule for G. Algorithm A is an approximation algorithm for the Two-stage Bin Packing problem with an *asymptotic approximation ratio* α and additive constant c, if for any instance G of the problem, it returns a solution in polynomial time such that $A(G) \leq \alpha \cdot \text{OPT}(G) + c$. The *(absolute) approximation ratio* α is defined analogously, with the restriction that $c = 0$.

3 Two Online Approximation Algorithms

The section studies the online setting of the Two-stage Bin Packing problem, where the jobs come in sequence and the current job needs to be packed without any knowledge of the future jobs. Two online algorithms named NF-PACKING and FF-PACKING are given, which are based on the rules NF and FF, respectively. Remark that NF-PACKING and FF-PACKING always assign the current job to the end of the job sequence in a specific flowshop. Thus a job fits into a flowshop if the flowshop can finish the new job sequence within the time limit 1. NF-PACKING assigns the current job to the open flowshop containing the last job if it fits, otherwise closes the flowshop and opens a new flowshop with the job as its first job. FF-PACKING assigns the current job to the first open flowshop that it fits, i.e., the lowest indexed open flowshop that is fit for the job. If no such flowshop exists, FF-PACKING opens a new flowshop and assigns the job to it.

It is easy to see for a coming job, NF-PACKING and FF-PACKING take runtime $O(1)$ and $O(n)$, respectively, to decide which flowshop the job should be assigned to, where n is the number of arrived jobs. For NF-PACKING and FF-FACKING, they have the properties formulated in the following lemma.

Lemma 1. *(I) For the schedule returned by* NF-PACKING, *the sum of loads of any two adjacent flowshops is greater than 1; (II) for the schedule returned by* FF-PACKING *with m flowshops, the sum of loads of any i $(2 \leq i \leq m)$ flowshops is greater than $i/2$.*

Given an instance G of the Two-stage Bin Packing problem, let NF(G) and FF(G) be the numbers of flowshops used by NF-PACKING and FF-PACKING for G, respectively. W.l.o.g, assume that both NF-PACKING and FF-PACKING use m flowshops for G. If $m \leq 1$, then the optimal schedule also uses m flowshop. By Lemma 1 (II), for FF-PACKING, the total load of the m flowshops is larger than $m/2$. As mentioned before, the total load of flowshop is at most 2. Thus OPT(G) is greater than $(m/2)/2 = m/4$. As a result, the algorithm FF-PACKING has the absolute approximation ratio

$$\mathrm{FF}(G)/\mathrm{OPT}(G) < \frac{m}{m/4} = 4.$$

For NF-PACKING, any two adjacent flowshops have a combined load larger than 1. We can group these adjacent flowshops into pairs. If m (the number of flowshops) is even, there are $m/2$ groups of flowshops, each of which has a combined load larger than 1. Thus for the m flowshops, the total load exceeds $m/2$. Combining it with the fact that the load flowshop in any schedule is at most 2, gives $\mathrm{OPT}(G) > (m/2)/2 = m/4$. Consequently,

$$\frac{\mathrm{NF}(G)}{\mathrm{OPT}(G)} < \frac{m}{m/4} = 4.$$

If m is odd, there are $(m-1)/2$ groups of adjacent flowshops. Using similar reasoning, we get $\mathrm{OPT}(G) > (m-1)/4$, i.e.,

$$m < 4 \cdot \mathrm{OPT}(G) + 1.$$

Therefore the algorithm NF-PACKING has an asymptotic approximation ratio 4.

Theorem 1. FF-PACKING *has an absolute approximation ratio 4, and* NF-PACKING *has an asymptotic approximation ratio 4.*

Next we analyze the lower bounds on the approximation ratios of FF-PACKING and NF-PACKING for the Two-stage Bin Packing problem.

Theorem 2. *The asymptotic approximation ratio of* FF-PACKING *is at least $19/6 \approx 3.166$, and that of the algorithm* NF-PACKING *is at least 4.*

4 An Improved Approximation Algorithm

As jobs in G arrive in an arbitrary order under the online setting, both FF-PACKING and NF-PACKING always append the current job to the end of the job sequence in a flowshop. This packing strategy may result in suboptimal job sequences that do not follow Johnson's Order, which is known to minimize the completion time for two-stage flowshop. To address this, we propose a slight relaxation of the online setting by allowing the flowshop to repack its assigned jobs according to Johnson's Order after each new job is added by FF rule. This leads to an approximation algorithm, called APPROX. Notably, APPROX still operates without any knowledge of future jobs; it only optimizes the current job sequence. Verifying whether there exists some F_j that is fit for job J_i with respect to Johnson's Order, and rescheduling the job J_i and the ones within F_j accordingly, can be done $O(n)$ time. Consequently, the time complexity of APPROX for processing n many jobs is $O(n^2)$.

Algorithm APPROX**:**

```
when a job J_i arrives;
if there exists some open flowshop that is fit for J_i with respect to Johnson's
Order then
  | assign J_i into the lowest indexed fit flowshop F_j;
  | repack the jobs in F_j in Johnson's Order;
else
  | open a new flowshop and assign J_i into it;
end
```

To simplify the analysis for the approximation ratio of APPROX, we assume that the flowshops in the schedule obtained by APPROX are ordered by their opening times, i.e., F_j is opened after F_i if $j > i$. Additionally, let X be the set of these flowshops, each of which contains exactly one job. It is worthy to point out that Lemma 1 (II) that holds for the FF-PACKING algorithm, also holds for APPROX. This is because that the jobs assigned to any flowshop in APPROX cannot be packed into previously open flowshops.

Lemma 2. *In the schedule returned by* APPROX*, if there is a flowshop F_i with a load at most $2/3$, then any flowshop F_j with $j > i$ either has a load larger than $2/3$ or contains exactly one job.*

Lemma 3. *In the schedule returned by* APPROX *for G, the optimal number of flowshops* $\mathrm{OPT}(G)$ *is at least $|X|$.*

Here we directly give the main result of the paper, Theorem 3, revealing the improvement on the approximation ratio due to the introduction of Johnson's Order.

Theorem 3. APPROX *has an asymptotic approximation ratio 3.061.*

The intuition behind the proof of Theorem 3 is as follows: If all the jobs in G have large processing times, then they are likely to be scheduled on the flowshops in a manner that is sufficiently close to an optimal schedule. On the other hand, if there is a job on a flowshop with a small processing time, then the loads of the previous flowshops must be large enough such that the job cannot be placed on any of them. Thus we divide the analysis into two cases, based on the processing times of the jobs in G.

Case 1. All jobs in G have their processing times larger than 2/5.

The open flowshops by APPROX can be classified into two sets by the number of their assigned jobs. Recall that the set X contains the flowshops with exactly one job. Let Y be the set of the rest flowshops with at least two jobs, i.e., the flowshops with load larger than 4/5. If there is at most one flowshop in X (i.e., $|X| \leq 1$), then $\text{APPROX}(G) \leq |Y|+1$ and $\text{OPT}(G) > (\frac{4|Y|}{5})/2$ by the definition of Y and the fact that the load of one flowshop is at most 2. These give

$$\text{APPROX}(G) \leq 2.5 \cdot \text{OPT}(G) + 1.$$

Otherwise,

$$\text{OPT}(G) > \left(\frac{|X|}{2} + \frac{4\,|Y|}{5}\right)/2, \tag{3}$$

as if $|X| \geq 2$, then the mean load of the flowshops in X is larger than 1/2 due to Lemma 1 (II). Thus $\text{OPT}(G) > \frac{|X|}{4} + \frac{2|Y|}{5} = \frac{2}{5}(|X| + |Y|) - \frac{3|X|}{20}$. Then,

$$\begin{aligned} \text{APPROX}(G) = |X| + |Y| &< \frac{5}{2}\left(\text{OPT}(G) + \frac{3\,|X|}{20}\right) \\ &\leq \frac{5}{2}\left(\frac{23 \cdot \text{OPT}(G)}{20}\right) = 2.875 \cdot \text{OPT}(G), \end{aligned} \tag{4}$$

where the "$\leq$" holds due to $|X| \leq \text{OPT}(G)$ by Lemma 3.

Case 2. At least one job in G has its processing time at most 2/5.

Let F_p $(1 \leq p \leq m)$ be the flowshop with the largest index that has an assigned job J_z with processing time at most 2/5, i.e., every job assigned to the last $m-p$ flowshops has its processing time larger than 2/5. Thus for each of the last $m-p$ flowshops, it has either exactly one assigned job or a load larger than 4/5 (i.e., at least two jobs). The following analysis is divided into two subcases based on the relation between the R-time r_z and T-time t_z of J_z.

4.1 Case 2.1. $r_z > t_z$

Now we analyze the status of the first $p-1$ flowshops.

Lemma 4. *In the schedule returned by* APPROX, *if there exists a flowshop* F_p *with an assigned job* J_z, *whose processing time* $r_z + t_z$ *is at most* 2/5, *and* $r_z > t_z$ *(i.e., Case 2.1 applies), then any previously open flowshop* F_i $(1 \leq i < p)$ *either has a load larger than* 4/5 *or has the load of its* R*-processor larger than* 3/5.

All flowshops except F_p in the schedule returned by APPROX are classified into three sets X, Y, and Z. Specifically, set X contains the flowshops each with exactly one assigned job, set Y contains the remaining flowshops with a load larger than 4/5, and set Z contains the rest flowshops. Thus

$$\text{APPROX}(G) = |X| + |Y| + |Z| + 1. \tag{5}$$

Combining the above discussion together with Lemma 4, we have that each flowshop in Z is opened before F_p, has a load at most 4/5, but with R-processor load larger than 3/5, and contains at least two jobs. Note that Inequality (3) $\text{OPT}(G) \geq (\frac{|X|}{2} + \frac{4|Y|}{5})/2$ also holds here. If $|Z| \leq 1$, then by Equality (5) and the analysis similar to that for Inequality (4), we have

$$\text{APPROX}(G) = 2.875 \cdot \text{OPT}(G) + 2. \tag{6}$$

In the following discussion, we assume $|Z| \geq 2$. By Lemma 2 and the definition of Z, at least $|Z| - 1$ flowshops in Z have loads larger than 2/3. We further classify the flowshops in X into two subsets: $X_{R \leq T}$ and $X_{R>T}$, based on the relationship between the R-time and T-time of the assigned job. Specifically, $X_{R \leq T}$ (resp., $X_{R>T}$) contains flowshops where the R-times of the assigned jobs are less than or equal to (resp., greater than) the corresponding T-time.

We first consider the properties of the flowshops in the subset $X_{R \leq T}$. Recall that by APPROX, packing the jobs from any two flowshops into one will cause the flowshop's completion time to exceed the time limit 1, even with rescheduling in Johnson's Order. Take a flowshop from $X_{R \leq T}$ and another from Z. W.l.o.g, let $S = \langle J_1, J_2, \ldots, J_c \rangle$ be the job sequence on the flowshop of Z (in Johnson's Order) , and let J_x be the job on the flowshop of $X_{R \leq T}$ (i.e., $x \notin \{1, \ldots, c\}$). Scheduling the jobs in S along with J_x in Johnson's Order gets a new sequence $S' = \langle J_1, \ldots, J_x, \ldots, J_c \rangle$ (note that J_x could be the first or the last job in S'), whose processing time would clearly exceed 1. If J_x is the critical job of S', then

$$r_1 + \cdots + r_x + t_x + \cdots + t_c > 1. \tag{7}$$

If J_x is scheduled after the critical job J_k, then

$$r_1 + \cdots + r_k + t_k + \cdots + t_x + \cdots + t_c > 1,$$

where we claim that $r_x \geq r_k$, because Johnson's Order sorts the jobs with R-times no larger than their corresponding T-times in non-decreasing order by R-time. That is

$$r_1 + \cdots + r_x + t_k + \cdots + t_x + \cdots + t_c > 1. \tag{8}$$

Additionally, by Johnson's Order each r_i in Inequalities (7) and (8) is not larger than t_i. Replacing all r_i with t_i except r_x in the two inequalities gets

$$t_1 + t_2 + \cdots + t_c + r_x + t_x = \left(\sum_{i=1}^{c} t_i\right) + r_x + t_x > 1. \tag{9}$$

Due to Lemma 4 and the definition of set Z, we also have that the load of R-processor of any flowshop in Z is larger than 3/5, i.e.,

$$r_1 + r_2 + \cdots + r_c = \sum_{i=1}^{c} r_i > 3/5. \tag{10}$$

Hence combining Inequalities (9) and (10) gives that if the job J_x is the critical job or scheduled after the critical job in sequence S', then the total load of the two flowshops exceeds $1 + 3/5 = 8/5$.

Now we repeatedly pair the flowshops from the sets $X_{R\le T}$ and Z such that each pair has a total load greater than 8/5. This implies that each individual flowshop has a mean load greater than 4/5. Consequently, we can add these flowshop pairs into set Y (recall that Y is the set of flowshops with a load larger than 4/5), continuing until no such pair of flowshops exists. If all the flowshops of Z can be paired, then all flowshops except F_p either belong to set Y with a load larger than 4/5 or belong to set X with exactly one assigned jobs. By the same reasoning for Inequality (3), we can get

$$\text{APPROX}(G) < 2.875 \cdot \text{OPT}(G) + 1, \tag{11}$$

where the constant is due to the flowshop F_p.

In the following, we assume that for any two flowshops, one from $X_{R\le T}$ and one from Z, in the sequence of the jobs assigned to the two flowshops that follows Johnson's Order, the job on the flowshop of $X_{R\le T}$ is scheduled before the critical job in the sequence.

Lemma 5. *In Case 2.1 of the schedule obtained by* APPROX, *(I). The load of R-processor on any flowshop in the set* $X_{R\le T}$ *is larger than 1/5; (II). If* $|X_{R\le T}| \ge 2$ *then the sum of loads of any* i $(2 \le i \le |X_{R\le T}|)$ *flowshops in the set* $X_{R\le T}$ *is larger than* $3i/5$.

Now we consider the properties of the flowshops in the set $X_{R>T}$.

Lemma 6. *In Case 2.1 of the schedule obtained by* APPROX, *if* $|X_{R>T}| \ge 3$, *then there exists a subset* $X' \subset X_{R>T}$ *with size at least* $|X_{R>T}| - 2$ *such that for any* i *flowshops in* X', *the sum of loads of their R-processors is larger than* $i/3$.

Finally, based on all the above discussion we can derive the main result for Case 2.1, formulated in the following lemma.

Lemma 7. *Given a two-stage job sequence* G, *if in Case 2.1, then the number of flowshops used by* APPROX *is bounded by* $3.061 \cdot \text{OPT}(G) + 5$.

4.2 Case 2.2. $r_z \le t_z$

Recall that F_p is the largest indexed flowshop containing a job J_z with $r_z + t_z \le 2/5$. In Case 2.2, we have $r_z \le t_z$ and aim to show that the asymptotic approximation ratio of APPROX is the same as that in Case 2.1, by converting the schedule for Case 2.2 to an equivalent schedule for Case 2.1 with the same performance.

Lemma 8. *Given a two-stage job sequence G, if in Case 2.2, then the number of flowshops used by* APPROX *is at most* $3.061 \cdot \mathrm{OPT}(G) + 5$.

Combining Inequality (4) for Case 1, and Lemmas 7 and 8 for Case 2, we can get the correctness of Theorem 3.

4.3 Lower Bounds for APPROX and Related Algorithms

Next we study the lower bounds on the approximation ratio of APPROX, and that of the algorithms incorporating a "repack" operation with NF rule.

Theorem 4. *The asymptotic approximation ratio of* APPROX *is at least* $\frac{8}{3}$.

Finally we consider the algorithms that incorporate a "repack" operation (not limited to Johnson's Order) with NF rule, i.e., the algorithms allow flowshop to repack its assigned jobs, after assigning a new job on it by NF. We observe that the counterexample for NF-PACKING also applies to the lower bound analysis of the asymptotic approximation ratio of these algorithms. Specifically, in the counterexample, the job sequence of each open flowshop is consistently maintained in Johnson's Order throughout the scheduling process, which is an optimal job sequence on a single flowshop. This implies that the repack operation does not improve the asymptotic performance in this scenario, as the counterexample demonstrates that the lower bound remains unaffected even when repacking is allowed. Therefore,

Theorem 5. *The asymptotic approximation ratio 4 of* NF-PACKING *is tight, even when job repacking is allowed in the open flowshop after a new job is added.*

5 Conclusion

We studied approximation algorithms for the Two-stage Bin Packing problem, which can be viewed as a two-stage generalization of the Bin Packing problem. To our best knowledge, these are the first approximation results for the Two-stage Bin Packing problem. Several issues are worthy for further investigation.

1. Is it possible to close the gap between the upper and lower bounds of the algorithm FF-PACKING for the Two-stage Bin Packing problem in the paper?
2. The *weight function* is a well-known technique studied in the context of the classical Bin Packing problem [14,19]. Could this technique be applied in the context of the Two-stage Bin Packing problem, in particular, improving the approximation ratio of the algorithm APPROX?
3. Is there a PTAS for the offline setting of the Two-stage Bin Packing problem?

References

1. Chen, J., Huang, M., Guo, Y.: Scheduling multiple two-stage flowshops with a deadline. Theor. Comput. Sci. **921**, 100–111 (2022)
2. Dong, J., et al.: An fptas for the parallel two-stage flowshop problem. Theor. Comput. Sci. **657**, 64–72 (2017)
3. Dong, J., Jin, R., Luo, T., Tong, W.: A polynomial-time approximation scheme for an arbitrary number of parallel two-stage flow-shops. Eur. J. Oper. Res. **281**(1), 16–24 (2020)
4. He, D., Kusiak, A., Artiba, A.: A scheduling problem in glass manufacturing. IIE Trans. **28**, 129–139 (1996)
5. Kovalyov, M.: Efficient ϵ-approximation algorithm for minimizing the makespan in a parallel two-stage system, Vesti Academii navuk Belaruskai SSR, Ser. Phiz.-Mat. Navuk **3**,119 (1985)
6. Tong, W., Xu, Y., Zhang, H.: A polynomial-time approximation scheme for parallel two-stage flowshops under makespan constraint. Theor. Comput. Sci. **922**, 438–446 (2022)
7. Wu, G., Chen, J., Wang, J.: Scheduling two-stage jobs on multiple flowshops. Theor. Comput. Sci. **776**, 117–124 (2019)
8. Wu, G., Chen, J., Wang, J.: On scheduling multiple two-stage flowshops. Theor. Comput. Sci. **818**, 74–82 (2020)
9. Zhang, X., van de Velde, S.: Approximation algorithms for the parallel flow shop problem. Eur. J. Oper. Res. **216**(3), 544–552 (2012)
10. Zhang, Y., Zhou, Y.: Separating computation and storage with storage virtualization. Comput. Commun. **34**(13), 1539–1548 (2011)
11. Armbrust, M., et al.: A view of cloud computing. Commun. ACM **53**(4), 50–58 (2010)
12. Cong, P., Xu, G., Wei, T., Li, K.: A survey of profit optimization techniques for cloud providers. ACM Comput. Surv. **53**(2), 1–35 (2020)
13. Greenberg, A., Hamilton, J., Maltz, D.A., Patel, P.: The cost of a cloud: research problems in data center networks. ACM SIGCOMM Comput. Commun. Rev. **39**(1), 68–73 (2009)
14. Coffman, E.G., Galambos, G., Martello, S., Vigo, D.: Bin Packing Approximation Algorithms: Combinatorial Analysis, pp. 151–207. Springer, US, Boston, MA (1999)
15. Garey, M.R., Johnson, D.S.: Computers and Intractability: A Guide to the Theory of NP-Completeness. W. H. Freeman & Co., USA (1979)
16. Hoberg, R., Rothvoss, T.: A logarithmic additive integrality gap for bin packing. In: Proceedings of the 28th Annual ACM-SIAM Symposium on Discrete Algorithms, USA, p. 2616–2625 (2017)
17. Rothvoß, T.: Approximating bin packing within o (log opt log log opt) bins. In: Proceedings of the 54th IEEE Annual Symposium on Foundations of Computer Science, Berkeley, pp. 20–29 (2013)
18. Karmarkar, N., Karp, R.M.: An efficient approximation scheme for the one-dimensional bin-packing problem. In: 23rd Annual Symposium on Foundations of Computer Science, pp. 312–320. IEEE (1982)
19. Johnson, D.S.: Near-optimal bin packing algorithms (1973)
20. Johnson, D.S.: Fast algorithms for bin packing. J. Comput. Syst. Sci. **8**(3), 272–314 (1974)

21. Johnson, D.S., Demers, A.J., Ullman, J.D., Garey, M.R., Graham, R.L.: Worst-case performance bounds for simple one-dimensional packing algorithms. SIAM J. Comput. **3**, 299–325 (1974)
22. Ullman, J.D.: The performance of a memory allocation algorithm, Technical report 100. Princeton University, Princeton, NJ (1971)
23. Baker, B.S., Coffman, E.G., Jr.: A tight asymptotic bound for next-fit-decreasing bin-packing. SIAM J. Algebraic Discrete Methods **2**(2), 147–152 (1981)
24. Dósa, G.: The tight bound of first fit decreasing bin-packing algorithm is ffd(i) $\leq$ 11/9 opt(i)+ 6/9. In: Proceedings of the International Symposium on Combinatorics, Algorithms, Probabilistic and Experimental Methodologies, pp. 1–11 (2007)
25. Yue, M.: A simple proof of the inequality ffd(l) $\leq$ 11/9 opt(l) + 1, $\forall l$ for the ffd bin-packing algorithm. Acta Math. Appl. Sin. **7**(4), 321–331 (1991)
26. Lee, C.C., Lee, D.T.: A simple on-line bin-packing algorithm. J. ACM **32**(3), 562–572 (1985)
27. Balogh, J., Békési, J., Dósa, G., Epstein, L., Levin, A.: A new and improved algorithm for online bin packing, in: Proceedings of the 26th Annual European Symposium on Algorithms, Vol. 112, Dagstuhl, Germany, pp. 5:1–5:14 (2018)
28. Balogh, J., Békési, J., Dósa, G., Sgall, J., van Stee, R.: The optimal absolute ratio for online bin packing. In: Proceedings of the 26th Annual ACM-SIAM Symposium on Discrete Algorithms, USA, p. 1425–1438 (2015)
29. Balogh, J., Békési, J., Dósa, G., Epstein, L., Levin, A.: A new lower bound for classic online bin packing. Algorithmica **83**, 2047–2062 (2021)
30. Johnson, S.M.: Optimal two- and three-stage production schedules with setup times included. Nav. Res. Logist. Q. **1**(1), 61–68 (1954)

An Improved Approximation Algorithm for the Minimum k-Star Partition Problem

Tong Xu, Wei Yu(✉), and Zhaohui Liu

School of Mathematics, East China University of Science and Technology, Shanghai 200237, China
y30231274@mail.ecust.edu.cn, {yuwei,zhliu}@ecust.edu.cn

Abstract. Given an undirected graph $G = (V, E)$, the minimum k-star partition problem is to find a collection of vertex-disjoint stars containing at most k vertices to cover all the vertices of V. The objective is to minimize the number of stars in the collection. In this paper, we give a local search algorithm which achieves an approximation ratio of $\frac{k}{2} - \frac{k-2}{k(k+1)}$ when $k \geq 5$ is even and $\frac{k}{2} - \frac{k-2}{2k^2}$ when $k \geq 5$ is odd. This improves on the previous best $\frac{k}{2}$-approximation algorithm implied by Hell and Kirkpatrick for each $k \geq 5$. In addition, we give examples to show that our analysis is tight.

Keywords: Approximation Algorithm · Star Partition · Path Partition · Local Search

1 Introduction

Given an undirected graph $G = (V, E)$, the minimum k-star partition problem (Min-kSP for short) is to find a minimum collection of vertex-disjoint stars, each of which has at most k vertices, to cover all the vertices of V. A star S in G is a tree containing at most one vertex of degree greater than one. If S contains more than two vertices, the vertex of maximum degree in S is called its center and the other vertices of S are its leaves. Clearly, in this case each leaf is connected by exactly one edge to the center and hence must be of degree one. If S contains precisely two vertices, either vertex can be chosen as the center and the other vertex is the leaf. If S consists of a single vertex, this vertex is the center of S but S has no leaves. By replacing stars with paths in the Min-kSP, we obtain the minimum k-path partition problem (Min-kPP for short). A star (path) is called a q-star (q-path) if it has exactly q vertices. A 1-star (1-path) is also called a singleton. A k^--star partition (k^--path partition) is a set of vertex-disjoint q-stars (q-paths) with $q \leq k$ that covers all the vertices.

The Min-kSP/Min-kPP has found applications in broadcasting problems in computer or communication network [20], facility location and network monitoring [2], vehicle routing problems [12,18], social networks [16], network controllability [19], postal delivery [10], and team formation problem [11], etc. When

F. V. Fomin and M. Xiao (Eds.): COCOON 2025, LNCS 15983, pp. 81–92, 2026.
https://doi.org/10.1007/978-981-95-0215-8_7

applied to the team formation problem [11], the graph $G = (V, E)$ represents the organizational network in a large global corporation, where each vertex in V corresponds to an employee and each edge in E between two vertices represents a communication relationship between the two corresponding employees. To reduce the operational cost, the corporation need to group all the employees into a small number of working teams. Moreover, each team has at most k employees: one leader and possibly $k-1$ ordinary employees and the leader must have communication relationship with each ordinary employee (if any) in his/her team. One can see that determining the minimum number of such teams is exactly the Min-kSP, where each team corresponds to a star in G consisting of a center (the leader) and at most $k-1$ leaves (the ordinary employees).

Since a q-star is equivalent to a q-path when $q \leq 3$, the Min-kSP is the same as the Min-kPP for $k \leq 3$. It is not difficult to see that the Min-2SP/Min-2PP is equivalent to the maximum matching problem, which is polynomially solvable [6]. When $k \geq 3$, the Min-kPP has been proved to be solvable in polynomial time when the input graphs are cographs [17] (with k being a fixed constant), trees [20] and bipartite permutation graphs [18] (with k being part of the input). However, the Min-kSP/Min-kPP has been shown to be NP-hard for any fixed $k \geq 3$ [8], even on bipartite graphs of maximum degree 3 [15]. As a consequence, we focus on the design of approximation algorithms for the Min-kSP/Min-kPP with $k \geq 3$.

Monnot and Toulouse [15] gave the first $\frac{3}{2}$-approximation algorithm for the Min-3PP using the matching method. This ratio was improved to $\frac{13}{9}$ by Chen et al. [4] using the local search approach. More elaborate local search algorithms by Chen et al. [3] and Chen et al. [5] further brought down the ratio to $\frac{4}{3}$ and $\frac{21}{16}$, respectively. Chen et al. [4] developed a polynomial algorithm to generate a 3^--path partition containing the minimum number of singletons, which implies a $\frac{k}{2}$-approximation algorithm for the Min-kPP (as in the proof of Theorem 1 in [2]). Chen et al. [2] obtained an improved $\frac{k+2}{3}$-approximation algorithm for the Min-kPP with $k \geq 7$ based on the maximum path-cycle cover approach. Li et al. [13] designed a simple local search algorithm for the Min-kPP with ratio $\frac{k^2}{3(k-1)} + \frac{(k-3)}{6(k-1)} \cdot (k \mod 3)$. By a reduction to a special case of the maximum traveling salesman problem (Max-TSP), Li et al. [14] proposed a $(\frac{k+12}{7} - \frac{6}{7k})$-approximation algorithm for the Min-kPP. They also obtained better approximation ratios of $\frac{31}{18}, \frac{17}{8}, \frac{7}{3}$ for the case of $k = 4, 5, 6$, respectively. The Max-TSP, which is a classical NP-hard problem [9], is to determine a maximum-weight Hamiltonian cycle for an edge-weighted undirected complete graph.

Compared to the above rich algorithmic results on the Min-kPP, few approximability results are devoted to the Min-kSP with $k \geq 4$. Prior to this work, Bao et al. [1] proposed a $\frac{19}{10}$-approximation algorithm based on the local search method for the Min-4SP. For the general Min-kSP, Hell and Kirkpatrick [7] gave a polynomial algorithm to produce a k^--star partition containing the minimum number of singletons. As in the case of Min-kPP, this yields a $\frac{k}{2}$-approximation algorithm for the Min-kSP. As far as we known, there exist no further approximation algorithms tailored for the Min-kSP with $k \geq 4$.

In this paper, we design a local search algorithm for the Min-kSP with $k \geq 5$, which achieves an approximation ratio of $\frac{k}{2} - \frac{k-2}{k(k+1)}$ when k is even and $\frac{k}{2} - \frac{k-2}{2k^2}$ when k is odd. We also give examples to show that these ratios are tight. Our algorithm starts with a k^--star partition containing the minimum number of singletons, and then iteratively applies local search operations to reduce the number of stars. The algorithm terminates when none of the designed local search operations is applicable and output the current solution as the approximate solution. One can see that our algorithm improves the $\frac{k}{2}$-approximation algorithm in [7] for every $k \geq 5$. Our algorithm for the Min-kSP with $k \geq 5$ is a generalization of the local search algorithm for the Min-4SP by Bao et al. [1]. The approximation ratio of our algorithm is proved through a more complex amortized strategy than that in [1]. Our results represent a solid step to further understand the approximability of the Min-kSP.

The rest of the paper is organized as follows. In Sect. 2, we give some basic notations used throughout the paper. In Sect. 3, we describe the local search algorithm for the Min-kSP in detail, analyze its approximation ratio and give tight examples. Finally, We conclude the paper in Sect. 4.

2 Preliminaries

Let $G = (V, E)$ be an undirected graph with vertex set V and edge set E, we define $n = |V|$ and $m = |E|$. Each $e = \{u, v\} \in E$ connects two vertices $u, v \in V$. For any subgraph S of G, let $V(S)$ and $E(S)$ denote the vertex set and edge set of S, respectively. The order of S is defined as the cardinality of $V(S)$. A q-star $S = u_1 - u_2 \cdots u_q$ in G is a subgraph of G with $V(S) = \{u_1, u_2, \ldots, u_q\}$ such that for each $i = 2, \ldots, q$ there is an edge $e_i = \{u_1, u_i\} \in E(S)$ connecting u_1 with u_i. u_1 is called the center of S while $u_2, \ldots, u_q$ are called the leaves of S. A 1-star, which contains only the center and no leaves, is also known as a singleton. We say that the q vertices $u_1, u_2, \ldots, u_q$ can be covered by a q-star if there exists a q-star in G containing exactly the vertices $u_1, u_2, \ldots, u_q$ (the center of this star may be any of these q vertices). A set $\mathcal{S} = \{S_1, S_2, \ldots, S_t\}$ of stars with $|V(S_i)| \leq k$ for each i is called a k^--star partition of G if $V = \bigcup_{i=1}^{t} V(S_i)$ and $V(S_i) \cap V(S_j) = \emptyset$ for any $i \neq j$.

Formally, the Minimum k-Star Partition Problem (Min-kSP) aims to find a k^--star partition $\mathcal{S}$ of G to minimize $|\mathcal{S}|$.

For an instance of the Min-kSP defined on G and a k^--star partition $\mathcal{Q}$, let $\mathcal{Q}_i$ $(i = 1, 2, \ldots, k)$ be the set of all i-stars in $\mathcal{Q}$. Then $\mathcal{Q}_1, \mathcal{Q}_2, \ldots, \mathcal{Q}_k$ form a partition of $\mathcal{Q}$. We use $\mathcal{Q}^*$ and to denote an optimal k^--star partition and therefore $|\mathcal{Q}^*|$ represents the corresponding optimal value. Analogously, $\mathcal{Q}_i^*$ $(i = 1, 2, \ldots, k)$ is the set of all i-stars in $\mathcal{Q}^*$ and $\mathcal{Q}_1^*, \mathcal{Q}_2^*, \ldots, \mathcal{Q}_k^*$ form a partition of $\mathcal{Q}$. Clearly, we have $|\mathcal{Q}| = \sum_{i=1}^{k} |\mathcal{Q}_i|$ and $|\mathcal{Q}^*| = \sum_{i=1}^{k} |\mathcal{Q}_i^*|$. We use SOL to indicate the k^--star partition obtained by the algorithm in discussion and sol is the size of SOL, i.e. $sol = |SOL|$. For a set $\mathcal{R}$ of stars, we define $V(\mathcal{R}) = \bigcup_{S \in \mathcal{R}} V(S)$ and $E(\mathcal{R}) = \bigcup_{S \in \mathcal{R}} E(S)$. In particular, we have $V(\mathcal{Q}_i) = \bigcup_{S \in \mathcal{Q}_i} V(S)$ and $V(\mathcal{Q}_i^*) = \bigcup_{S \in \mathcal{Q}_i^*} V(S)$.

3 A Local Search Algorithm for the Min-kSP

In this section, we propose a local search algorithm for the Min-kSP with $k \geq 5$. Then we will show that this algorithm achieves an approximation ratio of $\frac{k}{2} - \frac{k-2}{k(k+1)}$ when k is even and $\frac{k}{2} - \frac{k-2}{2k^2}$ when k is odd.

Our algorithm first invokes the algorithm in [7] to get a k^--star partition $\mathcal{Q}$ as the initial solution, which contains the minimum number of singletons, and then iteratively applies three local search operations to reduce the total number of stars in the current k^--star partition $\mathcal{Q}$. When further reduction is impossible, the algorithm terminates and outputs the current k^--star partition as the approximate solution.

Next, we describe the three local search operations of our algorithm in detail.

Operation 1

If there exist two 2-stars $x_1 - x_2, y_1 - y_2$ in $\mathcal{Q}_2$ and an i-star $z_1 - z_2 z_3 \cdots z_i$ in $\mathcal{Q}_i$ for some i with $2 \leq i \leq k-1$ such that the $i+4$ vertices $x_1, x_2, y_1, y_2, z_1, z_2, \ldots, z_i$ can be covered by a 3-star S_1, and an $(i+1)$-star S_2, update the current k^--star partition $\mathcal{Q}$ by replacing these two 2-stars and the i-star with S_1 and S_2 (see Fig. 1 for an illustration).

Operation 2

If there exist four 2-stars $x_1 - x_2, y_1 - y_2, z_1 - z_2, u_1 - u_2$ in $\mathcal{Q}_2$ and an i-star $v_1 - v_2 v_3 \cdots v_i$ in $\mathcal{Q}_i$ for some i with $3 \leq i \leq k-1$ such that the $i+8$ vertices $x_1, x_2, y_1, y_2, z_1, z_2, u_1, u_2, v_1, v_2, \ldots, v_i$ can be covered by three 3-stars S_1, S_2, S_3 and an $(i-1)$-star S_4, update the current k^--star partition $\mathcal{Q}$ by replacing these four 2-stars and the i-star with S_1, S_2, S_3, S_4 (see Fig. 2 for an illustration).

Operation 3 (This operation is applied only for even k)

If there exist $\frac{k}{2}$ 2-stars $x_{1,1} - x_{1,2}, x_{2,1} - x_{2,2}, \ldots, x_{k/2,1} - x_{k/2,2}$ in $\mathcal{Q}_2$ such that the k vertices $x_{1,1}, x_{1,2}, x_{2,1}, x_{2,2}, \ldots, x_{k/2,1}, x_{k/2,2}$ can be covered by a k-star S, update the current k^--star partition $\mathcal{Q}$ by replacing these $\frac{k}{2}$ 2-stars with S (see Fig. 3 for an illustration).

A high-level description of the complete algorithm, called Algorithm 1, for the Min-kSP is depicted in Fig. 4.

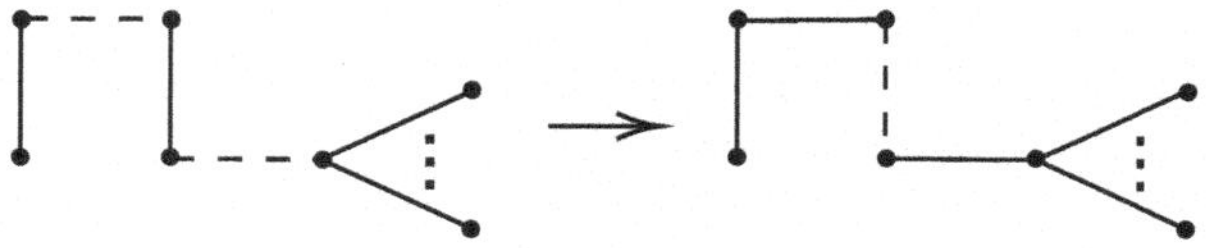

Fig. 1. Illustrating example for Operation 1. The solid edges are in $E(\mathcal{Q})$ and the dashed edges are in $E \setminus E(\mathcal{Q})$.

Step 1 runs in $O(nm)$ time, as shown by Hell and Kirkpatrick [7]. In Step 2, the algorithm tries to improve $\mathcal{Q}$ by Operations 1–3. If succeed, it updates $\mathcal{Q}$ accordingly and continues Step 2. Through an analysis similar to that in [3], one can show that when k is odd Step 2 takes $O(n^5)$ time and when k is even Step

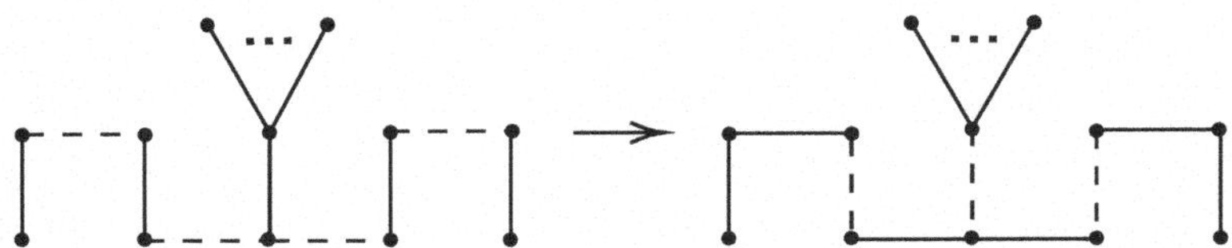

Fig. 2. Illustrating example for Operation 2. The solid edges are in $E(\mathcal{Q})$ and the dashed edges are in $E \setminus E(\mathcal{Q})$.

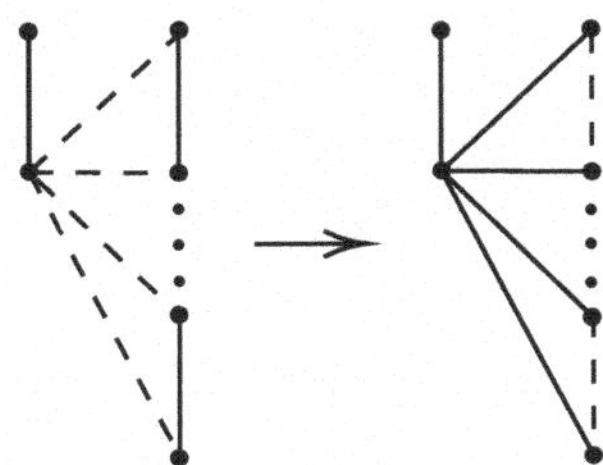

Fig. 3. Illustrating example for Operation 3. The solid edges are in $E(\mathcal{Q})$ and the dashed edges are in $E \setminus E(\mathcal{Q})$.

2 takes $O(n^5 + n^{k/2})$ time. Thus the overall time complexity of Algorithm 1 is $O(n^5)$ for odd k and $O(n^5 + n^{k/2})$ for even k.

Algorithm 1:

Input: An undirected graph $G = (V, E)$.
Output: A k^--star partition $\mathcal{Q}$ of G.
1. $\mathcal{Q}$ is initialized to be the k^--star partition with the minimum number of singletons computed by the algorithm in [7].
2. If one of Operations 1–3 is applicable, update $\mathcal{Q}$.
3. Return $\mathcal{Q}$.

Fig. 4. Description of Algorithm 1

Before analyzing the approximation ratio of Algorithm 1, we first introduce some notations to relate the size of $\mathcal{Q}_1, \ldots, \mathcal{Q}_k$ to those of $\mathcal{Q}_1^*, \ldots, \mathcal{Q}_k^*$. For a 2-star in $\mathcal{Q}_2$ containing two vertices u, v, we say that v is the *friend* of u and define $v = f(u)$. Clearly, u is also the *friend* of v and $u = f(v)$. Any two stars S_1^*, S_2^* in $\mathcal{Q}^*$, are called *adjacent*, if S_1^* is connected to S_2^* by a 2-star in $\mathcal{Q}_2$, i.e., there exists an edge $e = \{u, v\} \in E(\mathcal{Q}_2)$ with $u \in V(S_1^*)$ and $v \in V(S_2^*)$. In this occasion, we also say that S_1^* is adjacent to S_2^* via u or S_2^* is adjacent to S_1^* via v.

Next we classify the k-stars in $\mathcal{Q}_k^*$ into three types: $\mathcal{R}_1, \mathcal{R}_2, \mathcal{R}_3$. Given a k-star $S^* \in \mathcal{Q}_k^*$,

(i) $S^* \in \mathcal{R}_1$ if $V(S^*) \subseteq V(\mathcal{Q}_2)$ and $f(x) \notin V(\mathcal{Q}_1^*) \cup V(\mathcal{Q}_2^*)$ for each $x \in V(S^*)$.

(ii) $S^* \in \mathcal{R}_2$ if S^* is adjacent to some k-star in $\mathcal{R}_1$.
(iii) $S^* \in \mathcal{R}_3$ if S^* lies in neither $\mathcal{R}_1$ nor $\mathcal{R}_2$.

For a k-star $S^* \in \mathcal{R}_1$, let $f(S^*) = \{f(v) \mid v \in V(S^*)\}$.

Due to the local optimality of SOL and the choice of the initial solution, we get the following properties.

Lemma 1. $|\mathcal{Q}_1| \leq |\mathcal{Q}_1^*|$.

Proof. Since Algorithm 1 starts with a k^--star partition with the minimum number of 1-stars and all the local search operations do not change the number of 1-stars, it follows that SOL still contains the least number of 1-stars among all feasible solutions. Then we have $|\mathcal{Q}_1| \leq |\mathcal{Q}_1^*|$. □

Lemma 2. *For any k-star $S^* \in \mathcal{R}_1$, we have that (i) $|f(S^*) \setminus V(S^*)| \geq 2$ when k is even and $|f(S^*) \setminus V(S^*)| \geq 1$ when k is odd; (ii) S^* can not be adjacent to another k-star in $\mathcal{R}_1$; (iii) if S^* is adjacent to a star $S \in \mathcal{Q}^* \backslash (\mathcal{Q}_1^* \cup \mathcal{Q}_2^* \cup \mathcal{R}_1)$ via x and z is a vertex in $V(S)$ with $\{f(x), z\} \in E(S)$, then either $z \in V(\mathcal{Q}_k)$ or z is a leaf of some i-star in $\mathcal{Q}_i$ $(3 \leq i \leq k-1)$.*

Proof. (i) When k is even, since each 2-star contains two vertices, $|f(S^*) \backslash V(S^*)|$ cannot be odd. Now we show that $|f(S^*) \setminus V(S^*)| \neq 0$. Suppose by contradiction that $|f(S^*) \setminus V(S^*)| = 0$. Then we must have that $|f(S^*) \cap V(S^*)| = k$ and hence $f(S^*) = V(S^*)$. Therefore, $V(S^*)$ can be covered by $\frac{k}{2}$ 2-stars, say $S_1, S_2, \ldots, S_{k/2}$, in $\mathcal{Q}_2$. But this implies that we can apply Operation 3 for $S_1, S_2, \ldots, S_{k/2}$ to generate a k-star S^*, which is impossible. Therefore, $|f(S^*) \setminus V(S^*)| \geq 2$.

Similarly, when k is odd, $|f(S^*) \setminus V(S^*)|$ cannot be even. Thus, $|f(S^*) \setminus V(S^*)| \geq 1$.

(ii) Suppose that S^* is adjacent to a k-star $S \in \mathcal{R}_1$ via x. Then $f(x) \in V(S)$. Let u be any vertex in $V(S^*)$ with $\{x, u\} \in E(S^*)$. Denote by y any vertex in $V(S)$ with $\{f(x), y\} \in E(S)$. If $y \neq f(u)$, then Operation 1 can be applied for the three 2-stars $x - f(x)$, $u - f(u)$, $y - f(y)$, which is a contradiction (see Fig. 5(a) for an illustration). Otherwise, we have $y = f(u)$. Because S^* is a star and $\{x, u\} \in E(S^*)$, one of x, u is the center. We assume that u is the center. Since S^* is a k-star $(k \geq 5)$, there exists a leaf $v \neq x$ of S^* such that $\{u, v\} \in E(S^*)$. Now Operation 1 can be applied for the three 2-stars $x - f(x)$, $y - u$, $v - f(v)$. Again, this is a contradiction (see Fig. 5(b) for an illustration).

(iii) Since SOL contains the minimum number of singletons, any singleton in $\mathcal{Q}_1$ is not connected with a 2-star in $\mathcal{Q}_2$, hence $z \notin V(\mathcal{Q}_1)$. Let u be any vertex in $V(S^*)$ with $\{x, u\} \in E(S^*)$. By a similar argument as in the proof of conclusion (ii) we can deduce that $z \notin V(\mathcal{Q}_2)$. Thus know that $z \in V(\mathcal{Q}_i)$ $(3 \leq i \leq k)$. Suppose that $z \notin V(\mathcal{Q}_k)$, we shall shown that z is a leaf of some i-star in $\mathcal{Q}_i$ $(3 \leq i \leq k-1)$, which proves conclusion (iii). Assume by contradiction that z is the center of some i-star $S' \in \mathcal{Q}_i$ $(3 \leq i \leq k-1)$, then Operation 1 can be applied for two 2-stars $u - f(u)$, $x - f(x)$ in $\mathcal{Q}_2$ and the i-star S' in $\mathcal{Q}_i$ $(3 \leq i \leq k-1)$. But this is impossible due to the local optimality of SOL (see Fig. 5(c) for an illustration). □

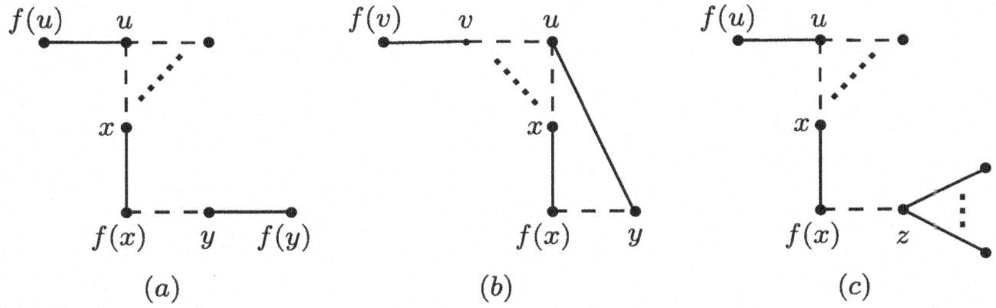

Fig. 5. Illustration of the proof of Lemma 2. The solid edges and the dashed edges are in $E(\mathcal{Q})$ and $E(\mathcal{Q}^*)$, respectively.

Lemma 2(ii) implies $\mathcal{R}_1 \cap \mathcal{R}_2 = \emptyset$. So we conclude that $\mathcal{R}_1$, $\mathcal{R}_2$, $\mathcal{R}_3$ form a partition of $\mathcal{Q}_k^*$.

Lemma 3. *For any k-star $S^* \in \mathcal{R}_2$ with $V(S^*) \cap V(\mathcal{Q}_k) = \emptyset$, S^* is adjacent to at most one k-star in $\mathcal{R}_1$.*

Proof. Assume that S^* is adjacent to two different k-stars S_1^* and S_2^* in $\mathcal{R}_1$ via x_1 and x_2, respectively. Then there exists some vertex $y_i \in V(S_i^*)$ with $\{f(x_i), y_i\} \in E(S_i^*)$ $(i = 1, 2)$. Let z_i $(i = 1, 2)$ be a vertex in $V(S^*)$ with $\{x_i, z_i\} \in E(S^*)$. If x_1 and x_2 are both leaf vertices of the S^*, we have that $z_1 = z_2$ must be a leaf of some i-star $S \in \mathcal{Q}_i$ $(3 \le i \le k-1)$ due to Lemma 2(iii) and the fact that $V(S^*) \cap V(\mathcal{Q}_k) = \emptyset$. Moreover, by Lemma 2(ii), we know that $y_2 \neq f(y_1)$, so $y_1 - f(y_1)$, $y_2 - f(y_2)$ are two distinct 2-stars in $\mathcal{Q}_2$. Thus Operation 2 can be applied for the four 2-stars $x_1 - f(x_1)$, $x_2 - f(x_2)$, $y_1 - f(y_1)$, $y_2 - f(y_2)$ in $\mathcal{Q}_2$ and the i-star $S \in \mathcal{Q}_i$ $(3 \le i \le k-1)$ (see Fig. 6(a) for an illustration), which is impossible. Otherwise, one of x_1, x_2 is the center vertex of the S^*. Without loss of generality, we suppose that x_1 is the center of S^* and x_2 is a leaf of S^*. Then $z_2 = x_1$ and therefore z_2 lies on the 2-star $x_1 - f(x_1)$, i.e., $z_2 \in V(\mathcal{Q}_2)$, which is a contradiction since Lemma 2(iii) implies that z_2 is a leaf of some i-star in $\mathcal{Q}_i$ with $3 \le i \le k-1$ (see Fig. 6(b) for an illustration). To sum up, the lemma holds true. □

Based on the above properties, we next show the relationship between $|\mathcal{Q}_i|$ and $|\mathcal{Q}_i^*|$ for $i \ge 2$.

Lemma 4. *(i) For any even $k \ge 8$, it holds that $\sum_{i=2}^{k}(k-i)|\mathcal{Q}_i| \le \frac{(k-1)(k^2-4)}{2(k+1)} \cdot opt$.*

(ii) For any odd $k \ge 7$, it holds that $\sum_{i=2}^{k}(k-i)|\mathcal{Q}_i| \le \frac{(k^2-1)(k-2)}{2k} \cdot opt$.

Lemma 5. *For $k = 5$, it holds that $3|\mathcal{Q}_2| + 2|\mathcal{Q}_3| + |\mathcal{Q}_4| \le \frac{(k^2-1)(k-2)}{2k} \cdot opt = \frac{36}{5} \cdot opt$.*

Lemma 6. *For $k = 6$, it holds that $4|\mathcal{Q}_2| + 3|\mathcal{Q}_3| + 2|\mathcal{Q}_4| + |\mathcal{Q}_5| \le \frac{(k-1)(k^2-4)}{2(k+1)} \cdot opt = \frac{80}{7} \cdot opt$.*

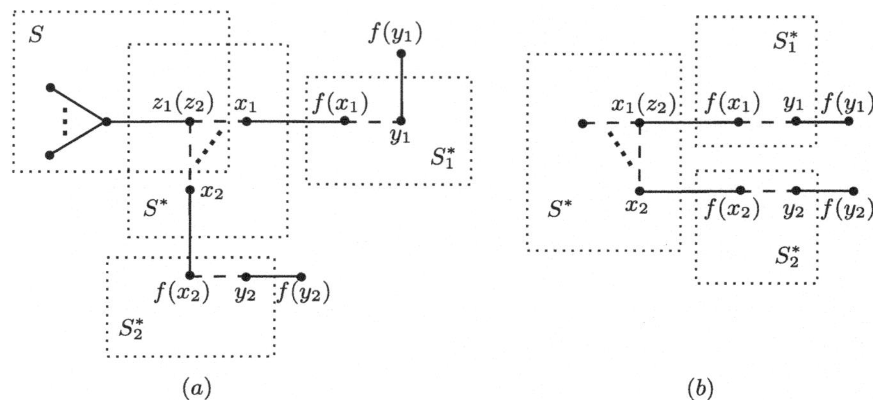

Fig. 6. Illustration of the proof of Lemma 3. As before, solid edges and dashed edges are in $E(\mathcal{Q})$ and $E(\mathcal{Q}^*)$, respectively.

Putting together the above lemmas, we obtain the main result as below.

Theorem 1. *For the Min-kSP with $k \geq 5$, Algorithm 1 is a $(\frac{k}{2} - \frac{k-2}{k(k+1)})$-approximation algorithm for even k and $(\frac{k}{2} - \frac{k-2}{2k^2})$-approximation algorithm for odd k, respectively.*

Proof. Consider the k^--star partition $\mathcal{Q}$ returned by Algorithm 1.

By Lemma 1, $|\mathcal{Q}_1| \leq |\mathcal{Q}_1^*|$. Since the total number of vertices in the graph can be counted with respect to either $\mathcal{Q}$ or $\mathcal{Q}^*$, we have

$$\sum_{i=1}^{k} i|\mathcal{Q}_i| = |V| = \sum_{i=1}^{k} i|\mathcal{Q}_i^*|. \tag{1}$$

By the definitions of *sol* and $\mathcal{Q}_i$, it follows that

$$\begin{aligned}
k \cdot sol &= \sum_{i=1}^{k} k|\mathcal{Q}_i| \\
&= \sum_{i=1}^{k} (k-i)|\mathcal{Q}_i| + \sum_{i=1}^{k} i|\mathcal{Q}_i| \\
&= (k-1)|\mathcal{Q}_1| + \sum_{i=1}^{k} i|\mathcal{Q}_i| + \sum_{i=2}^{k} (k-i)|\mathcal{Q}_i| \\
&\leq (k-1)|\mathcal{Q}_1^*| + \sum_{i=1}^{k} i|\mathcal{Q}_i^*| + \sum_{i=2}^{k} (k-i)|\mathcal{Q}_i|
\end{aligned}$$

$$\leq \begin{cases} (k-1)|\mathcal{Q}_1^*| + \sum\limits_{i=1}^{k} i|\mathcal{Q}_i^*| + \sum\limits_{i=1}^{k} \frac{(k-1)(k^2-4)}{2(k+1)} \cdot |\mathcal{Q}_i^*| & k \text{ is even} \\ (k-1)|\mathcal{Q}_1^*| + \sum\limits_{i=1}^{k} i|\mathcal{Q}_i^*| + \sum\limits_{i=1}^{k} \frac{(k^2-1)(k-2)}{2k} \cdot |\mathcal{Q}_i^*| & k \text{ is odd} \end{cases}$$

$$= \begin{cases} (k + \frac{(k-1)(k^2-4)}{2(k+1)}) \cdot |\mathcal{Q}_1^*| + \sum\limits_{i=2}^{k} (i + \frac{(k-1)(k^2-4)}{2(k+1)}) \cdot |\mathcal{Q}_i^*| & k \text{ is even} \\ (k + \frac{(k^2-1)(k-2)}{2k}) \cdot |\mathcal{Q}_1^*| + \sum\limits_{i=2}^{k} (i + \frac{(k^2-1)(k-2)}{2k}) \cdot |\mathcal{Q}_i^*| & k \text{ is odd} \end{cases}$$

$$\leq \begin{cases} \sum_{i=1}^{k} (k + \frac{(k-1)(k^2-4)}{2(k+1)}) \cdot |\mathcal{Q}_i^*| & k \text{ is even} \\ \sum_{i=1}^{k} (k + \frac{(k^2-1)(k-2)}{2k}) \cdot |\mathcal{Q}_i^*| & k \text{ is odd} \end{cases}$$

$$= \begin{cases} (k + \frac{(k-1)(k^2-4)}{2(k+1)}) \cdot opt & k \text{ is even} \\ (k + \frac{(k^2-1)(k-2)}{2k}) \cdot opt & k \text{ is odd,} \end{cases}$$

where the first inequality holds by Lemma 1 and Eq. (1), the second inequality follows from Lemmas 4, 5 and 6, and the last inequality follows from $i \leq k$. Dividing both sides of the above equation by k, we have

$$sol \leq \begin{cases} (\frac{k}{2} - \frac{k-2}{k(k+1)}) \cdot opt & k \text{ is even} \\ (\frac{k}{2} - \frac{k-2}{2k^2}) \cdot opt & k \text{ is odd.} \end{cases}$$

This completes the proof of the theorem. □

When k is even, Fig. 7 depicts a tight example. In this example, the input graph G consists of $\frac{k}{2}$ vertex-disjoint subgraphs $G_1, G_2, \ldots, G_{k/2}$ and $k-1$ edges $\{v_1, v_2\}$, $\{v_1, v_3\}$, …, $\{v_1, v_k\}$, where $G_1, G_2, \ldots, G_{k/2}$ are pairwise isomorphic and $v_{2i}, v_{2i+1} \in V(G_i)$ for $i = 1, 2, \ldots, \frac{k}{2}$ (set $v_{k+1} = v_1$). The vertices of each subgraph G_i can be covered by $k+1$ k-stars indicated by all the dashed edges. As a result, G admits a k^--star partition $\mathcal{Q}^*$ consisting of $\frac{k}{2} \cdot (k+1)$ k-stars, which is clearly optimal. On the other hand, all the vertices in $V(G_i) \setminus \{v_{2i}, v_{2i+1}\}$ can be covered by $\frac{k^2+k-2}{2}$ 2-stars represented by all the solid edges. Then G has a k^--star partition $\mathcal{Q}$ consisting of a single k-star $v_1 - v_2 v_3 \cdots v_k$ and $\frac{k}{2} \cdot \frac{k^2+k-2}{2}$ 2-stars. Since $\mathcal{Q}$ contains no singletons, it is a k^--star partition with the minimum number of singletons. Because the vertices of any three 2-stars in $\mathcal{Q}$ cannot be covered by two 3-stars and the vertices of any $\frac{k}{2}$ 2-stars in $\mathcal{Q}$ cannot be covered by a k-star, none of Operations 1–3 in Algorithm 1 can be applied with respect to $\mathcal{Q}$. So if Algorithm 1 chooses the k^--star partition $\mathcal{Q}$ as the initial feasible solution, it will output $\mathcal{Q}$ as the final k^--star partition SOL. Therefore, the approximation ratio for Algorithm 1 is at least

$$\frac{|\mathcal{Q}|}{|\mathcal{Q}^*|} = \frac{1 + \frac{k(k^2+k-2)}{4}}{\frac{k^2+k}{2}} = \frac{k}{2} - \frac{k-2}{k(k+1)}.$$

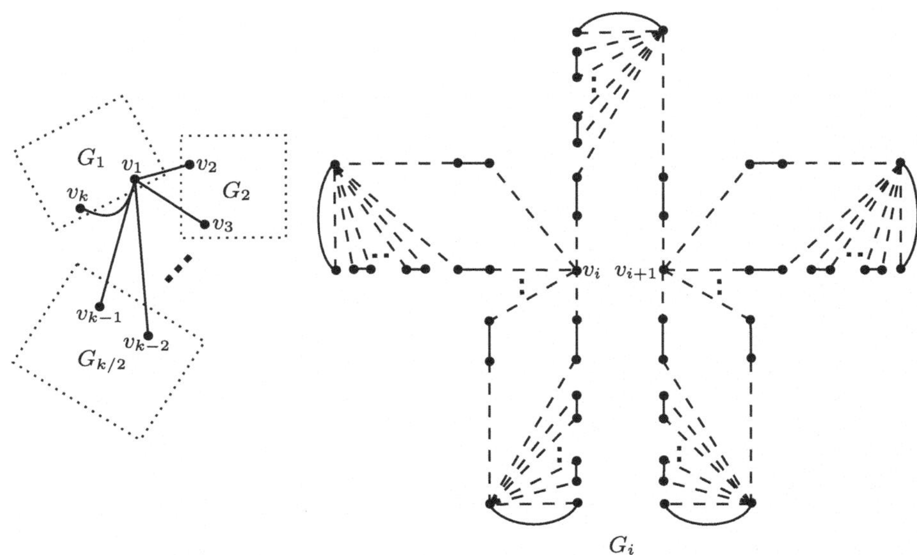

Fig. 7. A tight example for Algorithm 1 when k is even

When k is odd, Fig. 8 shows a tight example. In this example, the input graph G consist of k vertex-disjoint subgraphs $G_1, G_2, \ldots, G_k$ and $k-1$ edges $\{v_1, v_2\}$, $\{v_1, v_3\}$, ..., $\{v_1, v_k\}$, where $G_1, G_2, \ldots, G_k$ are pairwise isomorphic and $v_i \in V(G_i)$ for $i = 1, 2, \ldots, k$. The vertices of each subgraph G_i can be

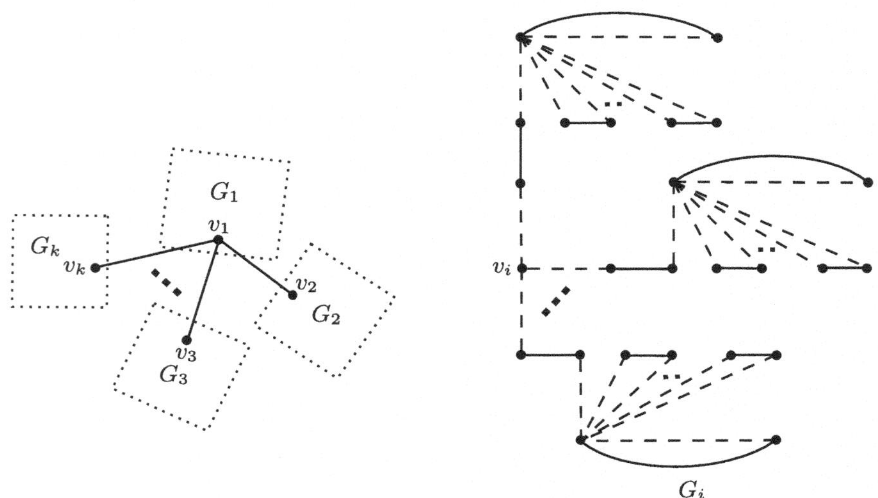

Fig. 8. A tight example for Algorithm 1 when k is odd

covered by k k-stars indicated by all the dashed edges. Also, all the vertices in $V(G_i) \setminus \{v_i\}$ can be covered by $\frac{k^2-1}{2}$ 2-stars represented by all the solid edges. Then, as before, G has a k^--star partition $\mathcal{Q}$ consisting of a single k-star $v_1 - v_2 v_3 \cdots v_k$ and $k \cdot \frac{k^2-1}{2}$ 2-stars, which represents SOL produced by Algorithm 1. On the other hand, G admits an optimal k^--star partition $\mathcal{Q}^*$ consisting of k^2 k-stars. Therefore, the approximation ratio for Algorithm 1 is at least

$$\frac{|\mathcal{Q}|}{|\mathcal{Q}^*|} = \frac{1 + \frac{k(k^2-1)}{2}}{k^2} = \frac{k}{2} - \frac{k-2}{2k^2}.$$

4 Conclusions

In this paper, we consider the Min-kSP for any $k \geq 5$ and design a local search algorithm. Algorithm 1 first computes a k^--star partition with the minimum number of singletons, and then iteratively applies local search operations to reduce the total number of stars until no improvements is possible. Then the algorithm terminates and outputs the current k^--star partition as the approximate solution. Through the amortization strategy, we prove that the approximation ratio of our algorithm is $\frac{k}{2} - \frac{k-2}{k(k+1)}$ when k is even and $\frac{k}{2} - \frac{k-2}{2k^2}$ when k is odd, which improve the best available $\frac{k}{2}$-approximation algorithm in [7]. We also showed that our analysis is tight.

For future work, one can try to improve our algorithm by introducing more elaborate local search operations for the Min-kSP.

Acknowledgments. This research is supported by the National Natural Science Foundation of China under grant number 12371317 and the Natural Science Foundation of Shanghai under grant number 24ZR1416900.

References

1. Bao, Q., Yu, W., Liu, Z., Chen, Y.: An improved approximation algorithm for the minimum 4-star partition problem (2024). submitted for publication
2. Chen, Y., Chen, Z., Kennedy, C., Lin, G., Xu, Y., Zhang, A.: Approximating the directed path partition problems. Inf. Comput. **297**, 105150 (2024)
3. Chen, Y., et al.: A local search 4/3-approximation algorithm for the minimum 3-path partition problem. J. Comb. Optim. **44**, 3595–3610 (2022)
4. Chen, Y., Goebel, R., Lin, G., Su, B., Xu, Y., Zhang, A.: An improved approximation algorithm for the minimum 3-path partition problem. J. Comb. Optim. **38**(1), 150–164 (2019). https://doi.org/10.1007/s10878-018-00372-z
5. Chen, Y., Goebel, R., Su, B., Tong, W., Xu, Y., Zhang, A.: A 21/16-approximation for the minimum 3-path partition problem. In: Proceedings of the 30th International Symposium on Algorithms and Computation, Article No. 46, pp. 461–4620 (2019)
6. Goldberg, A.V., Karzanov, A.V.: Maximum skew-symmetric flows and matchings. Math. Program. **100**, 537–568 (2004)

7. Hell, P., Kirkpatrick, D.G.: Packing by complete bipartite graphs. SIAM J. Algebraic Dis. Methods **7**(2), 199–209 (1986)
8. Kirkpatrick, D.G., Hell, P.: On the complexity of general graph factor problems. SIAM J. Comput. **12**(3), 601–609 (1983)
9. Karp, R.M.: Reducibility among combinatorial problems. In: Miller, R.E., Thatcher, J.W. (eds.) Complexity of Computer Computations, pp. 85–103 (1972)
10. Korpelainen, N.: A boundary class for the k-path partition problem. Electronic Notes Dis. Math. **67**, 49–56 (2018)
11. Lv B, Jiang J, Wu L, Zhao H, Team formation in large organizations: a deep reinforcement learning approac. Dec. Support Syst., 114343-114343 (2024)
12. Letchford, A.N., Salazar-Gonzalez, J.J.: The capacitated vehicle routing problem: stronger bounds in pseudo-polynomial time. Eur. J. Oper. Res. **272**, 24–31 (2019)
13. Li, S.M., Yu, W., Liu, Z.: A local search algorithm for the k-path partition problem. Optimizat. Lett. **18**, 279–290 (2024)
14. Li, S., Yu, W., Liu, Z.: Improved approximation algorithms for the k-path partition problem. J. Global Optimizat., 1–24 (2024)
15. Monnot, J., Toulouse, S.: The path partition problem and related problems in bipartite graphs. Oper. Res. Lett. **35**, 677–684 (2007)
16. Palsetia, D., Patwary, M.M.A., Hendrix, W., Agrawal, A., Choudhary, A.: Clique guided community detection. In: 2014 IEEE International Conference on Big Data (Big Data), pp. 500–509 (2014)
17. Steiner, G.: On the k-th path partition problem in cographs. Congr. Numer. **147**, 89–96 (2000)
18. Steiner, G.: On the k-path partition of graphs. Theoret. Comput. Sci. **290**, 2147–2155 (2003)
19. She, B., Mehta, S., Ton, C., Kan, Z.: Controllability ensured leader group selection on signed multiagent networks. IEEE Trans. Cybernet. **50**, 222–232 (2020)
20. Yan, J.H., Chang, G.J., Hedetniemi, S.M., Hedetniemi, S.T.: k-Path partitions in trees. Discret. Appl. Math. **78**, 227–233 (1997)

Doubly Constrained Fair Clustering for General p-Norms

Lunhao Zhang, Pengzhi Gao, and Peng Zhang(✉)

School of Software, Shandong University, Jinan 250101, Shandong, China
{zhanglunhao,gaopengzhi}@mail.sdu.edu.cn, algzhang@sdu.edu.cn

Abstract. Fairness in clustering has received significant attention. Dickerson et al. in 2023 first proposed the doubly constrained fair clustering problem that aims two fairness constraints, namely, (1) the Group Fairness (GF), which requires that different groups within each cluster have a certain degree of representation, and (2) the Diversity in Center Selection fairness (DS), which requires that the selected centers represent a diverse range of different groups. However, their algorithm only focuses on the k-center objective. In this paper, we generalize the doubly constrained fair clustering to ℓ_p norm objectives with general p, thus including k-Center, k-Median, and k-Means as special cases. We propose the first approximation algorithm for the doubly constrained fair clustering problem with general p-norms. In polynomial time, our algorithm finds an $O(\Delta^{\frac{1}{p}})$-approximate clustering that violates the GF constraint by an additive factor of 5 and satisfies the DS constraint, where Δ is the largest size of clusters in the solution. Our main contribution is a novel method to select centers using the min cost network flow approach. Finally, we conduct experiments to validate our algorithm. The experimental results show that the clustering cost of our algorithm, while simultaneously considering both of the GF and DS constraints, is nearly identical to that of the clustering algorithm which only considers the GF constraint.

Keywords: Fair Clustering · Network Flow · Approximation Algorithm · Combinatorial Optimization

1 Introduction

Clustering problems represent one of the most fundamental unsupervised learning problems. At the same time, it is also a classic operations research problem applied to facility location issues. In recent years, the fairness of clustering has become increasingly important. Similar to supervised learning, a series of mathematical constraints have been introduced to define the concept of clustering fairness. For instance, to comply with the disparate impact doctrine [6], measures are taken to prevent under-representation or over-representation of demographic groups within clusters [2,3]. Additionally, there are efforts to minimize

F. V. Fomin and M. Xiao (Eds.): COCOON 2025, LNCS 15983, pp. 93–105, 2026.
https://doi.org/10.1007/978-981-95-0215-8_8

the maximum clustering cost for each group among different clusters [7], as well as to ensure proportional demographic representation at selected cluster centers [11,14].

These constraints are all reasonable in specific domains, and the choice of which constraint to use depends on particular applications. The mainstream methods for fair clustering in the literature typically impose only one constraint in a given context. However, clustering may be applied to tasks in multiple different scenarios. This leads to the desire that coordinates different fairness constraints within a single clustering (rather than performing separate clusterings for different fairness concepts).

Consequently, Dickerson et al. [4] introduced a clustering method for the first time that aims to satisfy the two fairness constraints, that is, the group fairness (GF) constraint and the diversity in center selection fairness (DS) constraint. The group fairness constraint requires that different groups within each cluster have a certain degree of representation. The diversity in center selection fairness constraint requires that the selected centers represent a diverse range of different groups. However, the method in [4] primarily focused on the k-center objective. This naturally leads to the consideration of doubly constrained fair clustering for objectives other than k-center.

1.1 Our Results

In this paper, we consider the doubly constrained fair clustering problem with general ℓ_p norm objective, thus including k-Center, k-median, and k-Means as special cases.

We propose the first approximation algorithm for the doubly constrained fair clustering problem with general p-norms. In polynomial time, our algorithm finds an $O(\Delta^{\frac{1}{p}})$-approximate clustering that violates the GF constraint by an additive factor of 5 and satisfies the DS constraint, where Δ is the largest size of clusters in the solution. Our result extends the work of Dickerson et al. [4] from the k-center objective to general ℓ_p norm objective.

Our main contribution is a novel method to select centers using the min cost network flow approach. To deal with the diversity in center selection fairness constraint, we convert the problem of how to select center to the min cost network flow problem with lower and upper bounds, which in turn is solved by reducing to the classical min cost network flow problem. This improves upon the center selection method of Dickerson et al. [4], which only focuses on the k-center objective. Our center selection method ensures that no DS constraint violations occur.

Moreover, for the k-center objective, we achieve the same approximation guarantee as that established in [4]. This result will be given in the full version of the paper due to space limitation.

Finally, we validate our findings through experiments. The experimental results show that our clustering cost is almost equal to the clustering cost of solutions that only consider the GF constraint. Moreover, our method satisfies the DS constraint and violates the GF constraint by an additive factor of at most 5.

1.2 Related Work

Within the study of fairness in clustering, the group-level perspective is a crucial component. Chierichetti et al. [3] considered the requirement that the proportion of a group within any cluster should be similar to its proportion in the input data. They proposed a method of creating the so-called fairlets (which are minimal sets that satisfy fair representation while approximately preserving the clustering objective) and then applying general clustering algorithms on these fairlets. However, their work was limited to only two groups.

Bera et al. [2] extended the group fairness concept to multiple groups. They considered how points can be reassigned to centers so that each group has bounded representation within each cluster. Subsequently, Esmaeili et al. [5] further considered the problem of minimizing additive violations to group fairness while ensuring an upper bound on the clustering cost.

The concept of diversity in center selection was first explored by Kleindessner et al. [11]. The authors proposed an approximation algorithm with an approximation factor that varies exponentially with the number of groups. The subsequent work by Jones et al. [9] improved the approximation factor to a constant. Nguyễn et al. [12] extends this problem by requiring that the group representation of the centers falls within a desired range. Subsequently, Thejaswi et al. [14] considered the k-Median problem where the centers are required to satisfy lower-bound thresholds for diversity. Recently, Hotegni et al. [8] demonstrated that fair algorithms satisfying the DS constraint can be obtained for any ℓ_p-norm objective.

Doubly constrained fair clustering (with the k-center objective) was first studied by Dickerson et al. [4]. The authors [4] proved the following results. Given an α_{GF}-approximation algorithm for GF-constrained fair clustering problem, they can obtain a $2\alpha_{GF}$-approximation algorithm that violates the GF constraint by an additive violation of 2 and satisfies the DS constraint. Given an α_{DS}-approximation algorithm for the DS-constrained fair clustering problem, then they can obtain an $2(1+\alpha_{DS})$-approximation algorithm that violates the GF constraint by an additive factor of 3 and satisfies the DS constraint.

2 Preliminaries

Let $(\mathcal{X}, d)$ be a metric space with $n = |\mathcal{X}|$ points and distance $d(u, v)$ for every pair of points $u, v \in \mathcal{X}$. Given metric space $(\mathcal{X}, d)$ and an integer k, the goal of the (k, p)-Clustering problem is to find (a) a set of centers $S \subseteq \mathcal{X}$ with $|S|$ being at most k, and (b) an assignment $\phi : \mathcal{X} \to S$ that minimizes

$$\mathcal{L}_p(S, \phi) = \left(\sum_{v \in \mathcal{X}} d(v, \phi(v))^p \right)^{\frac{1}{p}}. \tag{1}$$

In (1), the assignment function ϕ indicates which center each point is assigned to. Obviously, the assignment function ϕ is very important to the problem. Let $u \in S$ be a center. We use $C(u) \subseteq \mathcal{X}$, called *cluster*, to denote the set of points that are assigned to v, that is, $C(u) = \{v \in \mathcal{X} \mid \phi(v) = u\}$. In general, we use $C \subseteq \mathcal{X}$ to denote a cluster whose center is neglected (but exists).

When $p = 1$ and $p = 2$ in (1), the (k, p)-CLUSTERING problem just corresponds to the k-MEDIAN problem and (a variant of) the k-MEANS problem, respectively. (Note that the classic k-MEANS has objective $\sum_v d(v, \phi(v))^2$.)

In the (k, p)-CLUSTERING problem, fairness is not considered. It will be okay that points in C are simply assigned to their nearest centers in S. Namely, in (k, p)-CLUSTERING the proximity of two points is measured by their distance.

In the fair clustering scenario, each point has a color representing its *demographic group information*. Let H be the set of all possible colors. We denote the set of points with color h by $C^h \subseteq \mathcal{X}$, and denote the number of points with color h by $n_h = |C^h|$.

Then, we will introduce two types of fairness usually considered in fair clustering, i.e., the group fairness and the diversity in center selection fairness. The *Group Fairness* (GF) [1–3] says that, for each center $u \in S$ and each color (i.e., demographic group) $h \in H$, it must hold that

$$\beta_h |C(u)| \leq |C^h(u)| \leq \alpha_h |C(u)|, \tag{2}$$

where $C^h(u) = C^h \cap C(u)$, and β_h and α_h are respectively the lower bound and upper bound for color h. Intuitively, the group fairness says that each cluster $C(u)$ should contain a certain number of points of color h, for each $h \in H$.

A clustering is of λ-additive violation on the GF constraint for some factor $\lambda \geq 0$ if $\forall u \in S$, $\forall h \in H$, it holds

$$\beta_h |C(u)| - \lambda \leq |C^h(u)| \leq \alpha_h |C(u)| + \lambda. \tag{3}$$

Let $k_h = |S \cap C^h|$ represents the number of centers of color h. The *Diversity in Center Selection* (DS) fairness [8,9,11,12,14] says that, it must hold that

$$l_h \leq k_h \leq r_h, \tag{4}$$

where l_h and r_h are the lower bound and upper bound respectively for the number of centers of color h. Intuitively, the diversity in center selection fairness says that each color group C^h ($h \in H$) should contain a certain number of centers.

Now we can give the definition of the doubly constrained fair (k, p)-CLUSTERING problem studied in the paper. Given a metric space $(\mathcal{X}, d)$ consisting of points, an integer k, a set of colors $H = \{1, 2, \ldots, m\}$, and bounds $\{\alpha_h, \beta_h\}_{h \in H}$ and $\{l_h, r_h\}_{h \in H}$, where each point $v \in \mathcal{X}$ has a color in H, the doubly constrained fair (k, p)-CLUSTERING problem asks to find a k-clustering of points in $\mathcal{X}$ so that the objective (1) is minimized, meanwhile the GF constraint (2) and the DS constraint (4) are satisfied.

The GF-constrained fair (k, p)-CLUSTERING problem is defined as follows. Given a metric space $(\mathcal{X}, d)$ consisting of points, an integer k, a set of colors

$H = \{1, 2, \ldots, m\}$, and bounds $\{\alpha_h, \beta_h\}_{h \in H}$, where each point $v \in \mathcal{X}$ has a color in H, the GF-constrained fair (k,p)-CLUSTERING problem asks to find a k-clustering of points in $\mathcal{X}$ so that the objective (1) is minimized, meanwhile the GF constraint (2) is satisfied.

For an integer $k > 0$, we use $[k]$ to denote the set $\{1, 2, \ldots, k\}$.

3 Algorithms

In this section, we give our algorithm for the doubly constrained fair (k,p)-CLUSTERING problem with general p-norms. The algorithm is shown as Algorithm G2GD (i.e., Algorithm 1). Algorithm G2GD consists of three stages, with step 1 constituting the first stage, step 2 constituting the second stage, and steps 3 to 5 constituting the third stage.

Algorithm 1. G2GD

Input: Instance $\mathcal{I}$ of the doubly constrained fair (k,p)-clustering problem.
Output: A k-clustering (S, ϕ).

1: Find a k-clustering $(\bar{S}, \bar{\phi})$ by Algorithm ALG-GF [2]. Let $\bar{k}$ be the size of $\bar{S}$, and $\bar{C}_1, \bar{C}_2, \ldots, \bar{C}_{\bar{k}}$ be the clusters defined by $\bar{\phi}$.
2: If there is no feasible solution for the DS constraint, then just return no feasible solution. Otherwise, given $(\bar{S}, \bar{\phi})$, find new centers $Q_1, \ldots, Q_{\bar{k}}$ using the Algorithm SNC in Section 3.1.
3: **for** $i \leftarrow 1$ **to** $\bar{k}$ **do**
4:     Call the algorithm DIVIDE in [4] on the input $(\bar{C}_i, Q_i)$ to partition $\bar{C}_i$ into sub-clusters $\{D(u)\}_{u \in Q_i}$ (that is, $\bar{C}_i = \bigcup_{u \in Q_i} D(u)$), where each sub-cluster $D(u)$ has a distinct center in Q_i.
5: **end for**
6: Let $S \leftarrow \bigcup_i Q_i$ be the set of all centers. Set the assignment function ϕ by $\forall u \in S, \forall v \in D(u), \phi(v) \leftarrow u$.
7: **return** (S, ϕ).

In the first stage of Algorithm G2GD, we call the algorithm in [2], denoted by Algorithm ALG-GF, on the input data set $(\mathcal{X}, d)$ to generate a k-clustering $(\bar{S}, \bar{\phi})$, where $\bar{S}$ is the set of cluster centers, and $\bar{\phi}$ is the assignment function mapping data points in $\mathcal{X}$ to centers in $\bar{S}$. By [2], $(\bar{S}, \bar{\phi})$ violates the GF constraint by an additive factor of 3.

In general, the center set $\bar{S}$ may not satisfy the DS constraint. That is to say, there would be some color group C^h which has representatives in $\bar{S}$ less than l_h or more than r_h. To address this, our strategy (the second stage of Algorithm G2GD) is to select new centers from each cluster $\bar{C}_i$ ($\bar{C}_i$ is defined by $\bar{\phi}$), but totally dropping $\bar{S}$. This is done by the algorithm (called SNC) in Sect. 3.1. Of course, an old center in $\bar{S}$ may be selected as a new center again by the Algorithm SNC. Let Q_i be the set of new centers selected by Algorithm SNC from cluster $\bar{C}_i$. The cluster centers found by our algorithm are just $S = \bigcup_i Q_i$, which satisfies the DS constraint.

In the third stage of Algorithm G2GD we construct a new clustering according to the center set S. This is done by breaking some of the old clusters ($\bar{C}_i$'s) into new pieces. If for some cluster $\bar{C}_i$ only one new center is chosen (i.e., $|Q_i| = 1$), we can simply assign the data points in $\bar{C}_i$ to this new center in Q_i, preserving the GF constraint satisfied by $\bar{C}_i$.

The difficult case comes from that for some cluster $\bar{C}_i$ we have $|Q_i| > 1$. In this case, we call the algorithm (called DIVIDE) by Dickerson et al. [4] to break $\bar{C}_i$ into $|Q_i|$ new pieces with each piece having a distinct cluster in Q_i. The GF constraint for the new pieces (clusters) is violated. However, the violation is slight and no more than an additive factor of 2 [4].

3.1 Algorithm SNC to Select New Centers

In this section we describe the Algorithm SNC to select new centers from a given clustering $(\bar{S}, \bar{\phi})$. Suppose that the clustering $(\bar{S}, \bar{\phi})$ has clusters $\{\bar{C}_1, ..., \bar{C}_{\bar{k}}\}$, where $\bar{k} = |\bar{S}| \leq k$. For clarity, we assume that the $\bar{k}$ centers in $\bar{S}$ are $u_1, u_2, \ldots, u_{\bar{k}}$. Based on $(\bar{S}, \bar{\phi})$, we will select new centers for each cluster $\bar{C}_i$ ($i \in [\bar{k}]$), denoted by $Q_i \subseteq \bar{C}_i$. The new centers $Q_1, Q_2, \ldots, Q_{\bar{k}}$ should satisfy the DS constraint (4). For our purpose, the incurred cost $\sum_{i=1}^{\bar{k}} \sum_{v \in Q_i} |\bar{C}_i| \, d(u_i, v)^p$ due to the new centers should be as minimum as possible.

We can formulate our problem of selecting new centers as the following mathematical programming (MP) (with constraints (5) and (6)). Constraint (6) is just a restatement of the DS constraint (4). Note that in (MP), the unknowns are $Q_1, Q_2, \ldots, Q_{\bar{k}}$.

$$\min \quad \sum_{i=1}^{\bar{k}} \sum_{u \in Q_i} |\bar{C}_i| \, d\left(u, u_i\right)^p \tag{MP}$$

$$\text{s.t.} \quad \bar{k} \leq \sum_{i=1}^{\bar{k}} |Q_i| \leq k \tag{5}$$

$$l_h \leq \sum_{i=1}^{\bar{k}} \left|Q_i \cap C^h\right| \leq r_h, \qquad \forall h \in H \tag{6}$$

$$\emptyset \neq Q_i \subseteq \bar{C}_i, \qquad \forall i \in [\bar{k}]$$

We solve (MP) by converting it to the minimum-cost flow problem with lower and upper bounds. The flow network (denoted by N) is illustrated in Fig. 1.

The nodes in the network N are as follows. The points in the center set $\bar{S}$ and the points in each cluster $\bar{C}_i$ ($i \in [\bar{k}]$) are all used as nodes in N. Note that the centers and points are different nodes in N. Each color in H has a node in N. Finally, there are a source node s and a sink node t.

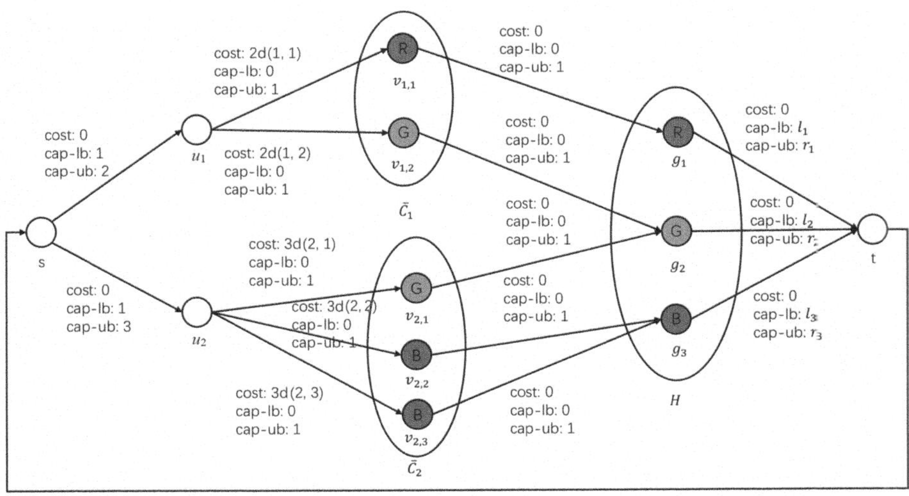

Fig. 1. An example of network N. Each point node $v_{i,j}$ in the cluster $\bar{C}_i$ ($i \in [\bar{k}]$) has a color depicted by a letter R, G, or B. Similarly, each color node g_h ($h \in H$) has also a letter depicting its color.

The (directed) edges in network N are as follows. For each center node u_i ($i \in [\bar{k}]$), there is an edge (s, u_i) having zero cost (i.e., distance), capacity lower bound of one, and capacity upper bound of $|\bar{C}_i|$.

For each center node u_i ($i \in [\bar{k}]$) (corresponding the center $u_i \in \bar{S}$) and each point node $v_{i,j}$ ($1 \le j \le |\bar{C}_i|$) (corresponding to a point in $\bar{C}_i$), there is an edge $(c_i, v_{i,j})$ having cost $|\bar{C}_i| \cdot d(c_i, v_{i,j})$, capacity lower bound of zero, and capacity upper bound of one. Here, for notational simplicity, we assume that we also have $c_i \in \bar{S}$ and $v_{i,j} \in \bar{C}_i$.

For each point node $v_{i,j}$ in every cluster $\bar{C}_i$ ($i \in [\bar{k}]$, $1 \le j \le |\bar{C}_i|$), there is an edge $(v_{i,j}, g_h)$, where we assume that the color of $v_{i,j}$ is h and g_h is the node corresponding color $h \in H$. The edge $(v_{i,j}, g_h)$ has zero cost, capacity lower bound of zero, and capacity upper bound of one.

For each color node g_h ($h \in H$), there is an edge (g_h, t) having a cost of zero and a capacity ranging from l_h to r_h. Finally, there is an edge (t, s) having cost of zero, capacity lower bound of $\bar{k}$, and capacity upper bound of k.

Then we compute a min cost flow in network N by converting the min cost flow problem with lower and upper bounds to the classical min cost flow problem. The approach is omitted here due to the limitation of space. It will be given in the full version of the paper.

If an edge $(v_{i,j}, g_h)$ connecting the point node $v_{i,j}$ and the color node g_h has flow value one, then the point $v_{i,j}$ is selected in Q_i. In this way, we obtain the center sets $Q_1, Q_2, \ldots, Q_{\bar{k}}$. One can see the cost of the flow is just the objective function $\sum_{i=1}^{\bar{k}} \sum_{v \in Q_i} |\bar{C}_i| \, d(u_i, v)^p$ of (MP).

The capacity lower bounds on edges from the source node s to the center nodes c_i's ensure that for each i at least one point in $\bar{C}_i$ is selected as the new center (i.e., $|Q_i| \geq 1$). The capacity lower bounds and upper bounds on edges from the color nodes g_h's to the sink node t ensure that condition (6) is met. The capacity lower bound and upper bound on edge (t, s) ensures condition (5).

3.2 Algorithm DIVIDE

Let C be a cluster of points and $Q \subseteq C$ be a subset of C whose points are used as centers for the points in C. Algorithm DIVIDE (proposed by Dickerson et al. [4]) partitions cluster C into $|Q|$ sub-clusters, each of which has a distinct center in Q.

For each color h appearing in C, Algorithm DIVIDE allocates the points of color h in set C evenly to the centers in Q. Specifically, for color h, Algorithm DIVIDE allocates $\left\lceil \frac{|C^h \cap C|}{|Q|} \right\rceil$ or $\left\lfloor \frac{|C^h \cap C|}{|Q|} \right\rfloor$ points of color h in C to each new center in Q, so that the numbers of points of color h assigned to the centers in Q differ at most one. See Fig. 2 for an illustration.

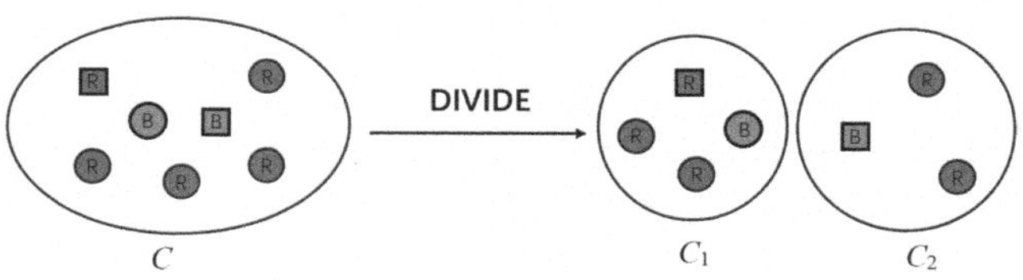

Fig. 2. Illustration of the sub-clusters generated by Algorithm DIVIDE. Rectangular points are centers, while round points are not centers. Each point has a letter depicting its color.

Algorithm 2. DIVIDE

Input: Point set C, subset $Q \subseteq C$ used as centers.
Output: Assignment function $\phi : C \to Q$.

1: **if** $|Q| = 1$ **then**
2: Assign all point in C to the single center in Q.
3: **else**
4: For each color h, assign $\left\lfloor \frac{|C^h \cap C|}{|Q|} \right\rfloor$ points in C of color h to a distinct center in Q.
5: If there are still points of color h unassigned, then assign each of them to a distinct center in Q. Ties are broken arbitrarily.
6: **end if**
7: **return** Function ϕ denoting the above assignment.

The even partitioning strategy of Algorithm DIVIDE has a property that the additive violation it introduces to the GF constraint will not exceed 2.

Lemma 1 ([4]). *Given a cluster C with radius R that violates the GF constraint by an additive violation of λ, and a subset of points $Q \subseteq C$, Algorithm DIVIDE generates a clustering of points in C with centers in Q such that the additive violation on the GF constraint of the new clustering is at most $\frac{\lambda}{|Q|} + 2$ and new radius is at most $2R$.*

Especially, if cluster C satisfies the GF constraint (i.e., the additive violation factor $\lambda = 0$), then Algorithm DIVIDE generates a clustering with additive violation factor on the GF constraint at most 2.

3.3 Analysis

In this section we analyze the performance of the k-clustering (S, ϕ) output by Algorithm G2GD. Recall that the clusters by (S, ϕ) are $C_1, C_2, \ldots, C_{|S|}$ (which are determined by the assignment function ϕ), with each cluster having a distinct center in S. We thus define Δ to be the maximum size of any cluster induced by (S, ϕ), i.e.,

$$\Delta = \max_{i \in [|S|]} |C_i| \,.$$

Lemma 2. *Algorithm G2GD finds a k-clustering (S, ϕ) with cost $\leq (1 + \Delta^{\frac{1}{p}})\mathcal{L}(\bar{S}, \bar{\phi})$, where $(\bar{S}, \bar{\phi})$ is the clustering generated by step 1 of Algorithm G2GD.*

Proof. By Algorithm G2GD, (S, ϕ) is built upon $(\bar{S}, \bar{\phi})$. Let $\bar{k} = |\bar{S}|$. So, we have

$$\begin{aligned}
\mathcal{L}(S, \phi) &= \left(\sum_{i=1}^{\bar{k}} \sum_{u \in Q_i} \sum_{v \in D(u)} d\left(u, v\right)^p \right)^{\frac{1}{p}} \\
&\leq \left(\sum_{i=1}^{\bar{k}} \sum_{u \in Q_i} \sum_{v \in D(u)} \left(d(u, u_i) + d(u_i, v)\right)^p \right)^{\frac{1}{p}} \\
&\leq \left(\sum_{i=1}^{\bar{k}} \sum_{u \in Q_i} \sum_{v \in D(u)} d\left(u, u_i\right)^p \right)^{\frac{1}{p}} + \left(\sum_{i=1}^{\bar{k}} \sum_{u \in Q_i} \sum_{v \in D(u)} d\left(u_i, v\right)^p \right)^{\frac{1}{p}}, \quad (7)
\end{aligned}$$

where the first inequality is due to the triangle inequality, and the second inequality is due to the Minkowski inequality.

We find that

$$\left(\sum_{i=1}^{\bar{k}}\sum_{u\in Q_i}\sum_{v\in D(u)} d\left(u,u_i\right)^p\right)^{\frac{1}{p}} = \left(\sum_{i=1}^{\bar{k}}\sum_{u\in Q_i} |D(u)| d\left(u,u_i\right)^p\right)^{\frac{1}{p}}$$
$$\leq \left(\Delta\sum_{i=1}^{\bar{k}}\sum_{u\in Q_i} d\left(u,u_i\right)^p\right)^{\frac{1}{p}} \leq \left(\Delta\sum_{i=1}^{\bar{k}}\sum_{v\in \bar{C}_i} d(u_i,v)^p\right)^{\frac{1}{p}}$$
$$= \Delta^{\frac{1}{p}}\mathcal{L}(\bar{S},\bar{\phi}), \tag{8}$$

where the last inequality holds since each Q_i is a subset of $\bar{C}_i$ ($i \in [\bar{k}]$).

Moreover, we have

$$\left(\sum_{i=1}^{\bar{k}}\sum_{u\in Q_i}\sum_{v\in D(u)} d\left(u_i,v\right)^p\right)^{\frac{1}{p}} = \left(\sum_{i=1}^{\bar{k}}\sum_{v\in \bar{C}_i} d\left(u_i,v\right)^p\right)^{\frac{1}{p}} = \mathcal{L}(\bar{S},\bar{\phi}). \tag{9}$$

By (7), (8), (9), we get $\mathcal{L}(S,\phi) \leq (1+\Delta^{\frac{1}{p}})\mathcal{L}(\bar{S},\bar{\phi})$, completing the proof. □

Lemma 3. *Suppose that the clustering $(\bar{S},\bar{\phi})$ generated by step 1 of Algorithm G2GD violates the GF constraint by an additive factor λ. Then Algorithm G2GD finds a clustering (S,ϕ) that violates the GF constraint by an additive factor at most $\lambda+2$, and satisfies the DS constraint.*

Proof. Algorithm SNC guarantees that the centers S satisfies the DS constraint. Since each cluster C_i ($i \in [|S|]$) determined by (S,ϕ) is generated by Algorithm DIVIDE, by Lemma 1, C_i violates the GF constraint by an additive factor at most $\lambda+2$. □

Theorem 1. *Algorithm G2GD finds an $O(\Delta^{\frac{1}{p}})$-approximation in polynomial time for the doubly constrained fair (k,p)-*Clustering *problem. The k-clustering output by Algorithm G2GD violates the group fairness (i.e., GF) constraint by an additive factor at most 5, and satisfies the diversity in center selection fairness (i.e., DS) constraint.*

Proof. In step 1, Algorithm G2GD calls the algorithm by Bera et al. [2] to obtain a $\bar{k}$-clustering $(\bar{S},\bar{\phi})$. By [2], $(\bar{S},\bar{\phi})$ is an $O(1)$-approximation for the GF-constrained fair (k,p)-Clustering problem that can be found in polynomial time. Let OPT_{GF} be the optimum of the GF-constrained fair (k,p)-Clustering problem. So, by Lemma 2, we have

$$\mathcal{L}(S,\phi) \leq (1+\Delta^{\frac{1}{p}})\mathcal{L}(\bar{S},\bar{\phi}) \leq (1+\Delta^{\frac{1}{p}})O(1)OPT_{GF}.$$

Let OPT_{GFDS} be the optimum of the doubly constrained fair (k,p)-Clustering problem. Obviously, we have $OPT_{GF} \leq OPT_{GFDS}$. So, we further have

$$\mathcal{L}(S,\phi) \leq (1+\Delta^{\frac{1}{p}})O(1)OPT_{GFDS} = O(\Delta^{\frac{1}{p}})OPT_{GFDS}.$$

Moreover, $(\bar{S}, \bar{\phi})$ violates the GF constraint by an additive factor of 3 [2]. So, by Lemma 3, The k-clustering (S, ϕ) output by Algorithm G2GD violates the GF constraint by an additive factor of 5.

Finally, it is straightforward that Algorithm G2GD runs in polynomial time. The theorem follows. □

4 Experiments

In this section, we present the experimental evaluation of our Algorithm G2GD. As our algorithm G2GD is for the doubly constrained fair clustering problem with general p-norms, in the experiment we focus on the case of $p = 2$, which corresponds to the doubly constrained fair clustering problem with the k-means objective. The source codes of our experiments can be downloaded from GitHub at https://github.com/zhanglh1023/dfc_clustering.

Tested algorithms. We compare the performance of our Algorithm G2GD with three existing algorithms. So, we have four tested algorithms in total. (i) Algorithm COLOR-BLIND is a k-medoids algorithm implemented in the PyClustering library [13]. (ii) Algorithm ALG-GF is a clustering algorithm considering only the GF constraints [2]. (iii) Algorithm ALG-GFDS is the algorithm proposed by Dickerson et al. [4] for the fair clustering problem that considers both the GF and DS constraints. Note that ALG-GFDS can be used for the clustering problem with the k-means objective. (Only analysis for the k-center objective was given in [4].) (iv) Algorithm G2GD is our proposed clustering algorithm, which satisfies the DS constraint, and violates the GF constraint by an additive factor of at most 5.

Data Sets. We evaluate our Algorithm G2GD together with the three existing algorithms using the data sets from the UCI repository [10]. Specifically, we employe the following three data sets. (i) The first is the *bank* dataset, which consists of 4,521 points, where marital status is used to define group membership. (ii) The second is a subset of the *adult* dataset, which contains 20,000 records, with race as the attribute for group membership. (iii) The third is a subset of the *creditcard* dataset, which comprises 20,000 records, where education level is used to determine group membership.

We employed Euclidean distance as the distance metric. For the GF constraint, we defined the lower and upper bounds for the proportion of each color h as $\beta_h = (1 - \delta)p_h$ and $\alpha_h = (1 + \delta)p_h$, respectively. Here, p_h represents the proportion of color h in the data set, and we set $\delta = 0.2$. For the DS constraint, we set $l_h = 0.8p_hk$ and $r_h = p_hk$, where k is the number of clusters.

Measurements. We evaluate the performance of the four tested algorithms using the following three metrics. (i) The first metric is the *price of fairness (PoF)* [2,5]. This metric quantifies the additional cost incurred to achieve fairness in the algorithm. Its general definition is

$$PoF = \frac{\text{Clustering cost subject to some constraints}}{\text{Clustering cost without fairness constraints}}.$$

In our experiments, PoF is computed as the ratio of the clustering cost of the algorithm considering fairness to the clustering cost of Algorithm COLOR-BLIND. (ii) The second metric is *GF-Violation*. This metric measures the maximum additive violation of the GF constraint, as defined in the formula (3). (iii) The third metric is *DS-Violation*. This metric captures the maximum additive violation of the DS constraint.

The experimental results show that the clustering cost of our Algorithm G2GD, while satisfying DS constraint and violating the GF constraint by an additive factor of only 5, is nearly identical to that of the clustering algorithm ALG-GF [2], which only considers the GF constraint. Due to the limitation of space, the details of the experiments will be given in the full version of the paper.

5 Conclusions

In this paper, we study the doubly constrained fair clustering problem with general ℓ_p norms, extending the work of Dickerson et al. [4] from k-center to general objectives including k-Median, k-Means, etc. We propose a simple and efficient approximation algorithm for the problem. The key contribution is a network flow-based algorithm to select clustering centers. Experimental results demonstrate that our algorithm, while effectively dealing with both the GF and DS constraints, achieves clustering costs nearly identical to those of only GF-constrained clustering. The superior experimental performance of the algorithm suggests that tighter approximation guarantees may exist, leaving room for further theoretical refinements and advances. Future work also include exploring additional fairness constraints to enhance the applicability of fair clustering.

Acknowledgments. This work is supported by the National Natural Science Foundation of China (62272280 and 61972228).

References

1. Ahmadian, S., Epasto, A., Kumar, R., Mahdian, M.: Clustering without over-representation. In: Proceedings of the 25th ACM SIGKDD International Conference on Knowledge Discovery & Data Mining (KDD), pp. 267–275 (2019)
2. Bera, S.K., Chakrabarty, D., Flores, N.J., Negahbani, M.: Fair algorithms for clustering. In: Proceedings of the 33rd International Conference on Neural Information Processing Systems (NeurIPS), pp. 4954–4965 (2019)
3. Chierichetti, F., Kumar, R., Lattanzi, S., Vassilvitskii, S.: Fair clustering through fairlets. In: Proceedings of the 31st International Conference on Neural Information Processing Systems (NeurIPS). pp. 5036–5044 (2017)
4. Dickerson, J., Esmaeili, S.A., Morgenstern, J., Zhang, C.J.: Doubly constrained fair clustering. In: Proceedings of the 37th Annual Conference on Neural Information Processing Systems (NeurIPS), pp. 13267–13293 (2023)
5. Esmaeili, S.A., Brubach, B., Srinivasan, A., Dickerson, J.P.: Fair clustering under a bounded cost. In: Proceedings of the 35th Conference on Neural Information Processing Systems (ICML), pp. 14345–14357 (2021)

6. Feldman, M., Friedler, S.A., Moeller, J., Scheidegger, C., Venkatasubramanian, S.: Certifying and removing disparate impact. In: Proceedings of the 21th ACM SIGKDD International Conference on Knowledge Discovery and Data Mining (KDD), pp. 259–268 (2015)
7. Ghadiri, M., Samadi, S., Vempala, S.: Socially fair k-means clustering. In Proceedings of the 2021 ACM Conference on Fairness, Accountability, and Transparency, pp. 438–448 (2021)
8. Hotegni, S.S., Mahabadi, S., Vakilian, A.: Approximation algorithms for fair range clustering. In: Proceedings of the 40th International Conference on Machine Learning (ICML), pp. 13270–13284 (2023)
9. Jones, M., Nguyễn, H.L., Nguyen, T.: Fair k-centers via maximum matching. In: Proceedings of the 37th International Conference on Machine Learning (ICML), pp. 4940–4949 (2020)
10. Kelly, M., Longjohn, R., Nottingham, K.: UCI machine learning repository. https://archive.ics.uci.edu
11. Kleindessner, M., Awasthi, P., Morgenstern, J.: Fair k-center clustering for data summarization. In: Proceedings of the 36th International Conference on Machine Learning (ICML), pp. 3448–3457 (2019)
12. Nguyễn, H.L., Nguyen, T., Jones, M.: Fair range k-center. arXiv e-prints arXiv:2207.11337 (2022)
13. Schubert, E., Rousseeuw, P.J.: Faster k-medoids clustering: improving the PAM, CLARA, and CLARANS algorithms. In: Proceedings of the 12th International Conference on Similarity Search and Applications (SISAP). Lecture Notes in Computer Science, vol. 11807, pp. 171–187 (2019)
14. Thejaswi, S., Ordozgoiti, B., Gionis, A.: Diversity-aware k-median: clustering with fair center representation. In: Proceedings of the European Conference on Machine Learning and Knowledge Discovery in Databases (ECML PKDD), Part II, pp. 765–780 (2021)

Combinatorial Optimization

Discrete Effort Distribution via Regret-Enabled Greedy Algorithm

Song Cao, Taikun Zhu, and Kai Jin(✉)

Shenzhen Campus of Sun Yat-sen University, Shenzhen, Guangdong, China
{caos6,zhutk3}@mail2.sysu.edu.cn, jink8@mail.sysu.edu.cn

Abstract. This paper addresses resource allocation problem with a separable objective function under a single linear constraint, formulated as maximizing $\sum_{j=1}^{n} R_j(x_j)$ subject to $\sum_{j=1}^{n} x_j = k$ and $x_j \in \{0, \ldots, m\}$. While classical dynamic programming approach solves this problem in $O(n^2m^2)$ time, we propose a regret-enabled greedy algorithm that achieves $O(n \log n)$ time when $m = O(1)$. The algorithm significantly outperforms traditional dynamic programming for small m. Our algorithm actually solves the problem for all k ($0 \le k \le nm$) in the mentioned time.

Keywords: Regret-enabled Greedy Algorithm · Discrete Effort Distribution · Resource Allocation · (max,+) convolution · Heaps

1 Introduction

We consider the effort distribution problem with separable objective function and one linear constraint. It can be formulated as follows.

$$\text{Maximize: } \sum_{j=1}^{n} R_j(x_j),$$

$$\text{subject to: } \sum_{j=1}^{n} x_j = k, \qquad x_j \in \{0, ..., m\}, m > 0 \text{ and } 1 \le k \le nm,$$

where,

n = the number of projects,
k = the total number of efforts,
m = the maximal number of efforts allowed to be allocated to each project,
x_j = the number of efforts allocated to j-th project, and
$R_j(x_j)$ = the revenue j-th project generates when it is allocated x_j efforts.
Assume that $R_j(0) = 0$.

This research is supported by Department of Science and Technology of Guangdong Province (Project No. 2021QN02X239) and Shenzhen Science and Technology Program (Grant No. 202206193000001, 20220817175048002).

F. V. Fomin and M. Xiao (Eds.): COCOON 2025, LNCS 15983, pp. 109–122, 2026.
https://doi.org/10.1007/978-981-95-0215-8_9

It can be solved by dynamic programming in $O(n^2m^2)$ time. Let $dp[j,k]$ be maximal revenue for the first j projects with k efforts. Then we have

$$dp[j,k] = \max_{0 \le x_j \le m} \{dp[j-1, k-x_j] + R_j(x_j)\}.$$

While prior works imposes concavity or near concavity constraints on R_j, our approach removes these constraints, requiring only separability of R_j. In this paper, we give an $O(n \log n)$ time algorithm based on a regret-enabled greedy framework, which solves the problem for all k ($0 \le k \le nm$) when $m = O(1)$.

Traditional greedy algorithms iteratively make locally optimal decisions to achieve global optimal solution, but they fail in our context. To overcome this we use a regrettable greedy mechanism—a paradigm that allows strategic revocation of prior decisions. Specifically, our algorithm adjusts allocations by (1) removing t (a parameter dependent on m) efforts from some projects, and (2) allocating $t+1$ efforts to some projects to maximize incremental revenue at each step.

The organization of this paper is as follows. In Sect. 2, we establish a crucial property of optimal solutions, and present our main algorithm in Sect. 3. Furthermore, for the special case where all revenue functions R_j are convex, we introduce two algorithms in Sect. 4: one computes optimal solutions for all k in $O(nm + \log n)$ time, while the other runs in $O(nm)$ time for a given k.

In Sect. 5, we define a class of functions called *oscillating concave functions* and demonstrate a computational property: if f is concave and g is oscillating concave, their (max,+) convolution can be computed in $O(n)$ time. Based on this property, we describe an algorithm for $m = 2$ that achieves $O(n)$ time after an initial sorting step.

Definition 1. *A distribution of k efforts to the n projects can be described by a vector $\mathbf{x} = (x_1, \ldots, x_n)$ where $x_i \in \{0, \ldots, m\}$ and $\sum_i x_i = k$. Such a vector is called a k-*profile. *For $k \ge 0$, denote by $\mathcal{P}_k$ the set of k-profiles.*

A k-profile is optimal if its revenue $\sum_i R_i(x_i)$ is the largest among $\mathcal{P}_k$.

1.1 Related Works

The problem we studied falls under the broad category of resource allocation. Resource allocation problem involves determining the cost-optimal distribution of constrained resources among competing activities under fixed resource availability. The multi-objective resource allocation problem (MORAP) has been formally characterized through network flow modeling by Osman *et al.* [1], establishing a generalized framework for handling different optimization criteria under resource constraints. Beyond that, resource allocation problem widely appears in manufacturing, computing, finance and network communication. Bitran and Tirupati [2] formulated two nonlinear resource allocation problems—targeting problem (TP) and balancing problem (BP)—for multi-product manufacturing systems. Bitran and Saarkar [3] later proposed an exact iterative algorithm for TP. Rajkumar *et al.* [4] presented an analytical model to measure quality of service (QoS) management, which referred to as QoS-based Resource Allocation

Model (Q-RAM). Bretthauer *et al.* [5] transferred various versions of stratified random sampling plan problem into resource allocation problems with convex objective and linear constraint. And they provided two branch-and-bound algorithms to solve these problems.

A fundamental variant known as the *simple resource allocation problem* [6] involves minimizing separable convex objective functions (or maximizing separable concave objective functions) with a single linear constraint, solvable via classical greedy algorithms [7,8]. Subsequent research has extended this framework along two directions: generalizing objective functions and complex constraints. For instance, Federgruen and Groenevelt [9] developed greedy algorithms for weakly concave objectives, while Murota [10] introduced M-convex functions–a specialized subclass of convex functions later studied by Shioura [11] for polynomial-time minimization. Nonlinear constraints were addressed by Bretthauder and Shetty [12], who proposed a branch-and-bound algorithm for separable concave objectives. Multi-objective scenarios were explored by Osman *et al.* [1] using genetic algorithms, and online stochastic settings were investigated by Devanur *et al.* [13] through a distributional model yielding an $1-O(\epsilon)$-approximation algorithm. Recent work by Deng *et al.* [14] further extended the framework to nonsmooth objectives under weight-balanced digraph constraints via distributed continuous-time methods. In this paper, we focus on generalizing the objective function by removing its concavity constraint.

From a computational perspective, our problem admits the computation of (max,+) convolution. Given two sequences $\{x_i\}_{i=1}^n$ and $\{y_i\}_{i=1}^n$, their (max,+) convolution computes $z_k = \max_{i=0}^k(x_i + y_{k-i})$, with (min,+) convolution defined analogously. Many problems occur to be computation of such convolution, such as the Tree Sparsity problem and Knapsack problem.

While naively computable in $O(n^2)$ time, Cygan *et al.* [15] put forward that there is no $O(n^{2-\epsilon})$ algorithm where $\epsilon > 0$ for (min,+) convolution. Subsequent improvements include Bremner *et al.*'s $O(n^2/\lg n)$ algorithm [16] and Bussieck *et al.*'s $O(n \log n)$ expected-time algorithm for random inputs [17]. Special sequence structures enable faster computation: When x and y are both convex, their (min,+) convolution can be easily computed in $O(n)$ time. For monotone integer sequences bounded by $O(n)$, Chan *et al.* [18] achieved $O(n^{1.859})$, later refined to $\tilde{O}(n^{1.5})$ upper bound by Chi *et al.* [19]. Bringmann [20] further considered Δ-near convex functions-those approximable by convex functions within additive error Δ-yielding an $\tilde{O}(n\Delta)$ algorithm. The conclusion we obtain in Sect. 5 slightly broadens the class of functions for which (max,+) convolution can be computed in $O(n)$ time.

Our resource allocation problem is closely related to the subset-sum and Knapsack problem [21,22]. Let W denote the maximum weight of the items, and P denote the maximum profit of the items. Pisinger [21] shows that (1) the subset-sum problem can be solved in $O(nW)$ time, improving over the trivial $O(n^2W)$ bound, and (2) the Knapsack problem can be solved in $O(nWP)$ time. Recently, an $\tilde{O}(n + W^2)$ time algorithm is given for the Knapsack problem by Bringmann *et al.* [22]. See more related work of the Knapsack problem (with the parameter W) in [22]. Note that the Knapsack problem is a special case of (and hence easier than) our resource allocation problem. An item with weight w can be seen as a project j; moreover, $R_j(w)$ is the profit of this item, where $R_j(x_j) = -\infty$ for $x_j \neq w$. Be aware that $m = W$ is the maximum weight of the items.

1.2 Preliminaries: Some Observations on Multisets

For convenience, in this paper a multiset refers to a multiset of $[m] = \{1, \ldots, m\}$.

A pair of multisets (A, B) is *reducible* if the sum of a nonempty subset of A equals the sum of a nonempty subset of B, and is *irreducible* otherwise.

Example 1. Reducible: $(A = \{1, 2, 2, 2\}, B = \{3, 3\})$, $(A, B) = (\{1, 3\}, \{2, 2\})$.
Irreducible: $(A, B) = (\{2, 2\}, \{3\})$, $(A, B) = (\{3\}, \{1, 1\})$.

Denote $\lambda_m = \begin{cases} \frac{(m-1)m(m+1)}{2}, & m > 1; \\ 1, & m = 1. \end{cases}$

For any multiset A, its sum of elements is denoted as $\sum A$.

Lemma 1. *A pair of multisets (A, B) is reducible if $\sum A \geq \lambda_m$ and $\sum B \geq \lambda_m$.*

Proof. The case $m = 1$ is trivial. We assume that $m > 1$ in the following. Note that $\lambda_m = (m - 1)(1 + \ldots + m)$ when $m > 1$.

Since $\sum A \geq \lambda_m$, exactly one of the following holds:
(1) There is $u \in [m]$ so that u appears at least m times in A.
(2) Each number in $[m]$ appears exactly $m - 1$ times in A.
Since $\sum B \geq \lambda_m$, exactly one of the following holds:
(1') There is $v \in [m]$ so that v appears at least m times in B.
(2') Each number in $[m]$ appears exactly $m - 1$ times in B.
If (1) and (1') hold, take v u's from A and u v's from B and we are done.
If (1) and (2') hold, take a u from both A and B and we are done.
If (2) and (1') hold, take a v from both A and B and we are done.
If (2) and (2') hold, take a 1 from both A and B and we are done. □

Remark 1. While Lemma 1 provides a foundational bound, its conclusion is not tight. Bringmann *et al.* [22] gave a better result with significantly increased analytical complexity:

Lemma 2. *[22] A pair of multisets (A, B) is reducible if*

$$|A| \geq 1500 \left(\log^3(2\,A|)\mu(A)m\right)^{1/2}$$

and

$$\sum B \geq 340000 \log(2\,A|)\mu(A)m^2/|A|,$$

where $\mu(A)$ denotes the maximal multiplicity of elements in A.

1.3 Irreducible Pair (A, B) with $\sum A - \sum B = 1$

Suppose we want to enumerate irreducible pairs (A, B) satisfying $\sum A - \sum B = 1$ (for some fixed small m) (which will be used in our algorithm). We only need to focus on (A, B) with $\sum B < \lambda_m$ (since otherwise $\sum A, \sum B \geq \lambda_m$, and (A, B) must be reducible by Lemma 1). Therefore we can enumerate all target pairs by brute-force programs (check all (A, B) where $\sum B < \lambda_m$ and $\sum A = \sum B + 1$).

Example 2. All irreducible pairs with $\sum A - \sum B = 1$ for $m = 2$ are:

$$A = \{1\}, \quad B = \emptyset;$$
$$A = \{2\}, \quad B = \{1\}.$$

Example 3. All irreducible pairs with $\sum A - \sum B = 1$ for $m = 3$ are:

$$A = \{1\}, \quad B = \emptyset;$$
$$A = \{2\}, \quad B = \{1\};$$
$$A = \{3\}, \quad B = \{2\};$$
$$A = \{3\}, \quad B = \{1, 1\};$$
$$A = \{2, 2\}, \quad B = \{3\}.$$

The number of irreducible pairs with $\sum A - \sum B = 1$ will be denoted by p_m, or p for simplicity. According to our brute-force programs,

$$p_1 = 1, \ p_2 = 2, \ p_3 = 5, \ p_4 = 11, \ p_5 = 27.$$

2 A Crucial Property of the Optimal k-Profiles

Definition 2. *Assume* $\mathbf{x} = (x_1, ..., x_n) \in \mathcal{P}_k$ *and* $\mathbf{y} = (y_1, ..., y_n) \in \mathcal{P}_{k+1}$. *Define* $\mathsf{diff}(\mathbf{x}, \mathbf{y}) = (A, B)$ *and call it the difference of* $(\mathbf{x}, \mathbf{y})$*, where*

$$A = \{y_i - x_i \mid i \in [n] \text{ and } y_i > x_i\}. \tag{1}$$
$$B = \{x_i - y_i \mid i \in [n] \text{ and } x_i > y_i\}. \tag{2}$$

Notice that $\sum A - \sum B = \sum_i (y_i - x_i) = (k + 1) - (k) = 1$.

Example 4. $\mathsf{diff}((1, 1), (3, 0)) = (\{2\}, \{1\})$. $\mathsf{diff}((2, 2, 2), (3, 1, 3)) = (\{1, 1\}, \{1\})$.

Lemma 3. *For any optimal k-profile* $\mathbf{x}$*, where* $k < nm$*, there exists an optimal* $(k + 1)$*-profile* $\mathbf{y}$ *such that* $\mathsf{diff}(\mathbf{x}, \mathbf{y})$ *is irreducible.*

Proof. First of all, take any $(k + 1)$-optimal profile $\mathbf{y}$. If $\mathsf{diff}(\mathbf{x}, \mathbf{y})$ is irreducible, we are done. Now, suppose to the opposite that $(A, B) = \mathsf{diff}(\mathbf{x}, \mathbf{y})$ is reducible.

For convenience, denote $I = \{i \in [n] \mid y_i > x_i\}$ and $J = \{j \in [n] \mid x_j > y_j\}$. We have $A = \{y_i - x_i \mid i \in I\}$ and $B = \{x_j - y_j \mid j \in J\}$ following (1) and (2).

As (A, B) is reducible, there exist nonempty sets $I_0 \subseteq I, J_0 \subseteq J$ such that $\sum_{i \in I_0}(y_i - x_i) = \sum_{j \in J_0}(x_j - y_j)$, which implies that

$$\sum_{i \in I_0} y_i + \sum_{j \in J_0} y_j = \sum_{j \in J_0} x_j + \sum_{i \in I_0} x_i.$$

Note that $I_0 \cap J_0 = \emptyset$ because $I \cap J = \emptyset$. We further obtain

$$\sum_{i \in I_0 \cup J_0} y_i = \sum_{i \in I_0 \cup J_0} x_i. \tag{3}$$

We claim that $\sum_{i \in I_0 \cup J_0} R_i(y_i) = \sum_{i \in I_0 \cup J_0} R_i(x_i)$. The proof is as follows. If $\sum_{i \in I_0 \cup J_0} R_i(y_i) < \sum_{i \in I_0 \cup J_0} R_i(x_i)$, we can see $\mathbf{y}$ is not $(k+1)$-optimal because by setting $y_i = x_i$ for $i \in I_0 \cup J_0$, the revenue of $\mathbf{y}$ is enlarged. Similarly, if $\sum_{i \in I_0 \cup J_0} R_i(y_i) > \sum_{i \in I_0 \cup J_0} R_i(x_i)$, we can see $\mathbf{x}$ is not k-optimal because by setting $x_i = y_i$ for $i \in I_0 \cup J_0$, the revenue of $\mathbf{x}$ is enlarged. Therefore, it must hold that $\sum_{i \in I_0 \cup J_0} R_i(y_i) = \sum_{i \in I_0 \cup J_0} R_i(x_i)$.

Following the claim above, the revenue of $\mathbf{y}$ is unchanged (and hence $\mathbf{y}$ is still optimal) if we modify $y_i = x_i$ for all $i \in I_0 \cup J_0$. Note that such a modification of $\mathbf{y}$ would decrease $\sum A$, and moreover $\sum A = \sum B + 1$ is always positive, therefore eventually $\mathbf{y}$ cannot be modified. This means that $\mathsf{diff}(x, y)$ becomes irreducible after several modifications of $\mathbf{y}$. So the lemma holds. □

As a side note, Lemma 3 implies that *for any optimal k-profile $\mathbf{x}$, where $k < nm$, there exists an optimal $(k+1)$-profile $\mathbf{y}$ such that $\sum_i |y_i - x_i| < 2\lambda_m$.* To see this, first find the optimal $(k+1)$-profile $\mathbf{y}$ with $\mathsf{diff}(\mathbf{x}, \mathbf{y}) = (A, B)$ irreducible. Observe that $\sum B < \lambda_m$. Otherwise, $\sum B \geq \lambda_m$ and $\sum A \geq \lambda_m$, and (A, B) is reducible by Lemma 1. Therefore $\sum_i |y_i - x_i| = \sum B + \sum A < 2\lambda_m$.

3 Algorithm for Finding Optimal k-Profile

It is sufficient to solving the following subproblem (for k from 0 to $nm - 1$):

Problem 1. Given a k-profile $\mathbf{x}^{(k)}$. Among all the $(k+1)$-profile $\mathbf{y}$ with $\mathsf{diff}(\mathbf{x}, \mathbf{y})$ being irreducible, find the one, denoted by $\mathbf{x}^{(k+1)}$, with the largest revenue.

Clearly, we can set $\mathbf{x}^{(0)}$ to be the unique (and optimal) 0-profile. Then, by induction, $\mathbf{x}^{(1)}, \ldots, \mathbf{x}^{(nm)}$ would all be optimal according to Lemma 3.

In what follows we solve this subproblem in $O(f(m) \log n)$ time, where $f(m)$ is some function of m, and factor $\log n$ comes from the application of heap.

For convenience, assume $\mathbf{x}^{(k)} = (x_1, \ldots, x_n)$.

Data Structures. Our algorithm uses $2m$ heaps.

For each $d \in [m]$, we build a max-heap DO_d whose items are those projects $i \in [n]$ for which $x_i + d \leq m$, and the value of item i is defined by $R_i(x_i + d) - R_i(x_i)$ – the increase of revenue when we distribute d more efforts into project i.

$$\mathsf{DO}_d = \{\langle i, R_i(x_i + d) - R_i(x_i)\rangle \mid i \in [n], x_i + d \leq m\}. \tag{4}$$

For each $d \in [m]$, we build a min-heap UNDO_d whose items are those projects $i \in [n]$ for which $x_i - d \geq 0$, and the value of item i is defined by $R_i(x_i) - R_i(x_i - d)$ – the lost of revenue when we withdraw d efforts from project i.

$$\mathsf{UNDO}_d = \{\langle i, R_i(x_i) - R_i(x_i - d)\rangle \mid i \in [n], x_i - d \geq 0\}. \tag{5}$$

Observe that if x_i is changed, we shall update the value of item i (calling UPDATE_VALUE) in each of the $2m$ heaps. (To be more clear, sometimes we may have to call DELETE or INSERT instead of UPDATE_VALUE, since the condition $x_i + d \leq m$ may change, so as $x_i - d \geq 0$ after the change of x_i.)

3.1 The Algorithm

Consider all irreducible pairs (A, B) with $\sum A - \sum B = 1$. Recall Examples 2 and 3. For convenience, denote them by $(A_1, B_1), \ldots, (A_p, B_p)$, where p is the number of such pairs. (Note: we can generate and store these p pairs by a brute-force preprocessing procedure, whose running time is only related to m.)

For each $c \in [p]$, denote by $\mathbf{y}^{(c)}$ the best $(k+1)$-profile among those satisfying $\mathsf{diff}(\mathbf{x}^{(k)}, \mathbf{y}) = (A_c, B_c)$. By Lemma 3, the best among $\mathbf{y}^{(1)}, \ldots, \mathbf{y}^{(p)}$ can serve as $\mathbf{x}^{(k+1)}$.

How do we compute $\mathbf{y}^{(c)}$ efficiently?

Let us first consider a simple case, e.g., $m = 2$ and $(A_c, B_c) = (\{2\}, \{1\})$. In this case computing $\mathbf{y}^{(c)}$ is equivalent to solving the following problem:

Find the indices i and j that maximize

$$R_i(x_i + 2) - R_i(x_i) - (R_j(x_j) - R_j(x_j - 1)),$$

subject to

$$i \neq j, x_i + 2 \leq m, x_j - 1 \geq 0.$$

We can find i so that $R_i(x_i + 2) - R_i(x_i)$ is maximized using heap DO_2, and find j so that $R_j(x_j) - R_j(x_j - 1)$ is minimized using heap UNDO_1. Clearly, $i \neq j$ because $x_i = 0$ whereas $x_j > 0$, and so the problem is solved.

Next, let us consider a more involved case: $m = 3$ and $(A_c, B_c) = (\{2\}, \{1\})$. If we do the same as in the above case, it might occur that $i = j$ (for those $x_i = 1$, item i is in DO_2 and UNDO_1 simultaneously when $m = 3$).

Nevertheless, utilizing the heaps, the above maximization problem can still be solved efficiently: Find the best i_1 and second best i_2 in DO_2, the best j_1 and second best j_2 in UNDO_1, and moreover, try every combination $(i, j) \in \{(i_1, j_1), (i_1, j_2), (i_2, j_1), (i_2, j_2)\}$. One of them must be the answer. (Indeed, we can exclude (i_2, j_2) from the trying set.)

With the experience on small cases, we now move on to the general case. For $d \in [m]$, let a_d denote the multiplicity of d in A_c, and b_d the multiplicity of d in B_c. Let $a = \sum a_d$ and $b = \sum b_d$ be the number of elements in A_c and B_c, respectively (which are bounded by λ_m according to the analysis in Sect. 1.3).

We now use a brute-force method to compute $\mathbf{y}^{(c)}$.

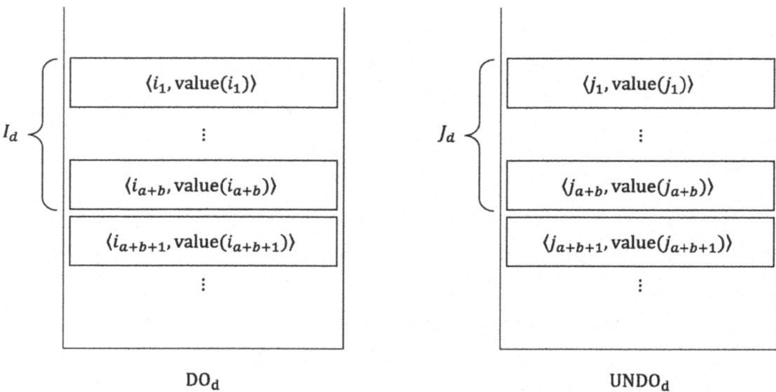

Fig. 1. An illustration of the algorithm.

1. For each $d \in [m]$, compute the set I_d that contains the best $a + b$ items in DO_d, and compute J_d that contains the best $a + b$ items in UNDO_d. See Fig. 1.
2. Enumerate $(I_1', \ldots, I_m'), (J_1', \ldots, J_m')$ such that

$$\begin{cases} I_d' \subseteq I_d \text{ and } I_d' \text{ has size } a_d, \\ J_d' \subseteq J_d \text{ and } J_d' \text{ has size } b_d. \end{cases}$$

When $I_1', \ldots, I_m', J_1', \ldots, J_m'$ are pairwise-distinct, we obtain a solution:

$$\text{increase } x_i \text{ by } d \text{ for } i \in I_d', \text{ and decrease } x_j \text{ by } d \text{ for } j \in J_d'.$$

Select the best solution and it is $\mathbf{y}^{(c)}$.

The enumeration to compute an $\mathbf{y}^{(c)}$ takes

$$g(m) = O\left(\binom{a_1}{a+b} \cdots \binom{a_m}{a+b} \binom{b_1}{a+b} \cdots \binom{b_m}{a+b}\right)$$

time. So Problem 1 can be solved in $O(pg(m) \log n)$ time, recall that p is the number of irreducible pairs (A, B) satisfying $\sum A - \sum B = 1$, entirely determined by m.

4 Separable Convex Objective Function

In this section, we consider a special case where all the separated objective function R_j are convex (the concave case has been studied extensively as mentioned in the introduction). We present two algorithms for this special case: One runs in $O(nm + n \log n)$ time and it finds the optimal k-profile for all k. The other runs in $O(nm)$ time and it finds the optimal solution for a given k.

Remark 2. If m is a constant, our first algorithm in this section runs in $O(n \log n)$ time, as the algorithm shown in Sect. 3. However, the constant factor of the algorithm in this section is much smaller.

Lemma 4. *There exists an optimal k-profile satisfies: At most one project receives more than 0 and less than m efforts.*

Proof. We prove it by contradiction. Suppose $\mathbf{x} = (x_1, \ldots, x_n)$ is an optimal k-profile. Assume $0 < x_i < m$, $0 < x_j < m$ for some $i \neq j$.

Assume $R_i(x_i+1) - R_i(x_i) \geq R_j(x_j+1) - R_j(x_j)$; otherwise we swap i and j. We can withdraw one effort from project j and give it to project i without decreasing the total revenue. By convexity of R_i and R_j, the inequality $R_i(x_i + 1) - R_i(x_i) \geq R_j(x_j+1) - R_j(x_j)$ still holds after such an adjustment, so this process can be repeated until $x_i = m$ or $x_j = 0$. □

First we sort the projects by $R_j(m)$ in descending order in $O(n \log n)$ time. Following Lemma 4, when $k \bmod m = 0$, the largest revenue equals $\sum_{j=1}^{k/m} R_j(m)$ (trivial proof omitted). Assume $k \bmod m \neq 0$ in the following. In this case, there must one project that is allocated with $k \bmod m$ efforts.

Denote $q(k) = \lfloor \frac{k}{m} \rfloor$.

For $i \leq q(k)+1$, denote by $\mathbf{x}^{(i)}$ the profile that allocates $k \bmod m$ efforts to project i, and m efforts to each project $j \in [q+1] \setminus \{i\}$.

For $i > q(k)$, denote by $\mathbf{x}^{(i)}$ the profile that allocates $k \bmod m$ efforts to project i, and m efforts to each project $j \in [q]$.

Lemma 5. *One of $\mathbf{x}^{(1)}, \ldots, \mathbf{x}^{(n)}$ is optimal. (It is trivial. Proof omitted.)*

Denote by $Ans[k]$ the largest revenue of k-profile.

Denote by $Ans_1[k] = \max(\text{revenue of } x^{(i)} : i \leq q(k)+1\}$.

Denote by $Ans_2[k] = \max(\text{revenue of } x^{(i)} : i > q(k)+1\}$.

It follows from Lemma 5 that $Ans[k] = \max(Ans_1[k], Ans_2[k])$.

We show how we compute the array Ans_1 altogether in $O(nm)$ time in the following. The array Ans_2 can also be computed in $O(nm)$ time using similar idea (details omitted).

For each $c \in [m] \setminus 0$, we compute $Ans_1[k]$ for k congruent to c (modulo m) in $O(n)$ time as follows, and thus obtain Ans_1 in $O(nm)$ time.

Define $r_j = R_j(m) - R_j(c)$. According to the definition of $x^{(i)}$ for $i \leq q(k)+1$,

$$Ans_1[k] = \sum_{j=1}^{q(k)+1} R_j(m) - \min(r_1, \ldots, r_{q(k)+1}), \tag{6}$$

As k increases by m, quotient $q(k)$ increases by 1, and we can compute the term $\sum_{j=1}^{q(k)+1} R_j(m)$ and $\min(r_1, \ldots, r_{q(k)+1})$ both in $O(1)$ time. Therefore it takes $O(1)$ time for each k congruent to c.

Now we move on the problem that asks $Ans[k]$ for a certain k.

In this problem, we do not have to sort $R_j(m)$. Instead, we only need to find out the largest $q(k)+1$ items of $R_1(m), \ldots, R_n(m)$, which takes $O(n)$ time through the algorithm for finding the K-th largest number in an array. Therefore, we cut off the term $O(n \log n)$ for this easier problem.

5 An Alternative Algorithm for $m = 2$

In this section, we describe an algorithm for $m = 2$ which costs $O(n)$ time after sorting. It also solves the problem for all k $(0 \leq k \leq mn)$.

5.1 Preliminaries: Some Observations on (max,+) Convolution

Definition 3. *Given a function $g : [n] \to \mathbb{R}$, we say g is* oscillating concave, *if it satisfies the following properties:*
For any k,

(1) $g(2k) - g(2k-2) \geq g(2k+2) - g(2k)$ (namely, $g(2k)$ is concave);
(2) $g(2k+2) - g(2k+1) \geq g(2k+1) - g(2k)$;
(3) $g(2k+1) - g(2k) \leq g(2k) - g(2k-1)$;
(4) $g(2k+1) - g(2k)$ is decreasing for k;
(5) $g(2k) - g(2k-1)$ is decreasing for k.

Lemma 6. *Let $f : [n] \to \mathbb{R}$ be a concave function and $g : [n] \to \mathbb{R}$ an oscillating concave function. The (max,+) convolution of f and g:*

$$h(k) = \max_{1 \leq i \leq k} \left(f(i) + g(k-i)\right), 1 \leq k \leq n,$$

can be computed in $O(n)$ time.

Proof. We demonstrate that:
Observation 1. For any fixed k $(1 \leq k \leq n-1)$, let i_k be a maximum point of $f(i) + g(k-i)$. Then $f(i) + g(k+1-i)$ attains its maximum at a point in $\{i_k - 1, i_k, i_k + 1\}$.

The above observation indicates that we can compute $h(k+1)$ by $h(k)$ in $O(1)$ time. We prove it by contradiction. Let i_{k+1} denote a maximum point of $f(i) + g(k-i)$, either $i_{k+1} > i_k + 1$, or $i_{k+1} < i_k - 1$.

By the optimality of i_k and i_{k+1}, we derive

$$f(i_k) + g(k - i_k) \geq f(i_{k+1} - 1) + g(k - i_{k+1} + 1) \tag{7}$$

and

$$f(i_{k+1}) + g(k + 1 - i_{k+1}) > f(i_k + 1) + g(k - i_k). \tag{8}$$

Notice that the equation doesn't hold in (8) because $i_k + 1$ is not a maximum point of $f(i) + g(k+1-i)$.

Combining (7) and (8) we have $f(i_{k+1}) - f(i_{k+1} - 1) > f(i_k + 1) - f(i_k)$. By the concavity of f, this implies $i_{k+1} < i_k + 1$, which contradicts to $i_{k+1} > i_k + 1$. So we can assume $i_{k+1} < i_k - 1$.

By the optimality of i_k and i_{k+1}, we derive

$$f(i_k) + g(k - i_k) \geq f(i_{k+1}) + g(k - i_{k+1}), \tag{9}$$

and

$$f(i_{k+1}) + g(k+1-i_{k+1}) > f(i_k) + g(k+1-i_k). \tag{10}$$

Notice that the equation doesn't hold in (10) because i_k is not a maximum point of $f(i) + g(k+1-i)$.

Combining (9) and (10) we derive

$$g(k+1-i_{k+1}) - g(k-i_{k+1}) > g(k+1-i_k) - g(k-i_k). \tag{11}$$

Similarly, by the optimality of i_k and i_{k+1}, we derive

$$f(i_k) + g(k-i_k) \geq f(i_{k+1}+1) + g(k-i_{k+1}-1). \tag{12}$$

and

$$f(i_{k+1}) + g(k+1-i_{k+1}) > f(i_k-1) + g(k+2-i_k). \tag{13}$$

Notice that the equation in (13) doesn't hold because $i_k - 1$ is not a maximum point of $f(i) + g(k+1-i)$.

Combine (12) and (13) together we derive

$$\begin{aligned} f(i_k) - f(i_k-1) + g(k-i_k) - g(k+2-i_k) > \\ f(i_{k+1}+1) - f(i_{k+1}) + g(k-i_{k+1}-1) - g(k+1-i_{k+1}). \end{aligned} \tag{14}$$

By the concavity of f and $i_{k+1} < i_k - 1$, we have $f(i_k) - f(i_k-1) < f(i_{k+1}+1) - f(i_{k+1})$. Further by (14) we have

$$g(k-i_k) - g(k+2-i_k) > g(k-i_{k+1}-1) - g(k+1-i_{k+1}). \tag{15}$$

We will use (11) and (15) to derive contradiction. For convenience, let $x = k - i_k, y = k+1-i_{k+1}$, and (11) can be simplified as

$$g(y) - g(y-1) > g(x+1) - g(x), \tag{16}$$

(15) can be simplified as

$$g(y) - g(y-2) > g(x+2) - g(x). \tag{17}$$

By $i_{k+1} < i_k - 1$ we know $y > x + 2$.

Case 1 (x, y are both even). By (17) and Definition 3.1, we know $y \leq x + 2$, which leads to a contradiction.

Case 2 (x, y are both odd). By Definition 3.2, we have

$$g(x+1) - g(x) \geq \frac{g(x+1) - g(x-1)}{2}. \tag{18}$$

and

$$\frac{g(y+1) - g(y-1)}{2} \geq g(y) - g(y-1). \tag{19}$$

By Definition 3.1 and $y \geq x+3$ we have

$$\frac{g(x+1)-g(x-1)}{2} \geq \frac{g(y+1)-g(y-1)}{2}. \tag{20}$$

Combine (18), (19) and (20) we derive $g(x+1)-g(x) \geq g(y)-g(y-1)$, which contradicts to (16).

Case 3 (x is even, y is odd). By Definition 3.4 and $y \geq x+2$ we have $g(y)-g(y-1) \leq g(x+1)-g(x)$, which contradicts to (16).

Case 4 (x is odd, y is even). By Definition 3.5 and $y \geq x+2$ we have $g(y)-g(y-1) \leq g(x+1)-g(x)$, which contradicts to (16).

□

5.2 Algorithm for Finding Optimal Solutions Based on (max,+) Convolution

Suppose the projects are sorted by $R_j(2)$ in descending order. For convenience, let $a_j = R_j(1)$, and $b_j = R_j(2) - R_j(1)$.

Divide all projects into two groups A, B. Group $A = \{j \mid a_j > b_j\}$, and group $B = \{j \mid a_j \leq b_j\}$. The number of elements in A is denoted as $|A|$, and $|B|$ analogously.

Definition 4. *The maximal revenue of allocating k efforts to group A projects is denoted as $f(k)$.*
The maximal revenue of allocating k efforts to group B projects is denoted as $g(k)$.

The following lemma indicates how to compute f and g.

Lemma 7 (Calculate f, g).

1. *$f(k)$ = sum of the kth largest a_i, b_i, where $i \in A$.*
2. $g(k) = \begin{cases} \sum_{i=1}^{\frac{k}{2}} R_i(2), k \text{ is even,} \\ \max\left(g(k-1) + \max_{\frac{k+3}{2} \leq i \leq |B|} a_i, g(k+1) - \min_{1 \leq i \leq \frac{k+1}{2}} b_i\right), k \text{ is odd,} \end{cases}$

   where $i \in B$.

Proof. 1. Proof is evident.

2. Proof is evident when k is even.

When k is odd, we demonstrate that for projects in group B, there exists an optimal k-profile, such that a unique project is allocated with one effort.

We prove it by contradiction. Assume there are two projects $i, j \in B$ receiving one effort separately. Without loss of generality, suppose $a_i \leq a_j$, then we have $a_i \leq a_j \leq b_j$. We can remove one effort from i-th project and allocate it to j-th project, without decreasing the total revenue. □

Denote the maximal revenue of allocating k efforts to all projects as $h(k)$, then $h(k)$ can be written as (max,+) convolution of f and g as follows:

$$h(k) = \max_{0 \leq i \leq 2|A|, 0 \leq k-i \leq 2|B|} f(i) + g(k-i).$$

The following lemma together with Lemma 6 ensure that we can compute $h(k)(1 \leq k \leq 2n)$ in $O(n)$ time.

Lemma 8 (Properties of f, g).

(1) $f(k)$ is an convex function. (Proof is evident by Lemma 7.1.)
(2) $g(k)$ is an oscillating concave function. (Proof is non-trivial and omitted here due to the space limit. See [23] for more details.)

6 Summary

We revisit the classic resource allocation problem with a separable objective function under a single linear constraint. A regret-enabled greedy algorithm is designed that achieves $O(n \log n)$ time for $m = O(1)$, outperforming dynamic programming algorithm for small m. The new algorithm is practical especially for very small m, and its analysis is not over complicated (see Lemma 3).

For the special case where all the separated objective function R_j are convex, we present fast algorithms that cost $O(nm + n \log n)$ time (for all k) or $O(nm)$ time (for one given k). For the special case where $m = 2$, we show that the main algorithm only costs linear time $O(mn) = O(n)$, after a sorting process that costs $O(n \log n)$ time. It arises an open question what is the lower bound for this allocation problem (for $m = 2$ or $m = O(1)$).

A more interesting open question (suggested by one reviewer) is that can we solve this resource allocation problem in time $O(n \log n \cdot \mathsf{poly}(m))$ or even $O(n \log n + \mathsf{poly}(m))$?

References

1. Osman, M.S., Abo-Sinna, M.A., Mousa, A.A.: An effective genetic algorithm approach to multiobjective resource allocation problems (MORAPs). Appl. Math. Comput. **163**(2), 755–768 (2005)
2. Bitran, G.R., Tirupati, D.: Tradeoff curves, targeting and balancing in manufacturing queueing networks. Oper. Res. **37**(4), 547–564 (1989)
3. Bitran, G.R., Sarkar, D.: Targeting problems in manufacturing queueing networks-an iterative scheme and convergence. Eur. J. Oper. Res. **76**(3), 501–510 (1994)
4. Rajkumar, R., Lee, C., Lehoczky, J., Siewiorek, D.: A resource allocation model for QoS management. In: Proceedings Real-Time Systems Symposium, pp. 298–307. IEEE (1997)
5. Bretthauer, K.M., Ross, A., Shetty, B.: Nonlinear integer programming for optimal allocation in stratified sampling. Eur. J. Oper. Res. **116**(3), 667–680 (1999)

6. Katoh, N., Ibaraki, T.: Resource allocation problems. In: Du, D.-Z., Pardalos, P.M. (eds.) Handbook of Combinatorial Optimization, pp. 905–1006. Springer US, Boston, MA (1999). https://doi.org/10.1007/978-1-4613-0303-9_14
7. Fox, B.: Discrete optimization via marginal analysis. Manage. Sci. **13**(3), 210–216 (1966). https://doi.org/10.1287/mnsc.13.3.210
8. Shih, W.: A new application of incremental analysis in resource allocations. J. Oper. Res. Soc. **25**(4), 587–597 (1974)
9. Federgruen, A., Groenevelt, H.: The greedy procedure for resource allocation problems: necessary and sufficient conditions for optimality. Oper. Res. **34**(6), 909–918 (1986)
10. Murota, K.: Discrete convex analysis. Math. Program. **83**, 313–371 (1998)
11. Shioura, A.: Minimization of an M-convex function. Discret. Appl. Math. **84**(1–3), 215–220 (1998)
12. Bretthauer, K.M., Shetty, B.: The nonlinear resource allocation problem. Oper. Res. **43**(4), 670–683 (1995)
13. Devanur, N.R., Jain, K., Sivan, B., Wilkens, C.A.: Near optimal online algorithms and fast approximation algorithms for resource allocation problems. In: Proceedings of the 12th ACM Conference on Electronic Commerce, pp. 29–38 (2011)
14. Deng, Z., Liang, S., Hong, Y.: Distributed continuous-time algorithms for resource allocation problems over weight-balanced digraphs. IEEE Trans. Cybern. **48**(11), 3116–3125 (2017)
15. Cygan, M., Mucha, M., Wegrzycki, K., Włodarczyk, M.: On problems equivalent to (min,+)-convolution. ACM Trans. Algorithms (TALG) **15**(1), 1–25 (2019)
16. Bremner, D., et al.: Necklaces, convolutions, and x+ y. In: Azar, Y., Erlebach, T. (eds) Algorithms–ESA 2006: 14th Annual European Symposium, Zurich, Switzerland, September 11-13, 2006. Proceedings 14, pp. 160–171. Springer, Berlin, Heidelberg (2006). https://doi.org/10.1007/11841036_17
17. Bussieck, M., Hassler, H., Woeginger, G.J., Zimmermann, U.T.: Fast algorithms for the maximum convolution problem. Oper. Res. Lett. **15**(3), 133–141 (1994)
18. Chan, T.M., Lewenstein, M.: Clustered integer 3SUM via additive combinatorics. In: Proceedings of the Forty-Seventh Annual ACM Symposium on Theory of Computing, pp. 31–40 (2015)
19. Chi, S., Duan, R., Xie, T., Zhang, T.: Faster min-plus product for monotone instances. In: Proceedings of the 54th Annual ACM SIGACT Symposium on Theory of Computing, pp. 1529–1542 (2022)
20. Bringmann, K., Cassis, A.: Faster 0-1-knapsack via near-convex min-plus-convolution. arXiv preprint arXiv:2305.01593 (2023)
21. Pisinger, D.: Linear time algorithms for knapsack problems with bounded weights. J. Algorithms **33**(1), 1–14 (1999)
22. Bringmann, K.: Knapsack with small items in near-quadratic time. In: Proceedings of the 56th Annual ACM Symposium on Theory of Computing, pp. 259–270 (2024)
23. Cao, S., Zhu, T., Jin, K.: Discrete effort distribution via regrettable greedy algorithm. arXiv preprint arXiv:2503.11107 (2025)

Improving Local Search for Weighted Partial MaxSAT by Initializing with Historical Information

Menghua Jiang[1], Rui Zhang[1], and Yin Chen[1,2(✉)]

[1] School of Computer Science, South China Normal University, Guangzhou, China
{jiangmenghua,2024023571}@m.scnu.edu.cn, ychen@scnu.edu.cn
[2] School of Artificial Intelligence, South China Normal University, Foshan, China

Abstract. Partial Maximum Satisfiability (PMS) is a generalisation of the well-known Maximum Satisfiability (MaxSAT), incorporating both hard and soft clauses. Weighted Partial Maximum Satisfiability (WPMS) further extends PMS by associating each soft clause with a positive integer weight. WPMS is particularly significant in practical applications, as it can encode numerous industrial optimisation problems involving hard constraints and soft constraints with varying priorities. Stochastic local search (SLS) algorithms have been extensively studied for solving WPMS, which has achieved significant advancements in recent years. In this work, we identify two issues in current SLS solvers and propose a corresponding solution. Firstly, we observe that current SLS solvers typically employ a fixed initialisation procedure at the start of each local search round, which may restrict the diversity of search directions. Secondly, current SLS solvers often fail to effectively utilise historical information. To address these issues, we propose a novel clause initialisation method that dynamically adjusts the weights of soft clauses based on both the current search state and historical information. Based on this method, we develop a new SLS solver for WPMS named HistLS. Extensive experiments on WPMS benchmarks from the incomplete track of MaxSAT Evaluations (MSEs) of the five recent years demonstrate that HistLS outperforms state-of-the-art SLS solvers.

Keywords: Weighted partial maximum satisfiability · Local search · Clause initialisation · Historical information

1 Introduction

Maximum Satisfiability (MaxSAT) problem is an optimization variant of the famous NP-complete Boolean Satisfiability (SAT) problem, which holds significant importance in various fields of computer science and operations optimization. Given a propositional formula in Conjunctive Normal Form (CNF), MaxSAT aims to find an assignment that maximizes the number of satisfied

The code is available at https://github.com/jmhmaxsat/HistLS.

F. V. Fomin and M. Xiao (Eds.): COCOON 2025, LNCS 15983, pp. 123–135, 2026.
https://doi.org/10.1007/978-981-95-0215-8_10

clauses in the CNF formula. Partial Maximum Satisfiability (PMS) problem extends MaxSAT by dividing clauses into hard and soft categories, seeking an assignment that satisfies all hard clauses while maximizing the number of satisfied soft clauses. Weighted Partial Maximum Satisfiability (WPMS) problem, an important generalization of MaxSAT, further extends PMS by associating a positive integer weight with each soft clause. Its goal is to satisfy all hard clauses while maximizing the total weight of the satisfied soft clauses. In practical scenarios, many real-world combinatorial problems involve both hard and soft constraints, where the soft constraints often have varying priorities. WPMS provides a more flexible encoding for such problems compared to SAT, MaxSAT, or PMS. WPMS has been successfully applied to solving various real-world problems across diverse domains, including planning [1], timetabling [2], routing [3] and combinatorial testing [4].

Algorithms for solving WPMS fall into two categories based on their ability to guarantee optimality: complete algorithms and incomplete algorithms. Complete algorithms either find and prove an optimal solution or establish that no such solution exists. Examples include branch-and-bound algorithms [5] and certain methods that iteratively call SAT solvers, commonly known as SAT-based algorithms [6]. SAT-based algorithms were first introduced for solving PMS in [7] and later extended to WPMS in [8]. These algorithms are well known for their strong performance on industrial instances, with significant recent efforts dedicated to their improvement [9–11].

Incomplete algorithms for solving WPMS are mainly stochastic local search (SLS) algorithms [12], which are very effective for many industrial problems with large instance sizes, high time costs, or limited global information. The objective of these problems is not to guarantee the optimality of the solutions but to find good-quality solutions within a reasonable time. Recently, an increasing number of techniques have been developed to improve SLS algorithms, including decimation-based initial solution generation methods [13], variable selection strategies [14–16] to guide the search direction, and a range of clause weighting schemes [17–20] designed to overcome local optima. The rapid advancement of these techniques has enabled SLS algorithms to achieve a level of industrial performance comparable to that of complete algorithms. In recent years, SLS algorithms have dominated all four categories in the incomplete track of MaxSAT Evaluations (MSEs)[1].

In this work, we focus on clause initialization methods, a relatively underexplored area distinct from the commonly studied techniques for improving local search. The initial values of soft clause weights play a critical role in the performance of SLS solvers for WPMS, as demonstrated in NuWLS [18]. However, existing SLS solvers typically initialize these weights uniformly at a fixed value before at the start of each local search round. This fixed initialization limits the diversity of search directions, often causing the algorithm to repeatedly encounter similar local optima from previous searches, ultimately reducing overall performance.

[1] https://maxsat-evaluations.github.io/.

In addition, current SLS solvers for WPMS tend to rely solely on the state of the best solution found so far when setting the initial weights of soft clauses, neglecting the historical search process. As a result, they fail to leverage information about how difficult certain soft clauses were to satisfy in past searches. This oversight prevents the algorithm from effectively distinguishing the relative importance or difficulty of soft clauses, reducing the effectiveness of weight adjustments.

To address these issues, we propose *hist_weight*(), which utilizes historical information to dynamically initialize soft clause weights. Specifically, during the local search process, whenever the SLS solvers updates the current solution to the best solution found so far, *hist_weight*() records the falsified soft clauses under this solution. At the end of the current local search round, it analyzes and identifies the soft clauses that remain falsified throughout the entire round. In the next round of local search, these consistently falsified soft clauses are assigned higher initial weights to increase their likelihood of being satisfied in the early stages of the search. This, in turn, helps guide the search towards a more optimal solution space.

Based on this method, we develop a new SLS solver named HistLS and compare it with state-of-the-art SLS solvers on WPMS benchmarks from the incomplete track of MSEs of the five recent years. The experimental results demonstrate that HistLS outperforms the current leading SLS solvers.

The remainder of this paper is organized as follows: Sect. 2 outlines the preliminary concepts, Sect. 3 describes the proposed HistLS solver in detail, Sect. 4 presents an evaluation of the performance of the proposed method, and finally, Sect. 5 concludes the paper with a summary.

2 Preliminaries

Given a set $V = \{x_1, x_2, \ldots, x_n\}$ of n Boolean variables, a literal is either a variable x_i (a positive literal) or its negation $\neg x_i$ (a negative literal). A positive literal x_i is satisfied when x_i is *True*, whereas a negative literal $\neg x_i$ is satisfied when x_i is *False*. A clause is a disjunction of literals, i.e., $c_i = l_{i1} \vee l_{i2} \vee \ldots \vee l_{ik}$, where k represents the number of literals in clause c_i. A clause is satisfied if at least one of its literals is satisfied; otherwise, it remains falsified. A propositional formula F in Conjunctive Normal Form (CNF) is a conjunction of all its clauses, i.e., $F = c_1 \wedge c_2 \wedge \ldots \wedge c_m$, where m is the total number of clauses in F. The set of variables in a formula F is denoted by $V(F)$. An assignment, represented as $\alpha : V(F) \rightarrow \{0, 1\}$, maps variables to Boolean values, where 1 indicates *True* and 0 indicates *False*. When α assigns values to all variables in $V(F)$, it is called a complete assignment.

Given a CNF formula F, the goal of MaxSAT is to find a complete assignment that satisfies the maximum number of clauses in F. PMS, a variant of MaxSAT, divides clauses into hard and soft, aiming to find a complete assignment that satisfies all hard clauses while maximizing the satisfaction of soft clauses. WPMS, an extension of PMS, introduces weights for each soft clause and seeks to find a complete assignment that satisfies all hard clauses while maximizing the total weight of the satisfied soft clauses. Obviously, PMS can be viewed as a special case of WPMS, where each soft clause has a weight of one.

Given a CNF formula F and a complete assignment α, we say that α is a solution to F. The solution α is feasible if it satisfies all hard clauses in F. The cost of a feasible solution α, denoted $cost(\alpha)$, is equal to the number of falsified soft clauses in PMS and the total weight of falsified soft clauses in WPMS under α. If α is infeasible, its cost is set to $+\infty$ for convenience. Additionally, a solution α_1 is better than another solution α_2, if $cost(\alpha_1) < cost(\alpha_2)$.

In SLS solvers for WPMS, flipping a variable involves changing its Boolean value from *True* to *False* or vice versa. For a variable x, the increase score of x, denoted by $score_{increase}(x)$, is the total weight of falsified clauses that would become satisfied if x is flipped. The decrease score of x, denoted by $score_{decrease}(x)$, is the total weight of satisfied clauses that would become falsified if x is flipped. The score of x, denoted by $score(x)$, represents the net change in the total weight of satisfied clauses if x is flipped, and can be understood as: $score(x) = score_{increase}(x) - score_{decrease}(x)$. If $score(x) > 0$, x is considered a good variable, as flipping x is likely to produce a positive net change in the total weight of satisfied clauses, making it a beneficial move in the search process.

Clause weighting techniques are commonly used in SLS solvers for WPMS, where each clause maintains two types of weights. The clause original weight, $w_{org}(c)$, represents the weight assigned to clause c in the CNF formula and is used to calculate the cost. The clause weight, $w(c)$, is dynamically adjusted using a clause weighting scheme and is used to compute each variable's score.

To enhance performance, SLS solvers typically employ a restart strategy when the current solution fails to improve after a certain number of steps. This strategy not only reinitializes the current assignment but also resets clause weights. At the beginning of each local search round, SLS solvers must initialize $w(c)$ for every clause c, where $w_{init}(c)$ represents its initial weight. Additionally, avg_{orgw} denotes the average original weight of soft clauses. Properly initializing $w_{init}(c)$ is crucial for effectively guiding the search process and enhancing solver performance.

3 Methodology

In this section, we first analyze the key questions that must be addressed when designing a clause initialization method. Based on this analysis, we propose a dynamic initialization method for soft clauses using historical information, called *hist_weight*(), which provides a simple yet effective solution to these questions. Building on *hist_weight*(), we develop a new SLS solver for WPMS named HistLS.

3.1 Analysis of Clause Initialisation Method

During the restart of an SLS solver, both the current assignment and the weights assigned to each clause are reinitialized. Note that these weights are used to compute variable scores, whereas the clause original weights are used to evaluate the cost of a solution. The initialization of clause weights is critical for determining the initial search direction and directly influences the search landscape. Therefore, when designing a clause initialisation method, the following key questions must be addressed:

- How should the initial weights of hard and soft clauses be set in the first round of local search?
- After the first round of local search, what should the initial weights of hard and soft clauses be? What is the relationship between their magnitudes?
- After the first round of local search, should the initial clause weights remain the same in each subsequent round? If not, how should they be set, and on what basis?

Different answers to these questions can significantly impact the performance of an SLS solver. Existing clause initialization methods typically adopt one of the following two strategies to address the first two questions: (1) The first strategy sets the initial weight of hard clauses to 1 and the initial weight of soft clauses to 0 in every round of local search [19]. (2) The second strategy sets the initial weight of hard clauses to 1 and the initial weight of soft clauses to 0 in the first round, while in subsequent rounds, the initial weight of hard clauses remains 1, and the initial weight of each soft clause c is set to $\frac{w_{org}(c)}{avg_{orgw}}$ [18,20].

However, when an SLS solver fails to improve the current solution after a certain number of steps, the core purpose of the restart strategy is to disrupt the current search landscape and prevent the algorithm from repeatedly exploring the same paths. Existing methods initialize each soft clause in the same way and set its initial weight to a fixed value after the first search round. This approach causes the algorithm to follow a similar search pattern after each restart, making it susceptible to repeatedly getting trapped in similar local optima, thereby limiting search diversity and efficiency.

In the following subsection, we propose a novel clause initialization method that offers a simple yet effective solution to the three questions mentioned above.

3.2 A Novel Clause Weight Initialization Method

The brief process of *hist_weight*() is illustrated in the figure1. This method interleaves initialization and local search, enabling the transfer of historical information between the two phases. *hist_weight*() operates in consecutive rounds, where each round initializes both hard and soft clauses before passing them to the subsequent local search phase for further optimization. Additionally, this method introduces a feedback mechanism that provides useful historical information extracted from the local search to guide the initialization in the next round. Specifically, the feedback consists of soft clauses that remained falsified during the previous local search when improvements to the best solution were made. *hist_weight*() operates as follows:

- For each hard clause c: set $w_{init}(c)$ to 1.

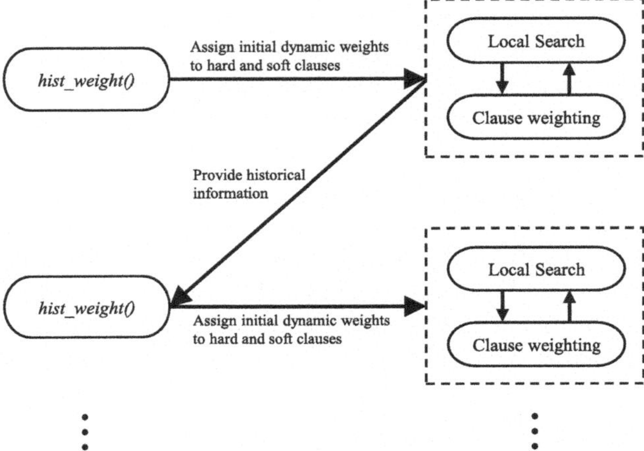

Fig. 1. The brief process of *hist_weight*().

- For each soft clause c: If the solver found a feasible solution in the previous local search round, then for the soft clauses that remained falsified during each improvement of the best solution in that round, set $w_{init}(c)$ to 1. Set $w_{init}(c)$ to 0 for all other soft clauses. Otherwise, set $w_{init}(c)$ to 0 for all soft clauses.

hist_weight() consists of two phases: historical information collection and dynamic initialization of clause weights. In the historical information collection phase, *hist_weight*() initially marks each soft clause c as a "difficult soft clause" before the SLS solver begins execution. During the local search process, whenever the solver finds a better solution, *hist_weight*() removes this mark from the soft clauses that are satisfied by the improved solution. This process continues until the local search round concludes.

In the clause weight initialization phase, *hist_weight*() sets the initial weight of all hard clauses to 1. For initializing the weights of soft clauses, if the best solution found so far or the local optimal solution from the previous local search round is infeasible, all soft clauses are initialized with a weight of 0. Otherwise, the "difficult soft clause" are assigned an initial weight of 1, while the remaining soft clauses are assigned a weight of 0. Once the initialization is complete, *hist_weight*() reassigns the "difficult soft clause" mark to each soft clause for use in the next round of historical information collection.

hist_weight() possesses the following characteristics:

- **Utilization of historical information for weight initialization.** By leveraging historical information accumulated from the previous search round, *hist_weight*() exhibits multiple advantages. Firstly, it expands the exploration scope of the solution space, enabling the solver to start from different states and effectively avoid repeatedly falling into the same local optimum, thereby enhancing search path diversity. Secondly, *hist_weight*() dynamically adjusts the initialization strategy based on feedback from the search process, enhancing its adaptability to different problem instances by adjusting clause priorities. Lastly, the *hist_weight*() strategy precisely handles "difficult soft clauses" by assigning them higher initial weights based on historical information, allowing the solver to focus on critical areas, reducing ineffective exploration.

- **Delayed activation strategy.** Specifically, *hist_weight*() applies dynamic initialization of soft clauses only when both the best solution found so far and the local optimal solution from the previous local search round are feasible. The rationale behind this design is that if the previous search fails to produce a feasible solution, *hist_weight*() lacks sufficient historical information to reasonably assess the difficulty of satisfying the soft clauses. In such cases, enforcing dynamic initialization may introduce noise and adversely affect subsequent search performance.
- **Integration with the clause weighting scheme Unified-SW in USW-LS** [19]. Since Unified-SW increases the weights of soft clauses only at feasible local optima, *hist_weight*() initializes hard clause weights to 1 and soft clause weights to 0 in the first round of local search. At this stage, the SLS solver focuses solely on satisfying all hard clauses, effectively transforming the problem into a SAT problem. By prioritizing hard clauses, the algorithm quickly identifies a feasible solution. Once a feasible solution is found, *hist_weight*() assigns equal initial weights to hard clauses and "difficult soft clauses", while setting the initial weights of other soft clauses to 0. This approach ensures that solver prioritizes "difficult soft clauses" without excessively focusing on soft clauses at the expense of hard clauses, thereby maintaining efficiency in finding feasible solutions.

3.3 The HistLS Algorithm

Based on the above *hist_weight*(), we develop a new SLS solver named HistLS. We outline the pseudo-code of HistLS in Algorithm 1 and describe the main process as follows. We denote the best solution found so far and its cost as α^* and $cost^*$, respectively. Throughout the local search, α represents the current assignment. Initially, α^* is empty, and $cost^*$ is initialized to $+\infty$.

HistLS iteratively calls the local search process until the *cutoff time* is reached (lines 2–20). In each round of local search, HistLS first generates an initial complete assignment α using UP-Decimation [17] (line 3) and initializes the clause weights using *hist_weight*() (line 4), and adds all variables x with $score(x) > 0$ to the set D (line 5). After initialization, HistLS proceeds with the search steps (lines 7–20). During the local search, if the solver discovers a solution with a lower cost than $cost^*$, then α^* and $cost^*$ are updated accordingly.

In each local search step, HistLS selects a variable and flips it to improve the current solution. When a local optimum is not reached, HistLS employs the BMS (Best from Multiple Selections) sampling strategy [21] to choose variables for flipping. This strategy is crucial because iterating through all elements in

Algorithm 1: HistLS

```
   Input: A WPMS instance F, cutoff time
   Output: The best solution found and its cost, or "No feasible solution found"
1  α* := ∅; cost* := +∞;
2  while running time < cutoff time do
3      α := an initial complete assignment by UP-Decimation;
4      hist_weight() initializes the weight of each clause under α;
5      Add all variables x with score(x) > 0 into the set D;
6      L := 10000000;
7      for step = 0; step < L; step++ do
8          if α is feasible and cost(α) < cost* then
9              α* := α; cost* := cost(α);
10             L := step + 10000000;
11         if D := {x | score(x) > 0} ≠ ∅ then
12             v := a variable in D selected by BMS strategy;
13         else
14             update clause weights by Unified-SW;
15             if ∃ falsified hard clauses then
16                 c := a random falsified hard clause;
17             else
18                 c := a random falsified soft clause;
19             v := the variable with highest score in c;
20         α := α with v flipped;
21 if α* ≠ ∅ then return α* and cost*;
22 else return "No feasible solution found";
```

the set D can be highly time-consuming; thus, BMS effectively streamlines the optimization process. Specifically, BMS randomly selects t variables from D (t is a parameter in BMS) and chooses the one with the highest score (lines 11–12). When D is empty, indicating that the search has reached a local optimum, HistLS updates the clause weights according to Unified-SW (line 14). Subsequently, it randomly selects a falsified clause c and selects the variable with the highest score from this chosen clause (lines 15–19).

Finally, if a feasible solution with cost $cost^*$ is found when the *cutoff time* is exceeded, HistLS reports α^* and $cost^*$; otherwise, it reports "No feasible solution found".

4 Experiments

In this section, we first introduce the benchmarks, competitors, parameter settings, and experimental setup used in our experiments. We then present the results from experiments conducted on a broad range of WPMS benchmarks to evaluate the efficiency of our HistLS algorithm. Finally, we perform ablation studies to validate the components in HistLS.

4.1 Experimental Preliminaries

Benchmarks. We evaluate HistLS on all WPMS benchmarks from the incomplete track of the five recent MSEs[2], denoted as WPMS_2020 (253 instances), WPMS_2021 (151 instances), WPMS_2022 (197 instances), WPMS_2023 (160 instances), and WPMS_2024 (229 instances). The benchmark instances mainly originate from classic graph theory problems (e.g., graph coloring, minimum vertex cover, and set covering) as well as real-world industrial applications (e.g., route planning, formal verification, and logistics optimization).

State-of-the-Art SLS Competitors HistLS is compared against seven state-of-the-art SLS solvers, with their source codes publicly available at:[3], [4], [5]. The parameter settings follow those specified in their respective papers.

- SATLike3.0 [17]: weighted runner-up in MSE 2020 and MSE 2021.
- BandMaxSAT [14]: weighted third place in MSE 2022.
- MaxFPS [15]: weighted third place in MSE 2022.
- NuWLS [18]: champion in all categories of MSE 2022.
- USW-LS [19]: champion in all categories of MSE 2023.
- SPB-MaxSAT [20]: champion in all categories of MSE 2024.
- NuWLS-BandHS [16]: a recently proposed SLS solver.

Parameter Settings. HistLS has two parameters: (1) h_{inc}, the increment of hard clause weights in the clause weighting scheme, and (2) t, the number of samples used in the BMS strategy. The parameter settings are as follows: $h_{inc} = 1$ and $t = 96$. These parameters are also used in the baseline solver USW-LS, and no additional tuning was performed.

Experimental Setup. HistLS is implemented in the programming language C++ and compiled using the g++ compiler with the -O3 optimization option. To ensure the fairness and validity of the experiments, both HistLS and its competitors are tested on a unified server with the same experimental environment. The server used in this study features an Intel Xeon Gold 6230 CPU @ 2.10GHz with 80 cores and 377GB of RAM, running the Ubuntu 22.04 Linux operating system.

In our experiments, we used the same evaluation methodology as employed in the incomplete track of MSEs, which is described as follows. Each solver performs one run on each instance. The *cutoff time* is set to 60 and 300 CPU seconds for each run, following the rules of the incomplete track of MSEs. We used the metrics "*#win*" and "*#score*" to evaluate performance. The "*#win*"

[2] https://maxsat-evaluations.github.io.
[3] http://lcs.ios.ac.cn/~caisw/Code/maxsat.
[4] https://github.com/JHL-HUST.
[5] https://github.com/filyouzicha.

metric indicates instances where the solver finds the best solution among all competing solvers in the same table, a widely used metric in comparing SLS solvers [14–17,20].

The “#*score*” metric used in the incomplete track of MSEs computes the $score(solver, i)$ for each instance i as follows: if the solver fails to find a feasible solution, its *score* is 0; otherwise, the *score* is given by $score(solver, i) = \frac{cost_{best}+1}{cost(\alpha^*)+1}$, where α^* denotes the feasible solution output by the solver, $cost_{best}$ is the lowest cost obtained for that instance among all solvers, and $cost(\alpha^*)$ is the cost of solution α^*. The overall *score* of a solver on a benchmark is the average *score* across all instances it solves, denoted by “#*score*”. The best results are highlighted in bold in the tables.

4.2 Comparison with SLS Competitors

Table 1. Comparison of HistLS with baseline solvers under two time limits of 60 s and 300 s.

Solvers	Metric	WPMS_2020 (253)		WPMS_2021 (151)		WPMS_2022 (197)		WPMS_2023 (160)		WPMS_2024 (229)	
		60 s	300 s	60 s	300 s	60 s	300 s	60 s	300 s	60 s	300 s
SATLike3.0l [17]	#*win*	28	37	20	27	17	18	9	9	24	29
	#*score*	0.6241	0.6472	0.5909	0.6135	0.5875	0.6029	0.5855	0.5895	0.6626	0.6653
BandMaxSAT [14]	#*win*	38	49	24	26	24	27	18	22	26	34
	#*score*	0.6431	0.6679	0.5950	0.6098	0.6202	0.6327	0.5986	0.6007	0.6685	0.6829
MaxFPS [15]	#*win*	43	47	17	20	10	17	9	8	23	25
	#*score*	0.6722	0.7021	0.5940	0.6358	0.6171	0.6312	0.6245	0.6220	0.6747	0.6827
NuWLS [18]	#*win*	65	69	25	23	36	44	24	24	48	65
	#*score*	0.7433	0.7635	0.6580	0.6731	0.7018	0.7066	0.6489	0.6596	0.7373	0.7475
USW-LS [19]	#*win*	94	101	36	**37**	56	64	54	**52**	88	91
	#*score*	0.8048	0.8207	0.7328	0.7372	0.7501	0.7582	0.7081	0.7170	0.8003	0.7988
SPB-MaxSAT [20]	#*win*	95	103	39	**37**	58	63	48	48	**90**	**94**
	#*score*	0.8135	0.8220	0.7318	0.7322	0.7590	0.7576	**0.7390**	**0.7454**	0.7856	0.7898
NuWLS-BandHS [16]	#*win*	64	71	29	31	40	43	28	31	50	64
	#*score*	0.7432	0.7688	0.6616	0.6748	0.6951	0.7027	0.6450	0.6573	0.7301	0.7426
HistLS	#*win*	**100**	**111**	**43**	**37**	**67**	**65**	**59**	48	86	90
	#*score*	**0.8183**	**0.8305**	**0.7480**	**0.7560**	**0.7736**	**0.7805**	0.7246	0.7215	**0.8174**	**0.8213**

The comparisons of HistLS with its competitors are summarized in Table 1. The results clearly demonstrate that HistLS consistently outperforms all seven SLS baselines across WPMS benchmarks, regardless of whether the time limit is set to 60 CPU seconds (60 s) or 300 CPU seconds (300 s). Specifically, with a 60 s cutoff, HistLS surpasses the second-best solver by 5.26% to 15.52% in terms of “#*win*” and by 0.59% to 4.05% in terms of “#*score*”. When the time limit is extended to 300 s, HistLS still maintains a competitive edge, outperforming the second-ranked solver by 1.56% to 7.77% in “#*win*” and by 1.03% to 3.99% in “#*score*”. Notably, HistLS achieves significantly higher “#*score*” values across the vast majority of benchmarks. This metric serves as the sole criterion for evaluating solver performance in the incomplete track of MSEs, further underscoring the effectiveness of our approach.

However, HistLS performs poorly on a very small number of benchmarks, such as in WPMS_2023, where the "#*score*" metric ranks second, falling short of SPB-MaxSAT. This indicates that HistLS exhibits slight deficiencies in performance on certain specific instances, which may be due to the characteristics of some instance families with lower structural complexity. These instances may have a small gap between the local and global optimal solutions, leading HistLS to be misled during exploration and consequently learn harmful historical information, resulting in a deviation from the local optimal state.

In summary, the proposed method combines our *hist_weight*() with the clause weighting scheme Unified-SW, helping HistLS outperform the state-of-the-art SLS solvers.

4.3 Ablation Study

Table 2. Comparison of HistLS with HistLS-alt1 and HistLS-alt2 under two time limits of 60 s and 300 s.

Solvers	Metric	WPMS_2020 (253)		WPMS_2021 (151)		WPMS_2022 (197)		WPMS_2023 (160)		WPMS_2024 (229)	
		60 s	300 s	60 s	300 s	60 s	300 s	60 s	300 s	60 s	300 s
HistLS-alt1	#*win*	**136**	121	67	51	83	75	70	73	120	116
	#*score*	0.8402	0.8339	0.7437	0.7381	0.7502	0.7457	0.7382	0.7374	0.8234	0.8192
HistLS-alt2	#*win*	132	125	69	64	95	90	**82**	76	123	110
	#*score*	0.8369	0.8377	0.7576	0.7578	0.7704	0.7716	0.7491	0.7530	0.8273	0.8237
HistLS	#*win*	130	**139**	**78**	**79**	**98**	**102**	81	**86**	**133**	**131**
	#*score*	**0.8447**	**0.8452**	**0.7740**	**0.7749**	**0.7879**	**0.7915**	**0.7534**	**0.7561**	**0.8361**	**0.8366**

To analyze the effectiveness and rationale of the proposed *hist_weight*() method in this paper, we compared HistLS with its two variants: HistLS-alt1 and HistLS-alt2. The first variant replaces the condition for updating soft clause weights in Unified-SW with SPB-Weighting, which is used in the current state-of-the-art SLS solver, SPB-MaxSAT [20]. This approach does not distinguish between feasible and infeasible local optima. The second variant, HistLS-alt2, sets the $w_{init}(c)$ of each "difficult soft clause" c to $\frac{w_{org}(c)}{avg_{orgw}}$, as done in these baseline solvers [18,20].

The comparison results of HistLS with these two variants are summarized in Table 2. HistLS outperforms HistLS-alt1 on almost all benchmarks, indicating the effectiveness of our dynamically initialized strategy, which is specifically designed for the characteristics of the Unified-SW structure. Furthermore, HistLS outperforms HistLS-alt2 as well, demonstrating the effectiveness of our clause weight initialization settings.

5 Conclusion

This paper addresses the limitations of existing SLS solvers in initialization by proposing a solution that utilizes historical information from local search for dynamic clause weight initialization. Based on this method, we developed a new SLS solver called HistLS. Extensive experiments demonstrate that HistLS outperforms current state-of-the-art SLS solvers. This research provides a more efficient tool for solving WPMS and offers new insights for improving SLS solvers.

Acknowledgments. This work was supported by the Scientific Research Innovation Project of Graduate School of South China Normal University (Grant No. 2024KYLX091).

References

1. Bonet, B., Frances, G., Geffner, H.: Learning features and abstract actions for computing generalized plans. In: Proceedings of AAAI, pp. 2703–2710 (2019)
2. Lemos, A., Monteiro, P.T., Lynce, I.: Minimal perturbation in university timetabling with maximum satisfiability. In: Proceedings of CPAIOR, pp. 317–333 (2020)
3. Khadilkar, H.: Solving the capacitated vehicle routing problem with timing windows using rollouts and max-sat. In: Proceedings of ICC, pp. 1–6 (2022)
4. Ansótegui, C., Manyà, F., Ojeda, J., Salvia, J.M., Torres, E.: Incomplete MaxSAT approaches for combinatorial testing. J. Heuristics **28**, 377–431 (2022)
5. Li, C.-M., Xu, Z., Coll, J., Manyà, F., Habet, D., He, K.: Combining clause learning and branch and bound for MaxSAT. In: Proceedings of CP, pp. 38–1 (2021)
6. Bacchus, F., Järvisalo, M., Martins, R.: Maximum satisfiabiliy. In: Handbook of Satisfiability. IOS Press, pp. 929–991 (2021)
7. Fu, Z., Malik, S.: On solving the partial max-sat problem. In: Proceedings of SAT, pp. 252–265 (2006)
8. Ansótegui, C., Bonet, M.L., Levy, J.: Solving (weighted) partial MaxSAT through satisfiability testing. In: Proceedings of SAT, pp. 427–440 (2009)
9. Ansótegui, C., Gabàs, J.: WPM3: an (in) complete algorithm for weighted partial MaxSAT. Artif. Intell. **250**, 37–57 (2017)
10. Berg, J., Demirović, E., Stuckey, P.J.: Core-boosted linear search for incomplete maxSAT. In: Proceedings of CPAIOR, pp. 39–56 (2019)
11. Nadel, A.: Anytime weighted MaxSAT with improved polarity selection and bit-vector optimization. In: Proceedings of FMCAD, pp. 193–202 (2019)
12. Cha, B., Iwama, K., Kambayashi, Y., Miyazaki, S.: Local search algorithms for partial MaxSAT. In: Proceedings of AAAI, pp. 263–268 (1997)
13. Cai, S., Luo, C., Zhang, H.: From decimation to local search and back: a new approach to MaxSAT. In: Proceedings of IJCAI, pp. 571–577 (2017)
14. Zheng, J., He, K., Zhou, J., Jin, Y., Li, C.-M., Manya, F.: BandMaxSAT: a local search MaxSAT solver with multi-armed bandit. In: Proceedings of IJCAI, pp. 1901–1907 (2022)
15. Zheng, J., He, K., Zhou, J.: Farsighted probabilistic sampling: a general strategy for boosting local search MaxSAT solvers. In: Proceedings of AAAI, pp. 4132–4139 (2023)
16. Zheng, J., He, K., Zhou, J., Jin, Y., Li, C.-M., Manyà, F.: Integrating multi-armed bandit with local search for MaxSAT. Artif. Intell. **338**, 104242 (2025)
17. Cai, S., Lei, Z.: Old techniques in new ways: clause weighting, unit propagation and hybridization for maximum satisfiability. Artif. Intell. **287**, 103354 (2020)

18. Chu, Y., Cai, S., Luo, C.: NuWLS: improving local search for (weighted) partial MaxSAT by new weighting techniques. In: Proceedings of AAAI, pp. 3915–3923 (2023)
19. Chu, Y., Li, C.-M., Ye, F., Cai, S.: Enhancing MaxSAT local search via a unified soft clause weighting scheme. In: Proceedings of SAT, pp. 8–1 (2024)
20. Zheng, J., Chen, Z., Li, C.-M., He, K.: Rethinking the soft conflict pseudo Boolean constraint on MaxSAT local search solvers. In: Proceedings of IJCAI, pp. 1989–1997 (2024)
21. Cai, S.: Balance between complexity and quality: local search for minimum vertex cover in massive graphs. In: Proceedings of IJCAI, pp. 747–753 (2015)

Regularized Submodular Maximization over Integer Lattice

Zhicheng Liu, Yang Lv, Yapu Zhang(✉), and Zhenning Zhang

Institute of Operations Research and Information Engineering, Beijing University of Technology, Beijing 100124, People's Republic of China
{manlzhic,lvyang}@emails.bjut.edu.cn,
{zhangyapu,zhangzhenning}@bjut.edu.cn

Abstract. In this paper, we delve deeply into the problem of regularized submodular maximization over the integer lattice. Our objective function, $f - c$, is simply the difference between a non-negative monotone submodular function f and a non-negative modular function c. While this problem has gained much attention in the set scenario recently, we broaden our focus to include the integer lattice. Our main contribution is an efficient algorithm for this problem, backed by strong approximation guarantees. We also test our algorithm in the real-world application of D-optimal design. To ensure fair comparisons, we created a greedy algorithm and calculated its approximation guarantees. The results show that our algorithm performs remarkably well with real datasets.

Keywords: lattice submodular · greedy · integer lattice · offline model

1 Introduction

Submodular set function maximization has garnered considerable attention both in practical applications and theoretical research [8,9,18]. For a given set N, a set function $f : 2^N \to \mathbb{R}$ is considered submodular if it meets the following key property: for any subsets $S, T \subseteq N$, $f(S) + f(T) \geq f(S \cap T) + f(S \cup T)$. Essentially, this means that adding an element e to a smaller set S gives at least as much gain as adding it to a larger set T containing S. This is expressed mathematically as $f(e|S) \geq f(e|T)$ when $S \subseteq$ T $\subset N$ and $e \in N \setminus S$, where $f(e|S) = f(e \cup S) - f(S)$. Furthermore, f is called monotone if adding any element always increases or maintains the function value, i.e., $f(e|A) \geq 0$ for all $e \in N \setminus A$. Additionally, f is normalized if its value is zero when the input set is empty, i.e., $f(\emptyset) = 0$.

The greedy algorithm is a powerful tool for submodular set maximization problems, achieving an approximation ratio of $1 - 1/e$ under a cardinality constraint and $1/(p+1)$ under p matroid constraints, respectively [3,11]. This makes the greedy approach an appealing choice for optimizing submodular functions within various constraint frameworks.

F. V. Fomin and M. Xiao (Eds.): COCOON 2025, LNCS 15983, pp. 136–147, 2026.
https://doi.org/10.1007/978-981-95-0215-8_11

In machine learning, many models and applications of submodular functions utilize set functions. While set functions have powerful modeling capabilities, sometimes we need to go beyond sets to incorporate more complex modeling and interactions. For instance, consider budget allocation problems, instead of simply deciding to include or exclude an advertisement, we may want finer control over the budget allocation, which also requires modeling submodular functions on the integer lattice. Similarly, in the sensor placement problem, which can be modeled as a set submodular function maximization problem where sensors are placed to cover an area. However, if sensors have different coverage levels based on their strength, we need to define submodularity on the integer lattice to determine the optimal sensor type for each location. Additionally, defining submodularity in the integer lattice is useful when we want to add the same element multiple times.

Recently, the optimization of submodular functions in an integer lattice has gained considerable attention. For a finite set S with $|S| = n$. A function $f : \mathbb{Z}^n \to \mathbb{R}$ is submodular if $f(x)+f(y) \geq f(x \vee y)+f(x \wedge y)$ for all $x, y \in \mathbb{Z}^n$, where $x \vee y$ and $x \wedge y$ denote the coordinate-wise maximum and minimum, respectively. A function $f : \mathbb{Z}^n \to \mathbb{R}$ is DR-submodular if $f(\chi_e|x) \geq f(2\chi_e|x)$ for arbitrary $x \in \mathbb{Z}^n$ and $e \in S$, where $f(y|x) = f(x + y) - f(x)$. In this paper, Our goal is to identify a vector $x \in \mathbb{Z}_+^n$ that maximizes the objective function defined as follows:

$$f(x) - c(x). \tag{1.1}$$

Here, $f : \mathbb{Z}_+^n \to \mathbb{R}_+$ represents a non-negative, monotone, submodular function. Meanwhile, $c : \mathbb{Z}_+^n \to \mathbb{R}_+$ is a non-negative modular function.

Problem (1.1) has various interpretations, potentially extending the current submodular framework to encompass a broader range of tasks. For example, the modular cost c can be introduced as a penalty term within existing submodular maximization problems, acting as a regularizer or soft constraint within the model. In the context of sensor placement, when f represents the financial benefit of installing sensors and c represents the total cost, solving Problem (1.1) involves maximizing profits by balancing revenue against costs.

1.1 Related Work

Below, we summarize some relevant theoretical work on submodular optimization, offering a novel guarantee for this paper's research.

Submodular function maximization over integer lattice under offline model. The classical problem involves maximizing a submodular function over the integer lattice, subject to various constraints. Soma et al. [19] employed a greedy algorithm to achieve a $1 - 1/e$ approximation ratio for monotone lattice submodular functions under a knapsack constraint. Gottschalk and Peis [4] used a double greedy approach to provide a 1/3-approximation algorithm for maximizing non-monotone lattice submodular functions over a bounded integer lattice. When the objective function becomes DR-submodular, they improved the approximation ratio to 1/2 using a random double greedy algorithm. Soma and

Yoshida [20] developed polynomial-time $1 - 1/e$ approximation algorithms for monotone lattice submodular functions with cardinality constraints. Additionally, Nong et al. [12] addressed the problem of maximizing a non-monotone, non-negative lattice submodular function on the bounded integer lattice without constraints, presenting a randomized algorithm with a 1/2 approximation guarantee.

Regularized submodular maximization. Buchbinder et al. [2] proposed a double greedy algorithm providing an approximation ratio of 1/2, provided $f(S) > c(S)$ holds holds for every set S. Harshaw et al. [5] studied the regularized monotone submodular maximization problem and use a distorted greedy algorithm to give a bifactor approximation ratio $(1 - 1/e, 1)$. Nikolakaki et al. [13] developed a framework that produces a set S such that$f(S) - c(S) \geq 1/2 f(OPT) - c(OPT)$ for matroid constraint under offline case and the same approximation guarantee for online unconstrained case. Kazemi et al. [7] give a bifactor approximation for this problem under streaming model when the constraint is a cardinality constraint. Jin et al. [6] proposed ROI-Greedy for the unconstrained regularized submodular maximization problem such that $f(S) - c(S) \geq f(OPT) - c(OPT) - \ln \frac{f(OPT)}{c(OPT)} c(OPT)$. Qiang and Liu consider dynamic DR-submodular maximization with linear costs over the integer lattice and they achieve a $(1/2, 1)$-approximation ratio. Besides, there are many research consider the case when submodular function is nonmonotone, authors can refer to Bodek and Feldman [1]; Lu and Yang [10]; Qi [14]; Sun et al. [21].

1.2 Our Contributions

Our first contribution in this research is demonstrating that a new greedy algorithm offers theoretical guarantees for solving the optimization problem $\max f(S) - c(S)$, where f is a non-negative, monotone, and submodular function, while c is a non-negative modular function. This approximation algorithm output a solution x_0, ensuring that

$$f(x_0) - c(x_0) \geq f(x^*) - c(x^*) - \ln \frac{f(x^*)}{c(x^*)} c(x^*).$$

This means the algorithm always guarantees a positive profit when the optimal solution x^* is profitable.

Our second contribution is an experiment in D-optimal design. For fair comparisons, we designed a Lattice-Return-On-Investment (LROI) algorithm and derived its approximation guarantees for monotone submodular functions f. Experimental results demonstrate that our algorithm performs well with real datasets.

1.3 Organization

The rest of this paper is organized as follows: In Sect. 2, we cover the essentials you need to fully understand our work. Next, in Sect. 3, we introduce the

Greedy algorithm and provide a detailed performance analysis. In Sect. 4, we present an experiment on the sensor placement problem. Finally, in Sect. 5, we summarize our findings and suggest a promising direction for future research.

2 Preliminaries

In this section, we define the concept of submodularity as follows:

Definition 1 *A function $f\colon \mathbb{Z}^n \to \mathbb{R}$ is submodular if for every $x, y \in \mathbb{Z}^n$, we have*

$$f(x) + f(y) \geq f(x \vee y) + f(x \wedge y).$$

Definition 2 *Let S be a finite set and $|S| = n$. A function $f\colon \mathbb{Z}^n \to \mathbb{R}$ is DR-submodular if for arbitrary $x \in \mathbb{Z}^n$ and $e \in S$, we have*

$$f(x + \chi_e) - f(x) \geq f(x + 2\chi_e) - f(x + \chi_e)$$

Property 1 *Assume S be a finite set and $|S| = n$. Let f be a submodular function. For arbitrary $x, y \in \mathbb{Z}^n$, we have*

$$f(x \vee y) \leq f(x) + \sum_{e \in \mathrm{supp}^+(y-x)} (f(x \vee y_e\chi_e) - f(x))$$

where $\mathrm{supp}^+(y - x)$ is the set of elements s in S such that $y(s) > x(s)$.

3 Regularized Submodular over Integer Lattice Without Constraint

In this section, we delve into the regularized monotone submodular maximization problem without constraint. Assume S be a finite set and $|S| = n$. Specifically, we aim to solve the optimization problem:

$$\max_{x \in \mathbb{Z}_+^n \cap \mathbb{B}} f(x) - c(x),$$

where $f : \mathbb{Z}_+^S \to \mathbb{R}_+$ is a monotone submodular function, $c : \mathbb{Z}_+^S \to \mathbb{R}_+$ is a normalized modular function and $\mathbb{B} = \{x | x \leq b, b \in \mathbb{Z}_+^S\}$ is a box constraint. This problem involves finding a vector $x \in \mathbb{Z}_+^S \cap \mathbb{B}$ that maximizes the difference between the monotone submodular function f and the modular function c.

3.1 Algorithm

LROI-Greedy is based on a simple idea: we start from 0 vector, and iteratively find $k \in \mathbb{Z}_+$ and $e \in S$ maximizing the average marginal return $\frac{f(k\chi_e|x_{i-1})}{kc_e}$ among all k, e, until none of the vectors can be added.

Algorithm 1

```
1: x_0 ← 0, i ← 0
2: while True do
3:     i ← i + 1
4:     (k_i, e_i) ← arg max f(kχ_e|x_{i−1}) / (k c_e)
5:     if f(k_i χ_{e_i}|x_{i−1}) − k_i c_{e_i} > 0 then
6:         x_i ← x_{i−1} + k_i χ_{e_i}
7:     else
8:         x_0 ← x_i
9:         Return x_0
10:    end if
11: end while
```

3.2 Analysis

Lemma 1 *For any i,*

$$f(k_i\chi_{e_i}|x_{i-1}) \geq \frac{k_i c_{e_i}}{c(x^*)}(f(x^*) - f(x_{i-1})).$$

Proof. According to the Line 4 of algorithm 1,

$$\begin{aligned}\frac{f(k_i\chi_{e_i}|x_{i-1})}{k_i c_{e_i}} &\geq \max_{s\in \mathrm{supp}^+(x^*-x_{i-1})} \frac{f(k\chi_e|x_{i-1})}{kc_e} \\ &\geq \frac{\sum_{s\in \mathrm{supp}^+(x^*-x_{i-1})} f(k_s\chi_{e_s}|x_{i-1})}{\sum_{s\in \mathrm{supp}^+(x^*-x_{i-1})} k_s c_{e_s}} \\ &\geq \frac{f(x^*\vee x_{i-1}) - f(x_{i-1})}{c(x^*)} \geq \frac{f(x^*) - f(x_{i-1})}{c(x^*)},\end{aligned}$$

where the first inequality is due to the Line 4 of algorithm 1 and the third inequality follows from the property of submodular.

Lemma 2 *For any i, assume $k_j c_{e_j} \leq c(x^*), \forall j \leq i$, then*

$$f(x_i) \geq \left(1 - \prod_{j=1}^{i}(1 - \frac{k_j c_{e_j}}{c(x^*)})\right) f(x^*).$$

Proof. From Lemma 1, the inequality holds for $i = 1$. Now, we assume $\forall t \leq i-1$, we have $f(x_t) \geq \left(1 - \prod_{j=1}^{t}(1 - \frac{k_j c_{e_j}}{c(x^*)})\right) f(x^*)$. So for any i, we have

$$\begin{aligned} f(x_i) &= f(k_i \chi_{e_i} | x_{i-1}) + f(x_{i-1}) \\ &\geq \frac{k_i c_{e_i}}{c(x^*)}(f(x^*) - f(x_{i-1})) + f(x_{i-1}) \\ &\geq \frac{k_i c_{e_i}}{c(x^*)} f(x^*) + (1 - \frac{k_i c_{e_i}}{c(x^*)}) f(x_{i-1}) \\ &\geq \frac{k_i c_{e_i}}{c(x^*)} f(x^*) + (1 - \frac{k_i c_{e_i}}{c(x^*)}) \left(1 - \prod_{j=1}^{i-1}(1 - \frac{k_j c_{e_j}}{c(x^*)})\right) f(x^*) \\ &= \left(1 - \prod_{j=1}^{i}(1 - \frac{k_j c_{e_j}}{c(x^*)})\right) f(x^*). \end{aligned}$$

By induction, the inequality holds.

Next, we will prove the main theorem from two cases.

Theorem 1 *Algorithm 1 returns a solution* x_0 *such that*

$$f(x_0) - c(x_0) \geq f(x^*) - c(x^*) - \ln \frac{f(x^*)}{c(x^*)} c(x^*).$$

Proof. **Case 1**: $c(x_0) < \ln \frac{f(x^*)}{c(x^*)} c(x^*)$. In this case, according to Line 5 of algorithm 1,

$$\begin{aligned} 0 \geq \max_{e \in supp^+(x^* - x_{i-1})} f(k\chi_e | x_0) - kc_e \geq \sum_{e \in supp^+(x^* - x_{i-1})} f(k_e \chi_e | x_0) - k_e c_e \\ \geq f(x^* \vee x_0) - f(x_0) - c(x^*) \geq f(x^*) - f(x_0) - c(x^*). \end{aligned}$$

So

$$f(x_0) \geq f(x^*) - c(x^*).$$

This leads to

$$f(x_0) - c(x_0) \geq f(x^*) - c(x^*) - c(x_0) \geq f(x^*) - c(x^*) - \ln \frac{f(x^*)}{c(x^*)} c(x^*).$$

Case 2: $c(x_0) \geq \ln \frac{f(x^*)}{c(x^*)} c(x^*)$. In this case, according to the monotone of function c, there exists a $i \in [1, n]$ such as $x_{i-1} \leq x_i \leq x_0$ and

$$c(x_i) \geq \ln \frac{f(x^*)}{c(x^*)} c(x^*) > c(x_{i-1}).$$

If $f(x_{i-1}) \geq f(x^*) - c(x^*)$, then we have

$$f(x_0) - c(x_0) \geq f(x_{i-1}) - c(x_{i-1}) \geq f(x^*) - c(x^*) - \ln \frac{f(x^*)}{c(x^*)} c(x^*).$$

If $f(x_{i-1}) \le f(x^*) - c(x^*)$ and $f(x_{i-1}) - c(x_{i-1}) \ge f(x^*) - c(x^*) - \ln \frac{f(x^*)}{c(x^*)} c(x^*)$, the theorem 1 also holds. Thus, in what follows, we only consider the case when $f(x_{i-1}) \le f(x^*) - c(x^*)$ and $f(x_{i-1}) - c(x_{i-1}) \le f(x^*) - c(x^*) - \ln \frac{f(x^*)}{c(x^*)} c(x^*)$. Before giving the proof of this case, we need some lemmas.

Lemma 3 *Assume $f(x_{i-1}) \le f(x^*) - c(x^*)$, then for any $j \le i-1$, $k_j c_{e_j} \le c(x^*)$.*

Proof. Suppose for contradiction that there exists a $j \le i-1$ such that $k_j c_{e_j} > c(x^*)$, then

$$f(x_j) = f(k_j \chi_{e_j} | x_{j-1}) + f(x_{j-1}) \ge \frac{k_j c_{e_j}}{c(x^*)} (f(x^*) - f(x_{j-1})) + f(x_{j-1}) > f(x^*).$$

Where the first inequality follows from Lemma 1 and the second inequality is due to the assumption. According to the monotone of function f, we have

$$f(x_{i-1}) \ge f(x_j) > f(x^*),$$

which contradicts with the assumption.

Lemma 4 *Let $a = \ln \frac{f(x^*)}{c(x^*)} c(x^*) - c(x_{i-1})$, we have*

$$\frac{a(f(x^*) - f(x_{i-1}))}{c(x^*)} + f(x_{i-1}) \ge f(x^*) - c(x^*).$$

Proof. First, we assume $a \le c(x^*)$. According to the Lemmas 2 and 3, we have

$$\begin{aligned}
&\frac{a(f(x^*) - f(x_{i-1}))}{c(x^*)} + f(x_{i-1}) \\
&= \frac{a}{c(x^*)} f(x^*) + (1 - \frac{a}{c(x^*)}) f(x_{i-1}) \\
&\ge \frac{a}{c(x^*)} f(x^*) + (1 - \frac{a}{c(x^*)}) \left(1 - \prod_{j=1}^{i-1} (1 - \frac{k_j c_{e_j}}{c(x^*)})\right) f(x^*) \\
&= \left(1 - (1 - \frac{a}{c(x^*)}) \prod_{j=1}^{i-1} (1 - \frac{k_j c_{e_j}}{c(x^*)})\right) f(x^*) \ge \left(1 - (1 - \frac{a + \sum_{j=1}^{i-1} c_{e_j}}{i c(x^*)})^i\right) f(x^*) \\
&= \left(1 - (1 - \frac{\ln \frac{f(x^*)}{c(x^*)}}{i})^i\right) f(x^*) \ge \left(1 - e^{-\ln \frac{f(x^*)}{c(x^*)}}\right) f(x^*) = f(x^*) - c(x^*)
\end{aligned}$$

where the first inequality is by Lemmas 1 and 3.

Second, considering the case when $a > c(x^*)$, we have

$$\frac{a(f(x^*) - f(x_{i-1}))}{c(x^*)} + f(x_{i-1}) > f(x^*) - f(x_{i-1}) + f(x_{i-1}) \ge f(x^*) - c(x^*),$$

which follows that $a > c(x^*)$ and $f(x_{i-1}) \le f(x^*) - c(x^*)$.

Finally, we continue to proof the Theorem 1.

Proof.

$$
\begin{aligned}
&f(x_0) - c(x_0) \geq f(x_i) - c(x_i) = f(k_i\chi_{e_i}|x_{i-1}) + f(x_{i-1}) - c(x_{i-1}) - k_i c_{e_i} \\
&\geq \frac{k_i c_{e_i}}{c(x^*)}(f(x^*) - f(x_{i-1})) + f(x_{i-1}) - c(x_{i-1}) - k_i c_{e_i} \\
&= \frac{(k_i c_{e_i} - a + a)}{c(x^*)}(f(x^*) - f(x_{i-1})) + f(x_{i-1}) - (k_i c_{e_i} - a) - \ln\frac{f(x^*)}{c(x^*)}c(x^*) \\
&= (k_i c_{e_i} - a)\left(\frac{f(x^*) - f(x_{i-1})}{c(x^*)} - 1\right) + \frac{a}{c(x^*)}(f(x^*) - f(x_{i-1})) + f(x_{i-1}) - \ln\frac{f(x^*)}{c(x^*)}c(x^*) \\
&\geq (k_i c_{e_i} - a)\left(\frac{f(x^*) - f(x_{i-1})}{c(x^*)} - 1\right) + f(x^*) - c(x^*) - \ln\frac{f(x^*)}{c(x^*)}c(x^*) \\
&\geq f(x^*) - c(x^*) - \ln\frac{f(x^*)}{c(x^*)}c(x^*).
\end{aligned}
$$

where the first inequality is from Lemma 1, the second inequality is due to Lemma 4 and the last inequality follows that $k_i c_{e_i} = c(x_i) - c(x_{i-1}) \geq a$.

4 Numerical Experiments

In experimental design, D-optimal design aims to optimize the process of collecting experimental data in order to maximize the fitting accuracy of regression models. To achieve this goal, we first define a monotone lattice submodular function $f : \mathbb{Z}_+^S \to \mathbb{R}$ and a cost function $c : \mathbb{Z}_+^S \to \mathbb{R}$, where c is a regularized submodular function. This framework facilitates solving the maximization problem over integer lattices. Specifically, the objective function of D-optimal experimental design can be expressed as:

$$f(x) = \log\det\left(\sum_{i=1}^{N} x_i a_i a_i^T\right),$$

where $x = [x_1, \ldots, x_N]$ represents the experimental points to be selected, a_i is a known column vector. This objective function aims to enhance the "informativeness" and numerical stability of the experimental design.

Meanwhile, the cost function $c(x)$ is defined as:

$$c(x) = \sum_{i=1}^{N} x_i \cdot c_i.$$

This cost function ensures that the problem possesses regularized DR submodular properties, which help control the number of experiments conducted.

In practical applications, the first experiment typically involves an additional setup cost, which is spread out over multiple repetitions. We can model this effect by considering the following transformation:

$$t_i(x_i) = \left\lfloor \frac{1}{c_i}\max(x_i - b_i,\ 0) \right\rfloor$$

where b_i represents the setup cost, and c_i is the repeating cost of the i-th experiment. The objective function then becomes:

$$f(x) = \log\det\left(I + \sum_i t_i(x_i) a_i a_i^T\right)$$

This function is a monotone lattice submodular function, and I denotes the identity matrix. However, it does not satisfy the diminishing returns condition. The cost function $c(x)$ remains unchanged.

The D-optimal design criterion has been widely applied in various fields, including environmental monitoring [17], and UAV swarm path optimization [16], among others. These applications typically involve optimal data collection or path planning tasks under specific constraints. Due to its strong theoretical properties and practical effectiveness, D-optimal design has become an important tool in the field of experimental design.

To address the above problem, we use random matrices to select experimental points $x = [x_1, \ldots, x_N]$.

To evaluate the effectiveness of the proposed algorithm, we compare it with the double-greedy algorithm and the classic greedy algorithm. The double-greedy algorithm expands the solution set in two directions: one starting from the empty set $\emptyset$, and the other starting from the full selection set $[2, 2, \ldots, 2]_{1\times N}$. The classic greedy algorithm incrementally selects the optimal experimental point until the marginal gain cannot be further improved.

By comparing with these algorithms, we aim to demonstrate the applicability and advantages of the proposed method in real-world scenarios. The code has been open-sourced on GitHub: https://github.com/lvymath1/Regularized_Lattice_submodular.

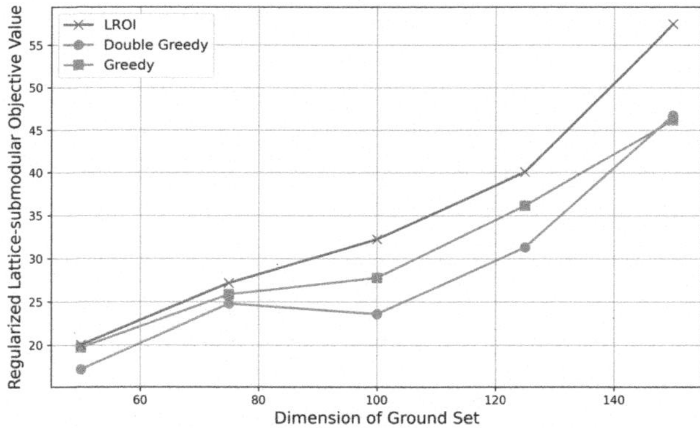

Fig. 1. Comparison of experimental results for the lattice submodular problem

For the lattice submodular problem, we employ a randomly generated matrix of size $x \in \mathbb{R}^{150\times 40}$ and vary the ground set size from 50 to 150. The objective function values obtained by different algorithms are evaluated, with the parameters set as follows: c is randomly selected from the range [2,6], m from [0.1,0.5], and b as an integer in [0,2]. The experimental results are illustrated in Fig. 1, showing that as the ground set size increases, all algorithms exhibit improved performance. It is noteworthy that the Double Greedy method demonstrates comparable performance to the Greedy method in large-scale cases. However, LROI consistently outperforms all competing algorithms across various settings, with its advantage becoming increasingly pronounced as the problem size grows.

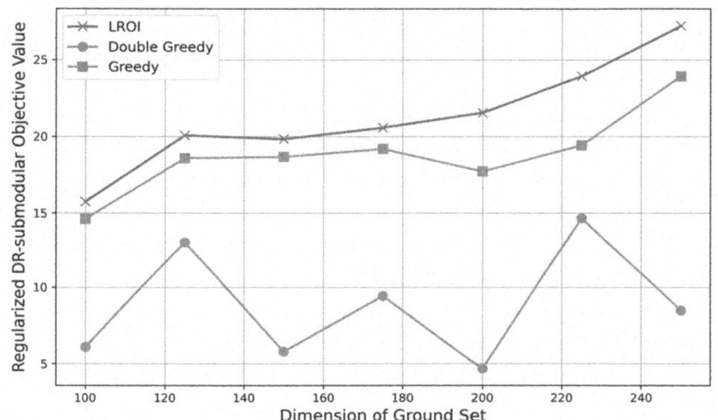

Fig. 2. Comparison of experimental results for the regularized DR submodular problem

For the regularized DR-submodular problem, we consider a matrix of size $x \in \mathbb{R}^{250\times 40}$ and vary the ground set size from 100 to 250. The objective function values obtained by different algorithms are compared, with the experimental results presented in Fig. 2. As the ground set size increases, the values of the blue and green curves exhibit a rising trend, indicating that the algorithms can identify superior solutions with a larger candidate set. Notably, the proposed LROI algorithm consistently outperforms all other methods across different ground set sizes, demonstrating its capability to achieve higher utility while utilizing lower-cost sensors. Furthermore, in small-scale settings, the performance of the Greedy algorithm is comparable to that of LROI; however, as the ground set size increases, the performance gap widens. This trend suggests that LROI exhibits superior scalability in large-scale problem instances, achieving a 12.9% improvement over other algorithms.

5 Conclusion

In this paper, we tackle the regularized submodular maximization problem in integer lattice, demonstrating that a new greedy algorithm possesses provable approximation guarantees for this problem, a promising direction for future research involves extending our approach to accommodate more generalized constraints, such as matroids or knapsacks.

Acknowledgements. The research is supported by Beijing Natural Science Foundation (No. IS24001), Beijing Natural Science Foundation Project (No. Z220004) and National Natural Science Foundation of China (No. 12401421).

References

1. Bodek, K., Feldman, M.: Maximizing sums of non-monotone submodular and linear functions: understanding the unconstrained case. ESA **23**(1–23), 17 (2022)
2. Buchbinder, N., Feldman, M., Naor, J., Schwartz, R.: A tight linear time (1/2)-approximation for unconstrained submodular maximization. SIAM J. Comput. **44**(5), 1384–1402 (2015)
3. Fisher, M.L., Nemhauser G.L., Wolsey, L.A.: An analysis of approximations for maximizing submodular set functions–ii. In: Balinski, M.L., Hoffman, A.J. (eds) Polyhedral Combinatorics. Mathematical Programming Studies, vol. 8, pp. 73–87. Springer, Berlin, Heidelberg (1978). https://doi.org/10.1007/BFb0121195
4. Gottschalk, O., Peis, B.: Submodular function maximization on the bounded integer lattice. In: WAOA, pp. 133–144 (2015)
5. Harshaw, C., Feldman, M., Ward, J., Karbasi, A.: Submodular maximization beyond non-negativity: guarantees, fast algorithms, and applications. In: Proceedings of the 36th International Conference on Machine Learning (ICML), pp. 2634–2643 (2019)
6. Jin, T., Yang, Y., Yang, R., Shi, J., Huang, K., Xiao, X.: Unconstrained submodular maximization with modular costs: tight approximation and application to profit maximization. Proc. VLDB Endow. **14**(10), 1756–1768 (2021)
7. Kazemi, E., Minaee, S., Feldman, M., Karbasi, A.: Regularized submodular maximization at scale. In: ICML, pp. 5356–5366 (2021)
8. Krause. A., Guestrin, A.: Near-optimal nonmyopic value of information in graphical models. In: UAI, p. 5 (2005)
9. Laitila, J., Moilanen, A.: New performance guarantees for the greedy maximization of submodular set functions. Optim. Lett. **11**, 655–665 (2017)
10. Lu, C., Yang, W., Gao, S.: Regularized nonmonotone submodular maximization. Optimization, 1–27(2024)
11. Nemhauser, G.L., Wolsey, L.A., Fisher, M.L.: An analysis of approximations for maximizing submodular set functions-i. Math. Program. **14**(1), 265–294 (1978)
12. Nong, Q., Fang, J., Gong, S., Du, D., Feng, Y., Qu, X.: A 1/2-approximation algorithm for maximizing a non-monotone weak-submodular function on a bounded integer lattice. J. Comb. Optim. **39**(4), 1208–1220 (2020). https://doi.org/10.1007/s10878-020-00558-4
13. Nikolakaki, S., Ene, A., Terzi, E.: An efficient framework for balancing submodularity and cost. In: KDD, pp. 1256–1266 (2021)
14. Qi, B.: On maximizing sums of non-monotone submodular and linear functions. Algorithmica **86**(4), 1080–1134 (2024)
15. Qiang, Y., Liu, B.: Dynamic DR-submodular maximization with linear costs over the integer lattice. In: AAIM (1), pp. 109–121 (2024)
16. Rong, Z., Hemin, S., Hao, L., Weilin, L.: TDOA and track optimization of UAV swarm based on D-optimality. J. Syst. Eng. Electron. **31**(6), 1140–1151 (2020)
17. Roy, V., Simonetto, A., Leus, G.: Spatio-temporal sensor management for environmental field estimation. Signal Process. **128**, 369–381 (2016)

18. Schulz, A.S., Uhan, N.A.: Approximating the least core value and least core of cooperative games with supermodular costs. Discrete Optim. **10**(2), 163–180 (2013)
19. Soma, T., Kakimura, N., Inaba, K., Kawarabayashi, K.: Optimal budget allocation: theoretical guarantee and efficient algorithm. In: ICML, pp. 351–359 (2014)
20. Soma, T., Yoshida, Y.: Maximizing monotone submodular functions over the integer lattice. Math. Program. (4), 539–563 (2018). https://doi.org/10.1007/s10107-018-1324-y
21. Sun, X., Han, C., Wu, C., Xu, D., Zhou Y.: The regularized submodular maximization via the lyapunov method. COCOON (2), 118–143 (2023)

Adaptive Weighting-Based Local Search for Route Number Minimization for Vehicle Routing Problem with Time Windows

Yuxuan Wang, Yunhao Li, Zhouxing Su(✉), Junwen Ding, Qingyun Zhang, and Zhipeng Lü

Huazhong University of Science and Technology, Wuhan, China
{smartlab_wyx,suzhouxing,junwending,qingyun_zhang,zhipeng.lv}@hust.edu.cn

Abstract. This paper presents an adaptive weighting-based local search (AWLS) algorithm for solving the vehicle routing problem with time windows (VRPTW), focusing on minimizing the number of routes. AWLS improves the ejection pool framework by adopting an adaptive weighting technique to diversify the search. First, it adjusts the penalties on customers when both inserting and ejecting them into and from the routes to prevent frequently moving the same customer. Second, we introduce route weighting in the route repairing procedure to encourage paying more attention to corrupted routes that persist for a long time. Third, we design a new objective function for identifying the best set of customers to eject, along with several acceleration strategies. Experimental results on 300 Gehring and Homberger's benchmarks show that our AWLS algorithm solves 247 instances to optimality, matches the state-of-the-art route minimization algorithms in significantly less runtime, and improves the best-known results for two instances in the literature.

Keywords: Vehicle routing problem with time windows · Large neighborhood search · Weighting-based local search

1 Introduction

The Vehicle Routing Problem (VRP) is a challenging NP-hard combinatorial optimization problem first introduced by Dantzig et al. [2]. Among various VRP variants, the Vehicle Routing Problem with Time Windows (VRPTW) is one of the most widely studied. Proposed in the early 1980s, VRPTW has been studied for over 40 years, yet many benchmark instances remain unsolved, making it highly challenging to design effective algorithms. Recently, VRPTW was featured as one of five tracks in the 2022 VRP Challenge held by the Center for Discrete Mathematics and Theoretical Computer Science (DIMACS). Several major companies (e.g., Google, Red Hat, Walmart, Huawei) participated in this VRPTW track. In the same year, a competition for both static and dynamic

F. V. Fomin and M. Xiao (Eds.): COCOON 2025, LNCS 15983, pp. 148–161, 2026.
https://doi.org/10.1007/978-981-95-0215-8_12

VRPTW was held in NeurIPS 2022. These indicate that VRPTW is widely considered by both the academia and industry.

VRPTW partitions customers into subsets, each forming a vehicle route, ensuring that every customer is served within a specified time window. Vehicles must arrive before the latest service time and may wait if they arrive early. All routes start and end at the depot, and customer demands must not exceed the vehicle capacity, without allowing split deliveries. The two main objectives of VRPTW are minimizing the number of routes and the total travel distance. Because the number of routes directly determines vehicle-related costs for logistics companies and, as a coarse-grained decision, is less sensitive to uncertainties than total travel distance, this study prioritizes minimizing route numbers.

In the early years of VRPTW studies, exact algorithms such as branch-and-cut-and-price [4], Lagrangian relaxation [12], and column generation [9] were developed, but they were effective only for instances with up to 100 customers. Heuristic approaches have been studied for decades, evolving from simple methods to complex metaheuristics with continual improvements on classic benchmark instances. One of the most popular frameworks for VRPTW is the large neighborhood search based on the ruin-and-recreate method. After Dees et al. [3] first proposed this method for routing problems [5,6], Schrimpf et al. [20] adapted this approach to tackle VRPTW. Following this idea, Christiaens et al. [1] improved the ruin-and-recreate method and proposed a slack induction by string removals (SISRs) method. Another well-known framework is the ejection pool method proposed by Lim et al. [14], which relaxes the constraints by temporarily ignoring a subset of customers. This approach improves the search diversification and makes metaheuristics more powerful on problems with strong constraints. Based on their studies, Nagata et al. [15] proposed a more effective ejection pool implementation and found 18 new best-known solutions.

In this paper, we propose an adaptive weighting-based local search (AWLS) based on the ejection pool framework for the route number minimization for VRPTW. Our main contributions can be summarized as follows:

- We introduce route weighting in the route repair phase to enhance neighborhood search and reduce redundant ejection-reinsertion loops.
- Our adaptive weighting strategy adjusts the weights of both inserted and ejected customers, and we propose a new objective function that maintains diversity while reducing the number of ejected customers.
- We design a new branch-and-bound algorithm for selecting customers to eject with pruning strategies based on service start time and ejected customer count, improving repaired solution quality.
- AWLS matches all best-known results in significantly less runtime and improves the results for two instances in the literature. Based on the proposed maximum clique method, we update the lower bounds for 10 benchmark instances. Finally, AWLS reaches the theoretical optimum for 247 of the 300 instances.

2 Problem Formulation and Reformulation

VRPTW is defined on a directed graph $G = (V, A)$ where $V = V' \cup \{0\}$ is the set of vertices composed of customers and the depot. The set of directed arcs is denoted by A and the distance from vertex v to vertex v' is denoted by $c(v, v')$. Each customer $v \in V'$ is equipped with quantity of demand q_v, time window $[e_v, l_v]$ and service duration st_v. The depot (denoted by vertex 0 without loss of generality) can be regarded as a special customer whose demand $q_0 = 0$ and service time $st_0 = 0$. A route is a permutation of customers $\{v_0, v_1, ..., v_n, v_{n+1}\}$ where the starting and ending vertex $v_0 = v_{n+1} = 0$ is the depot. Each route corresponds to a vehicle in a homogeneous fleet whose capacity is Q, i.e., $\sum_{i=1}^{n} q_{v_i} \leq Q$. Let $t(v_i) = \max\{e_{v_i}, t(v_{i-1}) + st_{v_{i-1}} + c(v_{i-1}, v_i)\}$ be the service time for customer v_i in a route, we need to satisfy the time window constraint $t(v_i) \leq l_{v_i}$.

The objective of VRPTW is to partition the customers into several routes and determine the order of customers in each route under the aforementioned constraints, so that the number of routes is minimized. Apparently, the route number minimization can be further decomposed into a series of decision subproblems with a fixed number of routes.

2.1 Penalty Function

One major challenge in local search metaheuristics for VRPTW is handling its strong constraints. A common approach is to relax certain constraints and penalize their violations, but choosing which constraints to relax has a significant impact on its performance. Some constraints can be naturally satisfied by appropriate neighborhood structures (e.g., consecutive tours can be preserved by vertex/arc exchange moves), while constraint violations can be shifted between constraints (e.g., capacity violations disappear if serving all customers is not mandatory). Therefore, it is beneficial to relax the hardest-to-satisfy constraints to reshape the search landscape and escape from local optima by transferring violations. Based on this insight, AWLS relaxes only the vehicle capacity and time window constraints, while strictly enforcing all other constraints.

Time-window violations can propagate: a single delay may cause successive delays along a route, resulting in a linear-time penalty. Therefore, we extend the penalty function [16] to enable incremental calculation of the penalty, reducing the complexity of evaluating each inter-route move to $O(1)$. Specifically, if a vehicle arrives at customer v after l_v, we reset it to l_v (capping the delay at the time-window limit). Let $P_t(r)$ and $P_c(r)$ denote the time window and capacity violation penalties for route r. We assign each route a weight W_r (initialized to

1). The total penalty for a solution S is given as follows:

$$P(S) = P_t(S) + \alpha P_c(S) \tag{1}$$

$$P_t(S) = \sum_{r \in S} W_r(r) \cdot P_t(r) \tag{2}$$

$$P_c(S) = \sum_{r \in S} \max\{0, \sum_{v \in r} q_v - Q\} \tag{3}$$

3 Adaptive Weighting-Based Local Search

In this section, we propose an adaptive weighting-based local search (AWLS) to solve the VRPTW for route number minimization. AWLS starts from a heuristic initial solution generation procedure (Sect. 3.1). Then, it repeatedly removes routes until the time limit is reached (Sect. 3.2). Each customer and route is associated with a weight which reflects its importance, respectively. As the search proceeds, the weights gradually increase so that the landscape of the solution space is reshaped adaptively. This diversification mechanism enables the algorithm to escape from the local optima, thus improving the effectiveness of AWLS. Follows are the details of the algorithm strategy.

3.1 Initial Solution

The goal of initialization is to construct a feasible solution with as few routes as possible. A solution is composed of several routes where each customer is served by exactly one route. Let $C(S) = \bigcup_{r \in S} r$ denote the set of customers already covered by S. Then, the ejection pool $EP = V' \setminus C(S)$ contains the remaining uncovered customers. Initially, $EP = V'$ since no customer is assigned. Then, we repeatedly pick a random customer v_{in} from EP and evaluate all its valid insertion positions in the current solution S. An insertion is valid if it does not violate the capacity or time window constraints for any customer in that route. For each feasible insertion gap in a route, let v_{in}^- and v_{in}^+ denote the predecessor and successor vertices, respectively; inserting v_{in} where the additional route length $\Delta D = c(v_{in}^-, v_{in}) + c(v_{in}, v_{in}^+) - c(v_{in}^-, v_{in}^+)$ is minimized. If no valid position exists, a new route $\{0, v_{in}, 0\}$ is created to serve v_{in}.

3.2 Customer Weighting-Based Route Removal

Once we obtain a feasible solution, we randomly remove a route and put all customers in the route into the ejection pool EP. The goal of the route removal procedure presented in Algorithm 1 is to empty EP by re-inserting customers into existing routes to form a complete feasible solution again.

The re-insertion follows a process similar to initialization: we iteratively insert a random customer $v_{in} \in EP$ to its best valid position in the current solution S. However, if no valid insertion exists, instead of creating a new route, we insert v_{in} at the position with the minimal penalty, producing an infeasible

Algorithm 1: RemoveRoute

Input: Feasible solution S
Output: Feasible solution with fewer routes than S
1 $S_{copy} \leftarrow S$, $S_{part} \leftarrow S$;
2 Initialize weights of customers $W_c(v_i) \leftarrow 1, \forall i \in V'$;
3 Randomly remove a route in S_{part} and put its customers into EP;
4 **while** $EP \neq \emptyset$ *and time limit is not reached* **do**
5   Randomly pick a customer v and remove v from EP, $W_c(v) \leftarrow W_c(v) + W_{in}$;
6   Insert v into the position with the minimal penalty in the current partial solution S_{part};
7   **if** $P(S_{part}) > 0$ **then**
8         $S_{part} \leftarrow$`Repair`(S_{part}) /* Algorithm 2 */;
9   **if** $P(S_{part}) > 0$ **then**
10             $(EJ,\ S_{part}) \leftarrow$`Eject`$(S_{part})$ /* Sect. 3.4 */;
11             $S_{part} \leftarrow S_{part} \setminus EJ$, $EP \leftarrow EP \cup EJ$;
12             $W_c(v_i) \leftarrow W_c(v_i) + W_{ej}, \forall v_i \in EJ$;
13             $S_{part} \leftarrow$`Shuffle`(S_{part}) /* Sect. 3.5 */;
14 **if** $EP = \emptyset$ **then**
15   **return** S_{part};
16 **else**
17   **return** S_{copy};

solution S^* (line 6). We then attempt to repair S^* using a route weighting-based local search (Algorithm 2 in Sect. 3.3). Once S^* is repaired, we move on to the next customer. If the local search fails (line 7), we perform an ejection procedure on one of the local optima solutions returned by the local search (Sect. 3.4). The ejection procedure removes a set of customers from the routes to reduce capacity consumption and travel time, thereby ensuring that solution S becomes feasible. After each insertion or ejection, we adjust the corresponding customer's weight by increasing it by W_{in} for insertions and by W_{ej} for ejections (line 12). To further improve diversification, we shuffle the current solution using a series of random non-worsening moves (line 13) as described in Sect. 3.5.

3.3 Route Weighting-Based Repairing Procedure

Algorithm 2 describes the route weighting-based local search to repair the infeasible solution, i.e., to reduce the route-weighted penalty value $P(S)$, by repeatedly exploring four kinds of neighborhood structures, which are 2-opt [10], 2-opt* [18], exchange [11], and out-relocate, respectively.

However, unlike the classic local search which employs the best-improvement policy, we do not evaluate the entire neighborhood. Instead, we randomly select a subset of neighborhood moves and perform the best move among them. S_{best} (line 8). If the best move improves the best solution S_{best} (line 5), we replace S_{best} with the best neighbor S^* (line 6) and reset the stagnation counter *stagnation* to 0 (line 6). Otherwise, we increase *stagnation* by one and the local search stops if we fail to improve the best solution in N_{step} consecutive local search iterations (line 3). In addition, if the best neighbor S^* is worse than the current solution S (line 9), the weight of each route with positive penalty will increase by one (line 10).

Algorithm 2: Repair

Input: Infeasible solution S
Output: Solution S_{best} where $P(S_{best}) \leq P(S)$

1 Initialize weights of routes $W_r(r_i) \leftarrow 1, \forall r_i \in S$;
2 $S^{best} \leftarrow S$, *stagnation* $\leftarrow 0$;
3 **while** $P(S) > 0$ *and stagnation* $< N_{step}$ **do**
4     $S^* \leftarrow$ the neighbor of S with minimal $P(S^*)$;
5   **if** $P(S^*) < P(S^{best})$ **then**
6         $S^{best} \leftarrow S^*$, *stagnation* $\leftarrow 0$;
7   **else**
8   *stagnation* $\leftarrow$ *stagnation* $+ 1$;
9   **if** $P(S^*) > P(S)$ **then**
10         $W_r(r_i) \leftarrow W_r(r_i) + 1, \forall r_i \in S, P(r_i) > 0$;
11     $S \leftarrow S^*$;
12 **return** S_{best};

3.4 Ejection Procedure

The ejection procedure starts from an infeasible solution that cannot be repaired by Algorithm 2, moving a subset of customers EJ with constraint violations to the ejection pool EP. The goal is to eliminate all constraint violations while minimizing the impact on the current solution structure. We prefer EJ with a small total customer weight ($\sum_{v \in EJ} W_c(v)$) and cardinality ($|EJ|$). The rationale for this preference is two fold: (1) the repair procedure has stagnated, so structural changes are needed to continue the search, and (2) while full customer coverage is required, we must avoid trivial ejections for efficiency. A trivial ejection occurs when too many customers are ejected or when the same customers are frequently removed, which does not improve feasibility. We combine these criteria into a single measure: $P_e(EJ) = \sum_{v \in EJ} W_c(v) \cdot |EJ|$.

Identifying the best EJ is time-consuming due to the large number of subsets. Previous work [15] constructed EJ route by route, limiting the number of customers to eject (K_{max}) and enumerating combinations in lexicographic order. We further improve this procedure in two ways: (1) we use a branch-and-bound algorithm to search for the best EJ set for each route, pruning nodes if their penalty exceeds the best found so far or if ejecting a customer does not advance the service start time, and (2) we adopt a tight maximum number of customers to eject ($K_{max} = 4$) to accelerate the search. If ejection of a few customers fails to restore feasibility, we apply a greedy strategy: ejecting the customer that produces the greatest penalty reduction until the route becomes feasible.

3.5 Shuffle Procedure

We apply a shuffle procedure after ejection by performing several neighborhood moves (as presented in Sect. 3.3). At each iteration, we evaluate up to 100 randomly sampled moves. If a move does not increase the total penalty $P(S)$ of the current solution S, we make the move. Otherwise, the shuffle procedure stops. To enhance diversification, the shuffle uses a tabu strategy to avoid reversing moves. We maintain a $|V| \times |V|$ tabu list matrix TL. For each move in the kth iteration that breaks arc (v_i, v_j), we set $TL_{ij} = k + rand(10)$. Then, for the k'th iteration, if $k' < TL_{ij}$, the arc (v_i, v_j) is forbidden to include.

3.6 Lower Bound of Route Number Minimization

We employ two methods for calculating the lower bound of the route number for VRPTW, which consider the capacity and time window constraints, respectively.

1. **Demand versus capacity.** By simply dividing the total demand by the vehicle capacity, we can obtain a simple lower bound $\lceil \sum_{v \in V} q_v / Q \rceil$ [9].
2. **Time window incompatibility.** By considering the incompatibility among customer time windows, we propose to model the lower bound of the route number by the maximum clique problem (MCP). In detail, we construct an undirected graph G', where there is an edge between nodes v_i and v_j if both routes $\{v_0, v_i, v_j, v_{n+1}\}$ and $\{v_0, v_j, v_i, v_{n+1}\}$ violate the time window constraints, i.e., $e_i + c(i, j) > l_j$ and $e_j + c(j, i) > l_i$. In other words, each edge represents the incompatibility between a pair of customers that must be in different routes. Then, the maximum clique size N_{mcp} of graph G' is the minimum route number since each customer in the clique must be in an independent route. To identify the maximum clique in graph G', we employ the state-of-the-art exact algorithm [13].

4 Experiments and Analysis

This section reports the experimental result of AWLS algorithm and compares its performance to the state-of-the-art algorithms on 300 benchmark instances [7]. This benchmark is categorized by customer distribution and instance characteristics. The distribution sets are R (random), C (clustered), and RC (mixed). Each set is divided into two types: Type 1 with small vehicle capacity and narrow time windows, and Type 2 with large capacity and loose time windows. This results in six groups (R1, R2, C1, C2, RC1, RC2), each containing five problem sizes (200, 400, 600, 800, 1,000 customers), with 10 instances per size.

4.1 Experimental Protocol

The proposed AWLS algorithm[1] is implemented in C++ and tested on Intel Xeon E5-2698v3 @ 2.30 GHz CPU. We performed 5 independent single-thread runs per instance, with time limits from 1 min up to 5 h to evaluate both efficiency and solution quality. For fairness in comparisons with the reference algorithms, competitor runtimes were scaled to an equivalent single-thread performance (using PassMark ratings[2] The parameter setting of the proposed algorithm is given in Table 1. Our preliminary experiments show that the proposed algorithm is not sensitive to the parameter values.

Table 1. The parameter settings of the proposed AWLS.

Parameter	Explanation	Value
α	Coefficient of capacity violation penalty	1
W_{in}	Customer weight increment on insertion	1
W_{ej}	Customer weight increment on ejection	2
N_{step}	Maximum stagnated step number in repairing	100

4.2 Comparison on Recent World Records

The best-known results for VRPTW are maintained by SINTEF[3] and the detailed updating history is recorded on combopt[4].

The updates show that further reducing route numbers on these well-studied instances is very challenging. The most recent improvement was made by SCR from Emapa on December 20, 2021. They reduced the route number for instances C2_8_7 and C2_10_9. AWLS matches these best-known route numbers under a 5-hour limit, even though SCR's results were achieved with unspecific hardware and time constraints.

Apart from Emapa, companies such as Quintiq, Nvidia, Huawei, and Cainiao also worked on this challenging problem. Overall, the industry leads the world record lists for VRPTW. Before this update, the last improvement was in 2016, highlighting the significant effort required to make even small advancements in VRPTW algorithm design. Regarding the academic society, the diversification criterion and squeeze procedure (DS) [15] keeps most world records. The DS algorithm was executed on AMD Opteron 2.4 GHz CPU. For a fair comparison, we scale down the runtime of DS by 936/1,918 = 48.8%.

Table 2 compares the improved results obtained by DS with ours. Column Best shows the best route number found by each algorithm, while columns

[1] https://github.com/HUST-Smart/DLLSMA-VRPTW.
[2] https://www.cpubenchmark.net/cpu_list.php.
[3] https://www.sintef.no/projectweb/top/vrptw/.
[4] http://combopt.org/history/.

Table 2. Comparison with recently improved results from DS and SCR. Bold numbers indicate results also reported by SCR, with unspecific hardware and time limits.

Instance	N	DS			AWLS				
		Best	Time (min)	ScaledT	Best	1 min	5 min	1 h	5 h
C2_4_8	400	11	60	29.3	11	0	5	5	5
RC2_4_5	400	8	10	4.9	8	2	5	5	5
C1_6_6	600	59	60	29.3	59	0	1	5	5
C1_6_7	600	57	60	29.3	57	1	5	5	5
RC2_6_5	600	11	10	4.9	11	5	5	5	5
C1_8_2	800	72	1	0.5	72	5	5	5	5
C1_8_6	800	79	10	4.9	79	1	5	5	5
C1_8_8	800	73	240	117.1	73	2	5	5	5
C2_8_6	800	23	60	29.3	23	0	1	5	5
C2_8_7	800	24	-	-	**23**	0	0	0	5
RC2_8_1	800	18	1	0.5	18	5	5	5	5
C1_10_6	1000	99	10	4.9	99	0	3	5	5
C1_10_7	1000	97	60	29.3	97	0	5	5	5
C1_10_8	1000	92	300	146.4	92	0	4	5	5
C2_10_3	1000	28	10	4.9	28	4	5	5	5
C2_10_6	1000	29	10	4.9	29	5	5	5	5
C2_10_7	1000	29	10	4.9	29	4	5	5	5
C2_10_8	1000	28	300	146.4	28	0	5	5	5
C2_10_9	1000	29	-	-	**28**	0	0	0	1
C2_10_10	1000	28	10	4.9	28	5	5	5	5
Average						1.95	3.95	4.50	4.80

Time and ScaledT report DS's runtime and scaled time in minutes, respectively. Columns 1min, 5min, 1h, and 5h show the number of hits to the best results achieved by AWLS within the corresponding time limits. The table demonstrates that AWLS achieves a 100% hit rate within one hour on all the 18 records held by DS, while DS requires several hours to match the best results on three instances. AWLS also reaches the best results within 5 min on all the 18 instances. For the two new records obtained by SCR with unknown hardware and time limits, AWLS matches the best-known results within 5 h, achieving a 100% hit rate for C2_8_7, highlighting AWLS's efficiency and robustness.

4.3 Comparison on the Complete Benchmark

To further assess the overall performance of AWLS, we conduct experiments on the entire benchmark dataset.

Table 3. Overall performance on all the 300 benchmark instances.

N	Type	PR	IIN	GD	LZ	DS				AWLS			SINTEF
						1 min	10 min	1 h	$\frac{\lceil N \rceil}{200}$ h	1 h	2 h	5 h	
200	R1	**18.2**	**18.2**	**18.2**	**18.2**	**18.2**	**18.2**	**18.2**	-	**18.2**	**18.2**	**18.2**	**18.2**
200	R2	**4.0**	**4.0**	**4.0**	**4.0**	**4.0**	**4.0**	**4.0**	-	**4.0**	**4.0**	**4.0**	**4.0**
200	RC1	**18.0**	**18.0**	**18.0**	**18.0**	**18.0**	**18.0**	**18.0**	-	**18.0**	**18.0**	**18.0**	**18.0**
200	RC2	**4.3**	**4.3**	**4.3**	**4.3**	**4.3**	**4.3**	**4.3**	-	**4.3**	**4.3**	**4.3**	**4.3**
200	C1	**18.9**	**18.9**	**18.9**	**18.9**	**18.9**	**18.9**	**18.9**	-	**18.9**	**18.9**	**18.9**	**18.9**
200	C2	**6.0**	**6.0**	**6.0**	**6.0**	**6.0**	**6.0**	**6.0**	-	**6.0**	**6.0**	**6.0**	**6.0**
CVN		x 694	**694**	**694**	**694**	**694**	**694**	**694**	-	**694**	**694**	**694**	**694**
Time (min)		7.7	-	53	10	1	10	60	-	60	120	300	-
#Run		10	1	5	2	1	1	1	-	1	1	1	-
400	R1	**36.4**	**36.4**	**36.4**	**36.4**	**36.4**	**36.4**	**36.4**	**36.4**	**36.4**	**36.4**	**36.4**	**36.4**
400	R2	**8.0**	**8.0**	**8.0**	**8.0**	**8.0**	**8.0**	**8.0**	**8.0**	**8.0**	**8.0**	**8.0**	**8.0**
400	RC1	**36.0**	**36.0**	**36.0**	**36.0**	**36.0**	**36.0**	**36.0**	**36.0**	**36.0**	**36.0**	**36.0**	**36.0**
400	RC2	8.5	8.6	8.6	8.5	8.5	**8.4**	**8.4**	**8.4**	**8.4**	**8.4**	**8.4**	**8.4**
400	C1	**37.6**	37.7	**37.6**	**37.6**	**37.6**	**37.6**	**37.6**	**37.6**	**37.6**	**37.6**	**37.6**	**37.6**
400	C2	12.0	12.0	11.9	11.7	12.0	11.8	**11.6**	**11.6**	**11.6**	**11.6**	**11.6**	**11.6**
CVN		1385	1387	1385	1382	1385	1382	**1380**	**1380**	**1380**	**1380**	**1380**	**1380**
Time (min)		15.8	-	89	20	1	10	60	120	60	120	300	-
#Run		5	1	5	4	1	1	1	1	1	1	1	-
600	R1	**54.5**	**54.5**	**54.5**	**54.5**	**54.5**	**54.5**	**54.5**	**54.5**	**54.5**	**54.5**	**54.5**	**54.5**
600	R2	**11.0**	**11.0**	**11.0**	**11.0**	**11.0**	**11.0**	**11.0**	**11.0**	**11.0**	**11.0**	**11.0**	**11.0**
600	RC1	**55.0**	**55.0**	**55.0**	**55.0**	**55.0**	**55.0**	**55.0**	**55.0**	**55.0**	**55.0**	**55.0**	**55.0**
600	RC2	11.6	11.6	11.7	11.5	11.6	**11.4**	**11.4**	**11.4**	**11.4**	**11.4**	**11.4**	**11.4**
600	C1	57.5	575.0	57.4	57.4	57.4	57.4	**57.2**	**57.2**	**57.2**	**57.2**	**57.2**	**57.2**
600	C2	17.5	**17.4**	17.5	**17.4**	**17.4**	**17.4**	**17.4**	**17.4**	**17.4**	**17.4**	**17.4**	**17.4**
CVN		2071	2070	2071	2068	2069	2067	**2065**	**2065**	**2065**	**2065**	**2065**	**2065**
Time (min)		18.3	-	105	30	1	10	60	180	60	120	300	-
#Run		5	1	5	6	1	1	1	1	1	1	1	-
800	R1	**72.8**	**72.8**	**72.8**	**72.8**	**72.8**	**72.8**	**72.8**	**72.8**	**72.8**	**72.8**	**72.8**	**72.8**
800	R2	**15.0**	**15.0**	**15.0**	**15.0**	**15.0**	**15.0**	**15.0**	**15.0**	**15.0**	**15.0**	**15.0**	**15.0**
800	RC1	73.0	72.4	**72.0**	**72.0**	72.1	**72.0**	**72.0**	**72.0**	**72.0**	**72.0**	**72.0**	**72.0**
800	RC2	15.7	15.7	15.8	15.6	15.6	**15.4**	**15.4**	**15.4**	**15.4**	**15.4**	**15.4**	**15.4**
800	C1	75.6	75.7	75.4	75.4	75.2	75.1	75.0	**74.9**	**74.9**	**74.9**	**74.9**	**74.9**
800	C2	23.7	23.4	23.5	23.4	23.4	23.4	23.3	23.3	23.3	23.2	**23.1**	**23.1**
CVN		2758	2750	2745	2742	2741	2737	2735	2734	2734	2733	**2732**	**2732**
Time (min)		22.7	-	129	40	1	10	60	240	60	120	300	-
#Run		5	1	5	8	1	1	1	1	1	1	1	-
1000	R1	92.2	**91.9**	**91.9**	**91.9**	**91.9**	**91.9**	**91.9**	**91.9**	**91.9**	**91.9**	**91.9**	**91.9**
1000	R2	**19.0**	**19.0**	**19.0**	**19.0**	**19.0**	**19.0**	**19.0**	**19.0**	**19.0**	**19.0**	**19.0**	**19.0**
1000	RC1	**90.0**	**90.0**	**90.0**	**90.0**	**90.0**	**90.0**	**90.0**	**90.0**	**90.0**	**90.0**	**90.0**	**90.0**
1000	RC2	18.3	18.3	18.5	18.3	18.4	**18.2**	**18.2**	**18.2**	**18.2**	**18.2**	**18.2**	**18.2**
1000	C1	94.6	94.5	94.3	94.4	94.1	94.0	93.9	**93.8**	**93.8**	**93.8**	**93.8**	**93.8**
1000	C2	29.7	29.4	29.5	29.3	29.3	28.9	28.9	28.8	28.8	28.8	**28.7**	**28.7**
CVN		3438	3431	3432	3429	3427	3420	3419	3417	3417	3417	**3416**	**3416**
Time (min)		26.2	-	162	50	1	10	60	300	60	120	300	-
#Run		5	1	5	10	1	1	1	1	1	1	1	-
CPU		P3.0G	P2.8G	O2.3G	P2.8G	O2.4C	O2.4G	O2.4G	O2.4G	O2.3G	O2.3G	O2.3G	-

Table 3 shows the detailed comparison among the proposed algorithm and the state-of-the-art algorithms in the literature, including PR [17], IIN [8], GD [19], and LZ [14]. We group the instances by scales and types, and report the average route numbers obtained by the reference algorithms for each group. To be consistent with the previous literatures, we also present the cumulative number of vehicles (CNV) in each instance scale. The result shows that AWLS outperforms or matches the state-of-the-art algorithms in one hour on all instances, while DS took 5 h to hit the best result on some instances. Moreover, when the time limit for AWLS is extended, better results can be obtained. For example, the CVN of AWLS under 2-hour time limit on the 800-customer instances is smaller than that of DS (and all other reference algorithms), and further improvement can be achieved if the time limit is extended to 5 h.

4.4 Results of Lower Bound Calculation

In this section, we present the results obtained by the proposed lower bound calculation methods. Combined with the upper bounds obtained by our algorithm, we can close most classical benchmark instances. The capacity-based bound, although obtained from a very simple argument, already matches the upper bound for 237 instances. For another 10 instances, the maximum-clique formulation tightens the lower bound up to the upper bound, so that 247 instances in total are now proven optimal.

Table 4. Optimal route numbers proved by the MCP model.

Instance	N	AWLS-UB	TimeWindow-LB	Capacity-LB	Runtime (min)
C1_2_1	200	20	**20**	18	< 1
C1_4_1	400	40	**40**	36	< 1
C1_6_1	600	60	**60**	56	< 1
C1_8_1	800	80	**80**	72	< 1
C1_8_2	800	72	72	72	< 1
C1_10_1	1000	100	**100**	90	< 1
R1_2_1	200	20	**20**	18	< 1
R1_4_1	400	40	**40**	36	< 1
R1_6_1	600	59	**59**	54	< 1
R1_8_1	800	80	**80**	72	< 1
R1_10_1	1000	100	**100**	91	< 1
RC2_2_5	200	4	4	4	< 1

Table 4 lists the instances whose optimal objective values are proven by the maximum clique model. Column AWLS-UB reports the upper bounds obtained by our AWLS. Column TimeWindow-LB presents the lower bounds obtained by the maximum clique model. Column Capacity-LB gives the lower bounds calculated by demand versus capacity. Column Runtime gives the computational time for building graph G' and solving the maximum clique problem in minutes. The numbers in bold highlight the lower bounds obtained by the maximum clique model which are better than those obtained by demand versus capacity. As shown in Table 4, the proposed lower bound calculation method is very efficient, and it is particularly effective on the instances with narrow time windows.

4.5 Effectiveness of Important Strategies

In order to validate the effectiveness of our strategies, we implemented four AWLS variants. The first variant, AWLS-i1e1r1, sets both W_{ej} and W_{in} to 1 (i.e., no extra penalty on ejection). The second variant, AWLS-i2e1r1, increases the insertion penalty by setting W_{in} to 2 to impose more penalties on insertion. The third, AWLS-i1e2r0, disables route weighting in the penalty function. The fourth variant, AWLS-OneObj, minimizes only the total ejected customer weight

Table 5. Comparison on different strategies of AWLS.

Instance	RN	AWLS			AWLS-i1e1r1			AWLS-i2e1r1			AWLS-i1e2r0			AWLS-OneObj		
		5 m	1 h	5 h	5 m	1 h	5 h	5 m	1 h	5 h	5 m	1 h	5 h	5 m	1 h	5 h
C2_4_8	11	5	5	5	4	5	5	4	5	5	0	0	4	4	5	5
RC2_4_5	8	5	5	5	5	5	5	5	5	5	0	1	1	5	5	5
C1_6_6	59	1	5	5	1	1	1	0	1	1	0	0	0	0	1	2
C1_6_7	57	5	5	5	4	5	5	2	5	5	0	0	4	4	5	5
RC2_6_5	11	5	5	5	5	5	5	5	5	5	0	2	4	5	5	5
C1_8_2	72	5	5	5	5	5	5	5	5	5	0	0	4	5	5	5
C1_8_6	79	5	5	5	0	5	5	0	5	5	0	0	0	2	5	5
C1_8_8	73	5	5	5	5	5	5	5	5	5	0	5	5	3	5	5
C2_8_6	23	1	5	5	4	5	5	3	5	5	0	1	3	5	5	5
C2_8_7	24	5	5	5	4	5	5	2	5	5	5	5	5	5	5	5
RC2_8_1	18	5	5	5	5	5	5	5	5	5	5	5	5	5	5	5
C1_10_6	99	3	5	5	1	5	5	0	5	5	0	0	0	0	5	5
C1_10_7	97	5	5	5	4	5	5	3	5	5	5	5	5	5	5	5
C1_10_8	92	4	5	5	2	5	5	0	5	5	0	1	4	1	5	5
C2_10_3	28	5	5	5	5	5	5	5	5	5	5	5	5	5	5	5
C2_10_6	29	5	5	5	5	5	5	5	5	5	2	5	5	5	5	5
C2_10_7	29	5	5	5	5	5	5	5	5	5	1	5	5	5	5	5
C2_10_8	28	5	5	5	5	5	5	1	5	5	5	5	5	5	5	5
C2_10_9	29	5	5	5	5	5	5	5	5	5	5	5	5	5	5	5
C2_10_10	28	5	5	5	5	5	5	5	5	5	5	5	5	5	5	5
Average		4.45	5.00	5.00	3.95	4.80	4.80	3.25	4.80	4.80	1.90	2.75	3.70	3.95	4.80	4.85

$\sum_{v \in EJ} W_c(v)$ rather than balancing it with the ejection pool size. All other components remain unchanged from the original AWLS.

Table 5 reports the hit counts to the route number (column RN) over 5 independent runs under various time limits. The original AWLS consistently achieves the highest hit rate under all time limits. These results indicate that penalizing ejections more than insertions is beneficial, because a large weight will prevent a customer from being ejected, but not encourage it to be inserted into some routes since the customer to insert is randomly picked. A large penalty with delayed influence (the weight has no effect until the ejected customer is inserted again) may improve the diversification of the search. Moreover, disabling route weighting degrades solution quality, underscoring its importance. Regarding the customer selection in the ejection procedure, AWLS comprehensively outperforms the simplified AWLS-OneObj variant. The reason might lie in the fact that customer weights can vary widely during the search, causing the algorithm to insert a single customer at the expense of ejecting many others. In other words, with the weighted penalty only, the search may prefer worse solutions (in terms of the number of ejected customers), which will slow down the convergence efficiency. However, multiplying the weighted penalty by the ejection pool size can calibrate excessive deviation. Based on the above analysis, the proposed strategies are crucial for AWLS's effectiveness and efficiency.

5 Conclusion

We present an adaptive weighting-based local search for the route number minimization of VRPTW. AWLS repeatedly removes routes of a greedy initial solution and adopts route weighting, customer weighting, and a customized objective to repair the time window and capacity violations. AWLS obtains competitive results on 300 classical benchmark instances, and improves the upper bounds reported in the literature for two instances for the first time, which were also recently obtained by an unpublished work from the industry. Combined with two lower bound calculation methods, AWLS reaches the theoretical optimum for 247 of the 300 instances.

Acknowledgments. This work was supported in part by the National Natural Science Foundation of China (NSFC) under Grant 62402191 and 62202192, and the Interdisciplinary Research Program of Hust under Grant 5003300129.

References

1. Christiaens, J., Vanden Berghe, G.: Slack induction by string removals for vehicle routing problems. Transp. Sci. **54**(2), 417–433 (2020)
2. Dantzig, G.B., Ramser, J.H.: The truck dispatching problem. Manage. Sci. **6**(1), 80–91 (1959)
3. Dees, W.A., Smith, R.J.: Performance of interconnection rip-up and reroute strategies. In: 18th Design Automation Conference, pp. 382–390. IEEE (1981)

4. Desrochers, M., Desrosiers, J., Solomon, M.: A new optimization algorithm for the vehicle routing problem with time windows. Oper. Res. **40**(2), 342–354 (1992)
5. Dong, B., Wu, W., Yang, Z., Li, J.: Software defined networking based on-demand routing protocol in vehicle ad-hoc networks. ZTE Commun. **15**(2), 11 (2017)
6. He, J., Cai, L.: Hybrid content distribution framework for large-scale vehicular ad hoc networks. ZTE Commun. **14**(3), 22 (2016)
7. Homberger, J., Gehring, H.: A two-phase hybrid metaheuristic for the vehicle routing problem with time windows. Eur. J. Oper. Res. **162**(1), 220–238 (2005). Logistics: From Theory to Application
8. Ibaraki, T., Imahori, S., Nonobe, K., Sobue, K., Uno, T., Yagiura, M.: An iterated local search algorithm for the vehicle routing problem with convex time penalty functions. Discrete Appl. Math. **156**(11), 2050–2069 (2008), in Memory of Leonid Khachiyan (1952 - 2005)
9. Kallehauge, B., Larsen, J., Madsen, O., Solomon, M.M.: Vehicle routing problem with time windows. In: Desaulniers, G., Desrosiers, J., Solomon, M.M. (eds.) Column Generation, pp. 67–98. Springer-Verlag, New York (2005). https://doi.org/10.1007/0-387-25486-2_3
10. Kindervater, G.A.: Vehicle routing: handling edge exchanges. Local Search in Combinatorial Optimization (1997), publisher: Princeton University Press
11. Kindervater, G.A., Savelsbergh, M.W.: Vehicle routing: handling edge exchanges. In: Local Search in Combinatorial Optimization, pp. 337–360. Princeton University Press (2018)
12. Kohl, N., Madsen, O.B.: An optimization algorithm for the vehicle routing problem with time windows based on Lagrangian relaxation. Oper. Res. **45**(3), 395–406 (1997)
13. Li, C.M., Jiang, H., Manyà, F.: On minimization of the number of branches in branch-and-bound algorithms for the maximum clique problem. Comput. Oper. Res. **84**, 1–15 (2017)
14. Lim, A., Zhang, X.: A two-stage heuristic with ejection pools and generalized ejection chains for the vehicle routing problem with time windows. INFORMS J. Comput. **19**(3), 443–457 (2007)
15. Nagata, Y., Bräysy, O.: A powerful route minimization heuristic for the vehicle routing problem with time windows. Oper. Res. Lett. **37**(5), 333–338 (2009)
16. Nagata, Y., Bräysy, O., Dullaert, W.: A penalty-based edge assembly memetic algorithm for the vehicle routing problem with time windows. Comput. Oper. Res. **37**(4), 724–737 (2010)
17. Pisinger, D., Ropke, S.: A general heuristic for vehicle routing problems. Comput. Oper. Res. **34**(8), 2403–2435 (2007)
18. Potvin, J.Y., Rousseau, J.M.: An exchange heuristic for routeing problems with time windows. J. Oper. Res. Soc. **46**(12), 1433–1446 (1995)
19. Prescott-Gagnon, E., Desaulniers, G., Rousseau, L.M.: A branch-and-price-based large neighborhood search algorithm for the vehicle routing problem with time windows. Networks **54**(4), 190–204 (2009)
20. Schrimpf, G., Schneider, J., Stamm-Wilbrandt, H., Dueck, G.: Record breaking optimization results using the ruin and recreate principle. J. Comput. Phys. **159**(2), 139–171 (2000)

Computational Complexity

Hunting a Rabbit Is Hard

Walid Ben-Ameur[1], Harmender Gahlawat[2(✉)], and Alessandro Maddaloni[1]

[1] SAMOVAR, Télécom SudParis, Institut Polytechnique de Paris, Palaiseau, France
{walid.benameur,alessandro.maddaloni}@telecom-sudparis.eu

[2] Université Clermont Auvergne, CNRS, Clermont Auvergne INP, Mines Saint-Étienne, LIMOS, 63000 Clermont-Ferrand, France
harmendergahlawat@gmail.com

Abstract. In the Hunters and Rabbit game, k hunters attempt to shoot an invisible rabbit on a given graph G. In each round, the hunters can choose k vertices to shoot at, while the rabbit must move along an edge of G. The hunters win if at any point the rabbit is shot. The hunting number of G, denoted $h(G)$, is the minimum k for which k hunters can win, regardless of the rabbit's moves. The complexity of computing $h(G)$ has been the longest standing open problem concerning the game and has been posed as an explicit open problem by several authors. The first contribution of this paper resolves this question by establishing that computing $h(G)$ is NP-hard even for bipartite simple graphs. We also prove that the problem remains hard even when $h(G)$ is $O(n^{\epsilon})$ or when $n-h(G)$ is $O(n^{\epsilon})$, where n is the order of G. Furthermore, we prove that it is NP-hard to additively approximate $h(G)$ within $O(n^{1-\epsilon})$. Finally, we give a characterization of graphs with loops for which $h(G) = 1$ by means of forbidden subgraphs, extending a known characterization for simple graphs.

Keywords: Hunters and rabbit · Complexity · Inapproximability

1 Introduction

The hunters and rabbit game has been studied under several different names. Although we hold nothing against rabbits, we choose to use the terminology of hunters and rabbit, as it is the most widely adopted. This game is played on an undirected graph G with a positive integer k representing the number of hunters. In each round or time step, the hunters shoot at k vertices, while the rabbit occupies a vertex unknown to the hunters (until the rabbit is possibly shot). The rabbit can start the game in a given subset of vertices (usually all of them) and, if the rabbit is not shot, it must move to an adjacent vertex after each round. The rabbit wins if it can ensure that its position is never shot; otherwise, the hunters win.

As an example consider a complete graph on n vertices: here $n-1$ hunters can shoot at the same $n-1$ vertices for two rounds and be sure the rabbit will be

F. V. Fomin and M. Xiao (Eds.): COCOON 2025, LNCS 15983, pp. 165–178, 2026.
https://doi.org/10.1007/978-981-95-0215-8_13

shot. On the other hand, on a path on $n > 2$ vertices $v_1, ..., v_n$, one hunter can win by subsequently shooting at all vertices from v_2, to v_{n-1} and then restart shooting backward $v_{n-1}, ..., v_2$.

The hunters and rabbit game was introduced in [10] for the case $k = 1$, where the authors show that one hunter wins on a tree if and only if it does not contain a 3-spider (H_1 in Fig. 4) as a subgraph. Further, it was also shown that when one hunter can win, he can win in a number of rounds linear in the number of vertices. Very similar results were obtained also in [16]. The minimum number of hunters needed to win on a given graph G is called the *hunting number* of G, denoted $h(G)$. In [1] it is proven that the hunting number is upper bounded by the graph pathwidth plus 1. It is also shown that the hunting number of an $(n \times m)$-grid is $\lfloor \frac{\min(n,m)}{2} \rfloor + 1$ and that the hunting number of trees is $O(\log(n))$. On the other hand, there are trees for which the hunting number is $\Omega(\log(n))$ [15]. In [8] the hunting number of the n dimensional hypercube is proven to be $1 + \sum_{i=1}^{n-2} \binom{i}{\lfloor \frac{i}{2} \rfloor}$. Here the authors also show that the graph degeneracy given by $\max_{S \subseteq V(G)} \delta(D[S])$ (i.e., the maximum through induced subgraphs of their minimum degree) is a lower bound for $h(G)$. In [12] the authors provide a polynomial-time algorithm to determine the hunting number of split graphs. They also show that computing the hunting number on any graph is FPT parameterized by the size of a vertex cover. Moreover they prove that, if a monotone capture is required, the number of hunters must be at least the pathwidth of the graph and it is not possible to additively approximate the monotone hunting number within $O(n^{1-\epsilon})$.

A more general version of the hunters and rabbit problem was also considered in [5,6] where the rabbit moves along the edges of a directed graph D that might also contain loops. It is shown there that it is NP-hard to decide whether the hunting number of a digraph is 1. Computing the hunting number is proved to be FPT parametrized in some generalization of the vertex cover. When the digraph is a tournament, tractability is achieved with respect to the minimum size of a feedback vertex set. The hunting number is also proved to be less than or equal to 1 + the directed pathwidth. An easy to compute lower bound is given by $\max_{S \subseteq V} \max(\delta^+(D[S], \delta^-(D[S])$ (i.e., the maximum through all induced subgraphs of the minimum indegree and the minimum outdegree). When a monotone capture is assumed, the hunting number is proved to be greater than or equal to the directed pathwidth, while pathwidth plus 1 is still a valid upper bound. Another result worth mentioning from [5] is related to the minimum number of shots (regardless of the number of hunters) required to shoot the rabbit. It is proved that this number is easy to compute, and that the rabbit can always be shot before time step n using this minimum number of shots. Some connections with the no-meet matroids of [7], as well as with the matrix mortality problem are also drawn in [6].

The hunters and rabbit game falls within the broader category of cops and robber games, where different versions are defined based on factors such as the available moves for the cops, the robber's speed, and the robber's visibility to the cops. These kind of games are widely studied, for a review see the book [9]. The first cops and robber game was defined in [20,23]. The difference between the hunters and rabbit and this game is that the cops must follow the edges of the graph and can always see the robber. Deciding whether k cops can catch a robber in this version is EXPTIME-complete when k is part of the input [18], but it is polynomial when k is fixed. Cops and robber games variants provide algorithmic interpretations of several graph (width) measures like treewidth [24], pathwidth [22], directed pathwidth [4], directed tree-width [17], DAG-width [3,21]. These games have been intensively studied also due to their applications in numerous fields such as multi-agent systems [2,25], robotics [11], database theory [14], distributed computing [19].

Contributions and paper organization.
In this paper we deal with the hunters and rabbit game on undirected graphs and we also consider the case when those graphs contain loops. Our main result states that computing $h(G)$ is NP-hard. This confirms the sentiment emerging from the literature (e.g. in [1,12]). In particular, we first provide a reduction from 3-PARTITION to the problem of computing the minimum number of hunters, $h_S(G)$, required when the rabbit starts in a subset S. We then provide a polynomial time reduction from computing $h_S(G)$ to computing $h(G)$ for graphs that may contain loops. We conclude by providing a reduction from $h(G)$ on graphs with loops to bipartite graphs. We also prove that the problem remains hard even if $h(G)$ or $n - h(G)$ is as small as $O(n^\epsilon)$, for any $\epsilon > 0$. Approximating the hunting number of a graph within an additive error of $O(n^{1-\epsilon})$ is shown to be NP-hard too, for every $\epsilon > 0$. Finally we extend, to graphs that can contain loops, the characterization from [10,16] of graphs G such that $h(G) = 1$.

The paper is organized as follows: notation and definitions are provided in Sect. 2; in Sect. 3 we derive preliminary properties; in Sect. 4 we prove our main result and the other hardness results; finally, in Sect. 5, we characterize graphs (with loops) where one hunter wins; the paper concludes with a few remarks in Sect. 6.

2 Notation and Some Terminology

Let us start with some notation. $G = (V, E)$ is an undirected graph where $V(G) := V$ (resp. $E(G) := E$) denotes the set of vertices (resp. edges) of G. All graphs considered in this paper contain neither isolated vertices nor parallel edges. One can then use uv to denote an edge whose endpoints are u and v. When it does not contain loops, the graph is said to be simple. The number of vertices is generally denoted by $n(G) := |V|$, or simply n when clear from the context. Let K_n be the complete graph on n vertices without loops, while K_n° denotes the complete graph with loops (so the number of edges is $n(n+1)/2$). Given $A \subset V$, $G[A]$ is the subgraph of G induced by A. Two disjoint subsets of vertices A and B are said to be *fully connected* if each vertex of A is adjacent to each vertex in B. For a vertex v, let $N(u) := \{w \mid uw \in E(G)\}$ and $N[u] := N(u) \cup \{u\}$. Two vertices u and v are said to be *twins* if $N[u] = N[v]$. We will use $[n]$ to denote the set of numbers $\{1, ..., n\}$.

The hunters and rabbit game is played on a graph $G = (V, E)$. We identify with $W_t \subseteq V$ the positions at which hunters are *shooting* at time t, while R_t denotes the *rabbit territory* (i.e., the set of possible positions of the invisible rabbit assuming that he was not yet shot). We have that $R_1 = V \setminus W_1$ and, for $t \geq 2$, $R_t = N(R_{t-1}) \setminus W_t$. A *hunter strategy* $(W_t)_{t \geq 1}$ is a *winning strategy*, if and only if, there exists a finite T such that $R_T = \emptyset$. With a slight abuse of notation, such a winning strategy could be simply identified with the first T sets $W_1, ..., W_T$. Observe that $(W_t)_{t \geq 1}$ is not a winning strategy, if and only if, there exists an *escape walk* of the rabbit (i.e., $(v_t)_{t \geq 1}$ such that $v_t \notin W_t$, and $v_t v_{t+1} \in E(G)$, $\forall t \geq 1$). If $R_t \cap A = \emptyset$, we say that A is *decontaminated* at time t. When $k = \max_{t \geq 1} |W_t|$, we say that the strategy $(W_t)_{t \geq 1}$ uses k hunters. The minimum integer k such that there exist an integer T and a winning strategy $W_1, ..., W_T$ such that $k = \max_{t \in [T]} |W_t|$ is called the *hunting number* of G and denoted by $h(G)$. When the set of possible initial positions of the rabbit is restricted to some subset S, the hunting number is noted $h_S(G)$ (so $h(G) = h_V(G)$). Note that $R_1 = S \setminus W_1$. If $h(G) \leq k$, G is said to be a *k-hunterwin* graph.

3 Preliminary Results

Let k represent the number of hunters ($|W_t| = k$, $\forall$ t ≥ 1) to be dealt with. From the definition of $h_S(G)$, we can ensure that if $k < h_S(G)$, then there are no winning strategies using k hunters if the rabbit starts at S. In other words, $R_t \neq \emptyset$ for any time t. The first lemma provides a reciprocal view that is mainly due to the undirected nature of the graph: using k hunters with $k < h_S(G)$ and assuming that the rabbit can start anywhere (not only in S), then the rabbit territory will always intersect with S (i.e., S cannot be decontaminated). Lemma 1 will be crucial to lift our hardness result for computing $h_S(G)$ to hardness of computing $h(G)$ in general.

Lemma 1. *If $h_S(G) > k$, then any strategy using k hunters in G is such that $R_\tau \cap S \neq \emptyset, \forall \tau \geq 1$.*

Proof. Let $(W_t)_{t \geq 1}$ be any strategy in G using k hunters. Let $\tau \geq 1$ be an integer and assume hunters shoot at $W_\tau, ..., W_1$ in this order: since $h_S(G) > k$, there must exist a rabbit walk $v_1, ..., v_\tau$ with $v_1 \in S$ that survives against $W_\tau, ..., W_1$ (i.e. $v_t \notin W_{\tau+1-t}$ for $t \in [\tau]$). But then $v_\tau, ..., v_1$ is a rabbit walk to S that survives against $W_1, ..., W_\tau$, namely $v_{\tau+1-t} \notin W_t$ for $t \in [\tau]$ and $R_\tau \cap S \neq \emptyset$. □

Let $G = (V, E)$ be a graph whose edge set might contain loops in addition to regular edges. Let B_G be the undirected bipartite graph built from G as follows: $V(B_G) = V \cup V'$ where V' is a copy of V and $E(B_G)$ contains edges $v'w$ and $w'v$ for any edge $vw \in E$. A loop vv of E is then represented by one edge $v'v$ in B_G. B_G can also be seen as the tensor product of G with K_2. Observe that B_G does not contain loops (see Fig. 1 for illustration). Next lemma states that G and B_G have the same hunting number.

Lemma 2. $h(G) = h(B_G)$.

Proof. Let $W_1, W_2, ..., W_T$ be a winning strategy in G using $h(G)$ hunters. A winning strategy W^B of length $2T$ is built in B_G as follows: for $1 \leq t \leq T$, let $W_t^B = \{v : v \in W_t\}$ if t is odd and $W_t^B = \{v' : v \in W_t\}$ for even time t. For $T+1 \leq t \leq 2T$, we take $W_t^B = \{v : v \in W_{t-T}\}$ if t is even and $W_t^B = \{v' : v \in W_{t-T}\}$ otherwise. If the rabbit was initially in V, then he will be shot during the first T iterations, otherwise this occurs between $T+1$ and $2T$. We consequently have $h(B_G) \leq h(G)$.

Conversely, if $(W_t^B)_{t\geq 1}$ is a winning strategy in B_G, then by simple projection on V we get a winning strategy in G: let $W_t = \{v : v \in W_t^B\} \cup \{v : v' \in W_t^B\}$. This implies that $h(B_G) \leq h(G)$. □

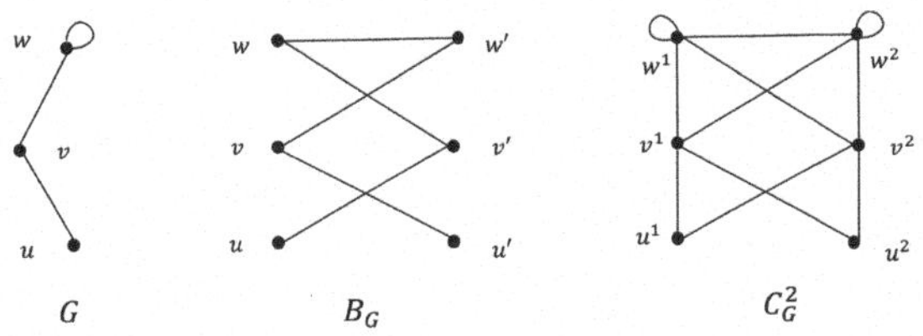

Fig. 1. Illustration of B_G and C_G^p (with $p = 2$)

For some number $p \geq 2$ and some graph $G = (V, E)$, let us build the graph C_G^p as follows: p copies $G^i = (V^i, E^i)_{1\leq i\leq p}$ of G are considered; if some edge $vw \in E$ then C_G^p contains also edges $v^i w^j$ for $1 \leq i, j \leq p$ (see Fig. 1 for illustration). The graph C_G^p can be seen as the tensor product of G with K_p°. We show that $h(C_G^p)$ is simply $p \times h(G)$.

Lemma 3. $h(C_G^p) = p \times h(G)$.

Proof. Consider a winning strategy $(W_t)_{t\geq 1}$ in G using $h(G)$ hunters. We build a strategy $(W_t^C)_{t\geq 1}$ in C_G^p using $p \times h(G)$ hunters as follows: $W_t^C = \bigcup_{1\leq i\leq p}\{v^i : v \in W_t\}$. This strategy is obviously a winning one showing that $h(C_G^p) \leq p \times h(G)$. Assume that $h(C_G^p) < p \times h(G)$. Consider a winning strategy $(W_t^C)_{t\geq 1}$ using $h(C_G^p)$ hunters. Observe that at each time step t, there exists at least one index i (denoted by $i(t)$) such that $|W_t^C \cap V^i| < h(G)$. Consider the strategy $(W_t)_{t\geq 1}$ (in G) defined by $W_t = \{v : v^{i(t)} \in W_t^C\}$. Observe that $|W_t| < h(G)$ implying the existence of a rabbit escape walk $(v_t)_{t\geq 1}$ allowing him to survive the hunter strategy. The rabbit escape walk in G can be transformed into an escape walk $(u_t)_{t\geq 1}$ in C_G^p where $u_t = v_t^{i(t)}$. This leads to contradiction since $(W_t^C)_{t\geq 1}$ was assumed to be a winning strategy. Hence, $h(C_G^p) = p \times h(G)$. □

Given two graphs $G = (V(G), E(G))$ and $H = (V(H), E(H))$, let $G \nabla H$ be the join graph obtained by considering the union of G and H and fully connecting

$V(G)$ and $V(H)$: $V(G\nabla \text{ H}) = V(H) \cup V(G)$ and $E(G\nabla H) = E(G) \cup E(H) \cup \{uv : \text{ u} \in V(G), v \in V(H)\}$. The next lemma states that $h(\cdot)$ is superadditive with respect to ∇ and provides an obvious upper bound.

Lemma 4. $h(G) + h(H) \leq h(G\nabla H) \leq \min(h(G) + n(H), h(H) + n(G))$.

Proof. Consider a winning strategy $(W_t)_{t\geq 1}$ in G using $h(G)$ hunters. Then $(V(H) \cup W_t)_{t\geq 1}$ is obviously a winning strategy in $G\nabla H$ showing that $h(G) + n(H)$ is an upper bound for $h(G\nabla H)$. By symmetry, $h(H)+n(G)$ is also an upper bound. The lower bound is proved using exactly the same technique already used in the proof of Lemma 3. More precisely, given any strategy $(W_t)_{t\geq 1}$ using strictly less than $h(G) + h(H)$ hunters, we have either $|W_t \cap V(G)| < h(G)$ or $|W_t \cap V(H)| < h(H)$ allowing to build a rabbit escape strategy. □

Observe that Lemma 4 implies that $h(G\nabla K_k^\circ) = h(G) + k$ while two inequalities can be obtained if $H = K_k$: $h(G) + k - 1 \leq h(G\nabla K_k) \leq h(G) + k$. Note that $h(K_k\nabla K_k) = h(K_{2k}) = 2k - 1 = h(K_k) + n(K_k) > h(K_k) + h(K_k)$ showing that the upper bound is sharp. On the other hand, let $G(a, p)$ be a graph obtained as the union of K_{a+1}° and a path $v_1...v_p$ on $p \geq 2$ vertices starting at the clique (so $v_1 \in K_{a+1}^\circ$). Observe that $n(G(a, p)) = a + p$ and $h(G(a, p)) = a + 1$. Now consider the join of two distinct copies of $G(a, p)$: we have $h(G(a, p)\nabla G(a, p)) \leq \max(2a + 2, a + p + 2)$. Indeed, two hunters can decontaminate the path on the first copy, while $a + p$ hunters are shooting at the second copy, then, for one step, $2(a + 1)$ hunters can cover the clique on both copies. At this point the rabbit's territory is reduced to $\{v_2, ..., v_p\}$ on the second copy. From there $a + p$ hunters shoot at the first copy, while two hunters decontaminate the path in the second copy. When $2 \leq p \leq a$, we have $h(G(a, p)\nabla G(a, p)) = 2a+2 = h(G(a, p)) + h(G(a, p)) < h(G(a, p)) + n(G(a, p))$, implying that the lower bound is also sharp.

4 Complexity

We start with a reduction from 3-PARTITION to the problem of computing $h_S(G)$. Then the latter is reduced to the problem of computing $h(G)$ in a graph having loops. Finally, using Lemma 2, we deduce that computing the hunting number is NP-hard in a bipartite graph.

Remember that an instance of 3-PARTITION is a multiset S of n positive integers $\{a_1, ..., a_n\}$ with $n = 3m$ for which we aim to decide whether there is a partition $S_1, ..., S_m$ of S such that the sum of the elements in each S_j equals $\beta = 1/m \sum_{i=1}^{n} a_i$. This problem is NP-hard even when the a_i are bounded by a polynomial in n and $\frac{\beta}{4} < a_i < \frac{\beta}{2}$ for $i = 1, ..., n$ [13]. Note that the latter condition implies that the S_j are triplets.

Proposition 1. *It is NP-hard to compute $h_S(G)$.*

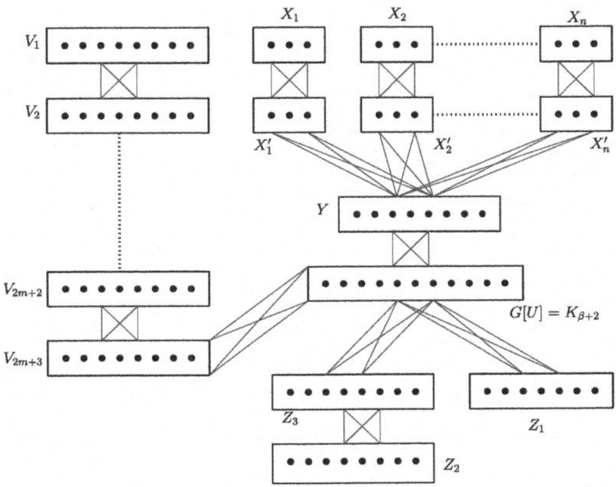

Fig. 2. Illustration for the construction used in the Proof of Proposition 1. Here, each block of vertices other than U is an independent set and U is a clique. Further, $Y, Z_1, Z_2, Z_3, V_1, \dots, V_{2m+3}$ contain β vertices each, U contains $\beta + 2$ vertices, X_i and X'_i contain a_i vertices (for $i \in [n]$). Finally, whenever two blocks, say A and B, are illustrated to be connected by a red connection, A and B are fully connected.

Proof. Let $\mathcal{S}$ be an instance of 3-PARTITION. We show that $\mathcal{S}$ admits a 3-PARTITION if and only if $h_S(G) = \beta$ where G and S are described below and shown in Fig. 2. The instance $\mathcal{S}$ defined by the numbers $(a_i)_{i\in[n]}$ is chosen as described above.

Construction of G. Let $Y, Z_1, Z_2, Z_3, V_1, \dots, V_{2m+3}$ each be a set of β independent vertices. Further, let U be a set of $\beta+2$ vertices that induces a clique in G. Finally, let $X_1, \dots, X_n, X'_1, \dots, X'_n$ be sets of independent vertices such that $|X_i| = |X'_i| = a_i$.

Now, for $i \in [2m + 2]$, fully connect V_i to V_{i+1} (i.e., $G[V_i \cup V_{i+1}]$ induces a complete bipartite graph with both partitions containing β vertices). Similarly, fully connect X_i to X'_i, and Z_2 to Z_3. Moreover, $Z_3 \cup Z_1 \cup V_{2m+3}$ and U are fully connected, and for $i \in [n]$, each X'_i is fully connected to U. Finally, let $S = Z_1 \cup Z_2 \cup V_1 \cup X_1 \cup \dots \cup X_n$ be the allowed starting positions of the rabbit.

In one direction, suppose $\mathcal{S}$ admits a 3-PARTITION $S_1, \dots, S_m$. Then, we claim that the hunter strategy $Z_1, Z_3, Y, S_1, Y, S_2, Y, S_3, Y, \dots, S_m, V_{2m+3}, V_{2m+2}, \dots, V_2$ is a winning one (here with a slight abuse of notation we are indicating with S_j the set $\bigcup_{i|a_i \in S_j} X'_i$). To ease the exposition, we provide a case by case analysis distinguished by the starting positions of the rabbit.

1. The rabbit starts in Z_1: The rabbit is shot in the first round since $W_1 = Z_1$.
2. The rabbit starts in Z_2: The rabbit is shot in the second round.
3. The rabbit starts in X_i for some $i \in [n]$: In this case, first we establish that the rabbit will be restricted to vertices in $X_1, \dots, X_n, X'_1, \dots, X'_n$ until round $2m + 2$. To this end, observe that if the rabbit needs to leave these vertices, it

needs to reach a vertex in Y, and it can only do so at an odd time step greater than 1. But this is not possible since $W_t = Y$ for odd rounds $3 \leq t \leq 2m+1$. Note that $G[X_1 \cup \ldots \cup X_n \cup X'_1 \cup \ldots \cup X'_n]$ is bipartite, thus $R_t \subseteq X_1 \cup \ldots \cup X_n$ when t is odd and $R_t \subseteq X'_1 \cup \ldots \cup X'_n$ when t is even, until round $2m+2$. Furthermore, observe that when $W_t \supseteq X'_i$, the set $X_i \cup X'_i$ becomes decontaminated and remains so for all subsequent rounds (until round $2m+2$) unless the rabbit moves to some vertex in Y, which is not possible. Finally, let the subset S_j, $j \in [m]$, contain numbers a_p, a_q, and a_r. Then, in the round $2j+2$, hunters shoot at all vertices in X'_p, X'_q, X'_r, and hence decontaminate $X_p \cup X'_p \cup X_q \cup X'_q \cup X_r \cup X'_r$ for all subsequent rounds (until round $2m+2$). Since $S_1, \ldots, S_m$ form a partition, after $2m+2$ rounds all vertices in $X_1, \ldots, X_n, X'_1, \ldots, X'_n$ will be decontaminated and as noted above, the rabbit is restricted to only these vertices for all these rounds. Hence, the rabbit gets shot.
4. The rabbit starts in V_1: Observe that the induced graph $G[\bigcup_{i\in[2m+3]} V_i]$ is bipartite, therefore, since $R_1 = V_1$ and $W_{2m+3} = V_{2m+3}$, we have $R_{2m+3} = \bigcup_{i\in[m+1]} V_{2i-1}$. Then at time $2m+4$, we have $W_{2m+4} = V_{2m+2}$ and $R_{2m+4} = \bigcup_{i\in[m]} V_{2i}$. A simple induction on t shows that, for $1 \leq t \leq m+1$, $R_{2m+2t} = \bigcup_{i\in[m-t+2]} V_{2i}$ and $R_{2m+2t+1} = \bigcup_{i\in[m-t+2]} V_{2i-1}$. Therefore, $R_{4m+3} = V_1$, $W_{4m+4} = V_2$ and thus $R_{4m+4} = \emptyset$.

In the other direction, suppose $\mathcal{S}$ does not admit a 3-PARTITION. We will show that the hunters do not have a winning strategy using only β hunters. Assume, by contradiction, that such a winning strategy exists. We begin by observing that once the rabbit reaches a vertex in U (i.e., $R_t \cap U \neq \emptyset$ for some $t > 0$), then the rabbit can never be shot since the hunting number of $G[U]$ is greater than β. Thus, to complete our proof, we only need to show that for any strategy of k hunters the rabbit has a walk that ensures $R_t \cap U \neq \emptyset$ for some $t > 0$. Furthermore, observe that all $\beta+2$ vertices of U are twins and there are at most β hunters, thus if $R_{t-1} \cap (Y \cup Z_3 \cup Z_1 \cup V_{2m+3}) \neq \emptyset$ for some $t > 1$, then $R_t \cap U \neq \emptyset$. As a consequence, we can safely assume that $W_1 = Z_1$ and $W_2 = Z_3$ (otherwise the rabbit can reach U in time step 2 and 3, respectively). Similarly, $W_3 = Y$, otherwise the rabbit has an escape strategy by starting on some X_i, moving to X'_i in the second time step, then to some $v \in Y \setminus W_3$ in the third time step, and finally moving to $U \setminus W_4$ in the fourth time step.

Let T be the first time step such that $R_T \cap \bigcup_{i\in[n]}(X_i \cup X'_i) = \emptyset$. For each odd time step t between 3 and T, $W_t = Y$ holds (otherwise the rabbit can reach U in the next time step). Observe that T is necessarily an even number. Moreover, observe that since hunters are shooting at Y in every odd time step, to decontaminate $X_i \cup X'_i$ the hunters must shoot at every vertex in X'_i in some even round. Furthermore, since hunters were able to decontaminate $\bigcup_{i\in[n]}(X_i \cup X'_i)$, they need to shoot at vertices in $\bigcup_{i\in[n]} X'_i$ for at least $m+1$ (even) time steps. This is obviously due to the non-feasibility of the 3-PARTITION instance implying that it is not possible to cover all vertices of $\bigcup_{i\in[n]} X'_i$ with only m subsets of size β. For $t \leq T$, let l_t be the largest index such that $R_t \cap V_{l_t} \neq \emptyset$ (i.e., the index of the lowest V_i in Fig. 2 that is not decontaminated at time t). We clearly have $l_1 = 1$, $l_2 = 2$ and $l_3 = 3$. Observe that if at some even time t the hunters shoot at $V_{l_{t-1}+1}$, then $l_t = l_{t-1} - 1$. Otherwise, if the hunters shoot at other vertices (i.e., $W_t \neq V_{l_{t-1}+1}$), for example, in $\bigcup_{i\in[n]} X'_i$, then $l_t = l_{t-1} + 1$. If t is odd, $W_t = Y$ implying that $l_t = l_{t-1} + 1$. Then, if we consider two consecutive time slots t and $t+1$, we either have $l_{t+1} = l_{t-1}$ or $l_{t+1} = l_{t-1} + 2$. Since hunters have to shoot at $\bigcup_{i\in[n]} X'_i$ for at least $m+1$ time steps, $l_{t+1} - l_{t-1}$ increases by 2, at least $m+1$ times (for some odd t). Since T is even and hunters are shooting at (even partially) $\bigcup_{i\in[n]} X'_i$ at time T, $l_{t+1} - l_{t-1}$ increases by 2, at least m times between 3 and $T-2$. One can then write that $l_{T-2} - l_2 = (l_{T-2} - l_{T-4}) + (l_{T-4} - l_{T-6}) + \cdots + (l_4 - l_2) \geq 2m$ implying that $l_{T-2} \geq 2m + 2$ and $l_{T-1} \geq 2m + 3$ (since $W_{T-1} = Y$). The rabbit can then reach U at time T implying that the hunter's strategy was not a winning one. □

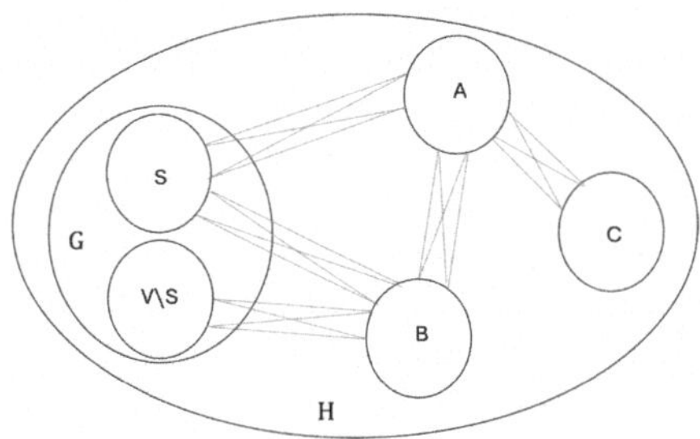

Fig. 3. Building H from G (Proposition 2): $1 \leq k \leq |S|$, $|A| = n-k$, $|B| = k$, $|C| = 2k$, $H[A] = K^\circ_{n-k}$, $H[B] = K^\circ_k$ and $H[C] = K^\circ_{2k}$ are complete graphs with loops.

Proposition 2. *It is NP-hard to compute $h(G)$ in a graph with loops.*

Proof. We prove the result using a reduction from the problem of computing $h_S(G)$. Consider a graph $G = (V, E)$ and a subset $S \subset V$ representing the possible initial positions of the rabbit. Let $n = |V|$ and let k be a number satisfying $1 \leq k \leq |S|$. We build a graph H as follows. The graph H contains G as a subgraph in addition to 3 complete subgraphs with loops: $H[A] = K^\circ_{n-k}$, $H[B] = K^\circ_k$ and $H[C] = K^\circ_{2k}$. The set A is fully connected to B, C and S while B is fully connected to V (and A) (see Fig. 3). Observe that $H[A \cup C] = K^\circ_{n+k}$ implying that $h(H) \geq n+k$. We claim that $h(H) = n+k$ if and only if $h_S(G) \leq k$.

Claim. $h_S(G) \leq k$, if and only if $h(H) = n + k$.

Proof. Assume that $h(H) = n + k$ and $h_S(G) > k \geq 1$. Let $W_1, ..., W_T$ be a winning strategy in H using $n + k$ hunters.

Let us first assume that there exists $t \geq 1$ such that $R_t \cap (A \cup B) \neq \emptyset$. Let δ be the last t such that $R_t \cap (A \cup B) \neq \emptyset$. Then, $W_{\delta+1} \supset (A \cup B)$ holds, implying $|W_{\delta+1} \cap (V \cup C)| \leq k$.

If $R_\delta \cap A \neq \emptyset$, then $R_{\delta+1} \cap C \neq \emptyset$ and $R_{\delta+1} \cap S \neq \emptyset$ since $|S| > k$ and $|C| > k$. Using that S (and even V) is connected to B and C is connected to A, we deduce that $W_{\delta+2}$ should also contain $A \cup B$ and $R_{\delta+2} \cap C \neq \emptyset$. By induction on $t \geq \delta + 1$, as long as $R_t \cap V \neq \emptyset$, we have $W_{t+1} \supset A \cup B$ and $R_{t+1} \cap C \neq \emptyset$. Since the game is supposed to end, there exists some T' between $\delta + 1$ and T such that $R_{T'} \cap V = \emptyset$. Observe that between rounds $\delta + 1$ and T' at most k hunters can shoot at vertices in V, while R_δ contains a vertex (of A) fully connected with S. Therefore, by considering the game played between rounds $\delta + 1$ and T' restricted to V, we obtain $h_S(G) \leq k$, which is a contradiction.

Assume now that $R_\delta \cap B \neq \emptyset$, then $R_{\delta+1} \cap S \neq \emptyset$ implying that $W_{\delta+2} \supset (A \cup B)$ (therefore, $|W_{\delta+2} \cap V| \leq k$). Since the rabbit can enter V (from B) at time $1 + \delta$ and $k < h_S(G)$, we can deduce from Lemma 1 that $R_{\delta+2} \cap S \neq \emptyset$. A simple induction on $t \geq \delta + 1$ shows that we will always have $W_{t+1} \supset A \cup B$ and $R_{t+1} \cap S \neq \emptyset$ implying that the game will never end.

Note that a crucial fact in the proof of the first (resp. second) case is that the rabbit can enter S (resp. V) at time $1 + \delta$, and at most $k < h_S(G)$ hunters can shoot at vertices of V from that time on.

Let us now assume that δ does not exist which is equivalent to say that $R_t \cap (A \cup B) = \emptyset \ \forall t \geq 1$. We consequently have $W_1 \supset (A \cup B)$, $R_1 \cap V \neq \emptyset$ and $R_1 \cap S \neq \emptyset$. The situation is then similar to the previous case where we had $R_\delta \cap B \neq \emptyset$ since the rabbit can be at any vertex of $V \backslash W_1$ at time 1. The same induction on t shows that $W_{t+1} \supset A \cup B$ and $R_{t+1} \cap S \neq \emptyset$ implying that the game is endless.

Finally, let us now show that $h(H) = n + k$ when $k \geq h_S(G)$. This can be done by providing a winning strategy. A possible one starts with $W_1 = B \cup V$ leading to $R_1 = A \cup C$. Assume that $W_1', W_2', ..., W_p'$ is a winning strategy (for some time p) allowing to shoot the rabbit in G if he starts at S. We can then take $W_2 = W_1' \cup A \cup B$, $W_3 = W_2' \cup A \cup B$, etc. So $W_{p+1} = W_p' \cup A \cup B$. At time $p + 1$ we have $R_{p+1} = C$. Therefore, by setting $W_{p+2} = A \cup C$, the whole graph is decontaminated. □

The claim immediately proves the result since one can compute $h_S(G)$ by computing the hunting number of $O(log(n))$ graphs of type H for different values of k. □

We are now able to state the main complexity result.

Theorem 1. *It is NP-hard to compute $h(G)$ in a bipartite graph.*

Proof. This is a consequence of Lemma 2 and Proposition 2. □

Let us focus now on the existence of polynomial-time approximation algorithms with additive guarantees. An obvious $O(n)$ additive guarantee is given by the upper bound n for the hunting number. We prove that it is not possible to do much better than $O(n)$.

Theorem 2. *It is NP-hard to additively approximate $h(G)$ within $O(n^{1-\epsilon})$ for any constant $\epsilon > 0$.*

Proof. Assume that we have a polynomial-time approximation algorithm with an $O(n^{1-\epsilon})$ additive guarantee. Given any graph G, let us build the graph C_G^p with $p = \lceil n^{2\frac{1-\epsilon}{\epsilon}} \rceil$. Using the approximation algorithm, we get an upper bound u satisfying inequalities $h(C_G^p) \leq u \leq h(C_G^p) + O((np)^{1-\epsilon})$. From Lemma 3, we know that $h(C_G^p) = p \times h(G)$ implying that $\frac{u}{p} - \frac{1}{p}O((np)^{1-\epsilon}) \leq h(G) \leq \frac{u}{p}$. Using that $p = \lceil n^{2\frac{1-\epsilon}{\epsilon}} \rceil$ leads to $\frac{u}{p} - O(\frac{1}{n^{1-\epsilon}}) \leq h(G) \leq \frac{u}{p}$. Since $h(G)$ is integer and $O(\frac{1}{n^{1-\epsilon}})$ is negligible, we get a polynomial-time algorithm to compute $h(G)$ (the construction is obviously polynomial for constant ϵ). Theorem 1 allows to conclude. □

Since graphs for which $h(G) = 1$ or $h(G) = n$ are well characterized, one might expect polynomial-time algorithms for either small values of h or large values of h. The next theorem states that this is not the case when either h or $n - h$ are upper bounded by a (small) power of n.

Theorem 3. *It is NP-hard to compute $h(G)$ for simple graph instances where $n - h(G) = O(n^\epsilon)$ (resp. $h(G) = O(n^\epsilon)$) for any constant $\epsilon > 0$.*

Proof. Given a simple graph G with $n = |V(G)|$, take k so that $n \leq k^\epsilon$ and consider the graph $G\nabla K_k$. Notice that $G\nabla K_k$ does not contain loops. From Lemma 4, we have $h(G) + k - 1 \leq h(G\nabla K_k) \leq h(G) + k$ implying that $h(G\nabla K_k) - k \leq h(G) \leq h(G\nabla K_k) - k + 1$. Furthermore $n(G\nabla K_k) - h(G\nabla K_k) \leq (n + k) - h(G) - k + 1 \leq n \leq (n + k)^\epsilon$, therefore $n(G\nabla K_k) - h(G\nabla K_k)$ is $O(n(G\nabla K_k)^\epsilon)$. A polynomial-time algorithm computing the hunting number if $n - h = O(n^\epsilon)$, applied on $G\nabla K_k$, can be used to get a lower bound and an upper bound for $h(G)$, whose difference is 1. In other words, we have a constant additive approximation for $h(G)$ contradicting Theorem 2.

To prove NP-hardness for instances where $h = O(n^\epsilon)$, one can start from a graph G and add a stable set of size $\Omega(n^{\frac{1}{\epsilon}})$ to get a new graph H for which $h(G) = h(H) = O(n(H)^\epsilon)$. □

5 Characterization of 1-Hunterwin Graphs with Loops

It is possible to linearly characterize graphs for which the hunting number equals 1 using the characterization in [10,16] and Lemma 2. Nonetheless we provide a characterization by means of forbidden subgraphs extending the one in [10,16] for loopless graphs. We will prove that cycles, connected loops and the 4 graphs H_1, H_2, H_3 and H_4 shown on Fig. 4 are precisely the forbidden subgraphs in a 1-hunterwin graph with loops. Notice that H_1 is the 3-spider graph used in [10,16] to characterize 1-hunterwin acyclic simple graphs.

Theorem 4. *A graph G is 1-hunterwin if and only if it does not contain cycles, two connected loops, H_1, H_2, H_3 or H_4.*

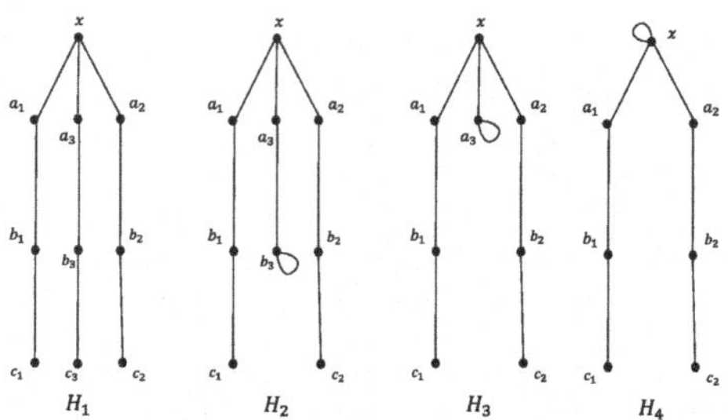

Fig. 4. The 4 forbidden graphs H_1, H_2, H_3 and H_4

Proof. First of all let us apply the construction of Lemma 2 and obtain the graph B_G. We will prove that G contains a forbidden subgraph among those listed above if and only if B_G contains a cycle or a 3-spider.

If G contains a cycle $v_1 v_2 ... v_k v_1$, then B_G contains the cycle $v_1 v_2' ... v_k' v_1$ if k is even and $v_1 v_2' ... v_k v_1' v_2 ... v_k' v_1$ if k is odd.

If G contains two loops on v_1 and v_k and a path $v_1 ... v_k$, then B_G contains the cycle $v_1 v_2' ... v_k' v_k ... v_2 v_1' v_1$ if k is even and $v_1 v_2' ... v_k v_k' ... v_2 v_1' v_1$ if k is odd.

If G contains paths $c_1 b_1 a_1 x a_2 b_2 c_2$ and $x a_3 b_3 c_3$ (i.e., H_1), then B_G will contain the 3-spider formed by $c_1' b_1 a_1' x a_2' b_2 c_2'$ and $x a_3' b_3 c_3'$. If G contains paths $c_1 b_1 a_1 x a_2 b_2 c_2$ and $x a_3 b_3$ with a loop on b_3 (i.e., H_2), then B_G contains the 3-spider formed by $c_1' b_1 a_1' x a_2' b_2 c_2'$ and $x a_3' b_3 b_3'$. If G contains paths $c_1 b_1 a_1 x a_2 b_2 c_2$ and $x a_3$ with a loop on a_3 (i.e., H_3), then B_G contains the 3-spider formed by $c_1' b_1 a_1' x a_2' b_2 c_2'$ and $x a_3' a_3 x'$. If G contains paths $c_1 b_1 a_1 x a_2 b_2 c_2$ and a loop on x (i.e., H_4), then B_G contains the 3-spider formed by $c_1' b_1 a_1' x a_2' b_2 c_2'$ and $x x' a_1 b_2'$.

Now suppose B_G has a cycle or a 3-spider S. If S contains two edges aa', bb', then S contains a path joining at least two of the endpoints of the above edges. Such a path projects into a walk joining a and b in G, implying that G contains two connected loops. We can thus suppose that S contains at most one edge of the form aa'. If S does not contain two copies v, v', then its projection on G is a cycle (resp. 3-spider), since S is a cycle (resp. 3-spider). If S contains two copies of the same vertex, let v, v' be a pair of such copies that are closest in S: if vv' is not an edge of S, then a shortest path of S joining v and v' has length at least 3 and does not contain two copies of the same vertex, thus it projects into a cycle of G. We can thus assume in what follows that S contains an edge vv' and no other edge of this form.

If S is a cycle, let S' be the subpath of S joining v and v' having length at least 3: if S' does not contain copies of the same vertex, then it projects into a cycle of G; if S' contains copies of the same vertex, let w, w' be a pair of such copies that are closest in S': there must exist a path in S' joining w and w' of length at least 3 not containing copies of the same vertex, this path projects into a cycle of G.

If S is a 3-spider, let P be the (possibly empty) path in S joining the center of the spider x and the endpoint of the edge vv' closest to x (note that the edge vv' does not belong to P). Let L_1, L_2 be the two legs of the spider that do not contain vv'. If in $P \cup L_1 \cup L_2$ there are no copies of the same vertex, then the projection of $P \cup L_1 \cup L_2 \cup vv'$ into G is a path $c_1b_1a_1xa_2b_2c_2$ and a path xa_3v with a loop on v or a path xv with a loop on v or a loop on x (when the length of P is respectively $2, 1$ or 0) corresponding respectively to H_2, H_3 and H_4. Otherwise, if there are copies of the same vertex in $P \cup L_1 \cup L_2$, let w, w' be the closest such pair: there must exist a path in $P \cup L_1 \cup L_2$, of length at least 3, joining w and w' not containing copies of the same vertex, this path projects into a cycle of G. □

6 Final Remarks

We know that deciding whether one hunter can win is polynomial, but the complexity of deciding whether the hunting number is less than some given constant k is unknown for $k \geq 2$ even in the case of trees.

Deciding whether $h(G) \leq k$ can be seen as a special case of the integer matrix mortality problem where given m binary square matrices of size n, one wants to determine whether any product obtained using these matrices results in the zero matrix; this is PSPACE-complete (see [6] for references). This implies that our problem is in PSPACE and might well be PSPACE-complete, since we have no evidence that it belongs to NP. In this direction, it would also be interesting to bound the number of rounds in a fastest winning hunter strategy.

Acknowledgments. We would like to thank Antoine Amarilli for preliminary discussions about the problem. This research benefited from the support of the FMJH Program Gaspard Monge for optimization and operations research and their interactions with data science, the IDEX-ISITE initiative CAP 20–25 (ANR-16-IDEX-0001), the International Research Center "Innovation Transportation and Production Systems" of the I-SITE CAP 20–25, and the ANR project GRALMECO (ANR-21-CE48-0004).

Disclosure of Interests. The authors declare that they do not have any competing interests.

References

1. Abramovskaya, T.V., Fomin, F.V., Golovach, P.A., Pilipczuk, M.: How to hunt an invisible rabbit on a graph. Eur. J. Comb. **52**, 12–26 (2016)
2. Alejandro Isaza, A.I., Lu, J., Bulitko, V., Greiner, R.: A cover-based approach to multi-agent moving target pursuit. In: Proceedings of the AAAI Conference on Artificial Intelligence and Interactive Digital Entertainment, vol. 4, issue (1), pp. 54–59 (2021)
3. Bang-Jensen, J., Larsen, T.M.: Dag-width and circumference of digraphs. J. Graph Theory **82**(2), 194–206 (2016)
4. Barát, J.: Directed path-width and monotonicity in digraph searching. Graphs Comb. **22**(1), 161–172 (2006)
5. Ben-Ameur, W., Maddaloni, A.: A cops and robber game and the meeting time of synchronous directed walks. Networks **84**(2), 238–251 (2024)

6. Ben-Ameur, W., Maddaloni, A.: Complexity results for a cops and robber game on directed graphs. Networks (2025)
7. Ben-Ameur, W., Kushik, N., Maddaloni, A., Neto, J., Watel, D.: The no-meet matroid. Discret. Appl. Math. **354**, 94–107 (2024)
8. Bolkema, J., Groothuis, C.: Hunting rabbits on the hypercube. Discret. Math. **342**(2), 360–372 (2019)
9. Bonato, A.: An invitation to pursuit-evasion games and graph theory. Am. Math. Soc. (2020)
10. Britnell, J.R., Wildon, M.: Finding a princess in a palace: a pursuit-evasion problem. Electr. J. Comb. **20**(1), (2013)
11. Chung, T., Hollinger, G., Isler, V.: Search and pursuit-evasion in mobile robotics. Auton. Robots **31**, 11 (2011). https://doi.org/10.1007/s10514-011-9241-4
12. Dissaux, T., Fioravantes, F., Galhawat, H., Nisse, N.: Further results on the hunters and rabbit game through monotonicity. Inf. Comput., 305 (2025)
13. Garey, M.R., Johnson, D.S.: Computers and Intractability: A Guide to the Theory of NP-Completeness. W.H. Freeman & Co Ltd. (1979)
14. Gottlob, G., Leone, N., Scarcello, F.: Robbers, marshals, and guards: game theoretic and logical characterizations of hypertree width. J. Comput. Syst. Sci. **66**(4), 775–808 (2003). Special Issue on PODS 2001
15. Gruslys, V., Méroueh, A.: Catching a mouse on a tree. arXiv e-prints, art. arXiv:1502.06591 (2015)
16. Haslegrave, J.: An evasion game on a graph. Discret. Math. **314**, 1–5 (2014)
17. Johnson, T., Robertson, N., Seymour, P., Thomas, R.: Directed tree-width. J. Comb. Theory. Ser. B **82**(1), 138–154 (2001)
18. Kinnersley, W.B.: Cops and robbers is EXPTIME-complete. J. Comb. Theory, Ser. B **111**, 201–220 (2015)
19. Nisse, N.: Network decontamination. In: Flocchini, P., Prencipe, G., Santoro, N., (eds.) Distributed Computing by Mobile Entities: Current Research in Moving and Computing, pp. 516–548 (2019)
20. Nowakowski, R., Winkler, P.: Vertex-to-vertex pursuit in a graph. Discret. Math. **43**(2), 235–239 (1983)
21. Obdržálek, J.: DAG-width: connectivity measure for directed graphs. In: Proceedings of the Seventeenth Annual ACM-SIAM Symposium on Discrete Algorithm, SODA '06, pp. 814–821, USA (2006). Society for Industrial and Applied Mathematics. ISBN 0898716055
22. Parsons, T.D.: Pursuit-evasion in a graph. Theory Appl. Graphs, LNCS **642**, 426–441 (1978)
23. Quillot, A.: Jeux et pointes fixes sur les graphes. PhD thesis, Université Paris VI (1978)
24. Seymour, P., Thomas, R.: Graph searching and a min-max theorem for tree-width. J. Comb. Theory, Ser. B **58**(1), 22–33 (1993)
25. Stern, R.: Multi-agent path finding – an overview. In: Osipov, G.S., Panov, A.I., Yakovlev, K.S. (eds.) Artificial Intelligence. LNCS (LNAI), vol. 11866, pp. 96–115. Springer, Cham (2019). https://doi.org/10.1007/978-3-030-33274-7_6

A Nearly-4 $\log n$ Depth Lower Bound for Formulas With Restriction on Top

Hao Wu(✉)

College of Information Engineering, Shanghai Maritime University, Shanghai, China
haowu@shmtu.edu.cn

Abstract. One of the major open problems in complexity theory is to demonstrate an explicit function which requires super logarithmic depth, a.k.a, the **P** versus $\mathbf{NC}^1$ problem. The current best depth lower bound is $(3 - o(1)) \cdot \log n$, and it is widely open how to prove a super-3 $\log n$ depth lower bound. Recently Mihajlin and Sofronova (CCC'22) [18] show if considering formulas with restriction on top, we can break the $3 \log n$ barrier. Formally, they prove there exist two functions $f : \{0,1\}^n \to \{0,1\}, g : \{0,1\}^n \to \{0,1\}^n$, such that for any constant $0 < \alpha < 0.4$ and constant $0 < \epsilon < \alpha/2$, their XOR composition $f(g(x) \oplus y)$ is not computable by an AND of $2^{(\alpha-\epsilon)n}$ formulas of size at most $2^{(1-\alpha/2-\epsilon)n}$. This implies a modified version of Andreev function is not computable by any circuit of depth $(3.2 - \epsilon) \log n$ with the restriction that top $0.4 - \epsilon$ layers only consist of AND gates for any small constant $\epsilon > 0$. They ask whether the parameter α can be push up to nearly 1 thus implying a nearly-3.5 $\log n$ depth lower bound.

In this paper, we provide a stronger answer to their question. We show there exist two functions $f : \{0,1\}^n \to \{0,1\}, g : \{0,1\}^n \to \{0,1\}^n$, such that for any constant $0 < \alpha < 2 - o(1)$, their XOR composition $f(g(x) \oplus y)$ is not computable by an AND of $2^{\alpha n}$ formulas of size at most $2^{(1-\alpha/2-o(1))n}$. This implies a $(4 - o(1)) \log n$ depth lower bound with the restriction that top $(2 - o(1)) \log n$ layers only consist of AND gates. We prove it by observing that one crucial component in Mihajlin and Sofronova's work, called the well-mixed set of functions, can be significantly simplified thus improved. Then with this observation and a more careful analysis, we obtain these nearly tight results. It is pointed out that actually similar result already follows from Komargodski, Raz and Tal's average-case lower bound [12]. The advantage of our work is not providing quantitatively better parameters. Instead, the main merit of our result is that our proof is based on a top-down approach and provides a new angle to questions about composition of functions.

Keywords: Circuit depth lower bound · Formula size lower bound · XOR composition

1 Introduction

One of the major open problems in complexity theory is to demonstrate an explicit function which requires super logarithmic depth, a.k.a, the **P** versus

F. V. Fomin and M. Xiao (Eds.): COCOON 2025, LNCS 15983, pp. 179–192, 2026.
https://doi.org/10.1007/978-981-95-0215-8_14

$\mathbf{NC^1}$ problem. The current best depth lower bound [8,20,21] is $(3 - o(1)) \cdot \log n$, and we still don't even know how to obtain a lower bound strictly larger than $3 \log n$. One promising approach to tackle this problem was suggested by Karchmer, Raz and Wigderson [11], they proposed that we should understand the complexity of (block)-composition of Boolean functions. Given two functions $f : \{0,1\}^m \to \{0,1\}$, $g : \{0,1\}^n \to \{0,1\}$, we define their composite function $f \diamond g : (\{0,1\}^n)^m \to \{0,1\}$ as: $f \diamond g\,(x_1, \ldots, x_m) = f\,(g\,(x_1)\,, \ldots, g\,(x_m))\,.$ Given any Boolean function f, we denote the depth complexity of f by $\mathsf{D}(f)$, that is the minimal depth of a circuit of AND, OR and NOT gates of fan-in 2 that computes f. And it is easy to see the depth complexity of $f \diamond g$ is upper-bounded by $\mathsf{D}(f) + \mathsf{D}(g)$ and it is natural to ask whether the depth complexity of $f \diamond g$ is far from this upper bound. Karchmer, Raz and Wigderson [11] conjectured that the depth complexity of $f \diamond g$ is not far from its upper bound:

Conjecture 1. Given two arbitrary non-constant Boolean functions $f : \{0,1\}^m \to \{0,1\}$ and $g : \{0,1\}^n \to \{0,1\}$, then $\mathsf{D}(f \diamond g) \approx \mathsf{D}(f) + \mathsf{D}(g)$.

If the conjecture is proved and the "approximate equality" is instantiated with proper parameters, by an argument of iterative composition [11], we will obtain an explicit function with super-logarithmic depth, which separates $\mathbf{P}$ from $\mathbf{NC^1}$. Recently, Cook and Mertz [2] even show that if the conjecture is true, $\mathbf{L}$ is not contained in $\mathbf{NC^1}$.

Many restricted cases of KRW conjecture have been proved to be true. For example, there are composition theorems when the inner function g satisfies certain property [3,5,8,20]. There are composition theorems about universal relation [4,6,9,13,17,22]. There are composition theorems where the composition itself is restricted such as monotone composition, semi-monotone composition [19] and strong composition [15]. There are also some variants [4,14,17] of original conjecture with the similar effect to the $\mathbf{P}$ versus $\mathbf{NC^1}$ problem, but we don't know how to prove them either. Maybe to prove the general form of KRW conjecture is far out of our reach now.

Note that we don't even know how to prove a super-$3 \log n$ depth lower bound, maybe we should consider following weaker conjecture which suffices to break the $3 \log n$ barrier in the first place.

Conjecture 2. There exist two non-constant Boolean functions $f, g : \{0,1\}^n \to \{0,1\}$ such that $\mathsf{D}(f \diamond g) \geq (1 + \epsilon)n$ for some small constant $\epsilon \in (0,1)$.

Unfortunately, we don't even know how to prove this weaker conjecture. Currently, the closest answer to Conjecture 2 is Meir's strong composition theorem [15], but we don't know how to prove it for the case of standard composition. Mihajlin and Sofronova [18] proposed we should consider proving depth lower bound against even weaker formulas by considering restriction on top of the formulas. They managed to prove a composition theorem for formulas with restriction on top via XOR composition. The so-called XOR composition, proposed by Mihajlin and Smal [17], is a useful special case of standard composition. In fact, the first nontrivial composition theorem [17] of a universal relation and some function is proved via XOR composition.

Given two functions $f : \{0,1\}^n \to \{0,1\}, g : \{0,1\}^n \to \{0,1\}^n$, their XOR composition $f \boxplus g$ is defined as :

$$f \boxplus g(x, y) = f(g(x) \oplus y)$$

where $\oplus$ denotes the bit-wise XOR of two binary strings. Mihajlin and Sofronova [18] proved following result.

Theorem 1 ([18]). *If we choose a function $f : \{0,1\}^n \to \{0,1\}$ randomly, with probability $1 - o(1)$, there exists a function $g : \{0,1\}^n \to \{0,1\}^n$, such that for any constant $0 < \alpha < 0.4$ and constant $0 < \epsilon < \alpha/2$, their XOR composition $f \boxplus g$ is not computable by an AND (or OR) of $2^{(\alpha-\epsilon)n}$ formulas of size at most $2^{(1-\alpha/2-\epsilon)n}$.*

This implies a super-3 $\log n$ depth lower bound for a modified version of the Andreev function against formulas with restriction on top.

Theorem 2 ([18]). *A modified version of Andreev function* **Andr′** *is not computable by any circuit of depth $(3.2 - \epsilon) \log n$ with the restriction that top $0.4 - \epsilon$ layers only consist of AND(or OR) gates for any small constant $\epsilon > 0$.*

They asked whether their result can be improved as asked by following question.

Question 1 ([18]). Is it possible to extend the range of parameter α in Theorem 1 to $0 < \alpha < 1$?

Before proceeding further, we want to emphasis that it is pointed out by Bathie and Williams [1], and the anonymous reviewers of this paper that result similar to Mihajlin and Sofronova's work could be obtained from Komargodski, Raz and Tal's average-case lower bound [12]. In fact, they prove that there is a feasible function which can not be computed by a large Majority gate on top of formulas of almost cubic size, and this implies a lower bound with a large AND gate on top. The main difference is that the result of Komargodski, Raz and Tal is obtained via restriction-based techniques while Mihajlin and Sofronova use a top-down approach which is more akin to the line of works using communication complexity. The merit is that the top-down approach is somewhat complete [7] for proving depth lower bound and no known obstacles are against this approach. Thus the real question here is that can we solve Question 1 in a top-down approach similar to that in Mihajlin and Sofronova's work. In this paper, we give a positive answer to the question with an even better result. In fact, we extend the range of parameter α to $0 < \alpha < 2 - o(1)$ which is nearly optimal.

1.1 Our Results

Our main result is an improved XOR composition theorem for formulas with restriction on top. Formally, we have following result.

Theorem 3. *Let $\mathsf{L}(f)$ be the formula size of any Boolean function f. For most functions $f : \{0,1\}^n \to \{0,1\}$, there exists a function $g : \{0,1\}^n \to \{0,1\}^n$, such that $f \boxplus g$ is not computable by an AND (or OR) of $2^{\alpha n}$ formulas of size at most $2^{\frac{n+\log \mathsf{L}(f) - \alpha n - 2\log\log n}{2}}$ for any $0 < \alpha < 1 + \frac{\log \mathsf{L}(f) - 2\log\log n}{n}$.*

This implies a nearly-$4\log n$ depth lower bound for formulas with restriction on top.

Theorem 4. *A modified version of Andreev function* **Andr′** *is not computable by any circuit of depth $(4-o(1))\log n$ with the restriction that top $(2-o(1))\log n$ layers only consist of AND (or OR) gates.*

Comparing to the results of Mihajlin and Sofronova [18], our results are nearly tight and our proof is much simpler.

Other Related Works. Besides Mihajlin and Sofronova's work, we note that Bathie and Williams [1] established a super-$3\log n$ depth lower bound against *uniform* circuits consisting of only NAND gates. Since their result is against uniform circuits, it is incomparable to ours. As pointed out in the introduction, we also note that result similar to Mihajlin and Sofronova's work already follows from Komargodski, Raz and Tal's average-case lower bound [12]. The advantage of our work is not providing quantitatively better parameter. Instead, the main merit of our result is that our proof is based on a top-down approach and more akin to the line of works using communication complexity, thus it provides a new angle to approach these questions. Further more, in a personal communication, Meir [16] pointed out that results similar to Mihajlin and Sofronova's work can also be obtained via techniques [15] from communication complexity but it is not clear whether such results are as tight as ours.

Our Approach. At first, we describe the intuitive idea about the proof in terms of communication complexity, a.k.a, Karchmer-Wigderson game. Although our proof does not follow exactly the terms of communication complexity, it shares the exact structure of a top-down approach. Now Alice and Bob want to solve $KW_{f\diamond g}$, the Karchmer-Wigderson game of $f \diamond g$. And for convenience, let's assume that both $D(f)$ and $D(g)$ are approximate n. To lower bound the communication complexity, firstly, we have make sure that to solve $KW_{f\diamond g}$, Alice and Bob have to solve both KW_f and KW_g, that is $KW_{f\diamond g}$ is some kind of direct sum [15] of KW_f and KW_g. Now assume the direct sum effect is true for $KW_{f\diamond g}$, we can show a lower bound about $1.5n$ for $KW_{f\diamond g}$ via following two-stage argument. In the first stage, Alice and Bob spent n bits communication where α bits communication is used for f and the other $n-\alpha$ bits communication for g. In the second stage, both KW_f and KW_g must be solved, thus we need another $\max\{n-\alpha, \alpha\}$ bits communication. Choose $\alpha = 0.5n$, the total number of bits required is about $n + 0.5n = 1.5n$. Unfortunately, we don't know how to carry above plan out for the general case of $KW_{f\diamond g}$. Instead, we successfully carry it out for the case of XOR composition of f, g with restriction on top, and the key is to define a measurement which characterizes the progress made to solve both KW_f and KW_g to be described next.

Now we give a concise description of the proof idea of Theorem 3. The whole proof strategy is similar to that in [18], we call such strategy as a double-measurement argument, a generalized form of the double-counting argument. One crucial component in such argument is the notion of well-mixed set of functions, and our improvement is mainly due to the simplification and improvement for such well-mixed set of functions.

Let $\mathcal{G}$ be the set of all functions from $\{0,1\}^n \to \{0,1\}^n$. Now given a hard function $f : \{0,1\}^n \to \{0,1\}$, we want to show there exists a function $g \in \mathcal{G}$, such that if $f \boxplus g$ can be computed by a formula $\phi_g = \bigwedge_{i=1}^{2^{\alpha n}} \phi_{g,i}$, there must be a sub-formula $\phi_{g,i}$ such that $\mathsf{L}(\phi_{g,i})$ is large. To show this, Mihajlin and Sofronova defined a sub-additive measure μ for Boolean functions of two arguments and $\mu(f \boxplus g)$ is large, thus by averaging, for every g, there exists some i_g such that $\mu(\phi_{g,i_g})$ is large enough. Note that for every g, i_g, ϕ_{g,i_g} computes some function h_{g,i_g}, and let $\mathcal{H}$ be the set of all such functions h_{g,i_g}. If the size of $\mathcal{H}$ is large, by a standard counting argument, there must be a hard function $h \in \mathcal{H}$ as required. But $\mathcal{H}$ may be a small set, to prevent this, we need to show for every $h \in \mathcal{H}$, there only exists a small subset of $\mathcal{G}_h \subseteq \mathcal{G}$ such that for every $g \in \mathcal{G}_h$, h_{g,i_g} is the same function as h. Formally, denote the $\{g | g \in \mathcal{G}, h_{g,i_g} = h\}$ by $\mathcal{G}_h$. Let $h_\star$ be the function such that the size of $\mathcal{G}_{h_\star}$ is maximum among all $\mathcal{G}_h$, it suffices to show $\mathcal{G}_{h_\star}$ is a small set. Intuitively, $\mu(h_\star)$ measures the progress made to solve both KW_f and KW_g. To proceed, we need to show an up bound of measurement $\mu(h_\star)$ in another way and this is why the notion of well-mixed set is involved.

Now consider this function $\mathcal{M}_{f,\mathcal{G}_{h_\star}}(x, y) = \bigvee_{g \in \mathcal{G}_{h_\star}} f(g(x) \oplus y)$. Let $\mathcal{M}^x_{f,\mathcal{G}_{h_\star}}$ be the function $\mathcal{M}_{f,\mathcal{G}_{h_\star}}$ with the first argument is fixed to be some $x \in \{0,1\}^n$, we want to show if the set $\mathcal{G}_{h_\star}$ is large, there are many xs such that $\mathcal{M}^x_{f,\mathcal{G}_{h_\star}}$ is almost a constant function, and it eventually implies $\mu(h_\star)$ is small which contradicts the fact that $\mu(h_\star)$ is already large. This is essentially the property that Mihajlin and Sofronova wanted for $\mathcal{G}_{h_\star}$, or in their terms, $\mathcal{G}_{h_\star}$ is well-mixed for function f. In their work, Mihajlin and Sofronova used a rather complicated probabilistic method to show that property.

We will show such complication is entirely unnecessary and the well-mixed property could be obtained by a simple counting argument if you choose the function f properly. For convenience, let $N = 2^n$, now choose a hard function $f : \{0,1\}^n \to \{0,1\}$ such that density$(f^{-1}(1)) = \frac{|f^{-1}(1)|}{N} \geq \delta$, typically, we set δ to be $\frac{1}{4}$. Note that given any fixed x, $\mathcal{M}^x_{f,\mathcal{G}_{h_\star}}(y) = 1$ if there is a function $g \in \mathcal{G}_{h_\star}$ such that $(g(x) \oplus y) \in f^{-1}(1)$. Given any $x \in \{0,1\}^n$, denote the set $\{z | \exists g \in \mathcal{G}_{h_\star}, g(x) = z\}$ by $\mathcal{G}_{h_\star}(x)$. Similarly, given $x, y \in \{0,1\}^n$, we denote the set $\{z | \exists g \in \mathcal{G}_{h_\star}, g(x) \oplus y = z\}$ by $\mathcal{G}_{h_\star}(x) \oplus y$. Given an x, if $|\mathcal{G}_{h_\star}(x)| > N(1-\delta)$, for any fixed y, we also have $|\mathcal{G}_{h_\star}(x) \oplus y| > N(1-\delta)$, since for any fixed y, $z \oplus y$ is a permutation function of z. When $|\mathcal{G}_{h_\star}(x) \oplus y| > N(1-\delta)$, $(\mathcal{G}_{h_\star}(x) \oplus y) \cap f^{-1}(1)$ is not empty, this means there exists $g \in \mathcal{G}_{h_\star}$ such that for that x, $f(g(x) \oplus y) = 1$.

Now we say x is bad, if $|\mathcal{G}_{h_\star}(x)| \leq \mathrm{N}(1-\delta)$. If $\mathcal{G}_{h_\star}$ is a dense subset of $\mathcal{G}$, the number of bad xs is small. Assume the size of $\mathcal{G}_{h_\star}$ is (at least) $|\mathcal{G}| \cdot (1-\delta)^P$, then there are at most P bad xs. If not, the number of functions in $\mathcal{G}_{h_\star}$ is less than

$$(N(1-\delta))^P \cdot N^{N-P} = N^N \cdot (1-\delta)^P = |\mathcal{G}| \cdot (1-\delta)^P$$

which is a contradiction. This means, given any x which is not bad, the function $\mathcal{M}^x_{f,\mathcal{G}_{h_\star}}(y) = 1$ for any y, thus $\mathcal{M}^x_{f,\mathcal{G}_{h_\star}}$ is a constant function. This eventually implies $\mu(h_\star)$ is small which leads to the desired contradiction. Finally, since $\mathcal{G}_{h_\star}$ has to be a small set, the set $\mathcal{H}$ must be a large set of distinct functions which contains a hard function $h \in \mathcal{H}$ as required. See more details in Lemma 2 and Theorem 6.

1.2 Organization of the Rest of the Paper

The rest of the paper is organized as follows. In Sect. 2, we provide necessary preliminaries. In Sect. 3, we prove Theorem 3, an improved XOR composition theorem of formulas with restriction on top. In Sect. 4, we present Theorem 4, a nearly-4 $\log n$ depth lower bound for formulas with restriction on top. In Sect. 5, we conclude and make some discussion about future directions.

2 Preliminaries

Definition 1 (De Morgan formula). *A De Morgan formula ϕ is a binary tree, its internal vertices are gates such as* AND($\wedge$) *or* OR($\vee$)*, its leaves are literals such as x_i or its negation $\neg x_i$. The depth of a formula is the depth of underling tree of the formula. The size of a formula is the number of its leaves.*

Definition 2. *The formula complexity of a boolean function $f : \{0,1\}^n \to \{0,1\}$, denoted $\mathsf{L}(f)$, is the size of the smallest formula that computes f. The depth complexity of f, denoted $\mathsf{D}(f)$, is the smallest depth of a formula that computes f.*

We will need following fact, given a large set of distinct functions, there is a function with large formula size in that set.

Fact 1 ([10], Theorem 1.23). *Let $\mathcal{F}$ be a set of distinct Boolean functions with input length n, then there exists a function $f \in \mathcal{F}$ such that $\mathsf{L}(f) \geq \frac{\log|\mathcal{F}|}{\log n+4}$.*

Definition 3 (XOR composition of two functions, [17]). *Let $f : \{0,1\}^n \to \{0,1\}, g : \{0,1\}^n \to \{0,1\}^n$ be two functions, their XOR composition $f \boxplus g : \{0,1\}^n \times \{0,1\}^n \to \{0,1\}$ is defined as follows:*

$$f \boxplus g(x,y) = f(g(x) \oplus y)$$

where $\oplus$ denotes the bit-wise XOR of two binary strings.

We recall a measure $\mu(h)$ of any Boolean function h of two arguments defined in [18].

Definition 4 **([18]).** *Let h be a Boolean function of two arguments, given any fixed x as the first argument, we define the function h^x by setting $h^x(y) = h(x,y)$ and define*

$$\mu(h) = \sum_{x \in X} \mathsf{L}(h^x).$$

Fact 2 **([18]).** $\mu(f \boxplus g) \geq 2^n \cdot \mathsf{L}(f)$ *and for every x,* $\mathsf{L}(h^x) \leq \mathsf{L}(h)$.

The measure μ is sub-additive in the following sense:

Lemma 1 **([18]).** *Let $h(x,y) = \circ(g_1, g_2, \ldots, g_k)(x,y)$, where $\circ$ is $\wedge$ or $\vee$. Then $\mu(h) \leq \mu(g_1) + \ldots + \mu(g_k)$.*

Matrix Representation for a Function of Two Arguments. For convenience, we follow a notation in [18] which treats a Boolean function of two arguments as a Boolean matrix.

Definition 5. *Set $X = \{0,1\}^n, Y = \{0,1\}^n$, given a function $h : X \times Y \to \{0,1\}$, define a corresponding matrix $\mathcal{M}_h$ such that*

- *the rows of $\mathcal{M}_h$ are indexed by $x \in X$ and the columns are indexed by $y \in Y$,*
- *and $\mathcal{M}_h(x,y) = h(x,y)$ for every x, y.*

Similarly, given two functions $f : \{0,1\}^n \to \{0,1\}, g : \{0,1\}^n \to \{0,1\}^n$, define a matrix $\mathcal{M}_{f,g}$ such that $\mathcal{M}_{f,g}(x,y) = f(g(x) \oplus y)$ for every x, y.

Furthermore, given a function $f : \{0,1\}^n \to \{0,1\}$ and a set $\mathcal{Z}$ of functions from $\{0,1\}^n \to \{0,1\}^n$, define a matrix $\mathcal{M}_{f,\mathcal{Z}}$ such that for every x, y,

$$\mathcal{M}_{f,\mathcal{Z}}(x,y) = \bigvee_{g \in \mathcal{Z}} f(g(x) \oplus y).$$

Finally, given a subset of indexes $A \subseteq X$ for rows in a matrix $\mathcal{M}$, the matrix $\mathcal{M}^A$ is a sub-matrix of $\mathcal{M}$ restricted to rows indexed by A.

Concentration of Measure

Theorem 5 (Chernoff bound). *Given n independent random variables $X_1, \ldots, X_n$ which distribute in $\{0,1\}$, let $X = \sum_{i=1}^n X_i$ be their sum and $\mathbb{E}[X] = \mu$. For any constant δ such that $0 < \delta < 1$, we have*

$$\Pr(X \geq (1+\delta)\mu) \leq e^{-\frac{\delta^2 \mu}{2+\delta}}$$

and

$$\Pr(X \leq (1-\delta)\mu) \leq e^{-\frac{\delta^2 \mu}{2}}.$$

By Chernoff bound, we have following fact.

Fact 3. *Let $N = 2^n$. If we choose a function $f : \{0,1\}^n \to \{0,1\}$ randomly, then $\Pr(\frac{|f^{-1}(1)|}{N} < \frac{1}{4}) \leq e^{-\Omega(N)}$ and $\Pr(\frac{|f^{-1}(0)|}{N} < \frac{1}{4}) \leq e^{-\Omega(N)}$.*

3 An Improved Xor Composition Theorem for Formulas With Restriction on Top

In this section, we prove Theorem 3. At first, let's recall the notion of well-mixed set of functions.

Definition 6 (Well-mixed set of functions, [18]**).** *A set of functions $\mathcal{G}$ from $\{0,1\}^n \to \{0,1\}^n$ is (Q,D,P)-well-mixed for f if $\forall \mathcal{Z} \subseteq \mathcal{G}, |\mathcal{Z}| \geq Q$, there exist a set $K \subseteq \{0,1\}^n, |K| \leq P$, such that $\mathcal{M}_{f,\mathcal{Z}}^{X\setminus K}$ has no more than D zeroes in total where $\mathcal{M}_{f,\mathcal{Z}}(x,y) = \bigvee_{g\in\mathcal{Z}} f(g(x)\oplus y)$.*

Now we show given any approximately balanced function f, the set $\mathcal{G}$ of all functions $\{0,1\}^n \to \{0,1\}^n$ is already a well-mixed set of functions for f.

Lemma 2. *Let $f : \{0,1\}^n \to \{0,1\}$ be a function, $\mathcal{G}$ be the set of all functions $\{0,1\}^n \to \{0,1\}^n$. For convenience, let $N = 2^n$ and $X = \{0,1\}^n$. Assume density$(f^{-1}(1)) = \frac{|f^{-1}(1)|}{N} \geq \delta$, then $\mathcal{G}$ is $(|\mathcal{G}|\cdot(1-\delta)^P, 0, P)$-well-mixed for f. Particularly, let $\mathcal{Z} \subseteq \mathcal{G}$ be a set of functions and density$(\mathcal{Z}) = \frac{|\mathcal{Z}|}{|\mathcal{G}|}$, there exists a set $K \subseteq \{0,1\}^n$ such that $|K| = P \leq \frac{\log density(\mathcal{Z})}{\log(1-\delta)}$ and all entries in $\mathcal{M}_{f,\mathcal{Z}}^{X\setminus K}$ are ones.*

Proof. Let $\mathcal{Z} \subseteq \mathcal{G}$ be a set of functions, given any $x \in \{0,1\}^n$, denote the set $\{z|\exists g \in \mathcal{Z}, g(x) = z\}$ by $\mathcal{Z}(x)$, and we say x is bad if $|\mathcal{Z}(x)| \leq N(1-\delta)$. Similarly, given $x, y \in \{0,1\}^n$, we denote the set $\{z|\exists g \in \mathcal{Z}, g(x)\oplus y = z\}$ by $\mathcal{Z}(x)\oplus y$.

Now assume there are P bad xs, the number of functions in $\mathcal{Z}$ is at most

$$(N(1-\delta))^P \cdot N^{N-P} = N^N \cdot (1-\delta)^P = |\mathcal{G}|\cdot(1-\delta)^P.$$

This implies density$(\mathcal{Z}) = \frac{|\mathcal{Z}|}{|\mathcal{G}|} \leq (1-\delta)^P$, this means $P \leq \frac{\log \text{density}(\mathcal{Z})}{\log(1-\delta)}$.

Now we show when x is not bad, for every $y \in \{0,1\}$, $\mathcal{M}_{f,\mathcal{Z}}(x,y) = \bigvee_{g\in\mathcal{Z}} f(g(x)\oplus y) = 1$. Since $|\mathcal{Z}(x)| > N(1-\delta)$, we have $|\mathcal{Z}(x)\oplus y| > N(1-\delta)$. Since $|\mathcal{Z}(x)\oplus y| > N(1-\delta)$, $(\mathcal{Z}(x)\oplus y)\cap f^{-1}(1)$ is not empty, there must be a $g \in \mathcal{Z}$ such that $g(x)\oplus y \in f^{-1}(1)$, thus $\bigvee_{g\in Z} f(g(x)\oplus y)$ must be 1 as required.

Now we are ready to prove Theorem 3 rephrased as follows. It is proved via a similar idea in [18] and a more careful analysis.

Theorem 6. *Let $f : \{0,1\}^n \to \{0,1\}$ be a function and density$(f^{-1}(1)) \geq \frac{1}{4}$, there exists a function $g : \{0,1\}^n \to \{0,1\}^n$, such that $f \boxplus g$ is not computable by an AND of $2^{\alpha n}$ formulas of size at most $2^{\frac{n+\log \mathsf{L}(f)-\alpha n-2\log\log n}{2}}$ for any $0 < \alpha < 1 + \frac{\log \mathsf{L}(f)-2\log\log n}{n}$.*

Proof (Proof of Theorem 6). We prove it by contradiction. Let $\mathcal{G}$ be the set of all functions $\{0,1\}^n \to \{0,1\}^n$. Assume the contrary that for all $g \in \mathcal{G}$ the XOR composition $f \boxplus g$ is computable by AND of small enough formulas. That is, for any $g \in \mathcal{G}$, there is a formula ϕ_g computing $f \boxplus g$ and ϕ_g is of following form

$$\phi_g = \bigwedge_{i=1}^{2^{\alpha n}} \phi_{g,i}$$

where the size of every $\phi_{g,i}$ is at most $2^{\frac{n+\log \mathsf{L}(f)-\alpha n-2\log\log n}{2}}$. Now let $h_{g,i}$ be the function that $\phi_{g,i}$ computes, thus $\mathsf{L}(h_{g,i}) \leq 2^{\frac{n+\log \mathsf{L}(f)-\alpha n-2\log\log n}{2}}$ and $f \boxplus g$ can be represented as $\bigwedge_{i=1}^{2^{\alpha n}} h_{g,i}$.

Recall that for every $g \in \mathcal{G}$, $\mu(f \boxplus g) \geq 2^n \cdot \mathsf{L}(f)$. By Lemma 1, there must be an $i_g \in [2^{\alpha n}]$ such that, the measure $\mu(h_{g,i_g})$ is large:

$$\mu(h_{g,i_g}) \geq 2^{(1-\alpha)n} \cdot \mathsf{L}(f).$$

Now we collect all such functions h_{g,i_g} and let $\mathcal{H}$ be the set of all such functions h_{g,i_g}. We want to show that the size of $\mathcal{H}$ is large, thus by a standard counting argument, there must be a function $h \in \mathcal{H}$ which requires large formulas which contradicts the hypothesis.

Given any $h \in \mathcal{H}$, denote the $\{g | g \in \mathcal{G}, h_{g,i_g} = h\}$ by $\mathcal{G}_h$. Let $h_\star$ be the function such that the size of $\mathcal{G}_{h_\star}$ is maximum among all $\mathcal{G}_h$. We will prove

$$-4 \log \text{density}(\mathcal{G}_{h_\star}) \cdot \mathsf{L}(h_\star) \geq \mu(h_\star) \geq 2^{(1-\alpha)n} \cdot \mathsf{L}(f),$$

before proving this, let's show it indeed leads to the contradiction required. Recall by assumption $\mathsf{L}(h_\star) \leq 2^{\frac{n+\log \mathsf{L}(f)-\alpha n-2\log\log n}{2}}$, this means

$$\text{density}(\mathcal{G}_{h_\star}) \leq 2^{-2^{\frac{n+\log \mathsf{L}(f)-\alpha n+2\log\log n-4}{2}}}.$$

Now we are ready to lower bound the size of $|\mathcal{H}|$, that is the number of distinct functions in $\mathcal{H}$. Since $|\mathcal{H}| \cdot |\mathcal{G}_{h_\star}| \geq |\mathcal{G}|$,

$$|\mathcal{H}| \geq \frac{1}{\text{density}(\mathcal{G}_{h_\star})} \geq 2^{2^{\frac{n+\log \mathsf{L}(f)-\alpha n+2\log\log n-4}{2}}}.$$

By Fact 1, there exists an $h \in \mathcal{H}$ such that

$$\begin{aligned}\mathsf{L}(h) &\geq \frac{2^{\frac{n+\log \mathsf{L}(f)-\alpha n+2\log\log n-4}{2}}}{\log 2n+4} \\ &> 2^{\frac{n+\log \mathsf{L}(f)-\alpha n-2\log\log n}{2}}, \text{when } n \text{ is large enough,}\end{aligned}$$

which is a contradiction to the assumption. Now we show $-4 \log \text{density}(\mathcal{G}_{h_\star}) \cdot \mathsf{L}(h_\star) \geq \mu(h_\star)$ by considering the matrix $\mathcal{M}_{f,\mathcal{G}_{h_\star}}$. By Lemma 2,

- there exists a set $K \subseteq \{0,1\}^n$ such that

$$\begin{aligned}|K| &\leq \frac{\log \text{density}(\mathcal{G}_{h_\star})}{\log(1-1/4)} \\ &\leq -4 \cdot \log \text{density}(\mathcal{G}_{h_\star}), \text{ since } \log\frac{3}{4} \approx -0.415 \text{ thus } 1 < \frac{\log\frac{3}{4}}{-0.25}.\end{aligned}$$

- And all entries in $\mathcal{M}_{f,\mathcal{G}_{h_\star}}^{X\setminus K}$ are ones.

Now we have following key fact about the function $h_\star$.

Fact 4. *For every $x, y \in \{0,1\}^n$, $\mathcal{M}_{f,\mathcal{G}_{h_\star}}(x,y) = 1$ implies $h_\star(x,y) = 1$.*

Proof. Recall that $\mathcal{M}_{f,\mathcal{G}_{h_\star}}(x,y) = \bigvee_{g\in\mathcal{G}_{h_\star}} (f \boxplus g)(x,y)$. If $\mathcal{M}_{f,\mathcal{G}_{h_\star}}(x,y) = 1$, there must be some $g \in \mathcal{G}_{h_\star}$ such that $(f \boxplus g)(x,y) = 1$. Furthermore, by assumption $f \boxplus g$ is simply $\bigwedge_{i=1}^{2^{\alpha n}} h_{g,i}$ and $h_\star = h_{g,i}$ for some i, thus $h_\star(x,y)$ must be 1 as well.

By Fact 4, given any $x \in X \setminus K$, for every $y \in \{0,1\}^n$, $h_\star^x(y) = 1$, in other words, $h_\star^x$ is a constant function and $\mathsf{L}(h_\star^x) = 0$. Now we are ready to up bound $\mu(h_\star)$ and have

$$\begin{aligned}\mu(h_\star) &= \sum_x L(h_\star^x) = \sum_{x\in K} L(h_\star^x) + \sum_{x\notin K} L(h_\star^x)\\ &= \sum_{x\in K} L(h_\star^x)\\ &\le |K| \cdot \mathsf{L}(h_\star)\\ &\le -4 \cdot \log \text{density}(\mathcal{G}_{h_\star}) \cdot \mathsf{L}(h_\star).\end{aligned}$$

as required.

Remark 1. We want to point out the AND($\wedge$) gate can be replaced with OR($\vee$) gate and since we use the counting argument to show there exists a hard function in the set $\mathcal{H}$, this result can be extended to the case of formula over full binary basis.

4 A Nearly-4 $\log n$ Depth Lower Bound for Formulas With Restriction on Top

We will prove a depth lower bound for a modified Andreev function defined in [18].

Definition 7 ([18]). *The modified Andreev function* $\mathbf{Andr}' : \{0,1\}^n \times \{0,1\}^{n\log n} \times \{0,1\}^{n\times 2\log n} \to \{0,1\}$ *is defined as follows:*

$$\mathbf{Andr}'(\mathtt{TT}_f, \mathtt{TT}_g, x_1, \ldots, x_{2\log n}) = (f \boxplus g)\,(\oplus(x_1), \ldots, \oplus(x_{2\log n}))$$

where $\mathtt{TT}_f$ *is a truth table of some function f from* $\{0,1\}^{\log n} \to \{0,1\}$,$\mathtt{TT}_g$ *is a truth table of some function g from* $\{0,1\}^{\log n} \to \{0,1\}^{\log n}$, *for every* $i \in [2\log n]$, x_i *is a binary string of length n and* $\oplus(\cdot)$ *is the parity function.*

Theorem 7. *There exist two parameters $\gamma = o(1), \epsilon = o(1)$, for every constant α such that $0 < \alpha < 2 - \gamma$, the modified Andreev function* $\mathbf{Andr}'$ *is not computable by an AND of n^α formulas of size at most $n^{3-\alpha/2-\epsilon}$.*

In terms of depth lower bound, the modified Andreev function $\mathbf{Andr}'$ *is not computable by any circuit of depth $(3+\alpha/2-\epsilon)\log n$ with the restriction that top α layers only consist of AND gates where $0 < \alpha < 2 - \gamma$. Choose $\alpha = 2 - o(1)$ properly, Theorem 4 follows.*

Theorem 7 is proved via the same idea from [18], the only differences here are details of parameters. For completeness of this paper, we present the proof here. At first, we recall the standard notion of random restriction.

Definition 8 (Restriction). *Given a Boolean function $f : \{0,1\}^n \to \{0,1\}$, a restriction $\rho \in \{0,1,*\}^n$ to function f is a vector of length n and for every $i \in [n]$, ρ_i is an element from $\{0,1,*\}$. Define $f|_\rho$ to be the function restricted according to ρ as follows: if ρ_i is $*$ then the i-th input bit of f is unfixed thus free to be 0 or 1; otherwise the i-th input bit of f is fixed to be ρ_i.*

Definition 9 (Random restriction). *Given $0 < p < 1$, the random restriction R_p randomly samples restrictions as follows: every ρ_i is sampled independently such that $\Pr[\rho_i = *] = p$ and $\Pr[\rho_i = 1] = \Pr[\rho_i = 0] = \frac{1-p}{2}$.*

In the rest of this section, we will set $p = \frac{2 \ln \log n}{n}$. Mihajlin and Sofronova proved following useful two lemmas about random restriction of the modified Andreev function implicitly in [18].

Lemma 3 (Implicit in[18]). *Let $\mathbf{Andr}'_{f,g}$ be the $\mathbf{Andr}'$ function hardwired with two fixed functions f, g. Then with probability $1 - o(1)$, the random restriction R_p will turn $\mathbf{Andr}'_{f,g}$ into $f \boxplus g$.*

Lemma 4 (Implicit in[18]). *Let α, β be two parameters such that $0 < \alpha < 2, 2 < \beta < 3$. Let ϕ be a formula of form $\bigwedge_{i=1}^{n^\alpha} \phi_i$ where the size of each ϕ_i is at most n^β. Then with probability $1 - o(1)$, the random restriction R_p will turn ϕ into a formula ϕ' of form $\bigwedge_{i=1}^{n^\alpha} \phi'_i$ where the size of each ϕ'_i is at most $n^{\beta-2+\delta}$ for some $\delta = o(1)$.*

Now we show how to prove Theorem 7 with above two lemmas.

Proof (Proof of Theorem 7). At first choose some function f from $\{0,1\}^{\log n} \to \{0,1\}$ and make sure that f is the function with maximum formula size and $\frac{|f^{-1}(1)|}{n} \geq \frac{1}{4}$. By Fact 1 and 3, we have $\mathsf{L}(f) \geq \log\left((1-o(1))2^{2^{\log n}}\right)/(\log\log n + 4) \geq n/2 \log\log n$ when n is large enough. By Theorem 6, we have following fact.

Fact 5. *Let f be the function chosen above, there exists a function $g : \{0,1\}^{\log n} \to \{0,1\}^{\log n}$, such that $f \boxplus g$ is not computable by an AND of n^α formulas of size at most $n^{1-\alpha/2-\frac{2\log\log\log n}{\log n}}$ for any $0 < \alpha < 2 - \frac{4\log\log\log n}{\log n}$ when n is large enough.*

Let δ be the same parameter from Lemma 4, $\gamma = \frac{4\log\log\log n}{\log n}$ and $\epsilon = \delta + \gamma/2$. Choose some α such that $0 < \alpha < 2 - \gamma$ and set $\beta = 3 - \alpha/2 - \epsilon$. Now assume Theorem 7 is false, that is $\mathbf{Andr}'$ is computable by formula ϕ of form $\bigwedge_{i=1}^{n^\alpha} \phi_i$ where the size of each ϕ_i is at most n^β, and so is $\mathbf{Andr}'_{f,g}$. By Lemma 3 and 4, $f \boxplus g$ is computable by a formula ϕ' of form $\bigwedge_{i=1}^{n^\alpha} \phi'_i$ where the size of each ϕ'_i is at most

$$n^{\beta-2+\delta} = n^{1-\alpha/2-\epsilon+\delta} = n^{1-\alpha/2-\frac{2\log\log\log n}{\log n}}$$

which contradicts Fact 5.

Remark 2. Note that the input length of the modified Andreev function $\mathbf{Andr}'$ is $n' = (3\log n+1)n$, writing the results in terms of n' doesn't change them essentially. For example, $(4-o(1))\log n > (4-o(1))\log \frac{n'}{4\log n} > (4-o(1))\log \frac{n'}{4\log n'} = (4-o(1) - \frac{\log\log n'+2}{\log n'})\log n' = (4-o(1))\log n'$. Similarly, in the restriction for top gates, AND could be replaced by OR.

5 Conclusion and Discussion

In this paper, we obtain a nearly-tight XOR composition theorem for formulas with restriction on top and with this composition theorem we have a nearly-$4\log n$ depth lower bound for formulas with restriction on top. Intuitively, in such the depth lower bound, we trade one unrestricted layer to nearly two layers of AND gates on top of the circuit and our trade-off is nearly optimal.

The next nature question as pointed out by Mihajlin and Sofronova [18] is to prove the case with $\mathbf{AC}^0$ formula on top. But it turns out even to prove the case of depth-2 formula with unbounded fan-in is difficult. The obstacle to extending current approach to the case of depth-2 formula on top is that we don't know how to find the sub-formula $h_\star$ such that the measure $\mu(h_\star)$ is large enough meanwhile $h_\star$ is correlated with $f \boxplus g$ properly like that in Fact 4. This difficulty also appears in the approach via communication complexity [16], since such result shares the same feature with a composition theorem of a depth-2 formula and a De Morgan formula, but we don't even know how to prove a composition theorem of two depth-2 formulas in general. New ideas are needed, maybe we should try to prove a general composition theorem of two depth-2 formulas in the first place.

Another interesting and even more important question is can we avoid the trade-off argument. For example, can we prove there exist two functions $f, g : \{0,1\}^n \to \{0,1\}$ such that the composed function $f \diamond g$ is not computable by circuits with following configuration: the top αn layers are restricted to be only AND gates and the rest $(1-o(1))n$ layers are unrestricted. In some sense, this kind of non-tradeoff result is more important, since it demonstrates that we can show true additional hardness in the composition rather than trading one kind of hardness to another. This kind of result already follows from Komargodski, Raz and Tal's average-case lower bound [12], so the question is that can we prove it with a top-down approach.

Acknowledgments. The author wants to thank the anonymous reviewers for helpful comments, especially for their suggestions about Komargodski, Raz and Tal's work. The author would like to thank Or Meir for sharing the idea about how to prove results similar to Mihajlin and Sofronova's work via techniques from communication complexity and other helpful discussions. The author would also like to thank Pei Wu for suggesting the question of the composition of two depth-2 formulas and other helpful discussions. This work is supported by National Natural Science Foundation of China (NSFC) under grant No.11601322.

References

1. Bathie, G., Williams, R.R.: Towards stronger depth lower bounds. In: 15th Innovations in Theoretical Computer Science Conference (ITCS 2024). Schloss-Dagstuhl-Leibniz Zentrum für Informatik (2024)
2. Cook, J., Mertz, I.: Tree evaluation is in space o(log n · log log n). In: Mohar, B., Shinkar, I., O'Donnell, R. (eds.) Proceedings of the 56th Annual ACM Symposium on Theory of Computing, STOC 2024, Vancouver, BC, Canada, 24–28 June 2024, pp. 1268–1278. ACM (2024). https://doi.org/10.1145/3618260.3649664
3. Dinur, I., Meir, O.: Toward the KRW composition conjecture: cubic formula lower bounds via communication complexity. Comput. Complex. **27**(3), 375–462 (2017). https://doi.org/10.1007/s00037-017-0159-x
4. Edmonds, J., Impagliazzo, R., Rudich, S., Sgall, J.: Communication complexity towards lower bounds on circuit depth. Comput. Complex. **10**(3), 210–246 (2001). https://doi.org/10.1007/s00037-001-8195-x
5. Filmus, Y., Meir, O., Tal, A.: Shrinkage under random projections, and cubic formula lower bounds for AC0 (extended abstract). In: Lee, J.R. (ed.) 12th Innovations in Theoretical Computer Science Conference, ITCS 2021, 6–8 January 2021, Virtual Conference. LIPIcs, vol. 185, pp. 89:1–89:7. Schloss Dagstuhl - Leibniz-Zentrum für Informatik (2021). https://doi.org/10.4230/LIPIcs.ITCS.2021.89
6. Gavinsky, D., Meir, O., Weinstein, O., Wigderson, A.: Toward better formula lower bounds: the composition of a function and a universal relation. SIAM J. Comput. **46**(1), 114–131 (2017). https://doi.org/10.1137/15M1018319
7. Göös, M., Riazanov, A., Sofronova, A., Sokolov, D.: Top-down lower bounds for depth-four circuits. In: 64th IEEE Annual Symposium on Foundations of Computer Science, FOCS 2023, Santa Cruz, CA, USA, 6–9 November 2023, pp. 1048–1055. IEEE (2023). https://doi.org/10.1109/FOCS57990.2023.00063
8. Håstad, J.: The shrinkage exponent of de Morgan formulas is 2. SIAM J. Comput. **27**(1), 48–64 (1998). https://doi.org/10.1137/S0097539794261556
9. Håstad, J., Wigderson, A.: Composition of the universal relation. In: Advances in Computational Complexity Theory, AMS-DIMACS (1993)
10. Jukna, S.: Boolean Function Complexity - Advances and Frontiers, Algorithms and Combinatorics, vol. 27. Springer, Cham (2012). https://doi.org/10.1007/978-3-642-24508-4
11. Karchmer, M., Raz, R., Wigderson, A.: Super-logarithmic depth lower bounds via the direct sum in communication complexity. Comput. Complex. **5**(3/4), 191–204 (1995). https://doi.org/10.1007/BF01206317
12. Komargodski, I., Raz, R., Tal, A.: Improved average-case lower bounds for de Morgan formula size: Matching worst-case lower bound. SIAM J. Comput. **46**(1), 37–57 (2017)
13. Koroth, S., Meir, O.: Improved composition theorems for functions and relations. Leibniz Int. Proc. Inform. LIPIcs **116**(48), 1–18 (2018). https://doi.org/10.4230/LIPIcs.APPROX-RANDOM.2018.48
14. Meir, O.: Toward better depth lower bounds: two results on the multiplexor relation. Comput. Complex. **29**(1), 1–25 (2020). https://doi.org/10.1007/s00037-020-00194-8
15. Meir, O.: Toward better depth lower bounds: a KRW-like theorem for strong composition. In: 64th IEEE Annual Symposium on Foundations of Computer Science, FOCS 2023, Santa Cruz, CA, USA, 6-9 November 2023, pp. 1056–1081. IEEE (2023). https://doi.org/10.1109/FOCS57990.2023.00064

16. Meir, O.: Personal Communication (2024)
17. Mihajlin, I., Smal, A.: Toward better depth lower bounds: the XOR-KRW conjecture. In: Kabanets, V. (ed.) 36th Computational Complexity Conference, CCC 2021, 20–23 July 2021, Toronto, Ontario, Canada (Virtual Conference). LIPIcs, vol. 200, pp. 38:1–38:24. Schloss Dagstuhl - Leibniz-Zentrum für Informatik (2021). https://doi.org/10.4230/LIPIcs.CCC.2021.38
18. Mihajlin, I., Sofronova, A.: A better-than-3log(n) depth lower bound for de Morgan formulas with restrictions on top gates. In: Lovett, S. (ed.) 37th Computational Complexity Conference, CCC 2022, 20–23 July 2022, Philadelphia, PA, USA. LIPIcs, vol. 234, pp. 13:1–13:15. Schloss Dagstuhl - Leibniz-Zentrum für Informatik (2022). https://doi.org/10.4230/LIPICS.CCC.2022.13
19. de Rezende, S.F., Meir, O., Nordström, J., Pitassi, T., Robere, R.: KRW composition theorems via lifting. In: Irani, S. (ed.) 61st IEEE Annual Symposium on Foundations of Computer Science, FOCS 2020, Durham, NC, USA, 16–19 November 2020, pp. 43–49. IEEE (2020). https://doi.org/10.1109/FOCS46700.2020.00013
20. Tal, A.: Shrinkage of de Morgan formulae by spectral techniques. In: 55th IEEE Annual Symposium on Foundations of Computer Science, FOCS 2014, Philadelphia, PA, USA, 18–21 October 2014, pp. 551–560. IEEE Computer Society (2014). https://doi.org/10.1109/FOCS.2014.65
21. Tal, A.: Computing requires larger formulas than approximating. Electron. Colloquium Comput. Complex. **TR16-179** (2016)
22. Wu, H.: An improved composition theorem of a universal relation and most functions via effective restriction. Electron. Colloquium Comput. Complex. **TR23-151** (2023)

Average-Case Deterministic Query Complexity of Boolean Functions with Fixed Weight

Yuan Li[1], Haowei Wu[1](✉), and Yi Yang[1,2]

[1] Fudan University, Shanghai 200438, China
{yuan_li,yyang1}@fudan.edu.cn, hwwu21@m.fudan.edu.cn
[2] Anqing Normal University, Anqing, Anhui 246011, China

Abstract. We study the *average-case deterministic query complexity* of boolean functions under a *uniform input distribution*, denoted by $\mathrm{D_{ave}}(f)$, the minimum average depth of zero-error decision trees that compute a boolean function f. This measure has found several applications across diverse fields, yet its understanding is limited.

We study boolean functions with fixed weight, where weight is defined as the number of inputs on which the output is 1. We prove $\mathrm{D_{ave}}(f) \leq \max\{\log \frac{\mathrm{wt}(f)}{\log n} + O(\log\log \frac{\mathrm{wt}(f)}{\log n}), O(1)\}$ for every n-variable boolean function f, where $\mathrm{wt}(f)$ denotes the weight. For any $4\log n \leq m(n) \leq 2^{n-1}$, we prove the upper bound is tight up to an additive logarithmic term for almost all n-variable boolean functions with fixed weight $\mathrm{wt}(f) = m(n)$. Håstad's switching lemma or Rossman's switching lemma [Comput. Complexity Conf. 137, 2019] implies $\mathrm{D_{ave}}(f) \leq n(1 - \frac{1}{O(w)})$ or $\mathrm{D_{ave}}(f) \leq n(1 - \frac{1}{O(\log s)})$ for CNF/DNF formulas of width w or size s, respectively. We show there exists a DNF formula of width w and size $\lceil 2^w/w \rceil$ such that $\mathrm{D_{ave}}(f) = n(1 - \frac{\log n}{\Theta(w)})$ for any $w \geq 2\log n$.

Keywords: average-case query complexity · decision tree · weight · switching lemma · criticality

1 Introduction

The *average-case deterministic query complexity* of a boolean function f under a *uniform input distribution*[1], denoted by $\mathrm{D_{ave}}(f)$, is the minimum average depth of zero-error decision trees that compute f. This notion serves as a natural average-case analogy of the classic deterministic query complexity $\mathrm{D}(f)$ and has

This work was supported by the National Natural Science Foundation of China under Grant No. 62402119 and the CCF-Huawei Populus Grove Fund (Yuan Li), and by the Scientific Research Compilation Plan Project of Anhui Province, China under Grant No. 2022AH051046 (Yi Yang).

[1] In [1], the complexity under distribution μ is denoted by $\mathrm{D}^{\mu}(f)$, and μ could be arbitrary. In this paper, we assume the input distribution μ is uniform.

F. V. Fomin and M. Xiao (Eds.): COCOON 2025, LNCS 15983, pp. 193–205, 2026.
https://doi.org/10.1007/978-981-95-0215-8_15

found applications in query complexity, boolean function analysis, learning algorithms, game theory, and percolation theory. Besides that, $\mathrm{D_{ave}}(f)$ is a measure with limited understanding, since $\mathrm{D_{ave}}(f)$ falls outside the class of polynomially-related measures, which includes $\mathrm{D}(f)$, $\mathrm{R}(f)$, $\mathrm{C}(f)$, $\mathrm{bs}(f)$, and $\mathrm{s}(f)$ (see the summaries in [2–4] and Huang's proof of the Sensitivity Conjecture [5]). This work is also inspired by Rossman's circuit lower bounds of detecting k-clique on Erdős–Rényi graphs in the average case [6,7]. Through this paper, we hope to initiate a comprehensive study on $\mathrm{D_{ave}}(f)$, exploring its implications and applications.

1.1 Background

To our knowledge, Ambainis and de Wolf were the first to introduce the concept of *average-case* query complexity [1]. They showed super-polynomial gaps between average-case deterministic query complexity, average-case bounded-error randomized query complexity, and average-case quantum query complexity.

Prior to the conceptualization by Ambainis and de Wolf, average-case query complexity had been studied implicitly since the early days of computer science. Yao [8] noticed that $\mathrm{D}^{\mu}_{\mathrm{ave}}(f)$ (with respect to any distribution μ) lower bounds the zero-error randomized query complexity, i.e., $\mathrm{D}^{\mu}_{\mathrm{ave}}(f) \leq \mathrm{R}_0(f)$. Furthermore, Yao's minimax principle says the maximum value of $\mathrm{D}^{\mu}_{\mathrm{ave}}(f)$ over all distributions μ equals $\mathrm{R}_0(f)$.

O'Donnell, Saks, Schramm, and Servedio established the OSSS inequality [9–11]: $\mathrm{D}^{\mu_p}_{\mathrm{ave}}(f) \geq \frac{\mathrm{Var}[f]}{\max_i \mathrm{Inf}_i[f]}$ for any boolean function f, where μ_p is the p-biased distribution and $\mathrm{Inf}_i[f]$ is the influence of coordinate i. By applying the inequality, O'Donnell et al. [9] showed that $\mathrm{R}_0(f) \geq \mathrm{D}^{\mu_p}_{\mathrm{ave}}(f) \geq (n/\sqrt{4p(1-p)})^{2/3}$ for any nontrivial monotone n-vertex graph property f with critical probability p. This result made progress on Yao's conjecture [8], which asserts that $\mathrm{R}_0(f) = \Omega(n)$ for every nontrivial monotone graph property. When $p = 1/2$, we have $\mathrm{R}_0(f) \geq \mathrm{D_{ave}}(f) \geq n^{2/3}$; Benjamini et al. proved that the lower bound $\mathrm{D_{ave}}(f) \geq n^{2/3}$ is almost tight [12].

While studying learning algorithms, O'Donnell and Servedio [13] discovered the OS inequality: $(\sum_i \hat{f}(\{i\}))^2 \leq \mathrm{D_{ave}}(f) \leq \log \mathrm{DT_{size}}(f)$ for any boolean function f, where $\hat{f}(\cdot)$ denotes f's Fourier coefficient and $\mathrm{DT_{size}}(f)$ denotes the decision tree size. The OS inequality plays a crucial role in learning monotone boolean functions (under the uniform distribution).

The most surprising connection (application) arose in game theory. Peres, Schramm, Sheffield, and Wilson [14] studied the *random-turn* HEX game, in which two players determine who plays next by tossing a coin before each round. They proved that the expected playing time (under optimal play) coincides with $\mathrm{D_{ave}}(f)$, where f is the $L \times L$ hexagonal cells connectivity function. Using the OS inequality and the results of Smirnov and Werner on percolation [15], Peres et al. proved a lower bound $L^{1.5+o(1)}$ on the expected playing time on an $L \times L$ board.

1.2 Our Results

The *weight* of a boolean function, defined as the number of inputs for which the output is 1, is related to its query complexity. For instance, Ambainis et al. [16] proved that the quantum query complexity of almost all n-variable functions with fixed weight m is $\Theta(\frac{\log m}{c+\log n-\log\log m}+\sqrt{n})$, where $c>0$ is a constant. In contrast, the hardest function with weight $m\geq 1$ has quantum query complexity $\Theta((n\cdot\frac{\log m}{c+\log n-\log\log m})^{1/2}+\sqrt{n})$. Ambainis et al. [16] also proved that almost all functions with fixed weight $m\geq 1$ have randomized query complexity $\Theta(n)$ as the hardest one.

Our first result proves that $\mathrm{D}_{\mathrm{ave}}(f)\leq\log\frac{m}{\log n}+O(\log\log\frac{m}{\log n})$ for any n-variable boolean function f with weight m.

Theorem 1. *For every boolean function $f:\{0,1\}^n\to\{0,1\}$, if the weight $\mathrm{wt}(f)\geq 4\log n$, then*

$$\mathrm{D}_{\mathrm{ave}}(f)\leq\log\frac{\mathrm{wt}(f)}{\log n}+O\left(\log\log\frac{\mathrm{wt}(f)}{\log n}\right). \qquad (1)$$

Otherwise, $\mathrm{D}_{\mathrm{ave}}(f)=O(1)$.

We prove Theorem 1 by developing a recursive query algorithm that attains the query complexity given in (1). The algorithm queries an arbitrary bit until the subfunction's weight becomes sufficiently small, or more specifically, smaller than the logarithm of its input length. Once this border condition is met, we invoke another algorithm which, on average, takes $O(1)$ bits to query the subfunction (Lemma 2).

Next, we prove Theorem 2, complementing our first result, which says that Theorem 1 is tight up to a lower order term for almost all fixed-weight functions.

Theorem 2. *Let $m:\mathbb{N}\to\mathbb{N}$ be a function such that $4\log n\leq m(n)\leq 2^{n-1}$. For almost all boolean functions $f:\{0,1\}^n\to\{0,1\}$ with the same weight $\mathrm{wt}(f)=m(n)$,*

$$\mathrm{D}_{\mathrm{ave}}(f)\geq\min_{x\in\{0,1\}^n}\mathrm{C}_x(f)\geq\log\frac{\mathrm{wt}(f)}{\log n}-O\left(\log\log\frac{\mathrm{wt}(f)}{\log n}\right), \qquad (2)$$

where $\mathrm{C}_x(f)$ denotes the size of the smallest certificate on input x.

Remark 1. For boolean functions with $\mathrm{wt}(f)\geq 2^{n-1}$, we can obtain a similar bound by replacing f with $\neg f$.

Beyond fixed-weight functions, we also examine CNFs, circuits, and formulas, studying the connection between $\mathrm{D}_{\mathrm{ave}}(f)$ and criticality.

Rossman introduced the notion of *criticality*, defined as the minimum value $\lambda\geq 1$ such that the following property holds: $\Pr_{\rho\sim\mathcal{R}_p}[\mathrm{D}(f|_\rho)\geq t]\leq(p\lambda)^t$ for any $p\in[0,1]$ and $t\in\mathbb{N}$. In terms of criticality, Håstad's switching lemma says every width-w CNF is $O(w)$-critical [17]; Rossman's switching lemma says every size-s

CNF is $O(\log s)$-critical [18,19]; Rossman proved depth-d size-s AC^0 circuits are $O(\log s)^{d-1}$-critical [19]; Harsha et al. proved depth-d size-s AC^0 formulas are $O(\frac{1}{d}\log s)^{d-1}$-critical [20].

For any λ-critical function f, applying a $(\frac{1}{2\lambda})$-random restriction and then querying the resulting subfunction using its optimal decision tree, we observe that $D_{ave}(f) \leq n(1-\frac{1}{\lambda}) + O(\sqrt{\frac{n}{\lambda}})$ (Lemma 5 from Sect. 4). Therefore, upper bounds on criticality imply upper bounds on average-case query complexity for CNFs, formulas, and circuits respectively.

For CNFs, circuits, or formulas, it is meaningful to understand whether the upper bounds on $D_{ave}(f)$ are tight or not. For example, consider a w-CNF f. By Lemma 5 from Sect. 4, we have $D_{ave}(f) \leq n(1-\frac{1}{O(w)})$. If the bound were indeed tight, it would suggest that the p-random restriction, with $p = \frac{1}{10w}$, is essentially an optimal query algorithm for generic w-CNFs. Otherwise, either a better query algorithm exists, or a stronger version of the switching lemma can be established. Either way, the answer would be interesting.

Along this line, we show that there exists a DNF formula of width w and size $\lceil 2^w/w \rceil$ with $D_{ave}(f) = n(1-\frac{\log n}{\Theta(w)})$. It indicates that even if there is a better query algorithm, the room for improvement is limited when w is large.

Theorem 3. *For any integer $w \in [2\log n, n]$, there exists a boolean function $f : \{0,1\}^n \to \{0,1\}$ computable by a DNF formula of width w and size $\lceil 2^w/w \rceil$ such that*

$$D_{ave}(f) = n\left(1 - \frac{\log n}{\Theta(w)}\right).$$

Lastly, we define *penalty shoot-out functions* in the full version of our paper, which are monotone balanced functions, such that the gap between $D(f)$ and $D_{ave}(f)$ is arbitrarily large. Moreover, unlike the worst-case measures $D(f)$, $R(f)$, $C(f)$, $bs(f)$, $s(f)$, which are known to be polynomially related [2–5], no such polynomial relation holds between *any* two of the average-case analogues[2] $D_{ave}(f)$, $R_{ave}(f)$, $C_{ave}(f)$, $bs_{ave}(f)$, $s_{ave}(f)$, even for monotone balanced functions[3].

2 Preliminaries

Let $f : \{0,1\}^n \to \{0,1\}$ be a boolean function. The worst-case deterministic query complexity of a boolean function f, denoted by $D(f)$, is the minimum depth of decision trees that compute f. The average-case deterministic query complexity of a boolean function f, denoted by $D_{ave}(f)$, is the minimum average depth of zero-error deterministic decision trees that compute f under a uniform input distribution.

[2] These average-case counterparts are defined in the uniform distribution.

[3] Super-polynomial gaps can be demonstrated using the threshold function [1], the tribes function, and $\mathrm{Maj} \circ \mathrm{AND}$, all of which are monotone. An extra trick can make them balanced: given a monotone f, let $g = \mathrm{Maj}(f, f^\dagger, z)$ for $z \in \{0,1\}$, where $f^\dagger$ denotes f's dual [21].

Definition 1. *The average-case deterministic query complexity of* $f : \{0,1\}^n \to \{0,1\}$ *under a uniform distribution is defined by*

$$\mathrm{D}_{\mathrm{ave}}(f) = \min_{T} \mathop{\mathbb{E}}_{x\in\{0,1\}^n} [\mathrm{cost}(T, x)],$$

where T *is taken over all zero-error deterministic decision trees that compute* f.

$\mathrm{D}_{\mathrm{ave}}(f)$ turns out to equal the average-case zero-error *randomized* query complexity, defined as $\min_{\mathcal{T}} \mathbb{E}_{x\in\{0,1\}^n}[\mathrm{cost}(\mathcal{T}, x)]$, where $\mathcal{T}$ is taken over all zero-error *randomized* decision trees that compute f (see Remark 8.63 in [21]).

3 Fixed-Weight Functions

3.1 Upper Bound

As a warm-up, the following proposition gives the exact value of $\mathrm{D}_{\mathrm{ave}}(f)$ for boolean functions f with weight 1, such as the AND function. For convenience, we say input x is a *black point* (with respect to f) if $f(x) = 1$.

Proposition 1. $\mathrm{D}_{\mathrm{ave}}(f) = 2(1 - \frac{1}{2^n})$ *for any* n*-variable boolean function* f *with* $\mathrm{wt}(f) = 1$.

Proof. Let $f : \{0,1\}^n \to \{0,1\}$ be a boolean function with a unique black point $z \in \{0,1\}^n$. We prove $\mathrm{D}_{\mathrm{ave}}(f) = 2(1 - \frac{1}{2^n})$ by induction on n. When $n = 1$, we have $\mathrm{D}_{\mathrm{ave}}(f) = 2(1 - \frac{1}{2^n}) = 1$.

Suppose an optimal query algorithm queries x_i first. If $x_i \neq z_i$, the algorithm outputs 0. Otherwise, the algorithm continues on the subfunction $f|_{x_i=z_i}$. Therefore,

$$\begin{aligned}
\mathrm{D}_{\mathrm{ave}}(f) &= 1 + \Pr_{x\in\{0,1\}^n}[x_i \neq z_i] \cdot 0 + \Pr_{x\in\{0,1\}^n}[x_i = z_i] \cdot \mathrm{D}_{\mathrm{ave}}(f|_{x_i=z_i}) \\
&= 1 + \frac{1}{2} \cdot 2\left(1 - \frac{1}{2^{n-1}}\right) \\
&= 2\left(1 - \frac{1}{2^n}\right),
\end{aligned}$$

where the second step is by the induction hypothesis. □

Next, we show a simple bound $\mathrm{D}_{\mathrm{ave}}(f) \leq \log \mathrm{wt}(f) + O(1)$ for any f. Say a query algorithm is *reasonable* if it terminates as soon as the subfunction becomes constant.

Lemma 1. $\mathrm{D}_{\mathrm{ave}}(f) \leq \log \mathrm{wt}(f) + 2$ *for any non-zero boolean function* f.

Proof. Let $m = \mathrm{wt}(f)$. We prove by induction on m and n. When $m = 1$, we have $\mathrm{D}_{\mathrm{ave}}(f) = 2(1 - \frac{1}{2^n}) \leq 2$ by Proposition 1. When $n = 1$, $\mathrm{D}_{\mathrm{ave}}(f) \leq 1$.

Suppose x_i is queried first. Let $m_0 = \mathrm{wt}(f|_{x_i=0})$ and $m_1 = \mathrm{wt}(f|_{x_i=1})$. If $m_b = 0$ for some $b \in \{0,1\}$, a reasonable algorithm will stop on a constant subfunction $f|_{x_i=b}$. Thus,

$$\mathrm{D_{ave}}(f) \le 1 + \frac{1}{2}(\log m + 2) \le \log m + 2$$

by the induction hypothesis. Otherwise, by the induction hypothesis and the AM-GM inequality, we have

$$\begin{aligned}\mathrm{D_{ave}}(f) &\le 1 + \frac{1}{2}(\log m_0 + 2) + \frac{1}{2}(\log m_1 + 2)\\ &= \log(2\sqrt{m_0 m_1}) + 2\\ &\le \log m + 2.\end{aligned}$$

□

We introduce concepts that will be used later. Suppose $f : \{0,1\}^n \to \{0,1\}$ has m black points $x^{(1)}, \ldots, x^{(m)} \in \{0,1\}^n$ in lexicographical order. We call $c_i = (x_i^{(1)}, \ldots, x_i^{(m)})$ the *column pattern* of coordinate i. Coordinates i, j are *positively (negatively) correlated* if $c_i = c_j$ ($c_i = \neg c_j$). An *equivalent coordinate set* (ECS) is a set of correlated coordinates.

Say a coordinate set $S \subseteq \{1, \ldots, n\}$ is *pure* if each c_i for $i \in S$ is either all-zero or all-one; otherwise, S is *mixed*. For example, in Table 1, the set $\{5, 9, 11\}$ is pure, since $c_5 = c_9 = (0,0)$ and $c_{11} = (1,1)$; the set $\{1,2,3\}$ is mixed, since $c_1 = c_3 = (0,1)$ and $c_2 = (1,0)$.

Table 1. A 11-variable boolean function with weight 2.

black points	x_1	x_2	x_3	x_4	x_5	x_6	x_7	x_8	x_9	x_{10}	x_{11}
$x^{(1)}$	0	1	0	1	0	1	0	0	0	0	1
$x^{(2)}$	1	0	1	0	0	1	0	1	0	1	1

Proposition 2. *If coordinates i, j are positively (negatively) correlated, then for any $x \in \{0,1\}^n$ with $f(x) = 1$, we have $x_i = x_j$ ($x_i \ne x_j$).*

Proposition 3. *Let S be a mixed ECS. For any coordinate $i \in S$ and $v \in \{0,1\}$, we have* $\mathrm{wt}(f|_{x_i=v}) < \mathrm{wt}(f)$.

Proposition 4. *Let $f : \{0,1\}^n \to \{0,1\}$ be a boolean function such that $n > k \cdot 2^{\mathrm{wt}(f)-1}$, then f has an ECS of size at least $k+1$.*

It is straightforward to prove Propositions 2 and 3. Proposition 4 follows from the pigeonhole principle, since there are 2^{m-1} distinct equivalence classes with respect to correlation. Using these facts, one can prove that any boolean function of weight $O(\log n)$ has constant $\mathrm{D_{ave}}(f)$.

Lemma 2. *Let* $f : \{0,1\}^n \to \{0,1\}$, *where* $\mathrm{wt}(f) < \log n$. *We have* $\mathrm{D_{ave}}(f) \le 5$.

Proof. Let $m = \mathrm{wt}(f) \ge 3$. (Lemma 2 follows directly from Lemma 1 if $\mathrm{wt}(f) < 3$.) We prove $\mathrm{D_{ave}}(f) \le 5$ by induction on n.

By Proposition 4, there exists a maximal ECS $I = \{i_1, \dots, i_k\}$ of size $k \ge 3$, since $n > 2^{\mathrm{wt}(f)} = 2 \cdot 2^{\mathrm{wt}(f)-1}$. Without loss of generality, assume that coordinates $i_1, \dots, i_k$ are positively correlated. By Proposition 2, we have $x_{i_1} = \cdots = x_{i_k}$ for any black point $x \in \{0,1\}^n$.

If $|I| = n$, then any black point x must satisfy $x_1 = \cdots = x_n$. Therefore, the only possible black points are the all-zero vector and the all-one vector, so there are at most 2 black points. By Lemma 1, $\mathrm{D_{ave}}(f) \le \log \mathrm{wt}(f) + 2 \le 3$.

From now on, we assume $|I| < n$, and thus there exists a coordinate $j \notin I$. Let J be a maximal ECS that contains j. Notice that I or J is mixed, since at most one of f's maximal ECSs can be pure.

For notational convenience, let $\rho_{u,v}$ denote the restriction fixing x_{i_1} and x_{i_2} to u, x_j to v, and leaving all other variables free. Our query algorithm T is defined as follows:

(1) query x_{i_1}, x_{i_2};
(2) output 0 if $x_{i_1} \ne x_{i_2}$;
(3) if $x_{i_1} = x_{i_2}$, then query x_j, and apply the query algorithm recursively on the subfunction.

The query algorithm T correctly computes f. Input x cannot be a black point if $x_{i_1} \ne x_{i_2}$, since i_1 and i_2 are positively correlated (Proposition 2).

For any $u, v \in \{0,1\}$, the number of inputs of $f|_{\rho_{u,v}}$ is $n-3$, and $\mathrm{wt}(f|_{\rho_{u,v}}) \le m-1$ since I or J is mixed (Proposition 3). Observe that $n-3 > 2^m - 3 > 2^{m-1} \ge 2^{\mathrm{wt}(f|_{\rho_{u,v}})}$ for any $m \ge 3$. By the induction hypothesis, we have $\mathrm{D_{ave}}(f|_{\rho_{u,v}}) \le 5$.

Let us analyze the average cost of T. First, notice that the probability of querying exactly 2 variables is $\frac{1}{2}$; the probability of querying at least 3 variables is $\frac{1}{2}$. Thus, we conclude that

$$\mathrm{D_{ave}}(f) \le \mathop{\mathbb{E}}_x [\mathrm{cost}(T,x)] \le \frac{1}{2} \cdot 2 + \Pr_x [\mathrm{cost}(T,x) \ge 3] \cdot (3+5) = 5.$$

□

Corollary 1. *Let* $f : \{0,1\}^n \to \{0,1\}$ *be a boolean function. If* $\mathrm{wt}(f) \le 4 \log n$, *then* $\mathrm{D_{ave}}(f) \le 40$.

Proof. We have $\mathrm{D_{ave}}(f_1(x) \vee \cdots \vee f_k(x)) \le \mathrm{D_{ave}}(f_1) + \cdots + \mathrm{D_{ave}}(f_k)$ for any $f_1, f_2, \dots, f_k$. This is because we can query $f_1(x)$, $f_2(x)$, …, $f_k(x)$ one by one and compute $f_1(x) \vee \cdots \vee f_k(x)$ afterward. The expected number of variables queried is at most the sum of the individual expectations.

Let $k = 8$ and B_f denote f's on-set (the set of inputs on which f outputs 1). Partition B_f into k disjoint sets $B_1 \dots, B_k$, where $|B_i| \le \lceil \frac{4 \log n}{8} \rceil$. Each B_i is the on-set of some function f_i. It can be verified that $f(x) = f_1(x) \vee \cdots \vee f_k(x)$ and that $\mathrm{wt}(f_i) = |B_i| \le \lceil \frac{4 \log n}{8} \rceil$. Note that $\mathrm{wt}(f_i) \le \lceil \frac{4 \log n}{8} \rceil < \frac{1}{2} \log n + 1 \le \log n$ when $n \ge 4$. (When $n < 4$, the corollary holds clearly.) Thus, by Lemma 2, $\mathrm{D_{ave}}(f_i) \le 5$ for any i, implying that $\mathrm{D_{ave}}(f) \le \sum_{i=1}^{8} \mathrm{D_{ave}}(f_i) \le 40$. □

To prove Theorem 1, we design a query algorithm and analyze its cost. Recall that in the proof of Lemma 1, we considered *any* reasonable query algorithm, which queries an arbitrary bit and terminates until the remaining function becomes constant. Similarly, to prove Theorem 1, we design a more sophisticated algorithm, which queries an arbitrary variable until the subfunction satisfies the following border condition: $D_{\text{ave}}(f) = O(1)$ if $\text{wt}(f) \leq 4\log n$ (Corollary 1); then the query algorithm used in the proof of Corollary 1 is invoked. The proof of Theorem 1 is in the full version of our paper.

3.2 Lower Bound

In this section, we prove Theorem 2, showing that Theorem 1 is tight up to an additive logarithmic term.

To illustrate the idea of the proof, let us take the XOR_n function as an example. Regardless of which variable is queried next, the black points are evenly partitioned, and the subfunction's weight is exactly halved. Since XOR_n has weight 2^{n-1} and the algorithm must continue until the subfunction becomes constant, it must query all n variables for every input. Thus, $D_{\text{ave}}(\text{XOR}_n) = n$. Similarly, the key idea of our proof is to show that most boolean functions exhibit a similar property: regardless of which variable is queried next, the black points are split into two roughly equal halves. In other words, for almost all $f \in \mathcal{B}_{n,m}$, where $\mathcal{B}_{n,m} = \{f : \{0,1\}^n \to \{0,1\} \mid \text{wt}(f) = m\}$, we shall prove $\text{wt}(f|_P)$ is "close" to $2^{-k}m$ for *any* tree path P querying $k = \epsilon \log m$ variables. (Ignoring the output of P, we view a tree path P as a restriction, with $f|_P$ representing the subfunction restricted to P.)

Now, we explain the proof strategy in more detail. To sample $f \in \mathcal{B}_{n,m}$ uniformly, a straightforward approach proceeds as follows: (1) randomly select m distinct inputs $x^{(1)}, \cdots, x^{(m)} \in \{0,1\}^n$; (2) set $f(x^{(i)}) = 1$ for all i; and (3) set the remaining inputs to 0. This can also be done by repeatedly drawing m vectors from $\{0,1\}^n$ without replacement and placing them into m vectors $y^{(1)}, \ldots, y^{(m)} \in \{0,1\}^n$. However, to estimate the probability that $\text{wt}(f|_P)$, where $f \in \mathcal{B}_{n,m}$, is close to $2^{-k}m$ for any length-k tree path, we adopt a different sampling method.

Fix a tree path P, viewed as a restriction. Instead of sampling a random $f \in \mathcal{B}_{n,m}$ and then estimating $\text{wt}(f|_P)$, we choose to sample $\text{wt}(f|_P)$ directly, where $f \in \mathcal{B}_{n,m}$, using the following method.

Fix a tree path P of length k, where

$$P = x_{i_1} \xrightarrow{v_1} x_{i_2} \xrightarrow{v_2} \cdots \to x_{i_k} \xrightarrow{v_k} c \tag{3}$$

and $c \in \{0,1\}$ is the output of the path. We denote the restriction $f|_{x_1=v_1,\cdots,x_k=v_k}$ by $f|_P$. In k rounds, for $j = 1, \ldots, k$, we sample without replacement from a box with 2^{n-j} 0's and 2^{n-j} 1's; place the numbers in the i_j-th position of each vector, and discard vectors with $(\neg v_j)$ at i_j-th position. (We can safely discard these vectors, because they are not counted in the weight of $f|_P$.) At the end of k rounds, $\text{wt}(f|_P)$ vectors remain.

Specifically, we sample $\mathrm{wt}(f|_P)$ as follows, given a fixed path P defined in (3), where $f \in \mathcal{B}_{n,m}$ uniformly:

(1) Let $y^{(1)}, \dots, y^{(m)} \in \{0, 1, \star\}^n$ be the m vectors, where all elements are set to $\star$ initially.
(2) In the first round, we sample $t_0 = m$ numbers without replacement from a box with 2^{n-1} zeros and 2^{n-1} ones. We then assign these numbers sequentially to the positions $y^{(1)}_{i_1}, \dots, y^{(m)}_{i_1}$, that is, the i_1-th position of all the m vectors. After that, discard the vectors with $(\neg v_1)$ at position i_1, that is, $y^{(p)}_{i_1} = \neg v_1$. Let t_1 be number of remaining vectors, where t_1 is a random variable equal to $\mathrm{wt}(f|_{x_{i_1}=v_1})$.
(3) In the second round, we sample t_1 numbers without replacement from a box with 2^{n-2} zeros and 2^{n-2} ones, since $f|_{x_{i_1}=v_1, x_{i_2}=0}$ and $f|_{x_{i_1}=v_1, x_{i_2}=1}$ have 2^{n-2} inputs. Assign these numbers sequentially to the i_2-th position of the remaining t_1 vectors, and discard the vectors with $(\neg v_2)$ at position i_2, i.e., $y^{(p)}_{i_2} = \neg v_2$. Let t_2 be number of remaining vectors, where t_2 is a random variable equal to $\mathrm{wt}(f|_{x_{i_1}=v_1, x_{i_2}=v_2})$.
(4) Proceed for k rounds. The number of remaining vectors t_k is a random variable equal to $\mathrm{wt}(f|_P) = \mathrm{wt}(f|_{x_{i_1}=v_1, \dots, x_{i_k}=v_k})$.

Recall that t_j is the number of vectors remaining after the j-th round, and $t_k = \mathrm{wt}(f|_P)$. If path P correctly computes f (on input $x \in \{0,1\}^n$ such that $x_{i_1} = v_1, \dots, x_{i_k} = v_k$), then we must have $\mathrm{wt}(f|_P) = 0$ or $\mathrm{wt}(f|_P) = 2^{n-k}$. Intuitively, in each round, $t_i \approx \frac{1}{2} t_{i-1}$ holds with high probability by Hoeffding's inequality, so it takes $\Omega(\log n)$ rounds to make $t_k = 0$. Thus, it is unlikely that a "short" path computes f.

Definition 2. **(δ-parity path).** *Let $P = x_{i_1} \xrightarrow{v_1} x_{i_2} \xrightarrow{v_2} \cdots \to x_{i_k} \xrightarrow{v_k} c$ be a path of length k. Let ρ_j denote the restriction that fixes x_{i_p} to v_p for $p = 1, \dots, j$, leaving other variables undetermined. The path P is called* δ-parity *with respect to $f : \{0,1\}^n \to \{0,1\}$ if*

$$\frac{1}{2}(1-\delta) \le \frac{\mathrm{wt}(f|_{\rho_j})}{\mathrm{wt}(f|_{\rho_{j-1}})} \le \frac{1}{2}(1+\delta)$$

for each $j = 1, \dots, k$.

The following lemma follows from a union bound and Hoeffding's inequality. See the full version of our paper for the proof.

Lemma 3. *Let $f : \{0,1\}^n \to \{0,1\}$ be a boolean function with $\mathrm{wt}(f) = m$. Let P be a decision tree path of length at most $\epsilon \log m$. For any $\delta \in (0, \frac{1}{2\epsilon \log m}]$ and $\epsilon \in (0,1)$,*

$$\Pr_{f \sim \mathcal{B}_{n,m}} [P \textit{ is not } \delta\textit{-parity for } f] < 2\epsilon \log m \cdot \exp\left(-\frac{1}{2} \cdot \delta^2 m^{1-\epsilon}\right).$$

Definition 3. (**(t, δ)-parity function**). *Let $f : \{0,1\}^n \to \{0,1\}$ be a boolean function. The function f is called* (t, δ)-parity *if any path of length at most t is a δ-parity path with respect to f.*

Lemma 4. *Let $f : \{0,1\}^n \to \{0,1\}$ be a (t, δ)-parity function with $\mathrm{wt}(f) \leq 2^{n-1}$ satisfying $1 \leq t \leq \log \mathrm{wt}(f) - 1$ and $\delta \leq \frac{1}{2t}$. Then, $\min_{x \in \{0,1\}^n} \mathrm{C}_x(f) \geq t$.*

The proof of Lemma 4 is in the full version of our paper. One can prove Theorem 2 using Lemmas 3 and 4; see the full version for the details.

4 DNFs, Circuits, and Formulas

In this section, we study $\mathrm{D}_{\mathrm{ave}}(f)$ of circuits that consist of AND, OR, NOT gates with unbounded fan-in.

As a warm-up, we show $\mathrm{D}_{\mathrm{ave}}(F) = O(s)$ for general size-s circuits F. The bound is tight up to a multiplicative factor, since $\mathrm{D}_{\mathrm{ave}}(\mathrm{XOR}_n) = n$, and XOR_n is computable by a circuit of size $O(n)$ and depth $O(\log n)$.

Proposition 5. *$\mathrm{D}_{\mathrm{ave}}(F) \leq 2s$ for every circuit F of size s.*

Proof. Notice that the average cost of evaluating each AND/OR/NOT gate does not exceed 2 (Proposition 1). Therefore, it takes at most $2s$ queries on average to evaluate s gates. □

Definition 4. *A* p-random restriction, *denoted by $\mathcal{R}_p$, is a distribution over restrictions leaving x_i unset with probability p and fixing x_i to 0 or 1 with equal probability $\frac{1}{2}(1-p)$ independently for each $i = 1, 2, \ldots, n$.*

Definition 5 ([18,19]). *A boolean function f is* λ-critical *if*

$$\Pr_{\rho \sim \mathcal{R}_p}[\mathrm{D}(f|_\rho) \geq t] \leq (p\lambda)^t$$

for any $p \in [0,1]$ and $t \in \mathbb{N}$.

The next lemma gives an upper bound on $\mathrm{D}_{\mathrm{ave}}(f)$ for λ-critical functions.

Lemma 5. *Let $f : \{0,1\}^n \to \{0,1\}$ be λ-critical. Then*

$$\mathrm{D}_{\mathrm{ave}}(f) \leq n\left(1 - \frac{1}{\lambda}\right) + 2\sqrt{\frac{n}{\lambda}}. \tag{4}$$

Proof. Let $\epsilon > 0$ and $p = \frac{1}{(1+\epsilon)\lambda}$. Since f is λ-critical, we have $\Pr_{\rho \sim \mathcal{R}_p}[\mathrm{D}(f|_\rho) \geq t] \leq (1+\epsilon)^{-t}$. Consider a query algorithm that queries each variable independently with probability $1-p$, and then applies a worst-case optimal query algorithm to $f|_\rho$. We have

$$\begin{aligned}
\mathrm{D}_{\text{ave}}(f) &\le \mathop{\mathbb{E}}_{\rho\sim\mathcal{R}_p}[|\operatorname{supp}(\rho)| + \mathrm{D}(f|_\rho)] \\
&= n(1-p) + \sum_{t=1}^{n} \Pr_{\rho\sim\mathcal{R}_p}[\mathrm{D}(f|_\rho) \ge t] \\
&\le n(1-p) + \sum_{t=0}^{\infty} \frac{1}{(1+\epsilon)^t} \\
&= n\left(1 - \frac{1}{(1+\epsilon)\lambda}\right) + \frac{1+\epsilon}{\epsilon} \\
&= n\left(1-\frac{1}{\lambda}\right) + \frac{n}{\lambda}\cdot\frac{\epsilon}{1+\epsilon} + \frac{1+\epsilon}{\epsilon}.
\end{aligned}$$

Let $\alpha = \frac{n}{\lambda} \ge 1$. The function $h(\epsilon) = \frac{\alpha\epsilon}{1+\epsilon} + \frac{1+\epsilon}{\epsilon}$ attains its minimum at $\epsilon = \frac{1}{\sqrt{\alpha}-1}$, where $h(\frac{1}{\sqrt{\alpha}-1}) = 2\sqrt{\alpha}$. Thus,

$$\mathrm{D}_{\text{ave}}(f) \le n\left(1-\frac{1}{\lambda}\right) + 2\sqrt{\frac{n}{\lambda}}.$$

□

Remark 2. Alternatively, one can prove $\mathrm{D}_{\text{ave}}(f) \le n(1-\frac{1}{2\lambda}) + O(1)$ by combining the OS inequality $\mathrm{D}_{\text{ave}}(f) \le \log \mathrm{DT}_{\text{size}}(f)$ [13] and the bound $\mathrm{DT}_{\text{size}}(f) \le O(2^{n(1-\frac{1}{2\lambda})})$ [19].

By combining Lemma 5 with the existing bounds on criticality for CNFs, bounded-depth circuits, and formulas [17–20, 22], the following upper bounds on $\mathrm{D}_{\text{ave}}(f)$ can be derived.

Corollary 2 ([17]). *Let $f : \{0,1\}^n \to \{0,1\}$ be computable by a CNF/DNF of width w. Then*

$$\mathrm{D}_{\text{ave}}(f) \le n\left(1 - \frac{1}{O(w)}\right).$$

Corollary 3 ([18, 19]). *Let $f : \{0,1\}^n \to \{0,1\}$ be computable by a CNF/DNF of size s. Then*

$$\mathrm{D}_{\text{ave}}(f) \le n\left(1 - \frac{1}{O(\log s)}\right).$$

Corollary 4 ([18, 22]). *Let $f : \{0,1\}^n \to \{0,1\}$ be computable by a circuit of depth d and size s. Then*

$$\mathrm{D}_{\text{ave}}(f) \le n\left(1 - \frac{1}{O(\log s)^{d-1}}\right).$$

Corollary 5 ([20]). *Let $f : \{0,1\}^n \to \{0,1\}$ be computable by a formula of depth d and size s. Then*

$$\mathrm{D}_{\mathrm{ave}}(f) \le n\left(1 - \frac{1}{O(\frac{1}{d}\log s)^{d-1}}\right).$$

It is natural to ask whether the upper bounds above are tight. A positive answer would suggest that random restrictions with the same probability p (as was used in the proof of the aforementioned results) are optimal. Toward this goal, we prove Theorem 3, which says there exists a DNF of width w and size $\lceil 2^w/w \rceil$ such that $\mathrm{D}_{\mathrm{ave}}(f) = n(1 - \frac{\log n}{\Theta(w)})$.

Here, we provide an outline of the proof and briefly explain how to find such a DNF formula, leaving the proof in the full version of our paper. In contrast to the $O(1)$ average cost to determine the output of the OR function under a uniform input distribution (Proposition 1), it costs $n(1-o(1))$ on average under a p-biased input distribution when $p = o(1/n)$ (Exercise 8.65 in [21]). Our approach is to employ a biased function g (given by Theorem 2) with $p = \Pr_{x\in\{0,1\}^n}[g(x) = 1]$ and $\mathrm{D}_{\mathrm{ave}}(g) = n(1 - o(1))$ as a "simulator" of p-biased variable. Then, we show the composition $\mathrm{OR} \circ g$ is hard to query under a uniform distribution and is computable by a somewhat small DNF formula. As such, Theorem 3 follows.

5 Conclusion

In this paper, we studied the average-case query complexity of boolean functions under the uniform distribution. We prove an upper bound on $\mathrm{D}_{\mathrm{ave}}(f)$ in terms of its weight; on the other hand, we prove that for almost all fixed-weight boolean functions, the upper bound is tight up to an additive logarithmic term. We show that, for any $w \ge 2\log n$, there exists a DNF formula of width w and size $\lceil 2^w/w \rceil$ such that $\mathrm{D}_{\mathrm{ave}}(f) = n(1-\frac{\log n}{\Theta(w)})$, which suggests that the criticality bounds $O(w)$ and $O(\log s)$ are tight up to a multiplicative $\log n$ factor.

Theorems 1 and 2 essentially relate $\mathrm{D}_{\mathrm{ave}}(f)$ to the zero-order Fourier coefficient $\widehat{f}(\{\emptyset\})$. Establishing an upper bound on $\mathrm{D}_{\mathrm{ave}}(f)$ in terms of higher-order Fourier coefficients (such as influences) would be valuable. For example, it is unclear whether the lower bound $\mathrm{D}_{\mathrm{ave}}(\mathrm{Hex}_{L\times L}) \ge L^{1.5+o(1)}$ is tight [14]; bounding $\mathrm{D}_{\mathrm{ave}}(f)$ in terms of Fourier coefficients might shed light on the open problem.

It remains open to prove tight upper bounds on $\mathrm{D}_{\mathrm{ave}}(f)$ for k-DNF, as well as for bounded depth formulas and circuits.

Acknowledgements. We are grateful to the anonymous reviewers for their valuable feedback.

References

1. Ambainis, A., de Wolf, R.: Average-case quantum query complexity. J. Phys. Math. Gen. **34**(35), 6741 (2001)

2. Buhrman, H., de Wolf, R.: Complexity measures and decision tree complexity: a survey. Theor. Comput. Sci. **288**(1), 21–43 (2002)
3. Aaronson, S., Ben-David, S., Kothari, R.: Separations in query complexity using cheat sheets. In: Proceedings of the forty-eighth annual ACM symposium on Theory of Computing, pp. 863–876 (2016)
4. Aaronson, S., Ben-David, S., Kothari, R., Rao, S., Tal, A.: "Degree vs. approximate degree and quantum implications of Huang's sensitivity theorem". In: Proceedings of the 53rd Annual ACM SIGACT Symposium on Theory of Computing, pp. 1330–1342 (2021)
5. Huang, H.: Induced subgraphs of hypercubes and a proof of the sensitivity conjecture. Ann. Math. **190**(3), 949–955 (2019)
6. Rossman, B.: "On the constant-depth complexity of k-clique". In: Proceedings of the fortieth annual ACM symposium on Theory of computing, pp. 721–730 (2008)
7. Rossman, B.: The monotone complexity of k-clique on random graphs. SIAM J. Comput. **43**(1), 256–279 (2014)
8. Yao, A.C.C.: Probabilistic computations: toward a unified measure of complexity. In: 18th Annual Symposium on Foundations of Computer Science (sfcs 977), pp. 222–227 (1977)
9. O'Donnell, R., Saks, M., Schramm, O., Servedio, R.A.: Every decision tree has an influential variable. In: 46th Annual IEEE Symposium on Foundations of Computer Science (FOCS'05), pp. 31–39 (2005)
10. Lee, H.K.: Decision trees and influence: an inductive proof of the OSSS inequality. Theor. Comput. **6**(1), 81–84 (2010)
11. Jain, R., Zhang, S.: The influence lower bound via query elimination. In: arXiv preprint arXiv:1102.4699 (2011)
12. Benjamini, I., Schramm, O., Wilson, D.B.: Balanced Boolean functions that can be evaluated so that every input bit is unlikely to be read. In: Proceedings of the thirty-seventh annual ACM symposium on Theory of computing. (2005)
13. O'Donnell, R., Servedio, R.A.: Learning monotone decision trees in polynomial time. In: 21st Annual IEEE Conference on Computational Complexity (CCC'06), pp. 213–225 (2006)
14. Peres, Y., Schramm, O., Sheffield, S., Wilson, D.B.: Randomturn hex and other selection games. Am. Math. Mon. **114**(5), 373–387 (2007)
15. Smirnov, S., Werner, W.: Critical exponents for twodimensional percolation. Math. Res. Lett. **8**, 729–744 (2001)
16. Ambainis, A.: Quantum query complexity of almost all functions with fixed on-set size. Comput. Complex. **25**, 723–735 (2016)
17. Håstad, J.: Computational limitations for small depth circuits. PhD thesis. Massachusetts Institute of Technology (1986)
18. Rossman, B.: An entropy proof of the switching lemma and tight bounds on the decision-tree size of AC0. (2017). URL: https://users.cs.duke.edu/~br148/logsize.pdf
19. Rossman, B.: Criticality of regular formulas. In: 34th Computational Complexity Conference (CCC 2019). vol. 137, pp. 1:1–1:28 (2019)
20. Harsha, P., Molli, T., Shankar, A.: Criticality of AC0-Formulae. In: 38th Computational Complexity Conference (CCC 2023) Leibniz International Proceedings in Informatics (LIPIcs), vol. 264, pp. 19:1–19:24 (2023)
21. O'Donnell, R.: Analysis of Boolean functions. Cambridge University Press (2014)
22. Håstad, J.: On the correlation of parity and small-depth circuits. SIAM J. Comput. **43**(5), 1699–1708 (2014)

Optimal Framework for Clustering with Noisy Queries

Jinghui Xia(✉) and Zengfeng Huang(✉)

School of Data Science, Fudan University, Shanghai, China
xiajh20@fudan.edu.cn, huangzf@fudan.edu.cn

Abstract. In clustering with noisy queries model, there are n vertices belonging to k unknown clusters. The algorithm is provided with an oracle that answers queries of whether two vertices belong to the same cluster with correct probability $\frac{1}{2}+\frac{\delta}{2}$. The goal is to recover the clusters with minimum number of queries. Most previous works give $O(\frac{nk\log n}{\delta^2}+\text{poly}(k,\frac{1}{\delta},\log n))$ query complexity, so we propose a framework that, given a black-box algorithm with the above query complexity, improves to $O(\frac{n(k+\log n)}{\delta^2}+\text{poly}(k,\frac{1}{\delta},\log n))$, which matches the lower bound of the problem up to an additive $\text{poly}(k,\frac{1}{\delta},\log n))$. With this framework, we propose two new algorithms, one recovers clusters with $O(\frac{n(k+\log n)}{\delta^2}+\frac{k^2\log^3 n}{\delta^4})$ queries, the other recovers clusters with $O(\frac{n(k+\log n)}{\delta^2}+\frac{k^4\log^3 n}{\delta^4}+\frac{k^3\log^8 n}{\delta^4})$ queries with polynomial time. The main idea is to utilize techniques from multi-armed bandit literature in the true cluster identification and verification procedure.

Keywords: Clustering · Multi-armed bandit · Complexity

1 Introduction

Clustering is a fundamental problem in data science and machine learning, with many applications in both theory and practice. Traditional clustering methods rely on full information of a similarity matrix, graph, or feature vectors, which are assumed to be accurate. However, in many real world problems, acquiring precise similarity data is either expensive or inherently noisy. Motivated by this challenge in many applications, e.g., crowdsourced entity resolution, the work of [14] proposed a theoretical model named clustering with noisy queries, also known as clustering with a faulty oracle. In this model, an algorithm seeks to recover an unknown k-clustering of n elements by making pairwise queries to an oracle, asking whether two elements are in the same cluster or not. The oracle's responses are only correct with probability $\frac{1}{2}+\frac{\delta}{2}$. The objective is to design algorithms that minimize the number of queries while achieving high-probability recovery of the true clustering.

On the theory side, the model is also closely related to the stochastic block model (SBM), which is one of the most basic models in graph clustering [1,

F. V. Fomin and M. Xiao (Eds.): COCOON 2025, LNCS 15983, pp. 206–217, 2026.
https://doi.org/10.1007/978-981-95-0215-8_16

10,16]. In SBM, n vertices belong to k unknown clusters, each intra-cluster edge is added independently with probability p and each inter-cluster edge is added independently with probability q, where $0 \leq q \leq p \leq 1$. Given a graph sampled from the above model, the goal is to recover hidden clusters. Consider a graph with edges between pairs of vertices that oracle gives positive response to the query, clustering with noisy queries model is equivalent to SBM with $p = \frac{1}{2} + \frac{\delta}{2}$ and $\frac{1}{2} - \frac{1}{\delta}$, except that SBM always observes all $O(n^2)$ edges while our goal is to use $o(n^2)$ queries. Recently in [17], the authors proposed an algorithm that breaks the small cluster barrier and obtained an improved algorithm for clustering with noisy queries as a byproduct.

Problem Definition. Given a set V of n vertices partitioned into k unknown, disjoint clusters $\{V_1, ..., V_k\}$ and each vertex i belongs to exactly one cluster $V_{c(i)}$. Let $\tau : V \times V \to \pm 1$ denote the ground-truth function where $\tau(u, v) = 1$ if u, v belong to the same cluster and $\tau(u, v) = -1$ otherwise. To infer cluster memberships, the algorithm may query an oracle, which provides noisy responses: for any pair (u, v), the oracle returns the correct value $\tau(u, v)$ with probability $\frac{1}{2} + \frac{\delta}{2}$ for some $\delta \in (0, 1)$. This noise can be formalized as receiving a perturbed output of τ, denoted as $\tilde{\tau}$:

$$\tilde{\tau}(u, v) = \sigma_{u,v} \cdot \tau(u, v),$$

where $\sigma_{u,v}$ is a $\{\pm 1\}$ random variable that attains $+1$ with probability $\frac{1}{2} + \frac{\delta}{2}$. We assume that the randomness in $\sigma_{u,v}$ is independent across distinct pairs and repeated queries for the same pair yield consistent results. The goal is to recover $\{V_1, ..., V_k\}$ with high probability while minimizing the number of queries.

1.1 Previous Results

In Table 1, we summarize previous results and our results.

[14] gave an quasi-polynomial time algorithm with $O(\frac{nk \log n}{\delta^2})$ queries, which recovers all clusters of size $\Omega(\frac{\log n}{\delta^2})$, and a polynomial time algorithm with $O(\frac{nk \log n}{\delta^2} + \min\{\frac{nk^2 \log n}{\delta^4}, \frac{k^5 \log^2 n}{\delta^8}\})$ queries, which recovers all clusters of size $\Omega(\frac{k \log n}{\delta^4})$. In the same paper, an information-theoretic lower bound of $\Omega(\frac{nk}{\delta^2})$ queries is proved. [11] focused on the two-cluster case and proposed a polynomial time algorithm with query complexity $O(\frac{n \log n}{\delta^2} + \frac{\log^2 n}{\delta^6})$. The minimum cluster size is trivial here since there are only two clusters. [18] focused on improving the dependency on δ, and gave a polynomial time algorithm with $O(\frac{nk \log n}{\delta^2} + \frac{k^{10} \log^2 n}{\delta^4})$ queries, which recovers all clusters of size $\Omega(\frac{k^4 \log n}{\delta^2})$. Compared to previous work, the dependence on δ in the second term is improved to $1/\delta^4$. [21] introduced a technique from multi-armed bandit literature and is the only work focusing on the main term of the query complexity since [14]. They gave a polynomial time algorithm with $O(\frac{n(k+\log n)}{\delta^2} + \frac{k^8 \log^3 n}{\delta^4})$ queries, which recovers all clusters of size $\Omega(\frac{k^4 \log n}{\delta^2})$. They also gave a lower bound of

Table 1. Query complexity upper and lower bounds, minimum size of clusters to be recovered and running time of corresponding algorithms.

Query Complexity	Cluster Size	Efficient?	Remark	Ref.
$\Omega(\frac{nk}{\delta^2})$	/	/	Lower bound	[14]
$O(\frac{nk\log n}{\delta^2})$	$\Omega(\frac{\log n}{\delta^2})$	No		
$O(\frac{nk\log n}{\delta^2}+\min\{\frac{nk^2\log n}{\delta^4},\frac{k^5\log^2 n}{\delta^8}\})$	$\Omega(\frac{k\log n}{\delta^4})$	Yes		
$O(\frac{n\log n}{\delta^2}+\frac{\log^2 n}{\delta^6})$	/	Yes	k=2	[11]
$O(\frac{nk\log n}{\delta^2}+\frac{k^4\log^2 n}{\delta^4})$	$\Omega(\frac{n}{k})$	Yes	Nearly-balanced	[18]
$O(\frac{nk\log n}{\delta^2}+\frac{k^{10}\log^2 n}{\delta^4})$	$\Omega(\frac{k^4\log n}{\delta^2})$	Yes		
$\Omega(\frac{n\log n}{\delta^2})$	/	/	Lower bound	[21]
$O(\frac{n(k+\log n)}{\delta^2}+\frac{k^4\log^2 n}{\delta^4})$	$\Omega(\frac{n}{k})$	Yes	Nearly-balanced	
$O(\frac{n(k+\log n)}{\delta^2}+\frac{k^8\log^3 n}{\delta^4})$	$\Omega(\frac{k^4\log n}{\delta^2})$	Yes		
$O(\frac{n^2\log^2 n}{\delta^2 s}+\frac{n^4\log^2 n}{\delta^4 s^4})$	$\Omega(s)$	Yes	$n\geq s\geq\frac{C\cdot\sqrt{n}\log^2 n}{\delta}$	[17]
$O(\frac{n(k+\log n)}{\delta^2}+\frac{k^2\log^3 n}{\delta^4})$	$\Omega(\frac{\log n}{\delta^2})$	No	Theorem 1	This work
$O(\frac{n(k+\log n)}{\delta^2}+\frac{k^4\log^3 n}{\delta^4}+\frac{k^3\log^8 n}{\delta^4})$	$\Omega(\frac{k\log^4 n}{\delta^4})$	Yes	Theorem 2	

$\Omega(\frac{n\log n}{\delta^2})$, which, combined with the previous $\Omega(\frac{nk}{\log n})$ bound, matches their first term of the upper bound. Very recently, [17] gave a polynomial time algorithm with $O(\frac{n^2\log^2 n}{\delta^2 s}+\frac{n^4\log^2 n}{\delta^4 s^4})$ queries, which recovers all clusters of size $\Omega(s)$, where s is an additional parameter that satisfies $n\geq s\geq\frac{C\cdot\sqrt{n}\log^2 n}{\delta}$.

1.2 Our Contributions

We propose a novel framework algorithm which, given a black-box algorithm that solves the clustering with noisy queries problem, solves the problem with improved query complexity. Our intuition comes from the multi-armed bandit algorithm as introduced in [21]. Then with this framework, we improve the query complexity of previous algorithms both without and with polynomial running time constraints. We propose a new algorithm with $O(\frac{n(k+\log n)}{\delta^2}+\frac{k^2\log^3 n}{\delta^4})$ queries, which is the first improvement for the inefficient setting, and may not be achieved by any current efficient algorithms.

Theorem 1. *There exists an algorithm that recovers all the clusters of size* $\Omega(\frac{\log n}{\delta^2})$ *with success probability* $1-o_n(1)$. *The total number of queries to the oracle is* $O(\frac{n(k+\log n)}{\delta^2}+\frac{k^2\log^3 n}{\delta^4})$.

Note that the lower bound suggests that the setting is meaningful only if $\frac{k}{\delta^2}\leq n$, as otherwise $\Omega(n^2)$ queries are required. Thus, $\frac{k^2}{\delta^4}\leq\frac{nk}{\delta^2}$, and if we ignore $\log n$, the dependence on both k and δ is optimal. We also show that, although our query complexity consists of two terms, it is always $O(\frac{nk\log n}{\delta^2})$, and is $o(\frac{nk\log n}{\delta^2})$ under certain conditions.

We also propose a new polynomial time algorithm with $O(\frac{n(k+\log n)}{\delta^2} + \frac{k^4 \log^3 n}{\delta^4} + \frac{k^3 \log^8 n}{\delta^4})$ queries.

Theorem 2. *There exists a polynomial time algorithm that recovers all the clusters of size* $\Omega(\frac{k \log^4 n}{\delta^2})$ *with success probability* $1 - o_n(1)$. *The total number of queries to the oracle is* $O(\frac{n(k+\log n)}{\delta^2} + \frac{k^4 \log^3 n}{\delta^4} + \frac{k^3 \log^8 n}{\delta^4})$.

This algorithm improves the dependence on k for the general case to match the previous ones in the nearly-balanced case, currently the best result under polynomial time constraints. While the query complexity is not strictly superior to that in [21] with respect to $\log n$, the improved dependence on k reduces the complexity for problems with $k = \Omega(\log n)$, which is a common case in practice.

1.3 Other Related Work

The clustering with noisy queries model captures some applications in entity resolution problem [8,15], the signed edges prediction problem in a social network [12] and the correlation clustering problem [4]. We refer to references [11,14,18] for more discussions of the motivations of this model.

In [19], the authors focus on a more general semi-random model that allows the oracle answers to be given adversarially with some probability.

Another line closely related to ours is the semi-supervised active clustering framework [3]. Their goal is to incorporate same-cluster queries in specific clustering problems to alleviate computational hardness, e.g., k-means [5,6,13], correlation clustering [2,20].

2 Preliminaries and Tools

Concentration Bound. We will use the well-known Chernoff-Hoeffding bound.

Theorem 3 (Chernoff-Hoeffding bound [9]). *Let* $X_1, ..., X_n$ *be i.i.d. random variables that can take values in* $\{0,1\}$, *with* $\mathbb{E}[X_i] = p$ *for* $1 \le i \le n$. *Let* $X := \frac{1}{n}\sum_{i=1}^{n} X_i$. *Then we have*

$$\mathbb{P}(X \ge p + \varepsilon) \le e^{-D(p+\varepsilon||p)n}$$
$$\mathbb{P}(X \le p - \varepsilon) \le e^{-D(p-\varepsilon||p)n},$$

where $D(x||y)$ *is the KL divergence of* x *and* y. *Recall the KL divergence between Bernoulli random variables* x, y $D(x||y) = x \ln(\frac{x}{y}) + (1-x)\ln(\frac{1-x}{1-y})$ *and* $D(x||y) \ge \frac{(x-y)^2}{2\max(x,y)}$.

With the Chernoff-Hoeffding bound, we give a general concentration bound for sampling subsets.

Lemma 1. *Let $p = \frac{1}{\log n}$ and $n_0 = \frac{128 \log n}{\delta^2}$. For all clusters $V_i \subset V$ of size $|V_i| \geq \frac{2n_0}{p}$, let $V' \subset V$ of size $|V'| = np$ where vertices in V' are selected independently of each other. Then with probability at least $1 - n^{-7}$ we have $|V_i \cap V'| \geq \frac{1}{2}|V_i| p \geq n_0$ for all cluster V_i.*

Proof. For each V_i, let $X_u = \mathbb{1}_{u \in V'}$ be the indicator function for each $u \in V_i$, then $X_u \sim \mathsf{Bernoulli}(p)$ and $|V_i \cap V'| = \sum_{u \in U} X_u$. We have $\mathbb{E}[|V_i \cap V'|] = |V_i| p$. Let $\varepsilon = \frac{1}{2}p$. By Chernoff-Hoeffding bound, with probability at least $1 - e^{D(\frac{1}{2}p||p)|V_i|} \geq 1 - e^{\frac{1}{8}p \cdot \frac{2n_0}{p}} \geq 1 - n^{-8}$,

$$|V_i \cap V'| \geq \frac{1}{2}|V_i| p \geq n_0.$$

By union bound, with probability at least $1 - n^{-7}$, it holds for all cluster V_i simultaneously.

Best Arm Identification. Following [21], we focus on a special variant of the Best Arm Identification (BAI) problem. In this setting, there are k different arms $a_1, ..., a_k$, where pulling arm a_i yields a binary reward $r \in \{0, 1\}$ from a_i. Assume the rewards from each a_i are i.i.d. $\mathsf{Bernoulli}(\mu_i)$ random variables with $\mu_i \in [0, 1], \forall i$. The objective is to identify the optimal arm $i^* = \arg\max_i \mu_i$ while minimizing the sample complexity, defined as the total number of arm pulls executed by the algorithm. In our specific application of BAI, all problem instances adhere to the following constraints: $i^* \in [k]$ is unique, $\mu_{i^*} \geq \frac{1}{2} + \frac{\delta}{2}$, and $\mu_i \leq \frac{1}{2} - \frac{\delta}{2}$ for all $i \neq i^*$. BAI is well-studied in the literature and the following result is used, which is a special case of Theorem 10 in [7].

Lemma 2. *Given $\delta \in (0, 1)$, $\alpha \in (0, 1)$ and a k-BAI instance defined above, there exists an algorithm $\mathsf{ME}(k, \delta, \alpha)$ that identifies the best arm with probability at least $1 - \alpha$, and its sample complexity is $O(\frac{k}{\delta^2} \log \frac{1}{\alpha})$.*

3 General Framework Algorithm

3.1 True Cluster Identification and Verification

In [21], the authors give an algorithm for the clustering with noisy queries problem with query complexity $O(\frac{n(k+\log n)}{\delta^2} + \frac{k^8 \log^3 n}{\delta^4})$. The idea of the algorithm is to first find some subsets of the clusters, then use best arm identification algorithm to identify the true cluster for each vertex with some probability independently and verify whether the cluster is correct for the vertex or not. Then the algorithm put together vertices of the same cluster and repeat the process until all vertices are correctly clustered. The true cluster identification and verification procedure is the crucial point for the algorithm to achieve the $O(\frac{n(k+\log n)}{\delta^2})$ term in query complexity. In this work, we use the same idea but with some modifications. In the original procedure, the identification and verification part are separate. We use the same verification algorithm.

Algorithm 1 ClusterVerify(v, B)

For each $u \in B$, query (v, u) and get answer $\tilde{\tau}(v, u)$
Compute $d(v, B) = \sum_{u \in B} \mathbb{1}_{\{\tilde{\tau}(v,u)=1\}}$
if $d(v, B) \geq \frac{1}{2}|B|$ **then**
  Output **TRUE**
else
  Output **FALSE**
end if

Algorithm 2 TrueClusterIV$(u, D_u = \{X_1, ..., X_k\}, D'_u = \{X'_1, ..., X'_k\}, \delta, \alpha)$

Simulate ME(k, δ, α):
Let $s_i = 0, i = 1, ..., k$ be the number of pulls for arm a_i.
repeat
  Receive an arm pull request from ME, say a_i
  Pick a random new vertex $v \in X_i$, and query (u, v) to get the answer $\tilde{\tau}(u, v)$ from the oracle. Let $s_i = s_i + 1$
  Return reward $r(a_i) := \frac{\tilde{\tau}(u,v)+1}{2}$ to ME
  if $s_i \geq \frac{64 \log n}{\delta^2}$ **then**
    Invoke ClusterVerify(u, X'_i)
    if ClusterVerify(u, X'_i) =**TRUE** **then**
      return Index i
    else
      Remove index i and Update $D_u = D_u \setminus \{X_i\}, D'_u = D'_u \setminus \{X'_i\}$,
      if $D_u = \emptyset$ **then**
        return -1
      end if
    end if
  end if
until ME terminates and outputs an arm a_{i^*}
if ClusterVerify(u, X'_{i^*}) =**TRUE** **then**
  return Index i^*
else
  return -1
end if

Lemma 3. *Let $B \subset V_i$ be a sub-cluster with size at least $\frac{64 \log n}{\delta^2}$ for some cluster V_i, then with probability at least $1 - n^{-6}$, the following holds for all $v \in V$ and all clusters V_i simultaneously,*
(1) if $v \in V_i$, ClusterVerify(v, B) *returns* ***TRUE;***
(2) if $v \notin V_i$, ClusterVerify(v, B) *returns* ***FALSE.***

Proof. This is a direct corollary of Lemma 3.4 in [21] with $\eta = \frac{1}{2}$. □

In our identification part, for each vertex u, we count the number of pulls s_i for each arm a_i. Once some $s_i \geq \frac{64 \log n}{\delta^2}$, we directly verify whether X_i is the correct cluster for u by ClusterVerify. If it is the correct cluster, then we output the index as the true cluster of u, otherwise we can remove X_i for u permanently.

If all sub-clusters are not the true cluster for u, that means the true cluster of u is not recovered yet, the algorithm returns -1 as a flag. With this modification, we can ensure that for each vertex u, we only need $O(\frac{\log n}{\delta^2})$ vertices from each cluster V_i in the whole true cluster identification and verification procedure, while in the previous algorithm, the BAI algorithm ME may require $O(\frac{k \log n}{\delta^2})$ pulls from one arm. In that case, we will need larger sub-clusters, which leads to more remaining clusters and vertices after the procedure and eventually higher order of k in query complexity for the algorithm. Note that the number of queries made by the extra verifications is strictly less than or equal to the number of queries made in the original identification part, so the query complexity remains the same (up to a constant of 2).

Recall that in the previous algorithm in [21], they set $\alpha = \frac{1}{4}$ for ME and use a recursive strategy since the number of remaining vertices after each round decreases by half, so the total number of queries over all rounds for algorithm 2 is $O(\frac{n(k+\log n)}{\delta^2})$. We also use the recursive strategy, but the condition is slightly different here. The number of vertices that are in the same clusters with $X_1, ..., X_h$ decreases by half after each round, but the vertices from other clusters will not find true cluster with $X_1, ..., X_h$, so they remain until the procedure stops. During the procedure, such vertex may query with $O(\frac{k \log n}{\delta^2})$ vertices of $X_1, ..., X_h$ at most. Suppose the number of these vertices is n', then they involve extra $O(\frac{n' k \log \text{ n}}{\delta^2})$ queries. Thus the total number of queries during the true cluster identification and verification procedure is

$$O(\frac{n(k + \log n)}{\delta^2} + \frac{n' k \log \text{ n}}{\delta^2})$$

3.2 Framework Algorithm

With the above true cluster identification and verification procedure Algorithm 2, we propose a novel framework Algorithm 3 that solves the clustering with noisy queries problem with $O(\frac{n(k+\log n)}{\delta^2} + \text{poly}(k, \frac{1}{\delta}, \log n))$ queries given an arbitrary black-box algorithm $\mathcal{A}$ that solves the problem with $O(\frac{nk \log n}{\delta^2} + \text{poly}(k, \frac{1}{\delta}, \log n))$ queries. The idea is to sample a subset $V' \subset V$ of size $|V'| = np$, where $p = \frac{1}{\log n}$ is the sampling ratio induced by the query complexity of $\mathcal{A}$, and invoke $\mathcal{A}$ to recover clusters of V', then use the recovered clusters as the sub-clusters of V. With probability at least $1 - o_n(1)$, $\mathcal{A}$ correctly recovers the sub-clusters $V'_1, ..., V'_{k'}$ of V', with $O(\frac{nk}{\delta^2} + \text{poly}(k, \frac{1}{\delta}, \log n))$ queries. Then with the true cluster identification and verification procedure, we can recover clusters of V with $O(\frac{n(k+\log n)}{\delta^2} + \text{poly}(k, \frac{1}{\delta}, \log n))$ queries. Suppose $\mathcal{A}$ can recover all clusters of size at least $n_0 = f(n, k, \delta)$. Let $n'_0 = \max(n_0, \frac{128 \log n}{\delta^2})$.

By Lemma 1, every cluster of size $\Omega(\frac{n'_0}{p}) = \Omega(n'_0 \log n)$ will have a sub-cluster in V' of size $\Omega(n'_0)$, which will be recovered by algorithm $\mathcal{A}(V', k, \delta)$. Then these clusters will be recovered in the true cluster identification and verification procedure with success probability at least $1 - o_n(1)$ by making $O(\frac{n(k+\log n)}{\delta^2} + \frac{n' k \log n}{\delta^2})$ queries. For the remaining vertices, they are all from smaller clusters

Algorithm 3 Framework$(V, k, \delta, \mathcal{A}, p, n_0)$

Let c be a sufficiently large constant. Initialize $U = V$.
Randomly sample $V' \subset V$ of size $|V'| = np$
Invoke $\mathcal{A}(V', k, \delta)$ to obtain sub-clusters $X_1, ..., X_k$
Keep $X_1, ..., X_h$ of size at least $\frac{128 \log n}{\delta^2}$
For each X_i, pick out arbitrary $X_i' \subseteq X_i$ of size $\frac{64 \log n}{\delta^2}$
For each $u \in V$, set $D_u = \{X_1, ..., X_h\}$, $D_u' = \{X_1', ..., X_h'\}$
while $|U| \geq \frac{2kn_0'}{p}$ **do**
  for each $v \in U$ **do**
    $j =$ TrueClusterIV$(v, D_v, D_v', \delta, \alpha)$
    if $j \neq -1$ **then**
      Update $X_j = X_j \cup \{v\}$, $U = U \setminus \{v\}$
    else
      Update D_v, D_v' in TrueClusterIV
    end if
  end for
end while
for each $v \in U$ **do**
  for each $X_j \in D_v$ **do**
    if ClusterVerify$(v, X_j') =$ **TRUE then**
      Update $X_j = X_j \cup \{v\}$, $U = U \setminus \{v\}$
    end if
  end for
end for
Invoke $\mathcal{A}(U, k - h, \delta)$ to obtain clusters $X_{h+1}, ..., X_k$
return k clusters $X_1, ..., X_k$

of size $O(n_0' \log n)$, so the total number will not exceed $n' = O(kn_0' \log n)$. We can again invoke algorithm $\mathcal{A}$, then any cluster of size $\Omega(n_0)$ will be recovered, with at most $O(k^2 n_0'^2 \log^2 n)$ queries, which is the number of all pairwise queries. Thus the minimum cluster size remains the same as in algorithm $\mathcal{A}$. If we have $n_0 = O(\text{poly}(k, \frac{1}{\delta}, \log n))$, which is true for previous algorithms e.g. [14,18], then the query complexity of our Algorithm 3 is $O(\frac{n(k+\log n)}{\delta^2} + \text{poly}(k, \frac{1}{\delta}, \log n))$. The correctness of the algorithm is easy to prove by union bound, since the success probability of both algorithm $\mathcal{A}$ and the identification and verification procedure is at least $1 - o_n(1)$.

Theorem 4. *If given a black-box algorithm that recovers all the clusters of size $\Omega(n_0)$ with success probability $1 - o_n(1)$ where $n_0 = O(\text{poly}(k, \frac{1}{\delta}, \log n))$ and the total number of queries to the oracle is $O(\frac{nk \log n}{\delta^2}) + \text{poly}(k, \frac{1}{\delta}, \log n)$, then there exists an algorithm that recovers all the clusters of size $\Omega(n_0)$ with success probability $1 - o_n(1)$ and the total queries to he oracle is $O(\frac{n(k+\log n)}{\delta^2}) + \text{poly}(k, \frac{1}{\delta}, \log n)$.*

In the following sections, we select specific algorithm $\mathcal{A}$ and analyze the query complexity upper bound under different problem settings.

4 Information Theoretic Optimal Algorithm

[14] gives an algorithm that satisfies Theorem 5. The running time of the algorithm is quasi-polynomial.

Theorem 5 (Theorem 2 of [14]). *There exists an algorithm that recovers all the clusters of size* $\Omega(\frac{\log n}{\delta^2})$ *with success probability* $1 - o_n(1)$. *The total number of queries to the oracle is* $O(\frac{nk \log n}{\delta^2})$.

Given this algorithm, our framework Algorithm 3 can recover all clusters of size $\Omega(\frac{\log n}{\delta^2})$ with probability at least $1 - o_n(1)$ by making $O(\frac{n(k+\log n)}{\delta^2} + \frac{k^2 \log^3 n}{\delta^4})$ queries.

Proof (Proof of Theorem 1). It suffices to calculate the query complexity. Here $n'_0 = O(\frac{\log n}{\delta^2})$. First, $\mathcal{A}$ takes $O(\frac{n}{\log n})$ vertices and makes $O(\frac{nk}{\delta^2})$ queries. Then TrueClusterIV makes $O(\frac{n(k+\log n)}{\delta^2} + \frac{k^2 \log^3 n}{\delta^4})$ queries since $n' = O(kn'_0 \log n) = O(\frac{k \log^2 n}{\delta^2})$. Finally, $\mathcal{A}$ takes $O(\frac{k \log^2 n}{\delta^2})$ vertices and makes $O(\frac{k^2 \log^3 n}{\delta^4})$ queries. Thus in total, the algorithm makes $O(\frac{n(k+\log n)}{\delta^2} + \frac{k^2 \log^3 n}{\delta^4})$ queries. □

Note that if the number of vertices from small clusters of size $O(\frac{log^2 n}{\delta^2})$ is $o(n)$, then the query complexity of our algorithm is $o(\frac{nk \log n}{\delta^2})$, which is better than the previous algorithm.

5 Computationally Efficient Algorithm

[17] gives an computationally efficient algorithm that satisfies Theorem 6. The running time of the algorithm is polynomial.

Theorem 6 (Theorem 1.6 of [17]). *There exists a polynomial time algorithm such that for any* $n \geq s \geq \frac{C \cdot \sqrt{n} \log^2 n}{\delta}$, *it recovers all clusters of size larger than* s *by making* $O(\frac{n^2 \log^2 n}{\delta^2 s} + \frac{n^4 \log^2 n}{\delta^4 s^4})$ *queries to the oracle.*

Note that this algorithm requires a parameter s. In order to recover all possible clusters, we may have to let $s = \frac{C \cdot \sqrt{n} \log^2 n}{\delta}$ even when $k \ll \frac{n}{s}$. We give an improved version with query complexity as a function of n, k, δ that does not require additional parameter.

Theorem 7. *There exists a polynomial time algorithm that recovers all the clusters of size* $\Omega(\frac{k \log^4 n}{\delta^2})$ *with success probability* $1 - o_n(1)$. *The total number of queries to the oracle is* $O(\frac{nk \log^2 n}{\delta^2} + \frac{k^4 \log^3 n}{\delta^4})$.

Proof. First we suppose k is known. If $\frac{n}{2k} \leq \frac{C \cdot \sqrt{n} \log^2 n}{\delta}$, run the algorithm with $s = \frac{C \cdot \sqrt{n} \log^2 n}{\delta}$. The algorithm recovers all clusters of size $\Omega(\frac{k \log^4 n}{\delta^2})$ by making $O(\frac{nk \log^2 n}{\delta^2} + \frac{k^4 \log^2 n}{\delta^4})$ queries, since $\frac{n}{s} = O(k)$.

Otherwise, let $s = \frac{n}{2k}$. Since there must be some clusters with size larger than $\frac{n}{2k}$, the algorithm will recover these clusters, and the number of remaining vertices is at most $\frac{n}{2}$. Remove the recovered clusters, update n and k, and recursively run the algorithm until $\frac{n}{2k} \leq \frac{C\cdot\sqrt{n}\log^2 n}{\delta}$. Then run the algorithm with $s = \frac{C\cdot\sqrt{n}\log^2 n}{\delta}$. The number of queries in the first round is $O(\frac{nk\log^2 n}{\delta^2} + \frac{k^4\log^2 n}{\delta^4})$. Since the remaining number of vertices decreases by half after each round, the number of queries is $O(\frac{nk\log^2 n}{2^{(i-1)}\delta^2} + \frac{k^4\log^2 n}{\delta^4})$ in the ith round, and there are at most $O(\log n)$ rounds. Thus the total number of queries is

$$\sum_{i=1}^{O(\log n)} O(\frac{nk\log^2 n}{2^{(i-1)}\delta^2} + \frac{k^4\log^2 n}{\delta^4}) = O(\frac{nk\log^2 n}{\delta^2} + \frac{k^4\log^3 n}{\delta^4}).$$

If k is unknown, let l be the guessed value of k. Let $l = 2$ first and run the above procedure with l, then double l every time. We stop doubling l when we meet the condition $\frac{n}{2l} \leq \frac{C\cdot\sqrt{n}\log^2 n}{\delta}$ in the procedure. When l is small, the algorithm may not recover any clusters, or recovers some cluster but fails to return any cluster when running with $s = \frac{n}{2l}$. But once we meet the condition $\frac{n}{2l} \leq \frac{C\cdot\sqrt{n}\log^2 n}{\delta}$ in the procedure, it is guaranteed that the algorithm recovers all possible clusters. Suppose $\max l = 2^t$. Note that l will never be larger than $2k$, as it must stop when running with $l \geq k$. Thus the number of queries in the last complete procedure is

$$O(\frac{nl\log^2 n}{\delta^2} + \frac{l^4\log^3 n}{\delta^4}) = O(\frac{nk\log^2 n}{\delta^2} + \frac{k^4\log^3 n}{\delta^4}).$$

And the total number of queries is

$$\sum_{j=1}^{t} O(\frac{n2^j\log^2 n}{\delta^2} + \frac{(2^j)^4\log^2 n}{\delta^4}) = O(\frac{n2^t\log^2 n}{\delta^2} + \frac{(2^t)^4\log^3 n}{\delta^4})$$

$$= O(\frac{nk\log^2 n}{\delta^2} + \frac{k^4\log^3 n}{\delta^4}).$$

The algorithm recovers all clusters of size $\Omega(\frac{k\log^4 n}{\delta^2})$. □

Now we prove theorem 2.

Proof (Proof of Theorem 2). Note that the first term of query complexity of the algorithm is $O(\frac{nklog^2n}{\delta^2})$, but it does not affect the structure of our algorithm. We only need to change the parameter. Let $p = \frac{1}{\log^2 n}$, Lemma 1 and Theorem 4 still hold.

Now it suffices to calculate the query complexity. Here $n_0' = O(\frac{k\log^4 n}{\delta^2})$. First, $\mathcal{A}$ takes $O(\frac{n}{\log^2 n})$ vertices and makes $O(\frac{nk}{\delta^2} + \frac{k^4\log^3 n}{\delta^4})$ queries. Then TrueClusterIV makes $O(\frac{n(k+\log n)}{\delta^2} + \frac{k^3\log^7 n}{\delta^4})$ queries since $n' = O(kn_0'\log^2 n) = O(\frac{k^2\log^6 n}{\delta^2})$. Finally, $\mathcal{A}$ takes $O(\frac{k^2\log^6 n}{\delta^2})$ vertices and makes $O(\frac{k^3\log^8 n}{\delta^4} + \frac{k^4\log^3 n}{\delta^4})$ queries. Thus in total, the algorithm makes $O(\frac{n(k+\log n)}{\delta^2} + \frac{k^4\log^3 n}{\delta^4} + \frac{k^3\log^8 n}{\delta^4})$ queries. □

6 Conclusion

In this paper, we study clustering with noisy queries. We propose a framework that improves the query complexity of a given algorithm for the problem. We provide new algorithms with improved query complexity. Some questions remain open are (1). whether the second term of query complexity can be further improved, both inefficient and efficient algorithms, and (2). whether there is a lower bound on the second term.

Disclosure of Interests. The authors have no competing interests to declare that are relevant to the content of this article.

References

1. Abbe, E.: Community detection and stochastic block models: recent developments. J. Mach. Learn. Res. **18**(177), 1–86 (2018)
2. Ailon, N., Bhattacharya, A., Jaiswal, R.: Approximate correlation clustering using same-cluster queries. In: Latin American Symposium on Theoretical Informatics, pp. 14–27. Springer, Cham (2018)
3. Ashtiani, H., Kushagra, S., Ben-David, S.: Clustering with same-cluster queries. In: Proceedings of the 30th International Conference on Neural Information Processing Systems, vol. 29, pp. 3216–3224. Curran Associates, Inc. (2016)
4. Bansal, N., Blum, A., Chawla, S.: Correlation clustering. Mach. Learn. **56**(1), 89–113 (2004)
5. Bressan, M., Cesa-Bianchi, N., Lattanzi, S., Paudice, A.: Exact recovery of mangled clusters with same-cluster queries. In: Proceedings of the 34th International Conference on Neural Information Processing Systems, vol. 33, pp. 9324–9334. Curran Associates, Inc. (2020)
6. Chien, I.E., Pan, C., Milenkovic, O.: Query k-means clustering and the double dixie cup problem. In: Proceedings of the 32nd International Conference on Neural Information Processing Systems, pp. 6650–6659. Curran Associates Inc., Red Hook, NY, USA (2018)
7. Even-Dar, E., Mannor, S., Mansour, Y.: Action elimination and stopping conditions for the multi-armed bandit and reinforcement learning problems. J. Mach. Learn. Res. **7**(39), 1079–1105 (2006)
8. Fellegi, I.P., Sunter, A.B.: A theory for record linkage. J. Am. Stat. Assoc. **64**(328), 1183–1210 (1969)
9. Hoeffding, W.: Probability inequalities for sums of bounded random variables. J. Am. Stat. Assoc. **58**(301), 13–30 (1963)
10. Holland, P.W., Laskey, K.B., Leinhardt, S.: Stochastic blockmodels: first steps. Soc. Netw. **5**(2), 109–137 (1983)
11. Larsen, K.G., Mitzenmacher, M., Tsourakakis, C.: Clustering with a faulty oracle. In: Proceedings of the Web Conference 2020, pp. 2831–2834. Association for Computing Machinery (2020)
12. Leskovec, J., Huttenlocher, D., Kleinberg, J.: Predicting positive and negative links in online social networks. In: Proceedings of the 19th International Conference on World Wide Web. WWW '10, pp. 641–650. Association for Computing Machinery, New York, NY, USA (2010)

13. Li, Y., Song, Y., Zhang, Q.: Learning to cluster via same-cluster queries. In: Proceedings of the 30th ACM International Conference on Information and Knowledge Management, pp. 978–987. Association for Computing Machinery, New York, NY, USA (2021)
14. Mazumdar, A., Saha, B.: Clustering with noisy queries. In: Proceedings of the 31st International Conference on Neural Information Processing Systems, pp. 5790–5801. Curran Associates Inc., Red Hook, NY, USA (2017)
15. Mazumdar, A., Saha, B.: A theoretical analysis of first heuristics of crowdsourced entity resolution. In: Proceedings of the Thirty-First AAAI Conference on Artificial Intelligence, pp. 970–976. AAAI Press (2017)
16. McSherry, F.: Spectral partitioning of random graphs. In: Proceedings of the 42nd IEEE Symposium on Foundations of Computer Science, pp. 529–537. IEEE Computer Society (2001)
17. Mukherjee, C.S., Peng, P., Zhang, J.: Recovering unbalanced communities in the stochastic block model with application to clustering with a faulty oracle. In: Proceedings of the 37th International Conference on Neural Information Processing Systems. NIPS '23. Curran Associates Inc., Red Hook, NY, USA (2023)
18. Peng, P., Zhang, J.: Towards a query-optimal and time-efficient algorithm for clustering with a faulty oracle. In: Proceedings of Thirty Fourth Conference on Learning Theory. Proceedings of Machine Learning Research, vol. 134, pp. 3662–3680. PMLR (2021)
19. Pia, A.D., Ma, M., Tzamos, C.: Clustering with queries under semi-random noise. In: Proceedings of Thirty Fifth Conference on Learning Theory. Proceedings of Machine Learning Research, vol. 178, pp. 5278–5313. PMLR (2022)
20. Saha, B., Subramanian, S.: Correlation clustering with same-cluster queries bounded by optimal cost. In: Proceedings of the 27th Annual European Symposium on Algorithms (ESA 2019). Leibniz International Proceedings in Informatics (LIPIcs), vol. 144, pp. 81:1–81:17. Schloss Dagstuhl–Leibniz-Zentrum fuer Informatik, Dagstuhl, Germany (2019)
21. Xia, J., Huang, Z.: Optimal clustering with noisy queries via multi-armed bandit. In: Proceedings of the 39th International Conference on Machine Learning. Proceedings of Machine Learning Research, vol. 162, pp. 24315–24331. PMLR (2022)

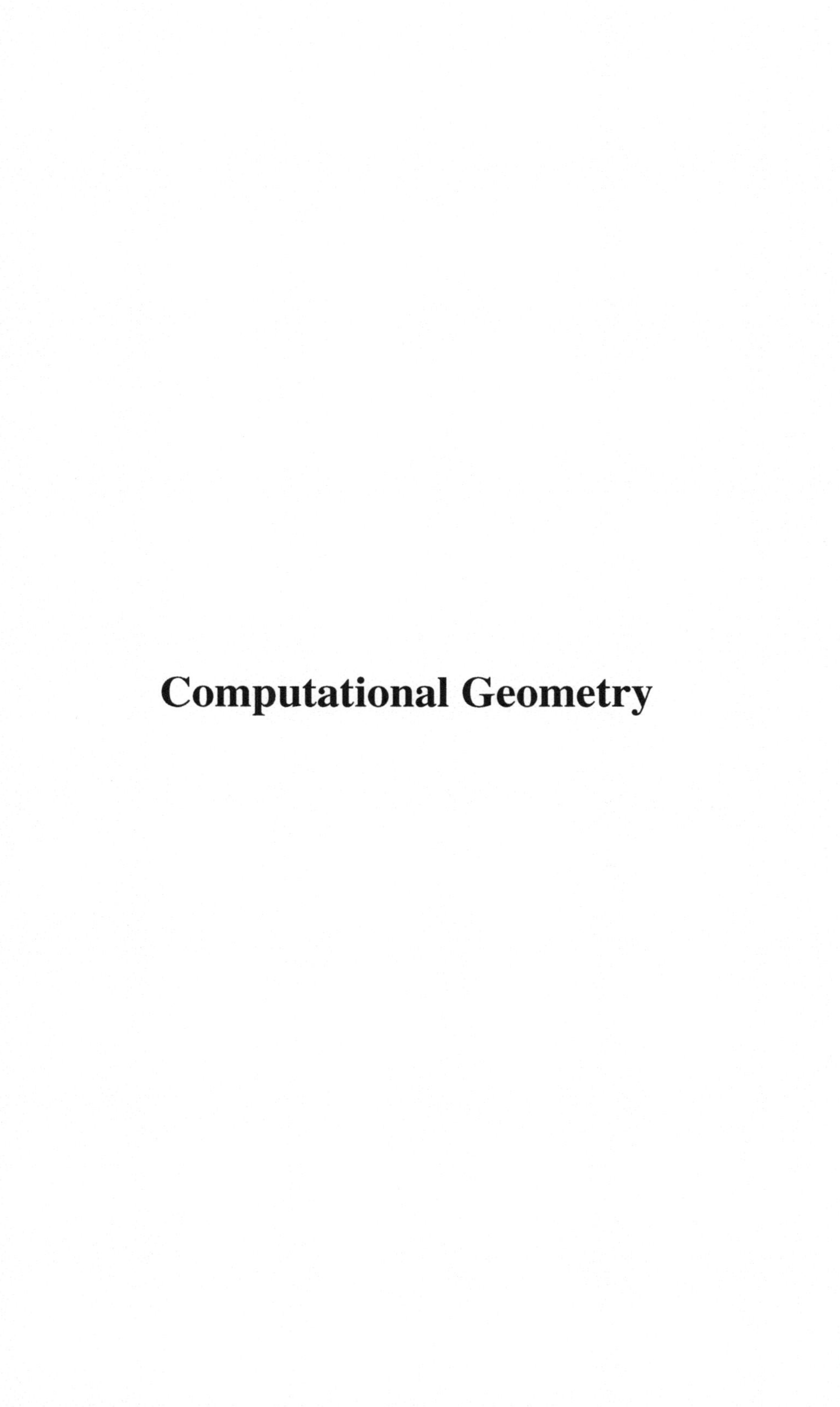

Computational Geometry

Minimum-Membership Geometric Dominating Set: Complexity and Algorithms

Bhavya Bansal[1], Raghunath Reddy Madireddy[1(✉)], and Supantha Pandit[2]

[1] Department of Computer Science and Information Systems, Birla Institute of Technology and Science, Pilani, Hyderabad Campus, Jawahar Nagar, Kapra Mandal, Medchal District, Telangana 500078, India
{f20211823,raghunath}@hyderabad.bits-pilani.ac.in

[2] Dhirubhai Ambani Institute of Information and Communication Technology, Gandhinagar, Gujarat, India

Abstract. We study the geometric version of the minimum-membership dominating set problem. In this problem, we are given a set O of geometric objects in the plane, and the goal is to find a dominating set S of O such that the maximum number of dominators in S for each object in O is the minimum. We call the maximum number of dominators in S of any object is called the depth of S. We give a polynomial-time exact algorithm for the problem when the objects are intervals on the real line. Further, the problem has polynomial-time algorithms when the objects are axis-parallel lines and axis-parallel strips. On the other hand, the discrete version of the problem is NP-hard, even for depth 1, for axis-parallel lines, axis-parallel strips, and axis-parallel rectangles. Finally, we show that the problem is NP-hard (for depth 1) when the objects are axis-parallel unit segments in the plane.

Keywords: Geometric dominating set · Minimum membership dominating set · NP-hard · APX-hard · Intervals · Lines · Segments

1 Introduction

The Dominating Set problem is a well-studied problem in computer science. Given a graph $G = (V, E)$, where V is the set of vertices and E is the set of edges, the goal is to find a minimum size subset $V' \subseteq V$ of vertices such that for each vertex $v \in V$ either $v \in V'$ or there exists a vertex $v' \in V'$ such that (v, v') is an edge in E. In the first case, v dominates itself and in the latter case, v' is called a dominator of v. The problem is known to be NP-hard [12], even for unit disk graphs [5]. Several variations of the dominating set problem are considered in the literature. One such variation is the minimum membership domination problem (MMDS) [1,8]. In this variation, given a graph $G = (V, E)$, we need to find a dominating set $V' \subseteq V$ of V such that the maximum number of dominators in V' for vertices in V is to be the minimum.

F. V. Fomin and M. Xiao (Eds.): COCOON 2025, LNCS 15983, pp. 221–233, 2026.
https://doi.org/10.1007/978-981-95-0215-8_17

In this paper, we study the geometric version of MMDS problem. For two objects, g and h, if g and h have a non-empty intersection, we say that h dominates g and vice versa. For a given set S of objects and an object h, we denote the number of dominators of h in S by $\mathsf{mem}(S, h)$[1], and we call this number as the membership value or depth of h with respect to S.

Definition 1. ***Minimum-Membership Geometric Dominating Set (MMGDS).*** *Given a set of objects O in the plane, the goal is to find a dominating set $S \subseteq O$ of O such that the maximum membership value of all objects in O with respect to S is minimum, i.e., $\max_{h \in O}\{\mathsf{mem}(S, h)\}$ is minimum. We call this number as the depth of S.*

We study the hardness and approximability of MMGDS problem with intervals, axis-parallel lines, axis-parallel unit segments, axis-parallel strips, and other related geometric objects. In addition, we examine the discrete version of the MMGDS problem. The formal definition of the problem is given below.

Definition 2. ***(Discrete MMGDS)*** *In the discrete minimum-membership geometric dominating set problem, we are given a set of points P and a set of objects O. We say that an object $g \in O$ dominates another object $h \in O$ if both g and h cover a common point from the set P. The goal is to find a dominating set S of O such that $\max_{h \in O}\{\mathsf{mem}(S, h)\}$ is minimum.*

In the rest of the paper, we use n to denote the number of objects in the instance of MMGDS problem. In the discrete version of MMGDS problem, we use m to denote the number of points in the instance.

1.1 Previous Work

Fellows and Hoover [8] introduced a variation of the dominating set problem called *perfect domination*, where every vertex has exactly one dominator in the solution. This problem is also studied under the name *perfect codes* [2,3,10,14]. The authors in [8] show that the problem is NP-hard even for planar graphs for degree three. Agarwal et al. [1] studied the problem of the minimum membership-dominating set problem on graphs. We note that the perfect dominating problem is a special case of the minimum membership domination problem with depth 1. The authors in [1] show that the minimum membership dominating set problem is NP-hard on planar bipartite graphs, with maximum degree four, even with depth 1. In addition, Sangam and Kare [21] investigated the computational complexity of the problem on various graph classes, including trees and planar graphs. They presented algorithmic results and demonstrated that the problem remains NP-hard even when restricted to planar graphs. More recently, Karthika et al. [13] strengthened these hardness results by proving that the MMDS problem remains NP-hard even on graphs with maximum degree three.

In the literature, a similar version of the set cover and hitting set problems is considered, called *Minimum Membership Set Cover* (MMSC) and *Minimum*

[1] If $h \in S$ then h will contribute to the value $\mathsf{mem}(S, h)$.

Membership Hitting Set (MMHS), respectively. The MMSC problem was first studied by Kuhn et al. [15], and they have proved that the problem is NP-complete and have shown that the problem cannot have a better approximation factor than $O(\log n)$ unless P=NP. In [20], Narayanaswamy et al. studied the MMHS problem. They have shown that the geometric version of MMHS problem is NP-hard when the objects are the axis-parallel segments in the plane. Mitchell and Pandit, [18] also studied the geometric versions of both MMSC and MMHS problems and proved that the problems are NP-hard when the objects are axis-parallel rectangles that intersect a horizontal line.

The Dominating Set problem is known to be NP-complete for unit disk graphs [5] and the problem has a PTAS for the unit disk graphs [11]. On the other hand, in [7], Erlebach and van Leeuwen have shown that the problem is APX-hard when the graphs denote the intersection graphs of objects like axis-parallel rectangles, ellipses, and other shapes. The discrete version of the Dominating Set problem is also known to be APX-hard [17] for various classes of geometric objects, such as axis-parallel rectangles, axis-parallel strips, axis-parallel ellipses, and unit balls in $\mathbb{R}^3$. Moreover, they established that the problem is NP-hard for unit disks intersecting a horizontal line and for axis-parallel unit squares intersecting a diagonal. Similar results are known for set cover and hitting set problems [4,16].

1.2 Our Contributions

- We give an $O(n^4)$ time exact algorithm for MMGDS problem when the objects are intervals on the real line (Theorem 1). When no interval is inside another interval in the problem instance, the problem admits the $O(n^2)$-time algorithm (Corollary 2). In addition, we show that there exists an $O(n^2)$ time algorithm to check if there exists an optimum solution with depth 1 for MMGDS problem with intervals (Corollary 1).
- Further, we show that the MMGDS problem admits polynomial time algorithms for axis-parallel infinite lines and axis-parallel strips (Theorem 2). On the other hand, we show that the *discrete* MMGDS problem is NP-hard for axis-parallel lines, strips, and some other geometric objects in the plane (Theorem 3 and Corollary 3).
- In the case of axis-parallel unit segments, we show that the MMGDS problem, even with depth 1, is NP-hard (Theorem 4).

We note that the membership value is always a positive integer. In our NP-hardness results, we have shown that determining the existence of a dominating set with depth 1 is NP-hard. Consequently, for all these NP-hard problems, unless P= NP, no polynomial-time approximation algorithm can achieve an approximation ratio smaller than 2.

1.3 Preliminaries

In this section, we give some definitions and terminologies that are required for the NP-hardness results in this paper.

Definition 3. *Positive exactly 1-in-*3SAT *(P1-in-3SAT). Given a* 3SAT *formula ϕ with variables $x_1, x_2, \ldots, x_n$ and clauses $C_1, C_2, \ldots, C_m$ where each clause contains exactly three unnegated (positive) literals, the goal is to find a satisfying assignment for ϕ such that in each clause there is exactly one true literal.*

The P1-in-3SAT problem is known to be NP-hard [6,9,22]. We now give a planar version of this problem that is known to be NP-hard [6,19].

Definition 4. *Rectilinear Positive Planar 1-in-3SAT (RPP1-in-3SAT) problem: Given a P1-in-3SAT instance ϕ, the variable-clause incidence graph, a bipartite graph where variables and clauses are represented as nodes and edges indicate variable occurrences in clauses, is planar. Furthermore, this planar graph admits a rectilinear embedding, where the variables are ordered and placed on a horizontal line, and clauses connect the variables by axis-parallel rectilinear legs (edges) from above and below the horizontal line. The problem asks whether there exists a truth assignment to the variables such that every clause contains at least one satisfied literal. Refer Fig. 1 for an instance of the RPP1-in-3SAT problem.*

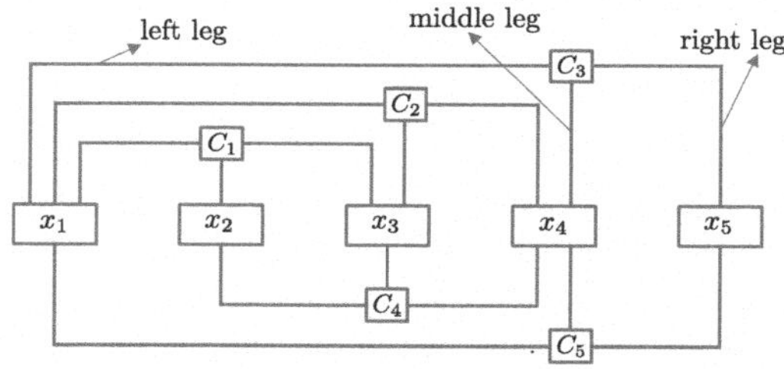

Fig. 1. An instance of RPP1-in-3SAT problem. The variables are on a horizontal line L, clauses C_1, C_2, and C_3 connect the variable from above, and clauses C_4 and C_5 connect the variable from below. Different legs are shown for C_3.

From the rectilinear planar embedding we have that, there are three different types of legs (edges). Consider a clause C_l that contains the literals x_i, x_j, and x_k in the order in which they appear in the RPP1-in-3SAT representation. Further, C_l connects these variable nodes from the above. We define the rectilinear edge connecting to x_i as the left leg, the edge connecting to x_j as the middle leg, and the edge connecting to x_k as the right leg. We can similarly define the left, middle, and right legs for a clause connecting to the three variables from below by rotating the RPP1-in-3SAT representation 180°.

2 MMGDSProblem with Intervals on the Real Line

In this section, we consider MMGDS problem with intervals on a real line (MMGDS-INT). We first show that the depth of any optimum solution is at most three, and using this bound, we give a sweep-line-based $O(n^4)$ time algorithm that returns an optimum solution to the instance of the problem.

Let $I = \{t_1, t_2, \ldots, t_n\}$ be the set of n intervals on the real line. For any interval t_i, let $\mathsf{left}(t_i)$ and $\mathsf{right}(t_i)$ denote the left and right endpoints of t_i, respectively. Without loss of generality, we assume that each interval in I has distinct left endpoints (resp. right endpoints).

2.1 Upper Bound on the Optimum Depth

In the following, we show that the depth of any optimum solution is at most three. Let $S_1 \subseteq I$ be the set of intervals such that each interval in S_1 completely covers some other interval in I but not completely covered by any other interval in I. Note that the intervals in S_1 may overlap, but one is not contained in another i.e., if $t_x, t_y \in S_1$ and $\mathsf{left}(t_x) < \mathsf{left}(t_y)$ then $\mathsf{right}(t_x) < \mathsf{right}(t_y)$. Let $I_1 \subseteq I$ be the set of intervals dominated by the intervals in S_1.

We note that S_1 is a dominating set for the set of intervals $S_1 \cup I_1$.

Observation 1: Let t_x, t_y, and t_z be intervals in S_1 such $\mathsf{left}(t_x) < \mathsf{left}(t_y) < \mathsf{left}(t_z)$. If t_y is completely covered by the union of t_x and t_z i.e., $\mathsf{left}(t_z) < \mathsf{right}(t_x)$, we can remove t_y from S_1 (since it is redundant), and still S_1 is a dominant set for $I_1 \cup S_1$ without increasing the depth of S_1.

Due to the above observation, we have the following claim.

Claim 1. *There exists a set $S'_1 \subseteq S_1$ that dominates all the intervals in $I_1 \cup S_1$ so that no three intervals in S'_1 dominate each other.*

Let $I'_1 \subseteq I$ be the set of intervals dominated by the intervals in S'_1.

Observation 2: Let $t_p \in I'_1$ be an interval dominated by four intervals t_w, t_x, t_y, and t_z in S_1^1 such $\mathsf{left}(t_w) < \mathsf{left}(t_x) < \mathsf{left}(t_y) < \mathsf{left}(t_z)$. Note that none of these intervals is fully covered by t_p, and none of these intervals fully covers the interval t_p (due to Claim 1). However, the union of t_p, t_w, and t_z completely covers the intervals t_x and t_y. So, we can replace t_x and t_y with t_p in S'_1, so that S'_1 is still a dominating set for I'_1 without increasing the depth of S'_1. In addition, t_p does not dominate any interval that is not dominated by intervals in S'_1.

Due to the above observation, we have the following claim.

Claim 2. *There exists a set $S^*_1 \subseteq I_1$ that dominates every interval in I_1 so that the depth of S^*_1 is at most three.*

Let $I_2 \subseteq I$ be the set of intervals that are not dominated by the intervals in S^*_1. The intervals in I_2 can overlap with each other, but no interval in I_2 contained another interval in I_2. We have the following claim.

Claim 3. *There exists a dominating set $S^*_2 \subseteq I_2$ for I_2 with depth at most 2.*

Proof. We can find S_2^* using a greedy strategy: (i) sort the intervals with respect to the left endpoint of the intervals in I_2 and (ii) pick the rightmost interval in I_2 into S_2^* that dominates the leftmost interval in I_2 that is not dominated by the intervals in S_2^* and repeat this step until all the intervals in I_2 are dominated by the intervals in S_2^*. We note that no two intervals in S_2^* dominate each other. Hence, the depth of S_2^* is at most 2. □

Thus, $S_1^* \cup S_2^*$ is a dominating set for I with depth at most three. Hence, we have the following lemma.

Lemma 1. *The depth of any optimum solution to the problem* MMGDS-INT *is at most three.*

2.2 Polynomial-Time Algorithm for MMGDS-INTProblem

In the following, we give a sweep-line based algorithm which returns an optimum solution to the given instance of MMGDS-INT problem in $O(n^4)$ time.

We first add two dummy intervals on the real line, t_0 and t_{n+1}, so that these intervals do not intersect with any given interval on the real line. More specifically, t_0 will be placed to the left of the intervals in I, and t_{n+1} will be placed to the right of all the intervals in I. In the rest of the section, we assume $I = \{t_0, t_1, t_2, \dots, t_n, t_{n+1}\}$.

We run a sweep line from left to right, starting from $x = \mathsf{left}(t_0)$ to $x = \mathsf{left}(t_{n+1})$. In a sweep-line position, we remember three intervals (due to the Lemma 1) and call the intervals active intervals. Whenever the sweep-line moves from one position to another position, we make sure that all the intervals whose left endpoints are within these two sweep-line positions are dominated by the active intervals in the two positions. To make the positions of the sweep line discrete, we always place the sweep line at the left endpoint of some interval in I. We do the following pre-processing steps:

Step (a) Sort the intervals in I with respect to their left endpoints. Say that the sorted order of intervals in I is $t_0, t_1, t_2, \dots, t_n, t_{n+1}$, that is, $\mathsf{left}(t_0) < \mathsf{left}(t_i) < \dots < \mathsf{left}(t_{n+1})$.

Step (b) We maintain two arrays γ_1, and γ_2, each of size $n + 1$. For each interval $t_i \in I$,

- $\gamma_1[i]$ is the largest index j such that $j < i$ and t_j is not dominated by t_i.
- $\gamma_2[i]$ stores the smallest index j such that $j \leq i$ and t_j are dominated by t_i.

We note that these arrays can be computed in $O(n^2)$ time.

Lemma 2. *Let t_x and t_y be two intervals such that $x < y$ and the intervals t_x and t_y do not intersect. If $\mathsf{left}(t_{\gamma_1[y]}) \leq \mathsf{right}(t_x)$, then every interval t_z with $\mathsf{left}(t_z) \in (\mathsf{right}(t_x), \mathsf{left}(t_y))$ is dominated by t_y.*

We construct a directed acyclic graph (DAG) $G = (V, E)$ where V is a set of vertices and E is an edge set. Here, each vertex in V, is of the form $(t_{past}, t_{prev}, t_{curr})$ such that $\mathsf{left}(t_{past}) < \mathsf{left}(t_{prev}) < \mathsf{left}(t_{curr})$ and $\mathsf{right}(t_{prev}) <\mathsf{right}(t_{prev}) < \mathsf{right}(t_{curr})$. We place a directed edge between two vertices $(t_{past}, t_{prev}, t_{curr})$ to $(t_{prev}, t_{curr}, t_{next})$ if the following conditions hold:

(a) $\mathsf{left}(t_{curr}) < \mathsf{left}(t_{next})$ and $\mathsf{right}(t_{curr}) < \mathsf{right}(t_{next})$,
(b) $\mathsf{left}(t_{next}) > \mathsf{right}(t_{prev})$, and
(c) If $\mathsf{right}(t_{curr}) < \mathsf{left}(t_{next})$, then all the intervals with left endpoint inside $(\mathsf{right}(t_{curr}), \mathsf{left}(t_{next}))$ are dominated by t_{next}. This can be checked by applying Lemma 2 with $x = curr$ and $y = next$ in $O(1)$ time.

The cost of the edge from $(t_{past}, t_{prev}, t_{curr})$ to $(t_{prev}, t_{curr}, t_{next})$ is assigned as follows:

- If some interval in the instance is dominated by all four intervals $t_{past}, t_{prev}, t_{curr}$, and t_{next}, then the cost of the edge is ∞. To see this, check if $\gamma_2[next]$ is dominated by t_{past}. If so, the edge cost is ∞.
- If some interval in the instance is dominated by three intervals, t_{prev}, t_{curr}, and t_{next}, then the cost of the edge is 3. To see this, check if $\gamma_2[next]$ is dominated by t_{prev} but not by t_{past}. If so, the edge cost is 3.
- If both t_{curr} and t_{next} intersect, or the intervals t_{curr} and t_{next} do not intersect, but some interval in the instance is dominated by both t_{curr} and t_{next}, then the cost of the edge is 2. To see the second part, we just check if $\gamma_2[next]$ is dominated by t_{curr}, but not dominated by t_{prev}.
- Otherwise, the cost of the edge is 1.

The graph G has $O(n^3)$ vertices, and there are at most $O(n)$ outgoing edges from each vertex in G. Thus, the number of edges in G is $O(n^4)$. In addition, $O(1)$ time is sufficient to check the feasibility of each edge and to calculate the cost of the edge. Thus, the graph G can be computed in $O(n^4)$ time.

From the construction, it is clear that the set of all intervals along any $t_0 - t_{n+1}$ path is a dominating set for the instance. The optimum solution to the given instance of MMGDS-INT problem is corresponding to a finite cost $t_0 - t_{n+1}$ path with the maximum edge cost on the path being the minimum. Furthermore, since G is DAG, such a path can be computed in linear time with respect to the number of edges in G, i.e., $O(n^4)$. Thus, we have the following theorem.

Theorem 1. *There exists an exact algorithm for MMGDS-INT problem that returns a solution in $O(n^4)$-time.*

Corollary 1. *There exists a $O(n^2)$ time algorithm to check if there is a depth 1 solution (perfect domination) for MMGDS-INT problem.*

Furthermore, if no interval in the given instance is inside another interval, it is clear that the depth of an optimum solution is at most two (from Claim 3). Thus, we have the following corollary.

Corollary 2. *There exists an $O(n^2)$ time exact algorithm for MMGDS-INT problem when no interval is contained within another interval in the given instance.*

3 MMGDSProblem with Lines and Strips

In this section, we consider the MMGDS problem with axis-parallel lines (MMDS-LINES) and axis-parallel strips (MMDS-STRIPS).

We first note that if the instance of MMDS-LINES problem (resp. MMDS-STRIPS) has both vertical and horizontal lines (resp. strips), then any combination of a vertical line (resp. strip) and a horizontal line (resp. strip) is an optimum solution with depth 2. If the instance has only vertical lines (the case with only horizontal lines is similar), then the given instance itself is an optimum solution with depth 1. If the instance contains only vertical strips (the horizontal strips case is symmetric), then the problem can be reduced to the MMGDS-INT problem by mapping every vertical strip to an interval on the real line, which is the intersection of the real line and the vertical strip, and hence the problem can be solved in polynomial time. Thus, we have the following result.

Theorem 2. *There exist polynomial time algorithms for MMDS-LINES and MMDS-STRIPS problems.*

In the following, we demonstrate that the discrete version of the MMGDS problem is NP-hard for axis-parallel lines, axis-parallel strips, and a few other geometric objects (the list is given in Corollary 3). To show the NP-hardness results, we provide polynomial-time reductions from the P1-in-3SAT problem (see Sect. 1.3 for the definition).

NP-Hardness Reduction for Axis-Parallel Lines: Let ϕ be an instance of the P1-in-3SAT problem consisting of n variables $x_1, x_2, \ldots, x_n$ and m clauses $C_1, C_2, \ldots, C_m$. We construct an instance $\mathcal{I}_\phi$ of the discrete MMDS-LINES problem as follows:

Variable Gadget: For each variable x_i, for $1 \leq i \leq n$, we define two axis-parallel lines: a vertical line v_i with the equation $x = i$ and a horizontal line h_i with the equation $y = -i$. Additionally, we place a point p_i at the intersection of v_i and h_i, which corresponds to the coordinate $(i, -i)$ (see Fig. 2(a)).

Clause Gadget and Interaction of Variable-Clause Gadgets: For each clause C_l (for $l = 1, 2, \ldots, m$), consisting of three positive literals x_i, x_j, and x_k, the corresponding gadget includes a vertical line c_l with the equation $x = n+l$ and three points: q_{li}, q_{lj}, and q_{lk}. The point q_{li} is placed at the intersection of the lines c_l and h_i, that is, at the coordinate $(n + l, -i)$. Similarly, q_{lj} is placed at the intersection of c_l and h_j (i.e., at $(n + l, -j)$), and q_{lk} is placed at the intersection of c_l and h_k (i.e., at $(n + l, -k)$) (see Fig. 2(b)).

This completes the construction. The construction can be completed in polynomial (in n and m) time. We now have the following theorem.

Theorem 3. *The discrete MMDS-LINES problem is* NP*-hard even for depth 1.*

By applying identical reductions as in the proof of Theorem 3, we establish the NP-hardness of strips and some other objects given in the following corollary.

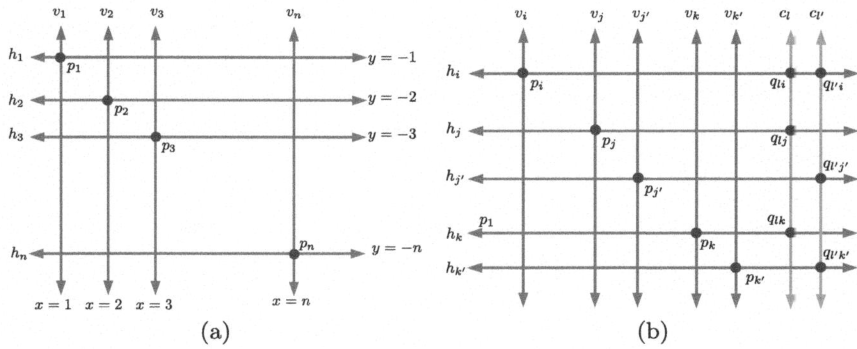

Fig. 2. (a) Variable gadget (b) Interaction of variable and clause gadget when clause $C_l = x_i \vee x_j \vee x_k$ and $C_{l'} = x_i \vee x_{j'} \vee x_{k'}$ such that $i < j < j' < k < k'$ and $l < l'$.

Corollary 3. *The discrete version of MMGDS problem in NP-hard for the following classes of objects:* ***(C1)*** *Axis-parallel strips,* ***(C2)*** *Axis-parallel rectangles inside a bounded region B, and one of the boundaries of each rectangle is attached to either the top boundary or the left boundary of B,* ***(C3)*** *Axis-parallel rectangles where every pair of rectangles intersect either zero or two times,* ***(C4)*** *Ais-parallel line segments with one of the endpoints on a diagonal line* ***(C5)*** *Axis-parallel rectangles anchored on a diagonal line, and* ***(C6)*** *Axis-parallel one-sided strips anchored on a diagonal.*

4 NP-Hardness of MMGDSProblem with Unit Segments

In this section, we prove that the MMGDS problem is NP-hard for axis-parallel unit segments, even with depth 1 (MMGDS-US problem). We give a polynomial-time reduction from a known NP-hard problem, the RPP1-in-3SATproblem [19] (see Sect. 1.3 for the definition).

Reduction: Let ϕ be an instance of the RPP1-in-3SAT problem consists of n variables $x_1, x_2, \ldots, x_n$ and m clauses $C_1, C_2, \ldots, C_m$. In the following, we construct an instance $\mathcal{I}_\phi$ of the MMGDS-US problem.

Variable Gadget: The variable gadget for each variable x_i consists of multiple components, including a gadget representing the variable node and a separate gadget for each leg (edge) that connects it to a clause gadget. We first describe each of these gadgets individually before explaining how they are combined to form the complete variable gadget.

Gadget for a Variable Node: For each variable node, we construct a gadget as illustrated in Fig. 3. The construction is described incrementally. Assume that $\tau = 6m$. First, we arrange 2τ segments, denoted $\alpha_1, \alpha_2, \ldots, \alpha_{2\tau}$, in a rectangular pattern, ensuring that no two segments intersect. Next, we introduce another set of 2τ segments, denoted $\beta_1, \beta_2, \ldots, \beta_{2\tau}$ where each β_i intersects orthogonally only with the corresponding α_i for $i = 1, 2, \ldots, 2\tau$. It is important to note that if α_i is horizontal, then β_i is vertical, and vice versa. Finally, we add another rectangular arrangement of segments, denoted $\gamma_1, \gamma_2, \ldots, \gamma_{2\tau}$ (shown in green), such that each γ_i intersects both β_i and β_{i+1} for $i = 1, 2, \ldots, 2\tau - 1$, and $\gamma_{2\tau}$ intersects both $\beta_{2\tau}$ and β_1.

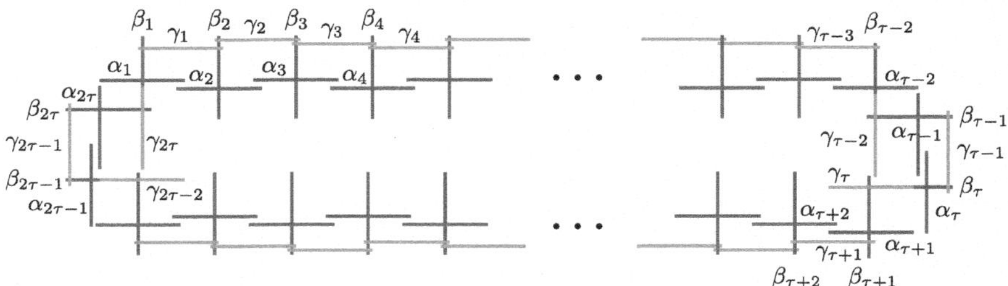

Fig. 3. The structure of a variable node

Gadget for a Leg: The gadget for a middle leg is shown in Fig. 4(a) and a left leg in Fig. 4(b). The structure of a right leg is the horizontal flip of the left leg. Each middle leg gadget has two special areas at the two ends, whereas the right leg, together with two special areas at the two ends, has one more special area at the corner of the leg (see the highlighted portions in the respective figures). At the lower end, we designate two blue segments, δ_2 and δ_3, which connect to the variable node gadget. Additionally, at the opposite end, there is a special segment, δ_5, that links to the clause gadget.

Attaching a Leg Gadget to the Variable Node Gadget: We now describe how a leg gadget can be attached to a variable node gadget. Let x_i be a variable connecting the clauses above the horizontal line L. Since the legs are pairwise nonintersecting, we can order the legs as $g_1, g_2, \ldots$ (which can have at most m g_i's) that connects the node of x_i in the left-to-right order. Let g_t be a leg. We now describe how g_t attaches to the node gadget of x_i. The segments α_{6t-3} and β_{6t-3} are removed from the variable node. The two green segments γ_{6t-4} and γ_{6t-3} become vertical, and they intersect their bottom endpoint with the top endpoint of β_{6t-4} and β_{6t-3}, respectively. The remaining gadget of the leg is above the variable node gadget, such that no segment, except the two green segments γ_{6t-4} and γ_{6t-3}, of the leg gadget, intersects with any segment of the variable node gadget.

Combining all the Gadgets: Together with the variable node rectangular structure and attaching at most $2m$ leg gadgets (a variable can connect to at most m clause either side of the horizontal line L) to it constitutes the variable gadget. It is important to note that the length of the leg gadgets, both vertically and horizontally, can be adjusted by either compressing or expanding the vertical and horizontal components of the structure. Further, assume that a variable gadget

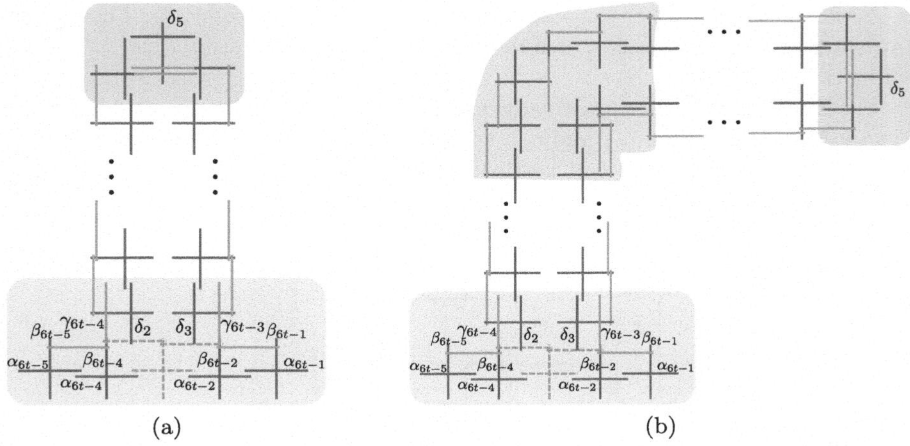

Fig. 4. Structure of (a) a middle leg, and (b) a left leg.

contains 6Δ segments for some Δ a multiple of 6. The segment intersection graph G_{var} of the variable gadget is depicted in Fig. 5.

We can now prove the following crucial result on G_{var}. Let $T_0 = \{\alpha_2, \alpha_4, \ldots, \alpha_{2\Delta}, \beta_1, \beta_3, \ldots, \beta_{2\Delta-1}\}$ and $T_1 = \{\alpha_1, \alpha_3, \ldots, \alpha_{2\Delta-1}, \beta_2, \beta_4, \ldots, \beta_{2\Delta}\}$. We have the following lemma.

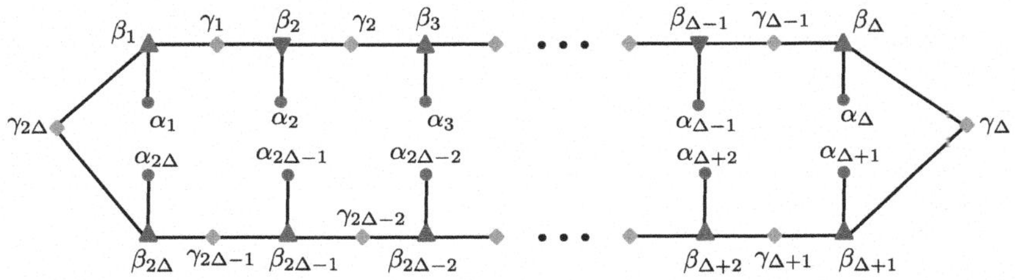

Fig. 5. Intersection graph of a variable gadget.

Lemma 3. *There exist exactly two sets, T_0 and T_1, of segments in G_{var}, each forming a dominating set with depth 1.*

The graph G_{var} is the segment intersection graph of the variable gadget for x_i. therefore, from Lemma 3, we say that the gadget of x_i has exactly two dominating set of segments T_0 and T_1, each one is a depth-1 solution We can correspondence the truth value of the variable x_i as *true* with T_0 and false with T_1.

Clause Gadget and Interaction of Variable-Clause Gadgets: Let C_l be a clause consisting of three positive literals x_i, x_j, and x_k. The literal x_i connects to C_l via the left leg, x_j via the middle leg, and x_k via the right leg. The gadget corresponding to C_l consists of nine segments: $c_{l,i}, c_{l,j}$, $c_{l,k}$, $c_l^1, c_l^2, c_l^3, c_l^4, c_l^5, c_l^6$, as illustrated in Fig. 6.

The interaction of the clause gadget with the variable gadget is established by horizontally intersecting $c_{l,i}$ and $c_{l,k}$ with the segment δ_5 of the variable gadget of x_i and x_k, respectively, and vertically intersecting the $c_{l,j}$ with the segment δ_5 of the variable gadget of x_j (ref. Fig. 6).

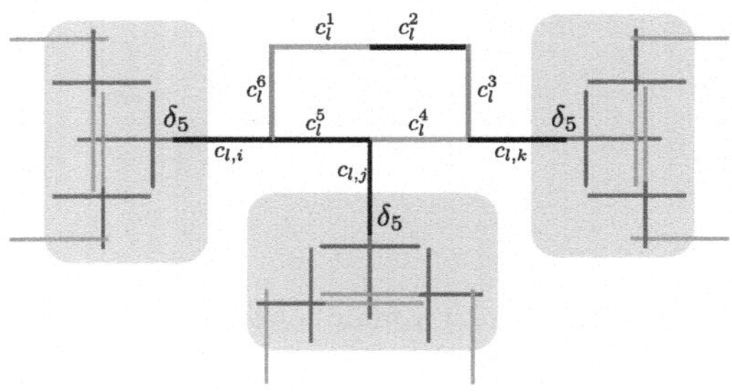

Fig. 6. A clause gadget and its interaction with the gadgets of the variables it contains.

This completes the construction. We now prove the following theorem.

Theorem 4. *The MMGDS-US problem is* NP*-hard even with depth 1.*

Acknowledgments.. We thank Sreeram Yerasani for helpful initial discussions on the problem.

References

1. Agrawal, A., Choudhary, P., Narayanaswamy, N.S., Nisha, K.K., Ramamoorthi, V.: Parameterized complexity of minimum membership dominating set. Algorithmica **85**(11), 3430–3452 (2023)
2. Biggs, N.: Perfect codes in graphs. Journal of Combinatorial Theory, Series B **15**(3), 289–296 (1973)
3. Cesati, M.: Perfect code is w[1]-complete. Inf. Process. Lett. **81**(3), 163–168 (2002)
4. Chan, T.M., Grant, E.: Exact algorithms and APX-hardness results for geometric packing and covering problems. Comput. Geom. **47**(2), 112–124 (2014)
5. Clark, B.N., Colbourn, C.J., Johnson, D.S.: Unit Disk Graphs. Discret. Math. **86**(1), 165–177 (1990)
6. Erik, D., Demaine, W.G., Hajiaghayi, M.: A Guide to Algorithmic Lower Bounds Computational Intractability. MIT press (2025)
7. Erlebach, T., van Leeuwen, E.J.: Domination in geometric intersection graphs. In: LATIN, pp. 747–758 (2008)
8. Fellows, M.R., Hoover, M.N.: Perfect domination. Australas. J Comb. **3**, 141–150 (1991)

9. Garey, M., Johnson, D.: The rectilinear steiner tree problem is NP-Complete. SIAM J. Appl. Math. **32**(4), 826–834 (1977)
10. Huang, H., Xia, B., Zhou, S.: Perfect codes in cayley graphs. SIAM J. Discret. Math. **32**(1), 548–559 (2018)
11. Hunt, H.B., Marathe, M.V., Radhakrishnan, V., Ravi, S.S., Rosenkrantz, D.J., Stearns, R.E.: NC-Approximation Schemes for NP- and PSPACE-Hard Problems for Geometric Graphs. J. Algorithms **26**(2), 238–274 (1998)
12. Karp, R.M.: Reducibility among Combinatorial Problems, pp. 85–103. Springer US, Boston, MA (1972)
13. Karthika, D., Muthucumaraswamy, R., Bentert, M., Bhyravarapu, S., Saurabh, S., Seetharaman, S.: On the complexity of minimum membership dominating set. In: SOFSEM 2025, pp. 94–107 (2025)
14. Kratochvíl, J.: Perfect codes over graphs. J. Combinatorial Theory, Series B **40**(2), 224–228 (1986)
15. Kuhn, F., von Rickenbach, P., Wattenhofer, R., Welzl, E., Zollinger, A.: Interference in cellular networks: the minimum membership set cover problem. In: Wang, L. (ed.) COCOON (2005)
16. Madireddy, R.R., Mudgal, A.: Approximability and hardness of geometric hitting set with axis-parallel rectangles. Inf. Process. Lett. **141**, 9–15 (2019)
17. Madireddy, R.R., Mudgal, A., Pandit, S.: Hardness results and approximation schemes for discrete packing and domination problems. In: COCOA, pp. 421–435 (2018)
18. Mitchell, J., Pandit, S.: Minimum membership covering and hitting. Theoret. Comput. Sci. **876**, 05 (2021)
19. Mulzer, W., Rote, G.: Minimum-weight triangulation is NP-hard. *J. ACM*, 55(2), 2008
20. Narayanaswamy, N.S., Dhannya, S.M., Ramya, C.: Minimum membership hitting sets of axis parallel segments. In: COCOON (2018)
21. Reddy, S.B., Kare, A.S.: Algorithms for minimum membership dominating set problem (2024)
22. Schaefer, T.J.: The complexity of satisfiability problems. In: STOC (1978)

New Lower Bound and Algorithm for Online Geometric Hitting Set Problem

Minati De[1(✉)], Ratnadip Mandal[1], and Satyam Singh[2]

[1] Department of Mathematics, Indian Institute of Technology Delhi, New Delhi, India
minati@maths.iitd.ac.in, maz218522@iitd.ac.in

[2] Department of Computer Science, Aalto University, Espoo, Finland
satyam.singh@aalto.fi

Abstract. The hitting set problem is a fundamental and well-studied combinatorial optimization problem in the offline setup. We study the online hitting set problem, where we know only the set of points beforehand, and objects are introduced individually. The objective is to maintain a minimum-sized hitting set by making irrevocable decisions. To hit homothetic axis-aligned hypercubes in $\mathbb{R}^d$ having widths in the range $[1, M]$ using points in $\mathbb{Z}^d$, we show that the competitive ratio of any algorithm, whether deterministic or randomized, is $\Omega(d \log M)$. Then, we propose a deterministic $\lfloor 4\alpha + 1 \rfloor^d \lfloor \log_2 2M \rfloor$ competitive algorithm to hit α-fat objects in $\mathbb{R}^d$ having widths in the range $[1, M]$ using points in $\mathbb{Z}^d$.

When the set of points is restricted to $(0, N)^d \cap \mathbb{Z}^d$ and the α-fat objects are in $(0, N)^d$, we obtain $\Omega(d \log N)$ and $O((4\alpha + 1)^d \log N)$, respectively, as the lower and upper bounds of the competitive ratio. This answers an open question raised by Alefkhani, Khodaveis, and Mari (WAOA 2023) by improving the best-known lower and upper bounds of $\Omega(\log N)$ and $O((4\alpha+1)^{2d} \log N)$, respectively, obtained by them for the problem. The techniques used are simple yet nontrivial, highlighting our paper's strength.

Keywords: Fat objects · Hitting set · Hypercubes · Online algorithm · Randomized lower bound

1 Introduction

The hitting set problem is one of the most fundamental problems in combinatorial optimization. Let $\Sigma = (\mathcal{P}, \mathcal{S})$ be a *range space*, where $\mathcal{P}$ is a set of *elements* and $\mathcal{S}$ is a collection of subsets of $\mathcal{P}$ known as *ranges*. A subset $\mathcal{H} \subseteq \mathcal{P}$ is called a *hitting set* of the range space $\Sigma = (\mathcal{P}, \mathcal{S})$ if the set $\mathcal{H}$ intersects every range s in $\mathcal{S}$. The objective of the *hitting set problem* is to find a hitting set

Work on this paper by M. De has been partially supported by SERB MATRICS Grant MTR/2021/000584, work by R. Mandal has been supported by CSIR, India, File Number- 09/0086(13712)/2022-EMR-I., and work by Satyam Singh has been supported by the Research Council of Finland, Grant 363444.

F. V. Fomin and M. Xiao (Eds.): COCOON 2025, LNCS 15983, pp. 234–247, 2026.
https://doi.org/10.1007/978-981-95-0215-8_18

with minimum cardinality. This problem is one of Karp's 21 classic NP-hard problems [20], and the best approximation factor achievable in polynomial time is $\Theta(\log |\mathcal{P}|)$ assuming P $\neq$ NP [12,17].

Due to several applications in VLSI design, database management, resource allocation, facility location, and wireless networks, researchers have considered the hitting set problem in the geometric setup (also known as *geometric hitting set problem*), where the set $\mathcal{P}$ is a collection of points from $\mathbb{R}^d$, and the set $\mathcal{S}$ is a finite family of geometric objects chosen from some infinite class (hypercubes, balls, etc.) [1,6,19,22]. By slightly misusing the notation, we use $\mathcal{S}$ to signify both the set of ranges and the set of objects that define these ranges. The hitting set problem remains NP-hard, even for simple objects such as unit disks and unit squares [18]. A PTAS is known when the geometric objects are half-spaces in $\mathbb{R}^3$, and when objects are r-admissible regions in a plane (this includes pseudo-disks) [22].

Alon et al. [3] initiated the study of the problem in an online setup where the entire range space $\Sigma = (\mathcal{P}, \mathcal{S})$ is known in advance, but the order of arrival of the elements of $\mathcal{S}$ is not known a priori. Note that since an online algorithm must maintain a feasible competitive solution by making irrevocable decisions, the problem is considerably difficult even if we know the whole range space in advance, but not the arrival order of elements in $\mathcal{S}$. In this setup, they proposed a deterministic $O(\log |\mathcal{P}| \log |\mathcal{S}|)$ competitive algorithm and obtained a lower bound of $\Omega\left(\frac{\log |\mathcal{P}| \log |\mathcal{S}|}{\log\log |\mathcal{P}| + \log\log |\mathcal{S}|}\right)$. In the same setup, Even and Smorodinsky [16] studied the problem for various geometric range spaces and achieved an optimal competitive ratio of $\Theta(\log |\mathcal{P}|)$ when objects are intervals, half-planes, and unit disks, and the set $\mathcal{P}$ is a finite set of points.

When $\mathcal{P}$ is not a finite set, the algorithm by Alon et al. [3] does not give any bound. De and Singh [10,11] considered the online hitting set problem when $\mathcal{P} = \mathbb{Z}^d$ is a discrete set of points which is not a finite set and the set $\mathcal{S}$ is not known in advance. In this setup, for unit hypercubes in $\mathbb{R}^d$, they proved that every deterministic algorithm has a competitive ratio of at least $d + 1$; whereas they obtained a randomized algorithm that achieves a competitive ratio of $O(d^2)$ for $d \geq 3$. For unit balls in $\mathbb{R}^d$, they proposed a deterministic algorithm having a competitive ratio of $O(d^4)$. In this direction, in this paper, we consider the problem when objects are similarly sized fat objects in $\mathbb{R}^d$ and $\mathcal{P} = \mathbb{Z}^d$.

Khan et al. [21] studied the problem, where, $\mathcal{P} \subseteq [0, N)^2 \cap \mathbb{Z}^2$ is a finite set for a given $N \in \mathbb{R}^+$ and the set $\mathcal{S}$ consists of axis-aligned integer squares $s \subseteq [0, N)^2$ in $\mathbb{R}^2$. Here, an integer square is a square whose all corners have integral coordinates. In this setup, the algorithm of Alon et al. [3] achieves a competitive ratio of $O(\log^2 N)$, while Khan et al. [21] obtained a tight $\Theta(\log N)$ competitive algorithm.

When $\mathcal{P} = (0, N)^d \cap \mathbb{Z}^d$ is a finite set and $\mathcal{S}$ is a family of α-fat objects in $(0, N)^d$, very recently, Alefkhani et al. [2] proposed a deterministic online algorithm with a competitive ratio of $(4\alpha + 1)^{2d} \log N$. They also proved that every deterministic algorithm has a competitive ratio of $\Omega(\log N)$ for this problem. They noted a large gap between the lower and upper bounds, and proposed to bridge this gap as an open problem. In this paper, we reduce this gap by obtaining an exponential factor improvement in the upper bound and giving an improved randomized lower bound. Note that the algorithm in [2] is not applicable for $\mathcal{P} = \mathbb{Z}^d$ as the value of N would be infinite; on the other hand, a family of α-fat objects in $(0, N)^d$ are fat objects of similarly sized as the width is bounded. This motivates us to ask whether one can achieve a bounded competitive ratio for similarly sized fat objects when $\mathcal{P} = \mathbb{Z}^d$. We answer this question positively.

1.1 Our Contributions and Comparison to Prior Works

As in [11], we study the online hitting set problem when the point set $\mathcal{P} = \mathbb{Z}^d$ is infinite, and the set $\mathcal{S}$ of objects is not known a priori; elements of this set are revealed one by one. The algorithm needs to maintain a hitting set for the revealed objects by choosing points from $\mathcal{P}$, and the decision to choose a point is irrevocable. In the following, we mention each contribution formally and discuss the context and comparison with prior work.

1. **A Randomized Lower Bound for Hypercubes.** First, we establish that the competitive ratio of any algorithm, whether deterministic or randomized, is $\Omega(d \log M)$ for hitting homothetic axis-aligned hypercubes having widths in the range $[1, M]$ using points in $\mathbb{Z}^d$ (Theorem 1). To obtain this lower bound, we use Yao's minimax principle (as in [7]). We first construct a complete binary tree $\mathcal{T}$ of height $O(d \log M)$, where each path from the root to a leaf of $\mathcal{T}$ represents a sequence of similarly sized homothetic axis-aligned hypercubes. If any path of $\mathcal{T}$ from the root node to one of the leaf nodes is presented at random as an input sequence, then we bound the competitive ratio of any deterministic algorithm for this random input sequence. We show that our lower bound construction provides a lower bound of $\Omega(d \log N)$ for hitting axis-aligned hypercubes in $(0, N)^d$ using points in $(0, N)^d \cap \mathbb{Z}^d$ and improves the result of Alefkhani et al. [2, Theorem 2] by a factor of d (Corollary 1).
2. **An Algorithm for Fat Objects.** Fat objects have several desirable properties and are well-studied due to their various applications and appearance in geometric problems [4,5,15]. We focus on the hitting set problem, where the set $\mathcal{P} = \mathbb{Z}^d$ and the set $\mathcal{S}$ consists of α-fat objects in $\mathbb{R}^d$ having widths in the range $[1, M]$. For this, we present a deterministic algorithm PruneAndChoose as follows. Upon the arrival of a new object σ having width w, if σ is not hit by the previously selected points, our algorithm selects any point from the set $(2^{i+1}\mathbb{Z})^d \cap \sigma$ to hit σ, where $i = \lfloor \log_2 w \rfloor$. We show that this algorithm achieves a competitive ratio of at most $\lfloor 4\alpha + 1 \rfloor^d \lfloor \log_2 2M \rfloor$ (Theorem 2). Our algorithm PruneAndChoose achieves a competitive ratio of $O((4\alpha + 1)^d \log N)$ for hitting α-fat objects in $(0, N)^d$ using points in $(0, N)^d \cap \mathbb{Z}^d$ (Corollary 3). This improves the upper bound of Alefkhani et al. [2, Theorem 1] by a factor of at least 5^d.

Table 1 summarizes the existing results and the results of the current paper.

Table 1. Summary of results for the online geometric hitting set problem. (∗) indicates that the set is known in advance. (#) indicates the randomized results.

Points	Objects	Lower Bound of the Competitive Ratio	Upper Bound of the Competitive Ratio
n points in $\mathbb{R}$	Intervals (∗)	$\Omega(\log n)$ [16]	$O(\log n)$ [16]
n points in $\mathbb{R}^2$	Half-planes and unit disks (∗)	$\Omega(\log n)$ [16]	$O(\log n)$ [16]
A finite subset of $[0, N)^2 \cap \mathbb{Z}^2$	Axis-aligned integer squares in $[0, N)^2$	$\Omega(\log N)$ [16]	$O(\log N)$ [21]
$(0, N)^d \cap \mathbb{Z}^d$	α-fat objects in $(0, N)^d$	$\Omega(\log N)$ [2]	$O((4\alpha + 1)^{2d} \log N)$ [2]
		$\Omega(d \log N)$ (#) (Corollary 1)	$O((4\alpha + 1)^d \log N))$ (Corollary 3)
$\mathbb{Z}^d$	Unit balls in $\mathbb{R}^d$	$d + 1$ for $d \leq 3$ [11]	$O(d^4)$ [11]
$\mathbb{Z}^d$	Unit hypercubes in $\mathbb{R}^d$	$\Omega(d + 1)$ [11]	$O(d^2)$ for $d \geq 3$ [11] (#)
$\mathbb{Z}^d$	α-fat objects in $\mathbb{R}^d$ having widths in the range $[1, M]$	$\Omega(d \log M)$ (#) (Theorem 1)	$O((4\alpha+1)^d \log M)$ (Theorem 2)

1.2 Other Related Works

The "continuous" version of the hitting set problem, where the point set $\mathcal{P} = \mathbb{R}^d$, is known as *piercing set problem*. When the set $\mathcal{S}$ consists of translated copies of a convex object, the corresponding piercing set problem is equivalent to *unit covering problem* [9]. Charikar et al. [8] considered the online unit covering problem and presented an $O(2^d d \log d)$ competitive algorithm for balls in $\mathbb{R}^d$, and they also gave a deterministic lower bound of $\Omega\left(\frac{\log d}{\log \log \log d}\right)$ for this problem. Later, Dumitrescu et al. [13] improved the upper and lower bounds of the competitive ratio to $O(1.321^d)$ and $\Omega(d + 1)$, respectively. In particular, they [13] obtained competitive ratios of 5 and 12 for balls in $\mathbb{R}^2$ and $\mathbb{R}^3$, respectively. Dumitrescu and Tóth [14] proved that the lower bound of the competitive ratio for the unit covering problem for hypercubes in $\mathbb{R}^d$ is at least 2^d, and this ratio is also optimal since it is achievable by a simple deterministic algorithm [7]. We refer to [9] for results related to the problem of the online piercing set of various other objects.

1.3 Notation and Preliminaries

We use $\mathbb{Z}^+$ and $\mathbb{R}^+$ to denote the set of positive integers and positive real numbers, respectively. For any $n \in \mathbb{Z}^+$, we use $[n]$ to represent the set $\{1, 2, \ldots, n\}$. For any $\beta \in \mathbb{Z}^+$, we use $\beta\mathbb{Z}$ to denote the set $\{\beta z \mid z \in \mathbb{Z}\}$, where $\mathbb{Z}$ is the set of integers.

Observation 1 *Let σ be an axis-aligned hypercube with side lengths between $\ell\beta$ and $r\beta$, where ℓ, $r \in \mathbb{R}^+$, $\beta \in \mathbb{Z}^+$, and $\ell < r$. Then, the hypercube σ contains at least $\lfloor \ell \rfloor^d$ and at most $\lfloor r+1 \rfloor^d$ points from $(\beta\mathbb{Z})^d$.*

From now onwards, we use the term hypercube to mean an axis-aligned hypercube. By an *object*, we refer to a compact set in $\mathbb{R}^d$ having a non-empty interior. For any object σ, we use $\partial\sigma$ to denote the *boundary* of the object, and $\text{int}(\sigma) = \{x \in \sigma \mid x \notin \partial\sigma\}$ refers the *interior* of the object. A set of the form $\sigma + \mathbf{v} = \{c + \mathbf{v} \mid c \in \sigma\}$, where $\mathbf{v} \in \mathbb{R}^d$, is called a *translated copy/translate* of σ. A set of the form $\lambda\sigma + \mathbf{v} = \{\lambda c + \mathbf{v} \mid c \in \sigma\}$, where $\mathbf{v} \in \mathbb{R}^d$ and $\lambda \in \mathbb{R}^+$, is called a *homothetic copy* of σ. A set $\mathcal{S}$ of objects is said to be *homothetic* if each object $\sigma \in \mathcal{S}$ is a homothetic copy of every other object $\sigma' \in \mathcal{S}$.

A number of different definitions of fatness (not extremely long and skinny) are available in the geometry literature [2,5,9]. We use a definition similar to Alefkhani et al. [2]. An object σ is called a *α-fat object*, for some $\alpha \geq 1$, if the ratio of the side length of the smallest hypercube containing σ and the side length of the largest hypercube contained in σ is at most α. Let H be the largest hypercube contained in σ. By *center* and *width* of σ, we denote the center and half of the side length of H, respectively. A set $\mathcal{S}$ of objects is said to be *similarly sized* when the ratio of the largest diameter of an object in $\mathcal{S}$ to the smallest diameter of an object in $\mathcal{S}$ is bounded by a fixed constant. For a fixed α, a set $\mathcal{S}$ of α-fat objects is *similarly sized fat objects* when the ratio of the largest width of an object in $\mathcal{S}$ to the smallest width of an object in $\mathcal{S}$ is bounded by a fixed constant.

2 Lower Bound for Similarly Sized Homothetic Hypercubes in $\mathbb{R}^d$

In this section, for hitting homothetic axis-aligned hypercubes having widths in the range $[1, M]$ using points in $\mathbb{Z}^d$, we prove that the lower bound of the competitive ratio of any deterministic or randomized algorithm is $\Omega(d \log M)$. Throughout this section, we use *width* of a hypercube to denote half of its side length. For simplicity of expression, we assume that M is an integral power of 4. Additionally, we assume that a hypercube σ is hit by some point p if p lies in the interior of σ. We can remove this assumption by replacing each constructed hypercube with an infinitesimally shrunken version of itself.

We use Yao's minimax principle (as in [7]) to prove the lower bound. To establish it, we create a complete binary tree $\mathcal{T}$ of height[1] $O(d \log M)$, where any path from the root to a leaf represents a sequence of axis-aligned hypercubes having widths in the range $[1, M]$, and the common intersection region of these hypercubes contains an integer point. Let $\mathcal{S}$ be a set consisting of all sequences corresponding to $2^{O(d \log M)}$ many paths from the root to each leaf node in $\mathcal{T}$. We show that if any of the sequences of $\mathcal{S}$ is chosen randomly with uniform

[1] The height of a tree is the maximum number of vertices lying in a path from the root node to any leaf node of the tree.

distribution, then the expected number of points placed by any deterministic algorithm is $\Omega(d \log M)$, whereas an offline optimum algorithm needs only one point. Consequently, the expected competitive ratio on any sequence from $\mathcal{S}$ is $\Omega(d \log M)$. A sequence in $\mathcal{S}$ can be chosen uniformly at random as follows. The root is introduced with certainty. Then, at each internal node, with equal probability, either the left or the right child is chosen.

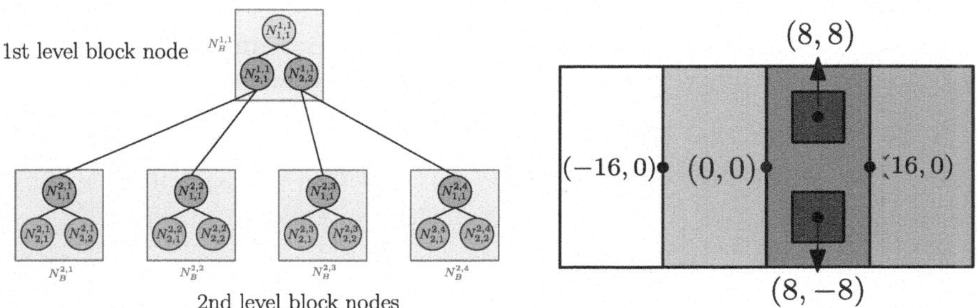

Fig. 1. Example of the tree $\mathcal{T}$ and $\mathcal{T}_B$ when $d = 2$ and $M = 16$. The red and blue colored squares represent the 1st and 2nd level block nodes of $\mathcal{T}_B$, respectively. The green, navy, brown, and magenta colored disks represent the 1st, 2nd, 3rd, and 4th level nodes of $\mathcal{T}$, respectively. Here, corresponding to the node $N_{1,1}^{1,1}$, we have an axis-aligned hypercube of width $M = 16$ centered at the point $c_{1,1}^{1,1} = (0, 0)$. Now, corresponding to the two children nodes $N_{2,1}^{1,1}$ and $N_{2,2}^{1,1}$ of $N_{1,1}^{1,1}$, we have two axis-aligned hypercubes of width $M = 16$ centered at the points $c_{2,1}^{1,1} = (16, 0)$ and $c_{2,2}^{1,1} = (-16, 0)$, respectively. Similarly, corresponding to the two children nodes $N_{1,1}^{2,1}$ and $N_{1,1}^{2,2}$ of $N_{2,1}^{1,1}$ we have two axis-aligned hypercubes of width $\frac{M}{4} = 4$ centered at the points $c_{1,1}^{2,1} = (8, 8)$ and $c_{1,1}^{2,2} = (8, -8)$, respectively. (Color figure online)

Now, we describe some terminologies related to the tree $\mathcal{T}$. Recall that $\mathcal{T}$ is a complete binary tree of height $O(d \log M)$. This tree $\mathcal{T}$ can be visualized as a complete 2^d-ary tree $\mathcal{T}_B$ of height $\log_4 M$ such that each node of $\mathcal{T}_B$, called *block node*, represents a complete binary tree of height d. See Fig. 1 for an example. For each $k \in [\log_4 M]$ and $l \in [2^{d(k-1)}]$, let $N_B^{k,l}$ denote the lth block node at the kth level[2] of $\mathcal{T}_B$, where at each level, we number the block nodes from left to right. So, the 2^d children block nodes of $N_B^{k,l}$ are $N_B^{k+1,2^d(l-1)+1}$, $N_B^{k+1,2^d(l-1)+2}$, $\ldots, N_B^{k+1,2^d(l-1)+2^d}$. Let $\mathcal{T}_B^{k,l}$ be a complete binary tree corresponding to a block node $N_B^{k,l}$. For each $i \in [d]$ and $j \in [2^{i-1}]$, let $N_{i,j}^{k,l}$ denote the jth node at the ith level of the tree $\mathcal{T}_B^{k,l}$, where at each level, we number the nodes from left to right. Now, for each $i \in [d-1]$ and $j \in [2^{i-1}]$, two children nodes of an internal node $N_{i,j}^{k,l}$ of $\mathcal{T}_B^{k,l}$ are $N_{i+1,2j-1}^{k,l}$ and $N_{i+1,2j}^{k,l}$, respectively. Next, for $k < \log_4 M$, we

[2] The level of a node of a tree is the number of vertices lying in the path from the root to that node. In particular, the root node is of level one.

describe the parent-child relation between a leaf of a tree $\mathcal{T}_B^{k,l}$ with the root of a tree corresponding to the child block node of $N_B^{k,l}$. For a leaf node $N_{d,j}^{k,l}$ of $\mathcal{T}_B^{k,l}$, where $j \in [2^{d-1}]$, two children are $N_{1,1}^{k+1,2^d(l-1)+2j-1}$ and $N_{1,1}^{k+1,2^d(l-1)+2j}$, respectively. Thus, $\mathcal{T}$ is a complete binary tree consisting of nodes $N_{i,j}^{k,l}$, where $k \in [\log_4 M]$, $l \in [2^{d(k-1)}]$, $i \in [d]$ and $j \in [2^{i-1}]$, such that $N_{1,1}^{1,1}$ is the root of it. As a result, the height h of $\mathcal{T}$ is $d \log_4 M = \frac{d}{2} \log_2 M$.

Corresponding to each node $N_{i,j}^{k,l}$, we have an axis-aligned hypercube having a width $\frac{M}{4^{k-1}}$ centered at a point $c_{i,j}^{k,l} \in \mathbb{R}^d$. Now, we explain the construction of the centers of all the hypercubes. This will be done recursively.

- First, we assume that $c_{1,1}^{1,1}$ is at the origin (see Fig. 1).
- Now, consider an internal node $N_{i,j}^{k,l}$ of $\mathcal{T}_B^{k,l}$, where $k \in [\log_4 M]$, $l \in [2^{d(k-1)}]$, $i \in [d-1]$ and $j \in [2^{i-1}]$. Recall that the two children nodes of $N_{i,j}^{k,l}$ are $N_{i+1,2j-1}^{k,l}$ and $N_{i+1,2j}^{k,l}$, respectively. We define $c_{i+1,2j-1}^{k,l} = c_{i,j}^{k,l} + \frac{M}{4^{k-1}}\mathbf{e}_i$ and $c_{i+1,2j}^{k,l} = c_{i,j}^{k,l} - \frac{M}{4^{k-1}}\mathbf{e}_i$, where $\mathbf{e}_i$ is the ith standard unit vector in $\mathbb{R}^d$. In other words, the center $c_{i+1,2j-1}^{k,l}$ (respectively, $c_{i+1,2j}^{k,l}$) corresponding to the left (respectively, right) child of $N_{i,j}^{k,l}$ is a translated copy of $c_{i,j}^{k,l}$, differing only in the ith coordinate. In particular, for the left (respectively, right) child, this translation is by $\frac{M}{4^{k-1}}\mathbf{e}_i$ (respectively, $-\frac{M}{4^{k-1}}\mathbf{e}_i$) (see Fig. 1).
- Next, consider a leaf node $N_{d,j}^{k,l}$ of $\mathcal{T}_B^{k,l}$, where $k < \log_4 M$, $l \in [2^{d(k-1)}]$ and $j \in [2^{d-1}]$. Recall that the two children nodes of $N_{d,j}^{k,l}$ are $N_{1,1}^{k+1,2^d(l-1)+2j-1}$ and $N_{1,1}^{k+1,2^d(l-1)+2j}$, respectively. To ensure that the hypercubes corresponding to the nodes $N_{1,1}^{k+1,2^d(l-1)+2j-1}$ and $N_{1,1}^{k+1,2^d(l-1)+2j}$ are contained inside the intersection of the hypercubes corresponding to the nodes lying in the path from the root $N_{1,1}^{k,l}$ to the leaf $N_{d,j}^{k,l}$, we define $c_{1,1}^{k+1,2^d(l-1)+2j-1} = \frac{c_{1,1}^{k,l}+c_{d,j}^{k,l}}{2} + \frac{M}{2^{2k-1}}\mathbf{e}_d$ and $c_{1,1}^{k+1,2^d(l-1)+2j} = \frac{c_{1,1}^{k,l}+c_{d,j}^{k,l}}{2} - \frac{M}{2^{2k-1}}\mathbf{e}_d$. In other words, the center $c_{1,1}^{k+1,2^d(l-1)+2j-1}$ (respectively, $c_{1,1}^{k+1,2^d(l-1)+2j}$) corresponding to the left (respectively, right) child of $N_{d,j}^{k,l}$ is a translated copy of the point $\frac{c_{1,1}^{k,l}+c_{d,j}^{k,l}}{2}$, differing only in the dth coordinate. In particular, for the left (respectively, right) child, this translation is by $\frac{M}{2^{2k-1}}\mathbf{e}_d$ (respectively, $-\frac{M}{2^{2k-1}}\mathbf{e}_d$) (see Fig. 1).

From the construction of hypercubes, it is easy to observe the following lemmas.

Lemma 1. *Let N, N', and N'' be three nodes of the tree $\mathcal{T}$ such that N' and N'' are the two children of N. Also, let σ' and σ'' be two hypercubes corresponding to the nodes N' and N'', respectively. Then, σ' and σ'' are interior disjoint.*

Proof. Let $N = N_{i,j}^{k,l}$ for some $k \in [\log_4 M]$, $l \in [2^{d(k-1)}]$, $i \in [d]$ and $j \in [2^{i-1}]$. First, consider the case when $N = N_{i,j}^{k,l}$ is an internal node of $\mathcal{T}_B^{k,l}$, i.e., $k \in [\log_4 M]$, $l \in [2^{d(k-1)}]$, $i \in [d-1]$ and $j \in [2^{i-1}]$. Then, we have $N' = N^{k,l}_{i+1,2j-1}$ and $N'' = N^{k,l}_{i+1,2j}$, and the two hypercubes σ' and σ'' are centered at points $c^{k,l}_{i+1,2j-1}$ and $c^{k,l}_{i+1,2j}$, respectively, having width $\frac{M}{4^{k-1}}$. Recall that $c^{k,l}_{i+1,2j-1} = c^{k,l}_{i,j} + \frac{M}{4^{k-1}}\mathbf{e}_i$ and $c^{k,l}_{i+1,2j} = c^{k,l}_{i,j} - \frac{M}{4^{k-1}}\mathbf{e}_i$. Thus, the centers of the two hypercubes σ' and σ'' only differ in the ith coordinate, and the distance between them is exactly $\frac{2M}{4^{k-1}}$. Since both σ' and σ'' are of width $\frac{M}{4^{k-1}}$, they are interior disjoint.

Now, consider the case when $N = N^{k,l}_{d,j}$ is a leaf node of $\mathcal{T}^{k,l}_B$, i.e., $k \in [\log_4 M - 1]$, $l \in [2^{d(k-1)}]$ and $j \in [2^{d-1}]$. Then, we have $N' = N^{k+1,2^d(l-1)+2j-1}_{1,1}$ and $N'' = N^{k+1,2^d(l-1)+2j}_{1,1}$, and the two hypercubes σ' and σ'' are centered at points $c^{k+1,2^d(l-1)+2j-1}_{1,1}$ and $c^{k+1,2^d(l-1)+2j}_{1,1}$, respectively, having width $\frac{M}{4^k}$. Recall that $c^{k+1,2^d(l-1)+2j-1}_{1,1} = \frac{c^{k,l}_{1,1}+c^{k,l}_{d,j}}{2} + \frac{M}{2^{2k-1}}\mathbf{e}_d$ and $c^{k+1,2^d(l-1)+2j}_{1,1} = \frac{c^{k,l}_{1,1}+c^{k,l}_{d,j}}{2} - \frac{M}{2^{2k-1}}\mathbf{e}_d$. Thus, the centers of the two hypercubes σ' and σ'' only differ in the dth coordinate, and the distance between them is exactly $\frac{M}{2^{2k-2}} = \frac{4M}{4^k}$. Since both σ' and σ'' are of width $\frac{M}{4^k}$, they are interior disjoint. □

Lemma 2. *Let $\mathcal{T}^{k,l}_B$ be a tree for some $k \in [\log_4 M]$ and $l \in [2^{d(k-1)}]$. Let $\mathcal{J} = (\sigma_u)^d_{u=1}$ be a sequence of hypercubes having width $w = \frac{M}{4^{k-1}}$ corresponding to nodes along a path from the root $N^{k,l}_{1,1}$ to a leaf $N^{k,l}_{d,j}$ of the tree $\mathcal{T}^{k,l}_B$ for some $j \in [2^{d-1}]$. Then, we have the following:*

(i) The common intersection region $Q = \cap^d_{u=1}\sigma_u$ contains two interior disjoint hypercubes H_1 and H_2 having width $\frac{w}{2}$ centered at the points $\frac{c^{k,l}_{1,1}+c^{k,l}_{d,j}}{2} + \frac{w}{2}\mathbf{e}_d$ and $\frac{c^{k,l}_{1,1}+c^{k,l}_{d,j}}{2} - \frac{w}{2}\mathbf{e}_d$, respectively.

(ii) Suppose $N^{k,l}_B$ is not a leaf node of $\mathcal{T}_B$ (i.e., $k < \log_4 M$). Let $\mathcal{T}^{k-1,l'}_B$ be a tree rooted at a child of the leaf node $N^{k,l}_{d,j}$ of the tree $\mathcal{T}^{k,l}_B$. Let $\Pi = (\sigma'_u)^d_{u=1}$ be a sequence of hypercubes corresponding to a path from the root to a leaf of the tree $\mathcal{T}^{k+1,l'}_B$. Then, the union of the hypercubes $\cup^d_{u=1}\sigma'_u$ is contained in the common intersection region $Q = \cap^d_{u=1}\sigma_u$.

Proof. First, from the construction of the hypercubes, recall that the center of the hypercube σ_1 corresponding to the node $N^{k,l}_{1,1}$ is $c^{k,l}_{1,1}$. Now, consider the next hypercube σ_2. Recall that the center of σ_2 is a translated copy of the center of σ_1 differing only in the 1st coordinate. Specifically, this translation is by $w\mathbf{e}_1$ (or $-w\mathbf{e}_1$) for left (or right) child. Hence, $c^{k,l}_{2,j'} = c^{k,l}_{1,1} + sign(\sigma_2)w\mathbf{e}_1$, where $w = \frac{M}{4^{k-1}}$ and the value of $sign(\sigma_2) = 1$ if σ_2 is the left child of σ_1, and -1 if σ_2 is the right child of σ_1. As a result of this, and σ_1 and σ_2 are both of width w, we also have $\sigma_2 = \sigma_1 + sign(\sigma_2)w\mathbf{e}_1$. Similarly, for each $u \in [d] \setminus \{1, 2\}$, we have $c^{k,l}_{u,j'} = c^{k,l}_{u-1,\lfloor \frac{j'}{2} \rfloor} + sign(\sigma_u)w\mathbf{e}_{u-1}$ and $\sigma_u = \sigma_{u-1} + sign(\sigma_u)w\mathbf{e}_{u-1}$, where the value of $sign(\sigma_u) = 1$ if σ_u is the left child of σ_{u-1}, and -1 if σ_u is the right child of σ_{u-1}. Therefore, for each $u \in [d] \setminus \{1\}$ and $j' \in [2^{u-1}]$, we have

$$c^{k,l}_{u,j'} = c^{k,l}_{1,1} + \sum_{v=2}^{u} sign(\sigma_v)w\mathbf{e}_{v-1} \tag{1}$$

and

$$\sigma_u = \sigma_1 + \sum_{v=2}^{u} sign(\sigma_v) w \mathbf{e}_{v-1}. \tag{2}$$

(i) Since $\sigma_2 = \sigma_1 + sign(\sigma_2)w\mathbf{e}_1$, we have that $\sigma_1 \cap \sigma_2$ is a hyperrectangle centered at the point $c_{1,1}^{k,l} + sign(\sigma_2)\frac{w}{2}\mathbf{e}_1$ whose side corresponding to the first coordinate is of length w, and all the remaining $(d-1)$ sides are of length $2w$. Since $\sigma_3 = \sigma_2 + sign(\sigma_3)w\mathbf{e}_2 = \sigma_1 + \sum_{v=2}^{3} sign(\sigma_v)w\mathbf{e}_{v-1}$ (due to Eq. 2), thus $\sigma_1 \cap \sigma_2 \cap \sigma_3$ is a hyperrectangle centered at the point $c_{1,1}^{k,l} + \sum_{v=2}^{3} sign(\sigma_v)\frac{w}{2}\mathbf{e}_{v-1}$ whose sides corresponding to the first two coordinates are of length w each, and the remaining $(d-2)$ sides are of length $2w$. Since for any $u \in [d] \setminus \{1\}$, the hypercube $\sigma_u = \sigma_{u-1} + sign(\sigma_u)w\mathbf{e}_{u-1} = \sigma_1 + \sum_{v=2}^{u} sign(\sigma_v)w\mathbf{e}_{v-1}$ (due to Eq. 2), we have $Q = \cap_{u=1}^{d}\sigma_u$ as a hyperrectangle centered at a point $c_{1,1}^{k,l} + \sum_{v=2}^{u} sign(\sigma_v)\frac{w}{2}\mathbf{e}_{v-1}$ whose sides corresponding to the first $(d-1)$ coordinates are of length w each, and the remaining side is of length $2w$.
As a result, the common intersection region Q will contain two interior disjoint hypercubes H_1 and H_2 of width $\frac{w}{2}$ centered at p_1 and p_2, respectively, where

$$\begin{aligned} p_1 &= c_{1,1}^{k,l} + \sum_{u=2}^{d} sign(\sigma_u)\frac{w}{2}\mathbf{e}_{u-1} + \frac{w}{2}\mathbf{e}_d \qquad \text{(due to Equation 1)} \\ &= \frac{c_{1,1}^{k,l}}{2} + \left(\frac{c_{1,1}^{k,l}}{2} + \frac{1}{2}\sum_{u=2}^{d} sign(\sigma_u)w\mathbf{e}_{u-1}\right) + \frac{w}{2}\mathbf{e}_d \\ &= \frac{c_{1,1}^{k,l}}{2} + \frac{c_{d,j}^{k,l}}{2} + \frac{w}{2}\mathbf{e}_d = \frac{c_{1,1}^{k,l} + c_{d,j}^{k,l}}{2} + \frac{w}{2}\mathbf{e}_d, \end{aligned}$$

and

$$\begin{aligned} p_2 &= c_{1,1}^{k,l} + \sum_{u=2}^{d} sign(\sigma_u)\frac{w}{2}\mathbf{e}_{u-1} - \frac{w}{2}\mathbf{e}_d \qquad \text{(due to Equation 1)} \\ &= \frac{c_{1,1}^{k,l}}{2} + \left(\frac{c_{1,1}^{k,l}}{2} + \frac{1}{2}\sum_{u=2}^{d} sign(\sigma_u)w\mathbf{e}_{u-1}\right) - \frac{w}{2}\mathbf{e}_d \\ &= \frac{c_{1,1}^{k,l}}{2} + \frac{c_{d,j}^{k,l}}{2} - \frac{w}{2}\mathbf{e}_d = \frac{c_{1,1}^{k,l} + c_{d,j}^{k,l}}{2} - \frac{w}{2}\mathbf{e}_d. \end{aligned}$$

(ii) From the construction of the centers of the hypercubes, we have that σ_1' is centered at the point $\frac{c_{1,1}^{k,l}+c_{d,j}^{k,l}}{2} + sign(\sigma_1')\frac{w}{2}\mathbf{e}_d$, where $sign(\sigma_1') = 1$ if σ_1' is the left child of σ_d, and $sign(\sigma_1') = -1$ if σ_1' is the right child of σ_d. W.l.o.g., assume that σ_1' is a hypercube corresponding to the left child of the node $N_{d,j}^{k,l}$. The other case, i.e., σ_1' is a hypercube corresponding to the right child of $N_{d,j}^{k,l}$, can be done in a similar way. From (i), we have that $\cap_{u=1}^{d}\sigma_u$ contains a hypercube H_1 having width $\frac{w}{2}$ centered at the point $\frac{c_{1,1}^{k,l}+c_{d,j}^{k,l}}{2}+\frac{w}{2}\mathbf{e}_d$. Notice that σ_1' and H_1 are concentric hypercubes having widths $\frac{w}{4}$ and $\frac{w}{2}$, respectively. Hence, $\sigma_1' \subset H_1$. In a similar way as we have Eq. 2, for each $u \in [d]\setminus\{1\}$, we also have $\sigma_u' = \sigma_1' + \sum_{v=2}^{u} sign(\sigma_v')\frac{w}{4}\mathbf{e}_{v-1}$, where $sign(\sigma_v') = 1$ if σ_v' is the left child of σ_{v-1}', and $sign(\sigma_v') = -1$ if σ_v' is the right child of σ_{v-1}'. Thus, the centers of the hypercubes σ_1' and σ_u' only differ in the first $(u-1)$th coordinates, and the difference is exactly $\frac{w}{4}$ at each coordinate for $u \in [d]\setminus\{1\}$. As a result of this, and σ_1' and H_1 are concentric hypercubes, and the hypercube σ_u' is of width $\frac{w}{4}$, we have that $\sigma_u' \subset H_1$ for each $u \in [d]\setminus\{1\}$. Therefore, we have $\cup_{u=1}^{d}\sigma_u' \subseteq H_1 \subseteq \cap_{u=1}^{d}\sigma_u$. Hence, the lemma follows.

□

Note that the width of a hypercube corresponding to a leaf node of $\mathcal{T}$ is 4 (by putting the value of $k = \log_4 M$ in $\frac{M}{4^{k-1}}$). As a result, due to Lemma 2, the intersection of hypercubes corresponding to a sequence $\mathcal{I} \in \mathcal{S}$ contains a hypercube having width 2. Since a hypercube having width 2 must contain an integer point, we have the following lemma.

Lemma 3. *Let $\mathcal{I} = (\sigma_i)_{i=1}^{h} \in \mathcal{S}$ be a sequence of hypercubes having widths in the range $[1, M]$ corresponding to nodes of a path from the root to a leaf of the tree $\mathcal{T}$, where h is the height of the tree $\mathcal{T}$. Then, their intersection $\cap_{i=1}^{h}\sigma_i$ must contain an integer point.*

Due to Lemma 3, for any sequence $\mathcal{I} \in \mathcal{S}$, the size of the hitting set produced by an optimum offline algorithm over the input sequence $\mathcal{I}$ is 1. Now, consider any online algorithm ALG. Let C_i be the cost of the algorithm ALG, i.e., the number of points selected by ALG, at the ith level. Clearly, $\mathbb{E}(C_1) = 1$. We will show that $\mathbb{E}(C_i) = \frac{1}{2}$ for $i \in [h] \setminus \{1\}$. While it is possible, in principle, for an algorithm to select several points at each level, it can be seen that this does not help in total cost reduction. W.l.o.g., we can assume that at each level, ALG selects only one point if the currently introduced hypercube is not hit by the solution set maintained by ALG. Due to Lemma 1, this point may lie either in the left or in the right child node, but not both. Thus, with probability $\frac{1}{2}$, the algorithm ALG chooses either a point or no point at the ith level. As a result, $\mathbb{E}(C_i) = \frac{1}{2} \times 1 + \frac{1}{2} \times 0 = \frac{1}{2}$. Therefore, we have that $\mathbb{E}(\text{ALG}) = \sum_{i=1}^{h}\mathbb{E}(C_i) = 1 + \frac{1}{2}(h-1) = \frac{1}{2}(1 + \frac{d}{2}\log_2 M)$. Hence, we have the following theorem.

Theorem 1. *The competitive ratio of any deterministic or randomized algorithm for hitting homothetic axis-aligned hypercubes in $\mathbb{R}^d$ having widths in the range $[1, M]$ using points in $\mathbb{Z}^d$ is $\Omega(d \log M)$.*

From the construction of the lower bound, it is easy to observe that all hypercubes are lying in $[-2M, 2M]^d$. So, with a change in coordinate, we can construct all hypercubes in such a way that they are lying in $(0, 4M+1)^d$. As a result, the lower bound is also applicable in the setup of Alefkhani et al. [2], and by considering $N = 4M + 1$, we have the following.

Corollary 1. *The competitive ratio of any deterministic or randomized algorithm for hitting axis-aligned hypercubes in* $(0, N)^d$ *using points in* $(0, N)^d \cap \mathbb{Z}^d$ *is* $\Omega(d \log N)$.

3 Algorithm for Similarly Sized Fat Objects in $\mathbb{R}^d$

In this section, we study the hitting set problem for a range space $\Sigma = (\mathcal{P}, \mathcal{S})$, where the point set $\mathcal{P} = \mathbb{Z}^d$ and the set $\mathcal{S}$ is a family of α-fat objects in $\mathbb{R}^d$ having widths in the range $[1, M]$ for $M > 1$. For this, we propose a deterministic online algorithm PruneAndChoose and prove that the algorithm achieves a competitive ratio of at most $\lfloor 4\alpha + 1 \rfloor^d \lfloor \log_2 2M \rfloor$. Throughout the section, if explicitly not mentioned, all distances are L_∞ distances. For simplicity, we use the term object to represent α-fat object.

Algorithm PruneAndChoose. For each $i \in \mathbb{Z}^+ \cup \{0\}$, let L_i be the ith layer containing objects of $\mathcal{S}$ having widths in the range $[2^i, 2^{i+1})$. Upon the arrival of an object σ having width w, we do the following. If σ is stabbed by the existing hitting set, we do nothing. Else, we first determine the layer L_i in which σ belongs, where $i = \lfloor \log_2 w \rfloor$. Now, to hit σ, our online algorithm selects any point from the set $(2^{i+1}\mathbb{Z})^d \cap \sigma$. The following lemma justifies the correctness of our algorithm.

Lemma 4. *For any object* σ *having width* w*, the set* $(2^{i+1}\mathbb{Z})^d \cap \sigma$ *is non-empty, i.e.,* $(2^{i+1}\mathbb{Z})^d \cap \sigma \neq \emptyset$*, where* $i = \lfloor \log_2 w \rfloor$.

Proof. Since the object σ is of width w, it contains a hypercube H of side length $2w$. As a result of this and also due to $w \geq 2^i$, the hypercube H, consequently the object σ, must contains a point from $(2^{i+1}\mathbb{Z})^d$, i.e., $(2^{i+1}\mathbb{Z})^d \cap \sigma \neq \emptyset$. Hence, the lemma follows. □

Theorem 2. *The deterministic algorithm* PruneAndChoose *achieves a competitive ratio of at most* $\lfloor 4\alpha + 1 \rfloor^d \lfloor \log_2 2M \rfloor$ *for hitting* α*-fat objects in* $\mathbb{R}^d$ *having widths in the range* $[1, M]$ *using points in* $\mathbb{Z}^d$.

Proof. Let $\mathcal{I}$ be a collection of similarly sized objects presented to the algorithm. Let $\mathcal{A}$ and $\mathcal{O}$ be hitting sets returned by the algorithm PruneAndChoose and an optimum offline algorithm, respectively, for the input $\mathcal{I}$. Let p be a point in $\mathcal{O}$. For each $i \in \{0, \ldots, \lfloor \log_2 M \rfloor\}$, let $\mathcal{I}_{p,i} \subseteq \mathcal{I}$ be the collection of all objects having widths in the range $\left[2^i, 2^{i+1}\right)$ that contains the point p. For each $i \in \{0, \ldots, \lfloor \log_2 M \rfloor\}$, let $\mathcal{A}_{p,i} \subseteq \mathcal{A}$ be the set of hitting points explicitly placed by PruneAndChoose to hit objects in $\mathcal{I}_{p,i}$. It is easy to see that $\mathcal{A} = \cup_{p \in \mathcal{O}} \cup_{i=0}^{\lfloor \log_2 M \rfloor} \mathcal{A}_{p,i}$. Now, we bound the cardinality of the set $\mathcal{A}_{p,i}$.

Let σ be an object in $\mathcal{I}_{p,i}$. Also, let $r \in (2^{i+1}\mathbb{Z})^d \cap \sigma$ be the point selected by our algorithm to hit σ. Since σ is an α-fat object having width strictly less than 2^{i+1}, the distance between any two points in σ is also strictly less than $\alpha 2^{i+1}$. Since $r, p \in \sigma$, we have that $d(r,p) < \alpha 2^{i+1}$. Hence, the interior of an axis-aligned hypercube of side length $\alpha 2^{i+3}$ centered at p contains all points in $\mathcal{A}_{p,i}$. Due to Observation 1, the interior of any axis-aligned hypercube of side length $2^{i+1}(4\alpha)$ contains at most $\lfloor 4\alpha + 1 \rfloor^d$ points from $(2^{i+1}\mathbb{Z})^d$. Thus, we have $|\mathcal{A}_{p,i}| \leq \lfloor 4\alpha + 1 \rfloor^d$. Therefore, we have $|\mathcal{A}| \leq \lfloor 4\alpha + 1 \rfloor^d \lfloor \log_2 2M \rfloor |\mathcal{O}|$. Hence, the theorem follows. □

For balls in $\mathbb{R}^d$, by putting the value of $\alpha = \sqrt{d}$ in Theorem 2, we have the following.

Corollary 2. *For hitting balls in $\mathbb{R}^d$ having radii in the range $[1, M]$ using points in $\mathbb{Z}^d$, the deterministic algorithm* PruneAndChoose *achieves a competitive ratio of at most $\lfloor 4\sqrt{d} + 1 \rfloor^d \lfloor \log_2 2M \rfloor$.*

Observe that any α-fat object lying in $(0, N)^d$ has a width at most $\frac{N}{2}$. As a result, due to Theorem 2, we have the following.

Corollary 3. *The deterministic algorithm* PruneAndChoose *achieves a competitive ratio of $O((4\alpha+1)^d \log N)$ for hitting α-fat objects in $(0, N)^d$ using points in $(0, N)^d \cap \mathbb{Z}^d$.*

4 Conclusion

In this paper, we have studied the online hitting set problem, where the point set $\mathcal{P} = \mathbb{Z}^d$. For hitting similarly sized homothetic axis-aligned hypercubes in $\mathbb{R}^d$, we show that no algorithm, be it deterministic or randomized, can have a competitive ratio better than $\Omega(d \log M)$. Then, for hitting similarly sized fat objects in $\mathbb{R}^d$, we present a deterministic algorithm PruneAndChoose having a competitive ratio of at most $\lfloor 4\alpha + 1 \rfloor^d \lfloor \log_2 2M \rfloor$. In particular, our results significantly improve the gap between the upper and lower bounds for the problem in the special setup of Alefkhani et al. [2]. There is room for further improvement. There is a huge gap between the lower and the upper bounds for hitting similarly sized fat objects in $\mathbb{R}^d$. We propose bridging this gap further as a future direction of research.

References

1. Agarwal, P.K., Pan, J.: Near-linear algorithms for geometric hitting sets and set covers. Discrete Comput. Geometry **63**(2), 460–482 (2019). https://doi.org/10.1007/s00454-019-00099-6
2. Alefkhani, S., Khodaveisi, N., Mari, M.: Online hitting set of d-dimensional fat objects. In: Approximation and Online Algorithms - 21st International Workshop, WAOA 2023, September 7-8, 2023, Proceedings. Lecture Notes in Computer Science, vol. 14297, pp. 134–144. Springer (2023). https://doi.org/10.1007/978-3-031-49815-2_10

3. Alon, N., Awerbuch, B., Azar, Y., Buchbinder, N., Naor, J.: The online set cover problem. SIAM J. Comput. **39**(2), 361–370 (2009). https://doi.org/10.1137/060661946
4. de Berg, M., van der Stappen, A.F., Vleugels, J., Katz, M.J.: Realistic input models for geometric algorithms. Algorithmica **34**(1), 81–97 (2002). https://doi.org/10.1007/S00453-002-0961-X
5. Chan, T.M.: Polynomial-time approximation schemes for packing and piercing fat objects. J. Algorithms **46**(2), 178–189 (2003). https://doi.org/10.1016/S0196-6774(02)00294-8
6. Chan, T.M., He, Q.: Faster approximation algorithms for geometric set cover. In: 36th International Symposium on Computational Geometry, SoCG 2020, June 23-26, 2020. LIPIcs, vol. 164, pp. 27:1–27:14. Schloss Dagstuhl - Leibniz-Zentrum für Informatik (2020). https://doi.org/10.4230/LIPIcs.SoCG.2020.27
7. Chan, T.M., Zarrabi-Zadeh, H.: A randomized algorithm for online unit clustering. Theory Comput. Syst. **45**(3), 486–496 (2009). https://doi.org/10.1007/s00224-007-9085-7
8. Charikar, M., Chekuri, C., Feder, T., Motwani, R.: Incremental clustering and dynamic information retrieval. SIAM J. Comput. **33**(6), 1417–1440 (2004). https://doi.org/10.1137/S0097539702418498
9. De, M., Jain, S., Kallepalli, S.V., Singh, S.: Online geometric covering and piercing. Algorithmica **86**(9), 2739–2765 (2024). https://doi.org/10.1007/S00453-024-01244-1
10. De, M., Singh, S.: Hitting geometric objects online via points in $\mathbb{Z}^d$. In: Computing and Combinatorics - 28th International Conference, COCOON 2022, October 22-24, 2022, Proceedings. Lecture Notes in Computer Science, vol. 13595, pp. 537–548. Springer (2022). https://doi.org/10.1007/978-3-031-22105-7_48
11. De, M., Singh, S.: Online hitting of unit balls and hypercubes in $\mathbb{R}^d$ using points from $\mathbb{Z}^d$. Theor. Comput. Sci. **992**, 114452 (2024). https://doi.org/10.1016/J.TCS.2024.114452
12. Dinur, I., Steurer, D.: Analytical approach to parallel repetition. In: Symposium on Theory of Computing, STOC 2014, May 31 - June 03, 2014, pp. 624–633. ACM (2014). https://doi.org/10.1145/2591796.2591884
13. Dumitrescu, A., Ghosh, A., Tóth, C.D.: Online unit covering in Euclidean space. Theor. Comput. Sci. **809**, 218–230 (2020). https://doi.org/10.1016/j.tcs.2019.12.010
14. Dumitrescu, A., Tóth, C.D.: Online unit clustering and unit covering in higher dimensions. Algorithmica **84**(5), 1213–1231 (2022). https://doi.org/10.1007/S00453-021-00916-6
15. Efrat, A., Katz, M.J., Nielsen, F., Sharir, M.: Dynamic data structures for fat objects and their applications. Comput. Geom. **15**(4), 215–227 (2000). https://doi.org/10.1016/S0925-7721(99)00059-0
16. Even, G., Smorodinsky, S.: Hitting sets online and unique-max coloring. Discret. Appl. Math. **178**, 71–82 (2014). https://doi.org/10.1016/j.dam.2014.06.019
17. Feige, U.: A threshold of ln n for approximating set cover. J. ACM **45**(4), 634–652 (1998). https://doi.org/10.1145/285055.285059
18. Fowler, R.J., Paterson, M., Tanimoto, S.L.: Optimal packing and covering in the plane are np-complete. Inf. Process. Lett. **12**(3), 133–137 (1981). https://doi.org/10.1016/0020-0190(81)90111-3
19. Ganjugunte, S.K.: Geometric Hitting Sets and Their Variants. Ph.D. thesis, Duke University, Durham, NC, USA (2011). https://hdl.handle.net/10161/4972

20. Garey, M.R., Johnson, D.S.: Computers and Intractability; A Guide to the Theory of NP-Completeness. W. H. Freeman & Co., USA (1990)
21. Khan, A., Lonkar, A., Rahul, S., Subramanian, A., Wiese, A.: Online and dynamic algorithms for geometric set cover and hitting set. In: 39th International Symposium on Computational Geometry, SoCG 2023, June 12-15, 2023. LIPIcs, vol. 258, pp. 46:1–46:17. Schloss Dagstuhl - Leibniz-Zentrum für Informatik (2023). https://doi.org/10.4230/LIPICS.SOCG.2023.46
22. Mustafa, N.H., Ray, S.: Improved Results on Geometric Hitting Set Problems. Discrete Comput. Geometry **44**(4), 883–895 (2010). https://doi.org/10.1007/s00454-010-9285-9

Erdős-Szekeres Maker-Breaker Games

Aleksa Džuklevski[1], Dömötör Pálvölgyi[2,3], Alexey Pokrovskiy[4], Csaba D. Tóth[5,6], Tomáš Valla[7], and Lander Verlinde[8](✉)

[1] University of Novi Sad, Novi Sad, Serbia
aleksa.dzuklevski@dmi.uns.ac.rs
[2] Eötvös Loránd University, Budapest, Hungary
domotor.palvolgyi@ttk.elte.hu
[3] Alfréd Rényi Institute of Mathematics, Budapest, Hungary
[4] University College London, London, UK
dralexeypokrovskiy@gmail.com
[5] California State University Northridge, Los Angeles, CA, USA
csaba.toth@csun.edu
[6] Tufts University, Medford, MA, USA
[7] Czech Technical University in Prague, Prague, Czech Republic
tomas.valla@fit.cvut.cz
[8] University of Auckland, Auckland, New Zealand
lver263@aucklanduni.ac.nz

Abstract. We present new results on Maker-Breaker games arising from the Erdős-Szekeres problem in planar geometry. This classical problem asks how large a set in general position has to be to ensure the existence of n points that are the vertices of a convex n-gon. Moreover, Erdős further extended this problem by asking what happens if we also require that this n-gon has an empty interior. In a 2-player Maker-Breaker setting, this problem inspires two main games. In both games, Maker tries to obtain an empty convex k-gon, while Breaker tries to prevent her from doing so. The games differ only in which points can comprise the winning k-gons: in the monochromatic version the points of both players can make up a k-gon, while in the bichromatic version only Maker's points contribute to such a polygon. Both settings are studied in this paper. We show that in the monochromatic game, Maker always wins. Even in a biased game where Breaker is allowed to place s points per round, for any constant $s \geq 1$, Maker has a winning strategy. In the bichromatic setting, Maker still wins whenever Breaker is allowed to place s points per round for any constant $s < 2$. This settles an open problem posed in 2019. Furthermore, we show that there are games that are not a lost cause for Breaker. Whenever $k \geq 8$ and Breaker is allowed to play 12 or more points per round, she has a winning strategy. We also consider the one-round bichromatic game (a.k.a. the offline version). In this setting, we show that Breaker wins if she can place twice as many points as Maker but if the bias is less than 2, then Maker wins for large enough set of points.

Keywords: Erdős-Szekeres Theorem · Maker-Breaker Game · Convex k-hole

F. V. Fomin and M. Xiao (Eds.): COCOON 2025, LNCS 15983, pp. 248–261, 2026.
https://doi.org/10.1007/978-981-95-0215-8_19

1 Introduction

The Erdős-Szekeres theorem [17] is a classical result in Ramsey theory. It states that every set of n points in the plane in general position contains a subset of $\Omega(\log n)$ points in convex position. A set of points in the plane is in *general position* if no three points are collinear, and in *convex position* if the points are the vertices of a convex polygon. The bound $\Omega(\log n)$ is the best possible; it has been refined to $(1-o(1))\log_2 n$ and significant efforts have been devoted to finding precise bounds [24,33]. The Erdős-Szekeres theorem made lasting impact on combinatorial geometry—many generalizations and variations has been studied.

For a point set $S \subset \mathbb{R}^2$ and an integer $k \geq 3$, a *k-hole* is a subset $H \subset S$ of size k such that $\text{conv}(H)$ is a convex k-gon and $\text{conv}(H) \cap S = H$, where $\text{conv}(H)$ denotes the convex hull of H. Let $h(k)$ denote the minimum integer n such that every set of n points in the plane in general position contains a k-hole. Horton [25] constructed point sets in general position that do not contain any 7-hole, which implies $h(k) = \infty$ for all $k \geq 7$. It is easy to see that $h(3) = 3$, $h(4) = 5$, and Harboth [21] showed that $h(5) = 10$. Gerken [19] and Nicolás [28] proved independently that $h(6) < \infty$. With a computer assisted proof, Heule and Scheucher [23] recently showed that $h(6) = 30$; completely answering a question posed by Erdős [16]. The minimum and maximum number of k-holes among n points in the plane were also studied [1,4,7,8,30], as well as the associated counting and enumeration problems [6,13,27].

Competitive games between two players to achieve a geometric structure were studied previously [2,22]. The natural competitive game arising from the Erdős-Szekeres problem is the endeavor of two players to achieve a k-hole for a given positive integer k by alternately placing points in general position in the plane. Depending on the goal of the game, three variants of the two-player game spawn from the Erdős-Szekeres problem.

1. Both players want to obtain a k-hole (*Maker-Maker*). Whoever creates the first k-hole wins.
2. Both players want to avoid a k-hole (*Avoider-Avoider*). Whoever creates the first k-hole loses.
3. The first player wants to obtain a k-hole and the second wants to prevent this (*Maker-Breaker*). The first player (Maker) wins if a k-hole is ever created by either player, and the second player (Breaker) wins if she is able to prolong the game indefinitely.

The *Erdős-Szekeres Maker-Maker game* was introduced by Valla [34]. Kolipaka and Govindarajan [26] studied the Avoider-Avoider game and showed that for $k = 5$ the second player can win within 9 rounds. Aichholzer et al. [3] proposed the Maker-Breaker game for $k \geq 7$. For $k \leq 6$, Maker inevitably wins as every sufficiently large point set in general position contains a k-hole. For $k = 7$, Maker can easily add one more point to extend a 6-hole to a 7-hole. Das and Valla [12] recently studied the Maker-Breaker game: they showed that Maker (i.e., the first player) has a winning strategy for all $k \leq 8$. If $r(k)$ denotes the minimum number of rounds for a winning strategy for Maker, then $r(3) = 2$ and

$r(4) = 3$ are obvious as Maker places the $(2r-1)$st point in round r; and they proved that $r(5) \leq 6$, $r(7) \leq 8$, and $r(8) \leq 12$. They asked what the maximum k is for which Maker has a winning strategy.

Aichholzer et al. [3] also introduced several *bichromatic* versions of Erdős-Szekeres games. In the bichromatic Maker-Breaker game, the set $S \subset \mathbb{R}^2$ is partitioned as $S = M \cup B$, where M and B denote the points played by Maker and Breaker, respectively. Maker wins if $M \cup B$ contains a k-hole H *and* $H \subset M$ (that is, the hole consists of Maker's points, and conv(H) contains neither Breaker points nor any additional Maker points).

The one-round game (the *offline version*) has also been considered. Conlon and Lim [10] proved that for every set M of m points in the plane in general position, there exists a set $S = M \cup B$ without 9-holes; however the set B is much larger than M in their construction. In the bichromatic variant, Maker places a set M of m points in the plane in general position, and then Breaker places a set B of points to prevent a k-hole $H \subset M$ with respect to $S = M \cup B$. For $k \in \{3, 4\}$, $|B| = 2m - O(1)$ points are always sufficient and sometimes necessary [11,32]; the current best lower bound is $|B| \geq \frac{2}{9}m - O(1)$ for $k = 5$ [9].

By default, each player places one new point per round in the online version. However, to balance the fortunes of the players, one might introduce *bias* by allowing one of the players to place more points than the other. The number of points that a player is allowed to place shall also be referred to as *speed* and we denote the respective integer speeds of Maker and Breaker by $s_M : s_B$. In particular, the default version, where the players have equal speed, is the $1:1$ game. In this paper we consider the bias $1 : s_B$ or $s_M : 1$, where both s_M and s_B are always rational. If a player has rational speed $a/b \in \mathbb{Q}$, then that player places a total of a points over a cycle of b consecutive rounds: each cycle starts with $\lfloor a/b \rfloor$ points per round and ends with $\lceil a/b \rceil$ points per round.

Our Results. We present three new strategies that can be used for different Maker-Breaker games. The first one is based on maintaining a structure which we shall call a *k-strip*: Intuitively, it can be seen as a set of k points on an x-monotone concave arc such that the region below the arc is empty (Sect. 2). We show that in the monochromatic $1:1$ game, Maker is able to construct such a k-strip of arbitrary size and give the algorithm to do so. Since such a k-strip is a specific type of k-hole, this proves that Maker always wins the monochromatic game (Theorem 1), settling the question of Das and Valla [12]. Furthermore, we show that even if bias is introduced in favor of Breaker, then Maker can still construct k-strips. Hence, Maker wins the monochromatic game even with speed $1 : s_B$ for any constant $s_B \geq 1$ (Theorem 2). A last adaption of the algorithm also shows that Maker has a winning strategy for the bichromatic game with bias $1 : s_B$ for any constant $s_B < 2$ (Theorem 3). Hence, this strategy is versatile, and with the right adjustments, Maker is able to use it to win multiple types of Maker-Breaker games.

Another strategy is based on *perturbed polygons* (Sect. 3.1). This is a strategy designed for the Breaker in the bichromatic game with large enough bias in her favor: Breaker places her points as vertices of a slightly perturbed regular polygon inscribed in sufficiently small circles around Maker's points. This strategy relies on a specific type of perturbation and a suitable choice of radii. Both elements of the strategy are explained in Sect. 3.1. We prove that whenever $k \geq 8$, Breaker is able to prevent Maker from constructing a k-hole in the bichromatic game with speed $1 : s_B$ for $s_B \geq 12$ (Theorem 4). We also provide a trade-off between k and the bias: Breaker wins the bichromatic game with $1 : 2\lambda$ bias for every integer $\lambda \geq 3$ if $k \geq 3\lceil 2\lambda/(\lambda - 2)\rceil - 1$ (Theorem 5).

Lastly, we propose a strategy for Maker for the one-round bichromatic game whenever the bias in favor of Breaker is strictly less than 2 (Sect. 3.3). Maker's strategy builds on the Density Hales-Jewett Theorem [18] from combinatorics. In particular, Maker starts with a section of the integer lattice $\mathbb{Z}^d$, for a sufficiently large dimension d, projected on a suitably chosen generic plane and perturbed to be in general position. Importantly, the perturbation ensures that some of the collinear t-tuples of points of the grid are mapped into t-holes (with respect to Maker's points). We prove that Maker wins the bichromatic one-round game with bias $1 : (2 - \varepsilon)$ for every $\varepsilon > 0$ (Theorem 6). This is the best possible, as Breaker wins the bichromatic one-round game with bias $1 : s_B$ for $s_B \geq 2$ (Proposition 1).

2 Monochromatic Maker-Breaker Games

When proving their result, Erdős and Szekeres [17] introduced a notion that later became known as *caps* and *cups*.

Definition 1. *A set $\{p_1, \ldots, p_k\}$ of k points, labeled in increasing order by their x-coordinates, forms a k-*cap *(k-*cup*) if the points are in convex position and their convex hull is bounded from below (above) by a single edge; see Fig. 1.*

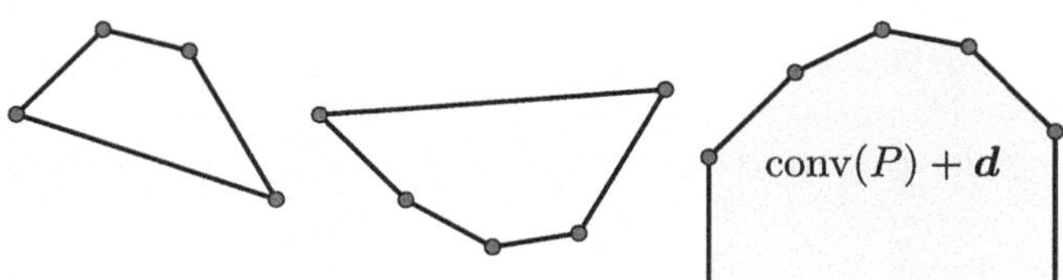

Fig. 1. Example of a 4-cap, a 5-cup and a 5-strip with respect to $\boldsymbol{d} = \overrightarrow{(0, -1)}$.

We extend these caps to a new notion of k-strips. Intuitively, a k-strip is a $(k + 1)$-hole with k points in $\mathbb{R}^2$ and one point "at infinity". Recall that the Minkowski sum of two sets $A, B \in \mathbb{R}^2$ is defined as $A+B = \{a+b : a \in A, b \in B\}$. For a nonzero vector v, let $\boldsymbol{v}$ denote the halfline from the origin in direction v. As a shorthand, $\boldsymbol{d} = \overrightarrow{(0, -1)}$ is a downward vertical halfline.

Definition 2. *For every* $k \in \mathbb{N}$, *a* k-strip *with respect to a point set* $S \subset \mathbb{R}^2$ *is a set* $P \subset S$ *of* k *points in convex position such that there exists a halfline* $\boldsymbol{v}$ *for which the Minkowski sum* $C = \mathrm{conv}(P) + \boldsymbol{v}$ *satisfies* $\mathrm{int}(C) \cap S = \emptyset$ *and* $\partial C \cap S = P$; *see an example in Fig. 1. We say that two* k-*strips,* P_1 *and* P_2, *are* parallel *if they are* k-*strips w.r.t. the same halfline* $\boldsymbol{v}$, *and* $C_1 = \mathrm{conv}(P_1) + \boldsymbol{v}$ *and* $C_2 = \mathrm{conv}(P_2) + \boldsymbol{v}$ *are disjoint.*

In particular, if P is a k-strip w.r.t. the downward halfline $\boldsymbol{d}$ (resp., upward halfline $\overrightarrow{(0,1)}$), then P is a k-cap (resp., k-cup).

2.1 Monochromatic Maker-Breaker Game Without Bias

In the $1:1$ game, we show by induction that Maker can create arbitrarily many k-strips for any $k \in \mathbb{N}$. As a k-strip is a k-hole, this immediately implies that Maker wins for any $k \in \mathbb{N}$.

Theorem 1. *In the monochromatic Maker-Breaker game (without bias), for every* $k, t \in \mathbb{N}$, *Maker can ensure that there are* t *parallel* k-*strips among the points placed by both players. This can be accomplished in* $t/2$ *rounds for* $k = 1$ *and at most* $(\frac{5}{3} \cdot 4^{k-2} - \frac{2}{3})t$ *rounds for* $k \geq 2$.

This settles an open problem recently posed by Das and Valla [12]. Furthermore, Theorem 1 generalizes: It is a special case of Theorem 2 in Sect. 2.2. (For completeness, we prove Theorem 1 in the full version of this paper [15].)

2.2 Monochromatic Maker-Breaker Game with Bias

We next consider a variation of the Maker-Breaker game in which Breaker has the advantage that she is allowed to place more points than Maker in each round. Nevertheless, Maker still has a winning strategy.

Theorem 2. *Consider the monochromatic Maker-Breaker game with* $1:s$ *bias. For every* $k, t \in \mathbb{N}$ *and* $s \in \mathbb{Q}_{\geq 1}$, *Maker can ensure that there are* t *parallel* k-*strips among the points placed by both players. This can be accomplished in at most* $t/(s+1)$ *rounds for* $k = 1$ *and at most* $((4s)^{k-2}/(s+1) + ((4s)^{k-2} - 1)/(4s-1))2t$ *rounds for* $k \geq 2$ *and* $s \in \mathbb{N}$.

Proof. Without loss of generality, assume that s is an integer. We proceed by induction on k. In the base case, $k = 1$, any t points in general position form t parallel 1-strips. Similarly, for $k = 2$, it is easy to see that any $2t$ points in general position form t parallel 2-strips. We may assume (by rotating the coordinate system, if necessary) that the points have distinct x-coordinates. Then we can partition the $2t$ points into t pairs of points with consecutive x-coordinates; and use downward 2-strips. Assume next that we are given $k, t \in \mathbb{N}$, $k \geq 2$, and the theorem holds for all $k', t' \in \mathbb{N}$ with $k' < k$. By the induction hypothesis, Maker can create $4st$ parallel $(k-1)$-strips with respect to the current point set S. As previously, we may assume (by rotating the coordinate system if necessary) all k-strips are w.r.t. a downward halfline $\boldsymbol{d} = \overrightarrow{(0, -1)}$, and all points in S have distinct x-coordinates. Denote the $(k-1)$-strips by $S_1, S_2, \ldots, S_{4st}$, from left to right, and the points in them as $s_{i,j}$, ordered according to increasing x-coordinate. For each $(k-1)$-strip S_i, we define the regions $Q_i = \text{conv}(S_i) + \boldsymbol{d}$ (see Fig. 2 for an illustration). Note that the regions Q_i are pairwise disjoint.

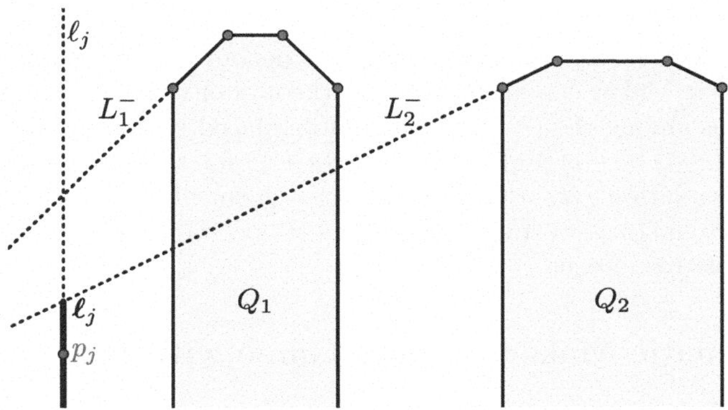

Fig. 2. A group of 4-strips for $s = 1$. The point p_j creates two possible k-strips.

For each $(k-1)$-strip S_i, we define a cone C_i^- as follows. Recall that $s_{i,1}$ is the leftmost point in S_i. Let $L(s_{i,1}) = s_{i,1} + \boldsymbol{d}$ be the downward halfline starting from $s_{i,1}$; and let L_i^- be the line passing through $s_{i,1}$ and $s_{i,2}$. Now let C_i^- be the cone swept by rotating the halfline $L(s_{i,1})$ clockwise about $s_{i,1}$ until it reaches L_i^- or it passes through a point in S. Note that for any point $p \in \text{int}(C_i^-)$, the set $S_i \cup \{p\}$ is a k-strip with respect to the point set $S \cup \{p\}$.

We partition the $4st$ parallel $(k-1)$-strips into $2t$ groups $\mathcal{S}_1, \ldots, \mathcal{S}_{2t}$, each consisting of $2s$ consecutive $(k-1)$-strips. For each group $\mathcal{S}_j$, we also define the region $\mathcal{Q}_j = \bigcup\{Q_i : S_i \in \mathcal{S}_j\}$, that is, the union of the regions Q_i of the corresponding $(k-1)$-strips. Note that the regions $\mathcal{Q}_j$ are pairwise disjoint. For each group $\mathcal{S}_j$, $j \in [2t]$, let ℓ_j be a vertical line that lies to the left of $\mathcal{S}_j$, but right of any other $\mathcal{S}_{j'}$, $j' < j$. Let $\boldsymbol{\ell}_j = \ell_j \cap \left(\bigcap_{S_i \in \mathcal{S}_j} C_i^-\right)$, and note that $\boldsymbol{\ell}_j$ is a downward halfline. Also, for any point $p_j \in \boldsymbol{\ell}_j$ and any $(k-1)$-strip $S_i \in \mathcal{S}_j$, the set $S_i \cup \{p_j\}$ is a k-strip with respect to the point set $S \cup \{p_j\}$; see Fig. 2.

In the next $2t$ rounds, Maker places arbitrary points $p_j \in \boldsymbol{\ell}_j$ for all $j \in [2t]$ such that the set of all points is in general position. For every $j \in [2t]$, point p_j creates $2s$ *candidates* for possible k-strips: $S_i \cup \{p_j\}$ for all $S_i \in \mathcal{S}_j$.

Meanwhile, Breaker plays $2st$ points. By the pigeonhole principle, there are at most t indices $j \in [2t]$ such that Breaker placed $2s$ or more points in $\mathcal{Q}_j$. Consequently, there are at least t indices $j \in [2t]$ such that Breaker placed fewer than $2s$ points in $\mathcal{Q}_j$. We show that each such group yields at least one k-strip.

Consider a group $\mathcal{S}_j$ such that Breaker placed at most $2s-1$ points in $\mathcal{Q}_j$. By the pigeonhole principle, there is a $(k-1)$-strip $S_i \in \mathcal{S}_j$ such that Breaker did not place any points in the region Q_i. If Breaker has not placed any points in $\operatorname{conv}(S_i \cup \{p_j\}) + \boldsymbol{d}$, either, then $S_i \cup \{p_j\}$ is a k-strip at this time. Otherwise, let C_i^* be the cone swept by rotating the halfline $L(s_{i,1})$ clockwise about $s_{i,1}$ until it passes through one of Breaker's points in $\operatorname{conv}(S_i \cup \{p_j\}) + \boldsymbol{d}$: let q be the first such point. In this case, $S_i \cup \{q\}$ is a k-strip at this time.

So in each group $\mathcal{S}_i$, there is a k-strip, adding to a total of t strips, as required.

Analyzing the Number of Rounds. Let $r(k,t)$ denote the minimum number of rounds such that Maker can ensure that at the end of round $r(k,t)$, there are t parallel k-strips among the $(s+1)r(k,t)$ points placed by both players. Clearly, we have $r(1,t) = t/(s+1)$ and $r(2,t) = 2t/(s+1)$ for all $t \in \mathbb{N}$; and we proved the recurrence relation $r(k,t) = r(k-1, 4st) + 2t$ for all $k \geq 3$ and $t \in \mathbb{N}$. The recursion solves to $r(k,t) = ((4s)^{k-2}/(s+1) + ((4s)^{k-2}-1)/(4s-1))2t$ for $k \geq 2$ when s is a positive integer. □

3 Bichromatic Maker-Breaker Game with Bias

In this section, we consider the bichromatic Maker-Breaker game with $1:s$ bias, first with $s < 2$ and later for $s \geq 2$. We show that in the first case, Maker can win; while for sufficiently large s (depending on k) Breaker can win the game.

Theorem 3. *The bichromatic Maker-Breaker game with* $1:s$ *bias, for any rational constant* $s < 2$ *and for any integer* $k \geq 3$*, is a win for Maker.*

We start by making a straightforward observation about how strips can be slightly changed in direction to avoid having a point in their interiors. This is now necessary since earlier, Breaker's points could also be used to extend a strip, while now this is no longer the case. Moreover, given several disjoint downward strips, rotate the downward halfline so that they remain parallel. Let $\operatorname{conv}(P)+\boldsymbol{d}$ and $\operatorname{conv}(Q)+\boldsymbol{d}$ be downward strips disjoint from a point set S. Then there exists a $\delta_0 > 0$ such that for all $\delta \in (-\delta_0, \delta_0)$ and the halfline $\boldsymbol{v} = \overrightarrow{(\delta, -1)}$, the following hold:

- both $\operatorname{conv}(P) + \boldsymbol{v}$ and $\operatorname{conv}(Q) + \boldsymbol{v}$ are strips disjoint from S;
- if $(\operatorname{conv}(P)+\boldsymbol{d}) \cap (\operatorname{conv}(Q)+\boldsymbol{d}) = \emptyset$, then $(\operatorname{conv}(P)+\boldsymbol{v}) \cap (\operatorname{conv}(Q)+\boldsymbol{v}) = \emptyset$;
- if $\operatorname{conv}(P) \cap \operatorname{conv}(Q) = \{s\}$ and $(\operatorname{conv}(P) + \boldsymbol{d}) \cap (\operatorname{conv}(Q) + \boldsymbol{d}) = \{s\} + \boldsymbol{d}$, then $(\operatorname{conv}(P) + \boldsymbol{v}) \cap (\operatorname{conv}(Q) + \boldsymbol{v}) = \{s\} + \boldsymbol{v}$.

Now we can prove the theorem.

Proof of Theorem 3. Let $\varepsilon = 2 - s$, that is, we assume $1:(2-\varepsilon)$ bias in favor of Breaker. We prove, by induction on k, that for all $k, r \in \mathbb{N}$, Maker can build r parallel k-strips disjoint from Breaker's points. The initial cases $k = 1, 2$ are easy, so we assume that $k \geq 3$ and Maker has built $2t$ parallel $(k-1)$-strips, where $t := \lceil r/\varepsilon \rceil$. As before, without loss of generality, we may assume that these are downward strips $S_1, \ldots, S_{2t}$, labeled in increasing order by x-projections of their convex hulls. Let S be the set of points played so far. We label the points in S_i by $s_{i,1}, \ldots, s_{i,k-1}$.

For each $j \in [2t-1]$, let $\boldsymbol{\ell}_j$ be a downward pointing halfline between S_j and S_{j+1}, similarly defined as above. Any point $p_j \in \boldsymbol{\ell}_j$ has the property that $S_j \cup \{p_j\}$ and $S_{j+1} \cup \{p_j\}$ are also downward strips disjoint from S. Note that for $i \neq j$, the strips $S_i \cup \{p_i\}$ and $S_{i+1} \cup \{p_i\}$ are disjoint from the strips $S_j \cup \{p_j\}$ and $S_{j+1} \cup \{p_j\}$.

By Observation 3, we find $\delta_0 > 0$ so that for all $\delta \in (-\delta_0, \delta_0)$ and for halfline $\boldsymbol{v} = \overrightarrow{(\delta, -1)}$, the regions $\text{conv}(S_i \cup \{p_i\}) + \boldsymbol{v}$ and $\text{conv}(S_{i+1} \cup \{p_i\}) + \boldsymbol{v}$ satisfy the following:

1. both $\text{conv}(S_i \cup \{p_i\}) + \boldsymbol{v}$ and $\text{conv}(S_{i+1} \cup \{p_i\}) + \boldsymbol{v}$ are strips;
2. $\text{conv}(S_i \cup \{p_i\}) + \boldsymbol{v}$ and $\text{conv}(S_{i+1} \cup \{p_i\}) + \boldsymbol{v}$ intersect only in the line $p_i + \boldsymbol{v}$;
3. for $i \neq j$, the strips $\text{conv}(S_i \cup \{p_i\}) + \boldsymbol{v}$ and $\text{conv}(S_{i+1} \cup \{p_i\}) + \boldsymbol{v}$ are disjoint from the strips $\text{conv}(S_j \cup \{p_j\}) + \boldsymbol{v}$ and $\text{conv}(S_{j+1} \cup \{p_j\}) + \boldsymbol{v}$; and
4. the strips $\text{conv}(S_i \cup \{p_i\}) + \boldsymbol{v}$ and $\text{conv}(S_{i+1} \cup \{p_i\}) + \boldsymbol{v}$ are disjoint from S.

Maker plays the points $p_1, \ldots, p_t$. During these moves Breaker plays points $b_1, \ldots, b_{(2-\epsilon)t}$. Pick a particular $\delta \in (-\varepsilon, \varepsilon)$ such that none of $b_1, \ldots, b_{(2-\epsilon)t}$ is on any line in direction $(\delta, -1)$ through $p_1, \ldots, p_t$; this can be done because there are finitely many points to avoid, but infinitely many choices for $\boldsymbol{v} = \overrightarrow{(\delta, -1)}$. This, in particular, ensures that each b_i is in at most one of the regions $\text{conv}(S_i \cup \{p_i\}) + \boldsymbol{v}$, $\text{conv}(S_{i+1} \cup \{p_i\}) + \boldsymbol{v}$, for $i = 1, \ldots, t$ (using (2) and (3)). By the pigeonhole principle, we can find at least $\varepsilon t = \varepsilon \cdot \lceil r/\varepsilon \rceil \geq r$ disjoint strips whose downward regions do not contain any Breaker points, as required. □

Note that for $s = 2$, Maker's strategy that we described no longer works. Indeed, in the inductive step, Breaker can place points to the left and right of each of the t points placed by Maker (with respect to the projection on x-coordinate) making sure that each k-strip contains one of her points either in the interior or as a vertex; see Fig. 3. We shall prove below that for $k \geq 8$, Breaker can win with speed $s \geq 12$, but it remains an open problem to determine the minimum speed that allows Breaker to win.

3.1 One-Round Bichromatic Game with Bias in Favor of Breaker

In the remainder of this section, we consider the case with $s \geq 2$. We start with an easy observation about the one-round bichromatic game.

Proposition 1. *For every $k, n \geq 3$, Breaker wins in the bichromatic one-round Maker-Breaker game with bias* $1:2$.

A proof is deferred to the full paper [15]. We give a brief summary here. In this game, Maker first places n points, and then Breaker can place $2n$ points. Breaker puts two points sufficiently close to every Maker point (above and below, but in general position). We show that each triangle Δabc spanned by Maker's points, with x-coordinates $a_1 < b_1 < c_1$, contains Breaker's point above or below b.

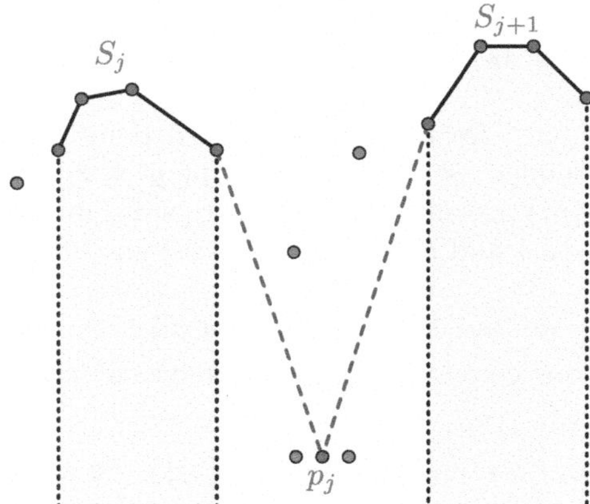

Fig. 3. If Breaker has a bias $s \geq 2$ in the bichromatic game, Maker's strategy to create k-strips no longer works. Breaker can place one point left and right of p_j.

3.2 Multi-round Bichromatic Game with Bias in Favor of Breaker

In this subsection, we address the multi-round bichromatic Maker-Breaker game, and show that Breaker wins with sufficient bias s in her favor. Breaker follows the basic strategy in Proposition 1, that is, placing points in some δ-neighborhood of Maker's points. However, the suitable value of δ may decrease as Maker adds new points. In the algorithm below, Breaker chooses $\delta_i > 0$ based on the first i points played by Maker. In round i, Breaker places q points in a δ_i-neighborhood of Maker's point and can choose up to one previous point p_j, $j < i$, played by Maker, and places new points in the smaller δ_i-neighborhood of p_j.

Breaker's Strategy for $s = 2\lambda$. Let $M_i = \{p_1, \dots, p_i\}$ denote the first i points placed by Maker. Breaker maintains a set $\mathcal{D}_i = \{D_1, \dots, D_i\}$ of pairwise disjoint closed disks, where D_j is centered at p_j for all $j \in \{1, \dots, i\}$. In round i, Maker first plays a new point p_i, and then Breaker plays up to 2λ points as follows. Breaker chooses two real numbers $0 < r_i < R_i$ (specified below), then creates a disk D_i of radius R_i and a circle C_i of radius r_i centered at p_i, and places λ points at the vertices of a regular λ-gon inscribed in C_i; followed by a perturbation that moves the points on the circle C_i to ensure that no three points are collinear. Specifically, when λ is odd, rotate the regular λ-gon such that none of its vertices is on a line spanned by p_i and a previous point. However, if λ is even, then p_i is the midpoint of the main diagonals of the regular λ-gon and we need to perturb the points more carefully.

If λ is even, we choose λ points on the circle C_i as follows. In round 1 (i.e., $i = 1$), we can place the λ points on C_1 arbitrarily in general position. For $i \geq 2$, let $\varepsilon_i \in (0, \frac{\pi}{10\lambda})$ be less than half of the smallest nonzero angle in $\{\angle p_x p_i p_y \mod \frac{2\pi}{\lambda} : 1 \leq x, y \leq i\}$. Initially, let $(q_1^0, \dots, q_\lambda^0)$ be the vertices of a regular λ-gon inscribed in C_i such that $\angle q_j^0 p_i p_{i'} > \varepsilon_i$ for all $1 \leq j \leq \lambda$ and

$1 \le i' < i$. See Fig. 4 for an illustration. For $j = 1, \ldots, \lambda$, we perturb q_j^0 to points q_j as follows: rotate q_j^0 along C_i counterclockwise by angle $\frac{1}{j}\varepsilon_i$. Note that the segments $q_j^0 q_j$ do not cross any line $p_i p_{i'}$, $i' < i$, and the perturbation guarantees $\angle q_j p_i q_{j+1} \le 2\pi/\lambda$ for $j = 1, \ldots, \lambda - 1$, and $2\pi/\lambda < \angle q_\lambda p_i q_1 < 2\pi/\lambda + \varepsilon_i$.

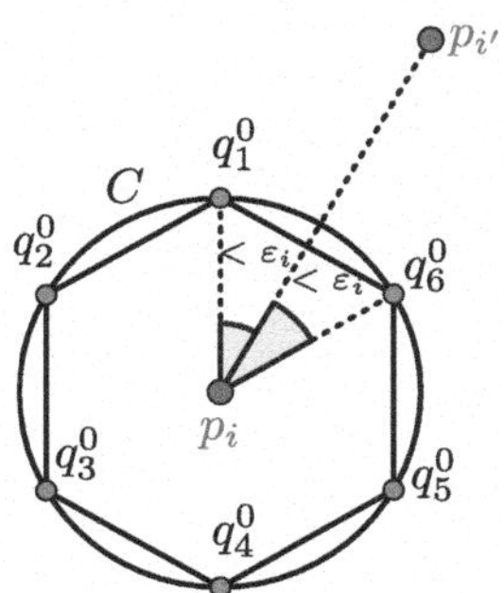

Fig. 4. Illustration on how to set up the to-be-perturbed λ-gon for $\lambda = 6$ and one point $p_{i'}$. We choose the points $(q_1^0, \ldots, q_\lambda^0)$ such that they are the vertices of a regular 6-gon and such that $\angle q_j^0 p_i p_{i'} > \varepsilon_i$ for all $j \in [\lambda]$. For the subsequent perturbation, we move q_j^0 counterclockwise along C by angle ε_i/j for all $j = 1, \ldots, \lambda$.

Furthermore, if $p_i \in D_j$ for some $j < i$, then Breaker chooses new values for r_j and R_j such that $0 < r_j < R_j$, and replaces the disk D_j and the circle C_j; and places λ new points on the new circle C_j (the new disk and circle will still be denoted by D_j and C_j, respectively). Overall, Breaker places at most $s = 2\lambda$ points in round i.

It remains to specify the radii $0 < r_i < R_i$, and the possible replacement radii $0 < r_j < R_j$. We first specify R_i and R_j. If $p_i \notin D_j$ for any $j < i$, then let R_i be any positive real such that D_i is disjoint from all previous disks D_j, $j < i$. Otherwise, $p_i \in D_j$ for a unique $j \in \{1, \ldots, i-1\}$. In this case, we set $R_j := |p_i p_j|/2$, so that the new disk D_j of radius R_j centered at p_j does not contain p_i, and then let $R_i > 0$ be any real such that D_i is disjoint from all previous disks D_j, $j < i$.

For choosing the radius r_i (and r_j if necessary), we proceed as follows. For all $a \in \{p_1, \ldots, p_{i-1}\}$, let $c_i(a)$ be the intersection point of the line ap_i and the boundary of D_i such that $\angle ap_i c_i(a) = \pi$; see Fig. 5. Now let $r_i > 0$ be sufficiently small so that the circle of radius r_i centered at p_i is disjoint from the lines $bc_i(a)$ and ab for all $\{a, b\} \subset \{p_1, \ldots, p_{i-1}\}$. In case the disk D_j has been replaced, then we recompute the points $c_j(a)$ for all $a \in \{p_1, \ldots, p_{j-1}\}$. This completes the description of Breaker's strategy for $s = 2\lambda$.

We prove in the full paper [15] that this is a winning strategy for Breaker.

Theorem 4. *In the bichromatic* $1:12$ *Maker-Breaker game with* $k \ge 8$, *Breaker has a winning strategy.*

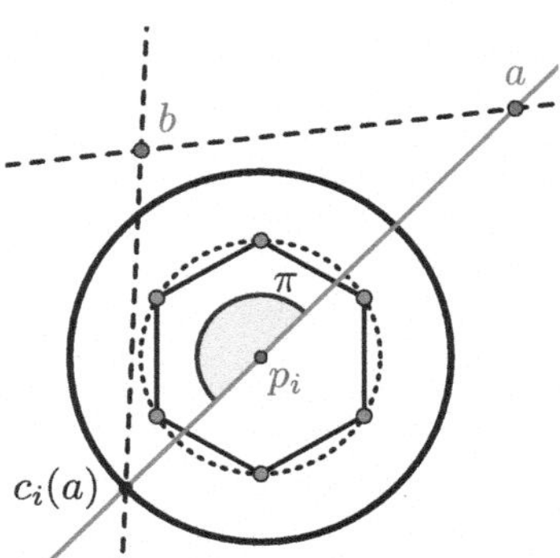

Fig. 5. Example on how to define $c_i(a)$ and r_i for $\lambda = 6$. We pick r_i such that lines $c_i(a)b$ and ab are both disjoint from the circle centered around p_i with radius r_i. The blue points represent Maker's points, while the orange points are Breaker's points. (Color figure online)

The proof of Theorem 4 generalizes and yields a trade-off between k and s, for all $s \geq 6$. We prove the following theorem in the full version of the paper [15].

Theorem 5. *For every integer $\lambda \geq 3$, Breaker wins the bichromatic Maker-Breaker game with $1 : 2\lambda$ bias if $k \geq 3\lceil 2\lambda/(\lambda - 2)\rceil - 1$. In particular, if $k \geq 17$ for $\lambda = 3$; $k \geq 11$ for $\lambda \in \{4, 5\}$; and $k \geq 8$ for $\lambda \geq 6$.*

3.3 One-Round Bichromatic Game with Bias in Favor of Maker

In contrast to Proposition 1, we show that Maker wins the one-round bichromatic game if she has a bias.

Theorem 6. *For every rational $\varepsilon > 0$ and integer $k \geq 3$, Maker wins in the bichromatic one-round Maker-Breaker game with bias $1 : (2 - \varepsilon)$.*

Maker's winning strategy uses the density Hales-Jewett theorem by Furstenberg and Katznelson [18] (see also [14,20,31] for simplified proofs). In the statement of the theorem below, we use the notation $[t] = \{1, 2, \ldots, t\}$.

Theorem 7. (Density Hales-Jewett Theorem *[18]***).** *For every integer $t \geq 2$ and every $\delta \in (0, 1]$, there exists an integer d_0 with the following property. If $R \subseteq [t]^{d_0} \subset \mathbb{Z}^{d_0}$ and $|R|/t^{d_0} \geq \delta$, then R contains t collinear points that form a combinatorial line, which means that we can form a sequence form the t points where each coordinate is either increasing, or constant.*

The proof of Theorem 6 is deferred to the full paper [15]; we give a brief summary. Maker projects the points of the high-dimensional grid $[t]^d$ to a generic plane, and carefully perturbs them such that some of the collinear t-sets become t-holes (w.r.t. Maker's points). Generic projections of a high-dimensional grid have previously been used for other problems in combinatorial geometry [5,

29]. Our contribution is an intricate perturbation that takes combinatorial lines to t-holes: We first create three "bundles" of collinear lines, the bundles are each almost parallel to three distinct directions (at angular distance $2\pi/3$ apart), and then perturb each bundle independently to bundles of parabolas. Each t-set on a parabola in a bundle forms a long and thin t-hole. Importantly, a typical Maker point p participates in such t-holes in three different directions, so Breaker needs at least two points in some small δ-neighborhood of p to block p from at least one t-hole in each bundle. Furthermore, any Breaker point that is not in a δ-neighborhood of a Maker point can block at most two such long and thin t-holes. Finally, an easy averaging argument shows that, with $1 : (2 - \varepsilon)$ bias, sufficiently many Maker points "survive" (in the sense that they each have at most one Breaker point in their δ-neighborhood) and can form a $\frac{t+1}{2}$-hole.

4 Conclusions and Open Problems

We have considered several Maker-Breaker games that arose from the Erdős-Szekeres problem in combinatorial geometry. Firstly, we considered the monochromatic game. In this game, Breaker's points can contribute to the k-hole that Maker is trying to obtain. We have shown that if both players play with unit speed (i.e., they each play one point per round), Maker always has a winning strategy. Furthermore, even if Breaker is allowed to play with increased (but constant) speed, Maker still wins. Due to a result of Conlon and Lim [10], we know that there exists a sufficiently fast growing function $f : \mathbb{N} \to \mathbb{N}$ such that if Breaker plays $f(i)$ points in round i, she wins; but determining the minimum growth rate of such a function f is an intriguing open problem.

The outcome of the game is less somber for Breaker whenever we consider the bichromatic version of the Maker-Breaker game. In this version, Breaker's points can only destroy k-holes and cannot contribute to them. When both players have low speed, Maker still wins. This is the case whenever the game is played with bias $1 : s$ for any constant $s < 2$ and $k \geq 3$. But if Breaker's speed is high enough, then she is able to win. We showed that for $k \geq 8$ and bias $1 : 12$ for Maker and Breaker, the latter has a winning strategy. This raises an interesting problem: What is the minimum speed $s(k)$ required for Breaker to win the $1 : s(k)$ game? We know that $2 \leq s(k) \leq 12$ for $k \geq 8$, and $2 \leq s(k) \leq 6$ for $k \geq 17$. However, we do not even know whether Breaker can win with any constant speed $s(k) > 1$ when $3 < k < 8$. Clearly, $s(k)$ is nonincreasing in k, but it remains an open problem to determine $\lim_{k\to\infty} s(k)$, that is, the limit of speeds that allow Breaker to prevent k-holes for sufficiently large values of k. Perhaps a first step to address these problems is the observation that Breaker wins the one-round game whenever she has speed two. Whether that also holds for multiple round games is still open, though. The geometric arguments that support Maker's winning strategy using k-strips with bias $1 : s(k)$, for $s(k) < 2$, no longer work for $s(k) \geq 2$. Showing that Maker can still win when Breaker is playing with speed higher than two seems to require novel ideas and a new strategy. In contrast, in the one-round game, when the respective speeds for Maker and Breaker are

1 and $2 - \varepsilon$, Maker has a winning strategy based on the Density Hales-Jewett Theorem.

Lastly, another interesting research direction is to determine how efficient the above winning strategies are for Maker. These proposed strategies give an exponential upper bound on the number of rounds required to form a k-hole. It is still unclear whether there exist strategies that allow Maker to win in fewer rounds. Proving the optimality of a strategy, and thus proving the exact number of points required, seems to be an even more challenging open problem.

Acknowledgments. Collaboration on this project started at the *Novi Sad Workshop on Foundations of Computer Science* (*NSFOCS*), held June 22-26, 2024, in Novi Sad, Serbia. The authors thank Mirjana Mikalački, Miloš Stojaković, Marko Savić and Jelena Stratijev for their hospitality and for organizing this successful workshop.

Research on this project was supported in part by the NRDI EXCELLENCE-24 grant no. 151504 Combinatorics and Geometry, the ERC Advanced Grant no. 101054936 ERMiD (Pálvögyi), the NSF award DMS-2154347 (Tóth), the Czech Science Foundation Grant no. 24-12046S (Valla), and the New Zealand Marsden Fund (Verlinde).

Disclosure of Interests. The authors have no competing interests to declare that are relevant to the content of this article.

References

1. Aichholzer, O., et al.: A superlinear lower bound on the number of 5-holes. J. Comb. Theory A **173**, 105236 (2020)
2. Aichholzer, O., et al.: Games on triangulations. Theoret. Comput. Sci. **343**(1–2), 42–71 (2005)
3. Aichholzer, O., Díaz-Báñez, J.M., Hackl, T., Martín, D.O., Pilz, A., Ventura, I., Vogtenhuber, B.: Erdős-Szekeres-type games. In: EuroCG, pp. 23:1–7 (2019)
4. Aichholzer, O., et al.: On k-gons and k-holes in point sets. Comput. Geom. **48**(7), 528–537 (2015)
5. Alon, N.: A non-linear lower bound for planar epsilon-nets. Discr. Comput. Geom. **47**(2), 235–244 (2012)
6. Bae, S.W.: Faster counting empty convex polygons in a planar point set. Inf. Process. Lett. **175**, 106221 (2022)
7. Balko, M., Scheucher, M., Valtr, P.: Tight bounds on the expected number of holes in random point sets. Random Struct. Algorithms **62**(1), 29–51 (2023)
8. Bárány, I., Valtr, P.: Planar point sets with a small number of empty convex polygons. Stud. Sci. Math. Hung. **41**(2), 243–266 (2004)
9. Cano, J., Olaverri, A.G., Hurtado, F., Sakai, T., Tejel, J., Urrutia, J.: Blocking the k-holes of point sets in the plane. Graphs Comb. **31**(5), 1271–1287 (2015)
10. Conlon, D., Lim, J.: Fixing a hole. Discr. Comput. Geom. **70**, 1551–1570 (2023)
11. Czyzowicz, J., Kranakis, E., Urrutia, J.: Guarding the convex subsets of a point-set. In: CCCG. Fredericton, NB (2000)
12. Das, A.K., Valla, T.: On Erdős-Szekeres maker-breaker games. In: CCCG (2024). https://cosc.brocku.ca/~rnishat/CCCG_2024_proceedings.pdf
13. Dobkin, D.P., Edelsbrunner, H., Overmars, M.H.: Searching for empty convex polygons. Algorithmica **5**(4), 561–571 (1990)

14. Dodos, P., Kanellopoulos, V., Tyros, K.: A simple proof of the density Hales-Jewett theorem. Int. Math. Res. Not. **2014**, 3340–3352 (2014)
15. Džuklevski, A., Pálvölgyi, D., Pokrovskiy, A., Tóth, C.D., Valla, T., Verlinde, L.: Erdős-Szekeres maker-breaker games. In preparation, available on arXiv (2025)
16. Erdős, P.: Some applications of graph theory and combinatorial methods to number theory and geometry. In: Algebraic Methods in Graph Theory, vol I, II (Szeged, 1978), pp. 137–148. North-Holland, Amsterdam (1981)
17. Erdős, P., Szekeres, G.: A combinatorial problem in geometry. Compos. Math. **2**, 463–470 (1935)
18. Furstenberg, H., Katznelson, Y.: A density version of the Hales-Jewett theorem. Journal d'Analyse Mathématique **57**, 54–119 (1991)
19. Gerken, T.: Empty convex hexagons in planar point sets. Discr. Comput. Geom. **39**(1–3), 239–272 (2008)
20. Gowers, W.T.: Polymath and the density Hales-Jewett theorem. In: An Irregular Mind, Bolyai Society Mathematical Studies, vol. 21, pp. 659–687. Springer (2010)
21. Harborth, H.: Konvexe Fünfecke in ebenen Punktmengen. Elem. Math. **33**, 116–118 (1978)
22. Hefetz, D., Krivelevich, M., Stojaković, M., Szabó, T.: Positional Games. OS, vol. 44. Springer, Basel (2014). https://doi.org/10.1007/978-3-0348-0825-5_9
23. Heule, M., Scheucher, M.: Happy ending: An empty hexagon in every set of 30 points. In: TACAS. LNCS, vol. 14570, pp. 61–80. Springer, Cham (2024)
24. Holmsen, A.F., Mojarrad, H.N., Pach, J., Tardos, G.: Two extensions of the Erdős-Szekeres problem. J. Eur. Math. Soc. **22**(12), 3981–3995 (2020)
25. Horton, J.D.: Sets with no empty convex 7-gons. Canad. Math. Bull. **26** (1983)
26. Kolipaka, P., Govindarajan, S.: Two player game variant of the Erdős-Szekeres problem. Discret. Math. Theor. Comput. Sci. **15**(3), 73–100 (2013)
27. Mitchell, J., Rote, G., Sundaram, G., Woeginger, G.J.: Counting convex polygons in planar point sets. Inf. Process. Lett. **56**(1), 45–49 (1995)
28. Nicolás, C.: The empty hexagon theorem. Disc. Comput. Geom. **38**, 389–397 (2007)
29. Pach, J., Tardos, G., Tóth, G.: Indecomposable coverings. Canad. Math. Bull. **52**(3), 451–463 (2009)
30. Pinchasi, R., Radoičić, R., Sharir, M.: On empty convex polygons in a planar point set. J. Comb. Theory A **113**(3), 385–419 (2006)
31. Polymath, D.: A new proof of the density Hales-Jewett theorem. Ann. Math. **175**, 1283–1327 (2012)
32. Sakai, T., Urrutia, J.: Covering the convex quadrilaterals of point sets. Graphs Combinatorics **23**, 343–357 (2007)
33. Suk, A.: On the Erdős-Szekeres convex polygon problem. J. AMS **30**(4), 1047–1053 (2017)
34. Valla, T.: Ramsey theory and combinatorial games. Master's thesis, Charles University, Prague (2006)

Minimum Membership Geometric Set Cover in the Continuous Setting

Sathish Govindarajan(✉), Mayuresh Patle, and Siddhartha Sarkar

Indian Institute of Science, Bengaluru, India
gsat@iisc.ac.in

Abstract. We study the minimum membership geometric set cover, i.e., MMGSC problem [SoCG, 2023] in the continuous setting. In this problem, the input consists of a set P of n points in $\mathbb{R}^2$, and a geometric object t, the goal is to find a set $\mathcal{S}$ of translated copies of the geometric object t that covers all the points in P while minimizing $\mathsf{memb}(P, \mathcal{S})$, where $\mathsf{memb}(P, \mathcal{S}) = \max_{p \in P} |\{s \in \mathcal{S} : p \in s\}|$.

For unit squares, we present a simple $O(n \log n)$ time algorithm that outputs a 1-membership cover. We show that the size of our solution is at most twice that of an optimal solution. We establish the NP-hardness on the problem of computing the minimum number of non-overlapping unit squares required to cover a given set of points. This algorithm also generalizes to fixed-sized hyperboxes in d-dimensional space, where an 1-membership cover with size at most 2^{d-1} times the size of a minimum-sized 1-membership cover is computed in $O(dn \log n)$ time. Additionally, we characterize a class of objects for which a 1-membership cover always exists. For unit disks, we prove that a 2-membership cover exists for any point set, and the size of the cover is at most 7 times that of the optimal cover. For arbitrary convex polygons with m vertices, we present an algorithm that outputs a 4-membership cover in $O(n \log n + nm)$ time.

Keywords: Computational Geometry · Minimum-Membership Geometric Set Cover · Minimum Ply Covering · Approximation Algorithms

1 Introduction

The minimum membership geometric set cover problem has attracted significant interest in computational geometry due to its relevance in applications such as wireless networks, where minimizing interference is crucial [2,3,9,10,14]. Traditionally, much of the research on set cover problems has focused on discrete settings, where both the points to be covered and the covering objects are confined to predefined positions. However, many real-world scenarios require continuous flexibility in the placement of covering objects, leading to the study of the continuous variant of the problem.

The continuous geometric set cover problem for unit disks is a well-studied, classical problem in computational geometry. Its objective is to cover a given

F. V. Fomin and M. Xiao (Eds.): COCOON 2025, LNCS 15983, pp. 262–276, 2026.
https://doi.org/10.1007/978-981-95-0215-8_20

set of points with the smallest number of unit disks. In particular, a well-known PTAS exists for this problem based on the Hochbaum-Maass shifting strategy [15]. Furthermore, there are numerous practical, fast constant-factor approximation algorithms [5,6,11,13].

Definition 1 *(Ply). Given a set $\mathcal{S}$ of geometric objects, the ply of $\mathcal{S}$, denoted by* $\mathsf{ply}(\mathcal{S})$*, is* $\max_{p \in \mathbb{R}^2} |\{s \in \mathcal{S} : p \in s\}|$.

Definition 2 *(Membership). Given a set P of points and a set $\mathcal{S}$ of geometric objects, the membership of P with respect to $\mathcal{S}$, denoted by* $\mathsf{memb}(P, \mathcal{S})$*, is* $\max_{p \in P} |\{s \in \mathcal{S} : p \in s\}|$.

1.1 Our Contribution

The Minimum-Membership Geometric Set Cover (MMGSC) problem is well studied in the discrete setting where both points and objects are given as input. In this paper, we initiate the study of the minimum membership geometric set cover (MMGSC) problem in the continuous setting. In this problem, the input consists of a set P of n points in $\mathbb{R}^2$, and a geometric object t, the goal is to find a set $\mathcal{S}$ of translated copies of t that covers all the points in P, while minimizing $\mathsf{memb}(P, \mathcal{S})$, where $\mathsf{memb}(P, \mathcal{S}) = \max_{p \in P} |\{s \in \mathcal{S} : p \in s\}|$. We present the following results on this problem.

1. 1-membership Hypercube Cover: For unit intervals in one dimension, we give an exact algorithm that constructs a 1-membership cover in $O(n \log n)$ time. Using this algorithm, we construct a 1-membership cover for unit squares, and show that the size of the cover is a 2-approximation to the optimum size. This algorithm also generalizes to (translates of) hyperboxes in d-dimension, where a 1-membership cover with size at most 2^{d-1} times the size of a minimum-sized 1-membership cover is computed in $O(dn \log n)$ time.
2. We show that the problem of computing the minimum-size 1-membership unit square cover is NP-hard by a reduction from PLANAR3SAT.
3. We show that a 1-membership cover can be constructed if the geometric object t tiles the plane. Moreover, for objects that do not tile the plane we show a point set for which a 1-membership cover does not exist.
4. For unit disks, we construct a 2-membership cover, and show that the size of the cover is a 7-approximation to the optimum.
5. For convex polygons, we leverage homothetic approximations to achieve a 4-membership cover.

In this paper, we prove the bounds on ply, which implies the same bounds for membership. For example, in Sect. 2, we construct a 1-ply hypercube cover. Since a 1-ply cover is a set of non-overlapping objects, the membership of any point is at most 1. Hence, this cover is a 1-membership cover.

1.2 Related Work

Minimum Membership Geometric Set Cover (MMGSC) problem in the discrete setting (both points and objects are given as input) was introduced by Erlebach et al. [10], who presented NP-hardness and approximation results. A related problem is Minimum Ply Geometric Set Cover (MPGSC), introduced by Biedl et al. [3]. They prove that the problem is NP-hard for unit squares and unit disks. Also, they gave a polynomial-time 2-approximation when the minimum ply for the instance is a constant. Durocher et al. presented the first constant approximation algorithm for the MPGSC problem with unit squares [9]. Bandyapadhyay et al. introduced the *generalized* MMGSC problem, which is a generalization of both MMGSC and MPGSC. They gave a polynomial-time 144-approximation algorithm for unit squares [2]. Govindarajan and Sarkar later improved the approximation factor to 16 [14]. The Unique Coverage problem is another related problem, which was introduced by Demaine et al. for the set systems [8]. Erlebach and van Leeuwen gave the first set of results for geometric unique coverage with unit squares and unit disks [10]. They showed that the unique coverage of unit disks is NP-hard and presented an 18-approximation algorithm with $O(n^3m^8)$ runtime. For unit squares, they gave a 4-approximation algorithm. Later, van Leeuwen established NP-hardness of the unique coverage of unit squares as well, and gave a 2-approximation algorithm for the problem [19].

2 1-Ply Hypercube Cover

In this section, we construct a 1-ply cover for hypercubes in d dimension using the 1-ply cover in $(d-1)$ dimension.

2.1 Unit Interval Cover

First, we consider the setting in one dimension where P is a set of points on the x-axis. We show that a set S of disjoint unit-length intervals that cover all points in P can always be found.

Lemma 1. *Given a set P of n points on the x-axis, a 1-ply cover with minimum number of unit intervals can be computed in $O(n \log n)$ time.*

Proof. Any two distinct input points on the x-axis are separated by a non-zero distance. We construct a 1-ply interval cover by sweeping the x-axis from left to right. Whenever an uncovered point $p \in P$ is encountered, add a unit interval s to the solution set $\mathcal{S}$, where p is the left endpoint of s. This algorithm, referred to as $\mathsf{Separate}(P)$, produces an ordered set $\mathcal{S}$ of non-overlapping unit intervals covering P, from left to right, in $O(n \log n)$ time. The optimality of this greedy algorithm can be proved by a standard *stay ahead* argument (refer to Chapter 4.1 in [16]). □

2.2 Unit Square Cover

Given a set P of points in the plane, the goal is to produce a set S of axis-aligned unit squares that cover all the points in P while minimizing the ply. Unless stated otherwise, a square refers to a unit square.

Theorem 1. *Given a set P of n points in the plane, a 1-ply unit square cover can be computed in $O(n \log n)$ time.*

Proof. To generate a 1-ply unit square cover for a given set P of n points in the plane, apply the Separate algorithm (defined in the proof of Lemma 1) on the x-coordinates of the points, which distributes the points into non-overlapping vertical strips; see Fig. 1(a). Again, apply the Separate algorithm with unit intervals on the y-coordinates of the points within each strip to split it into squares; see Fig. 1(b).

This algorithm is referred to as SquareCover. Suppose, k vertical strips are generated and the i-th vertical strip contains n_i points such that $\sum_{i\in[k]} n_i = n$. Then the running time of SquareCover, ignoring multiplicative constants, is $n \log n + \sum_{i\in[k]} n_i \log n_i \leq n \log n + \log n \sum_{i\in[k]} n_i = 2 \cdot n \log n$. □

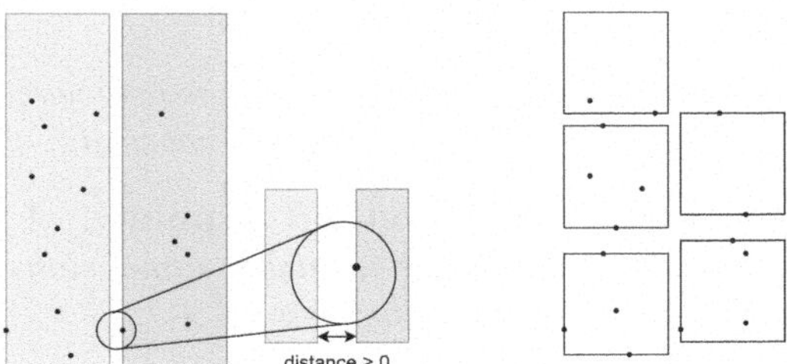

(a) Partitioning into vertical strips. (b) 1-ply unit square cover.

Fig. 1. (a) Partitioning into vertical strips. (b) 1-ply unit square cover.

Theorem 2. *The size of the cover generated by the* SquareCover *algorithm is at most twice the size of any minimum-sized 1-ply unit square cover for a given set P of points in the plane.*

Proof. The Separate algorithm divides the points into vertical strips of width 1. These strips are enumerated from left to right as $V_1, V_2, V_3, \ldots$. Let O be the union of the odd-indexed strips, and E be the union of the even-indexed strips. Since there is a gap between V_i and V_{i+1}, as well as between V_{i+1} and V_{i+2}, any unit square that covers a point in V_i cannot cover a point in V_{i+2}. Therefore, if only the points in O (resp. E) were given, an optimal cover could be found by finding the optimal covers for each strip in O (resp. E) individually. For a vertical strip, the problem reduces to the 1-ply unit interval cover. So applying the Separate algorithm on the y-coordinates of points in a strip gives an optimal cover for a vertical strip. Thus, we have an optimal cover for the points in O (resp. E). Let S_O and S_E be the sets of squares obtained by SquareCover for the points in O and E, respectively. Let OPT be a minimum-sized 1-ply unit square cover for P. Since the set of points contained in O (resp. E) is a subset of P, therefore, $|S_O| \leq |OPT|$, and $|S_E| \leq |OPT|$. Thus $|S_O| + |S_E| \leq 2|OPT|$. Therefore, the set cover $S_O \cup S_E$ is at most twice as large as OPT. □

2.3 Hypercube Cover

Given a set P of points in the d-dimensional space, the goal is to produce a set S of axis-aligned unit hypercubes that cover all points in P while minimizing the ply.

Theorem 3. *Given a set P of n points in d-dimensional space, a 1-ply unit hypercube cover, which is at most 2^{d-1} times the size of any minimum-sized 1-ply cover, can be generated in $O(d \cdot n \log n)$ time.*

This theorem can be derived by generalizing the results of the minimum ply unit square cover problem inductively. We prove two lemmas from which Theorem 3 follows directly. First, we define some terms.

Definition 3 (d-dimensional wall). *For $x \in \mathbb{R}$, an orthogonal range of the form $(-\infty, \infty)^{d-1} \times [x, x+1]$, is called a d-dimensional wall.*

Definition 4 (d-dimensional projection of a point). *A d-dimensional point, obtained by ignoring the coordinates in higher dimensions (if any) while preserving the coordinates in the first d dimensions.*

Lemma 2. *Given a set P of n points in d-dimensional space, a 1-ply unit hypercube cover can be generated in $O(d \cdot n \log n)$ time.*

Proof. Consider the HypercubeCover algorithm that takes the point set P and the number of dimensions d as input and returns a set $\mathcal{S}$ of axis-aligned d-dimensional unit hypercubes that cover the d-dimensional projection of P.

The base case for $d = 2$ generates a 1-ply unit square cover. For $d > 2$, we assume that HypercubeCover$(P, d-1)$ generates a 1-ply $(d-1)$-dimensional unit hypercube cover. We observe that I_d, which is a 1-ply unit interval cover for the dth dimension, assigns each point to its respective wall. For a wall corresponding to $drange \in I_d$, containing subset $P' \subseteq P$ of points, we have a 1-ply cover obtained using HypercubeCover$(P', d-1)$, which is S_{d-1}. Thus, we conclude that $\{box \times drange \mid \forall box \in S_{d-1}\}$ will be a 1-ply unit hypercube cover for the d-dimensional wall. No two walls intersect or touch since I_d is a 1-ply cover. Hence, the combined solution of all walls is a 1-ply hypercube cover for P.

Let $T_d(|P|)$ denote the time complexity of HypercubeCover(P, d). By Theorem 1, we have the base case, $T_2(n) = 2 \cdot n \log n$. Suppose, there are k d-dimensional walls and the i-th one contains n_i points such that $\sum_{i \in [k]} n_i = n$. Each recursive

Algorithm 1: HypercubeCover(P, d)

if $d = 2$ **then**
  Return SquareCover(P)
$\mathcal{S} \leftarrow \emptyset$; ▷ Set of hypercubes in the cover
Sort P in non-decreasing order based on the dth coordinate (for efficient identification of points in the same wall);
$P_d \leftarrow$ Set of dth coordinates of points in P;
$I_d \leftarrow$ Separate(P_d);
for *drange* $\in I_d$ **do**
  $P' \leftarrow$ All points in P having their d^{th} coordinate in *drange*;
  $S_{d-1} \leftarrow HypercubeCover(P', d-1)$;
  for *box* $\in S_{d-1}$ **do**
    Insert *box* $\times$ *drange* into $\mathcal{S}$;
Return $\mathcal{S}$;

invocation of the HypercubeCover algorithm includes a sorting operation, leading to the subsequent recurrence.

$$T_d(n) = n \log n + \sum_{i \in [k]} T_{d-1}(n_i) = d \cdot n \log n \tag{1}$$

The second equality follows from the base case of $d = 2$. Thus, the overall time complexity of HypercubeCover(P, d) is $O(d \cdot n \log n)$, where $|P| = n$. □

Lemma 3. *The cover generated by the* HypercubeCover *algorithm is at most* 2^{d-1} *times the size of any minimum-sized* 1*-ply unit hypercube cover for a given set* P *of points in d-dimensional space.*

Proof. We prove this claim by induction on the number of dimensions d. The claim is true for $d = 2$, as proved in Theorem 2. Assume that the claim is true for $d = i - 1$, and let us prove it for $d = i$.

The algorithm for the unit hypercube cover in d-dimensional space distributes the points into several disjoint walls. Let $W_1, W_2, W_3, \ldots$ denote the walls enumerated in increasing order of their ranges in d^{th} dimension, and let O be the union of odd-indexed walls and E be the union of even-indexed walls.

A wall has d^{th}-dimension range with unit length. Thus, for points in a single wall, the size of the smallest d-dimensional 1-ply hypercube cover is the same as the size of the smallest $(d-1)$-dimensional 1-ply hypercube cover for $(d-1)$-dimensional projections of those points. Thus, by construction and inductive hypothesis, we have a 2^{d-2} approximation for the points in each of the d-dimensional walls. Now, by the separation of walls due to gaps in S_d, and by the presence of a wall W_{i+1} between W_i and W_{i+2}, any hypercube cannot cover two points such that one lies in W_i and the other lies in W_{i+2}. Hence, we have a 2^{d-2} approximation for the points in O and E, respectively.

Let OPT_O and OPT_E be the optimal solutions for points in O and E, respectively, and let OPT be the optimal solution for P. Since P contains points in $O \cup E$, we have $|OPT_O| \leq |OPT|$ and $|OPT_E| \leq |OPT|$.

Let S_O and S_E be the sets of hypercubes generated by the HypercubeCover algorithm for the regions corresponding to O and E, respectively. Thus,

$$|S_O| \leq 2^{d-2}|OPT_O| \leq 2^{d-2}|OPT|, |S_E| \leq 2^{d-2}|OPT_E| \leq 2^{d-2}|OPT|$$
$$\implies |S_O \cup S_E| = |S_O| + |S_E| \leq 2^{d-1}|OPT|.$$

Therefore, the hypercube cover generated by the HypercubeCover algorithm is at most 2^{d-1} times the size of any minimum-sized 1-ply unit hypercube cover. □

Remark. This theorem can be further extended to d-dimensional axis-aligned hyperbox cover, with dimensions $l_1 \times l_2 \times \cdots \times l_d$, by determining the l_i-length interval cover while separating along the ith dimension.

2.4 NP-Hardness of Minimum Size 1-Ply Unit Square Cover

Let us first define the minimum size 1-ply unit square cover problem formally.

Definition 5 (Minimum Size 1-Ply Set Cover of Unit Squares). *Given a set of n points P on $\mathbb{R}^2$, the goal in* MS1P-SC-US *is to cover P with the minimum number of non-overlapping unit squares.*

We prove that the above problem is NP-hard via a reduction from PLANAR3SAT, which is known to be NP-hard [20]. Recall that 3SAT asks whether there exists a truth assignment to the variables of a given 3SAT formula φ that satisfies it. PLANAR3SAT adds a geometric constraint that the variable-clause incidence graph of φ must be planar. There are many ways to embed the incidence graph of a PLANAR3SAT formula on the plane without edge crossings. Knuth and Raghunathan show how the graph can be laid out (in polynomial time) such that variables correspond to points on the x-axis and clauses correspond to non-crossing three-legged "combs" above or below the x-axis [17]. First, place all the variable nodes along a horizontal line, in order. Then, for each clause, connect its three variable nodes with rectilinear (i.e., right-angled) non-crossing line segments. Visually, for each clause, the rectilinear connections form a "three-legged comb". Refer to Fig. 2. See [12] for a relevant reduction.

Theorem 4. MS1P-SC-US *is NP-hard.*

Proof. Given a PLANAR3SAT instance φ with n variables and m clauses, we construct in polynomial time an instance P_φ of MS1P-SC-US such that the following holds: P_φ can be covered by at most k non-overlapping unit squares if and only if φ is satisfiable. We fix k later. First, we place some *guiding* unit squares on the plane that, in turn, decide the points in P_φ. These squares will be colored red or blue. The squares of the same color will be pairwise disjoint. All but m

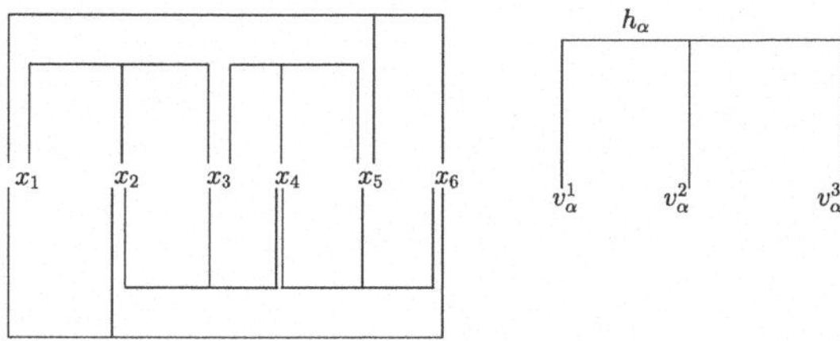

Fig. 2. The left figure shows a planar embedding of a PLANAR3SAT instance. The right figure shows a comb structure for a clause α.

points in P_φ will be placed in the intersection regions of the overlapping unit squares. The placement of the *guiding* unit squares and the points in P_φ is done in steps, leading to some gadgets, as described below.

Variable Gadget. A variable gadget is a horizontal chain of an even number of unit squares. There could be two types of unit squares in a variable gadget, namely, 'variable squares' and 'separating squares'. Let x_i be a variable in φ that appears in k_i clauses. We put $k_i + 2(k_i - 1) = 3k_i - 2$ pairs of unit squares sequentially, forming a horizontal chain, where every two consecutive squares intersect. Essentially, between every two 'variable square' pairs, we have a quadruple of 'separating squares'. To be precise, the chain starts with a pair of 'variable squares'. These are immediately followed by a quadruple of 'separating squares'. The pattern continues as alternating between a pair of 'variable squares' and a quadruple of 'separating squares', until we place k_i pairs of 'variable squares'. The separating squares ensure enough spacing among the vertical chains of squares (to be placed later). See Fig. 3(a).

The unit squares in the chain alternate colors, with the leftmost being red. Two points are placed within each rectangular intersection region of two consecutive squares (i.e., a red and a blue square): one at the bottom-left and the other at the top-right corner. These points are termed the *variable points*. Furthermore, certain variable squares are moved vertically to appropriately connect with the *clause squares*.

Clause Gadget. For every clause α in φ, there is a 3-legged comb, i.e. a horizontal segment h_α and three vertical segments $v_\alpha^1, v_\alpha^2, v_\alpha^3$, in the planar embedding (see Fig. 2). We place squares along $h_\alpha, v_\alpha^1, v_\alpha^2$, and v_α^3 and color them red and blue alternately. Every two adjacent squares intersect. Along h_α, we place two subchains of squares with a gap in the junction of h_α and v_α^2 (i.e., the middle leg of the comb). Thus, a clause gadget has two horizontal chains and three vertical chains of guiding unit squares. These squares are called *clause squares*. See Fig. 3(b). As in the variable gadget, two points are placed within each rectangular intersection region of two intersecting squares. These points are termed as *clause points*. We place a point p_α near the intersection of h_α with v_α^2 in a specific way, to be made precise shortly. The point p_α is called the *central clause point* of α. If a variable x_i appears in its positive form in α, then a blue square

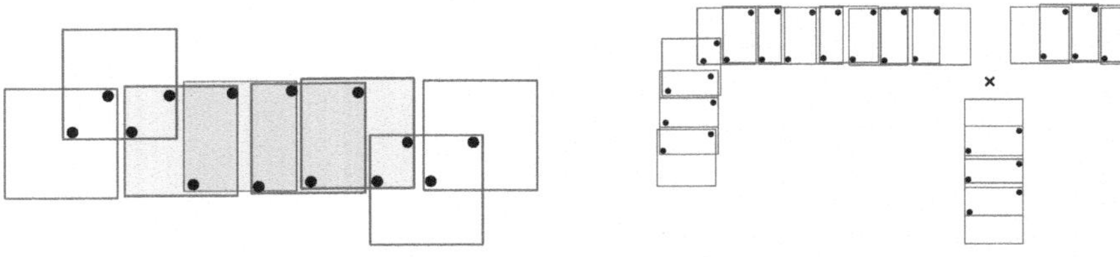

(a) A variable gadget for a variable that appears in two clauses. The *separating* squares are shaded.

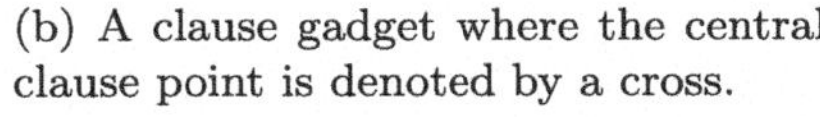

(b) A clause gadget where the central clause point is denoted by a cross.

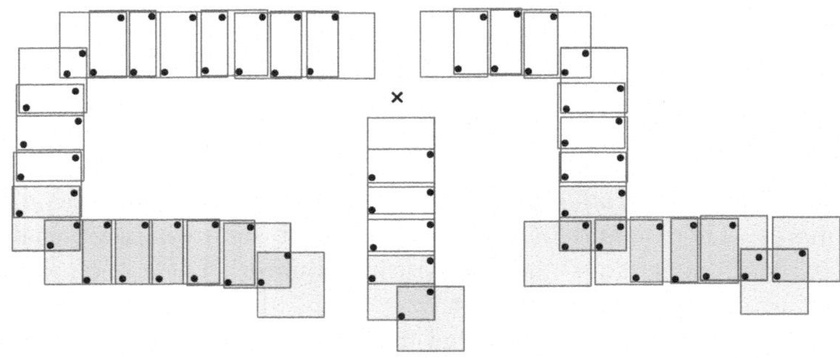

(c) Connecting a clause gadget with the corresponding variable gadgets. The shaded squares correspond to the variable gadgets.

Fig. 3. Depicting a clause $(\neg x_i \vee \neg x_j \vee x_k)$. The variable gadgets for x_i, x_j, x_k are drawn from left to right; x_i appears in its positive form in some other clause; x_j appears only in α; and x_k appears in its negative form in some other clause.

of the variable gadget of x_i intersects a red square of the corresponding vertical chain of α, as shown in Fig. 3(c). If a variable appears in its negated form, then a red square of the variable gadget intersects a blue square of the corresponding vertical chain, as shown in Fig. 3(c). At this point, we do not worry about the exact number of squares in a chain. The *central clause squares* are defined as the three squares nearest to p_α, each located at the ends of the corresponding chains of clause squares for α. We ensure that the positioning of the point p_α and the central clause squares respects the following properties.

- For each clause α, any square (on the plane) containing p_α intersects at least one of the corresponding central clause squares.
- A square containing p_α can be drawn, intersecting only one central clause square and no other guiding squares.

This completes the construction of the set P_φ of points. It is easy to see that the reduction takes polynomial time.

We now prove that the PLANAR3SAT formula φ is satisfiable if and only if there exists a set of at most k non-overlapping unit squares covering P_φ. Here,

$k = m + c$, where m is the number of clauses in φ and c is half the number of guiding squares placed during the reduction.

First, consider the forward direction. Suppose that φ is a satisfiable formula. We place k non-overlapping unit squares that cover the point set P_φ as follows. Consider a satisfying truth assignment δ_φ for φ. For each variable x_i which is set to True according to δ_φ, select the red squares in the corresponding variable gadget. For each of the remaining variables, select the blue squares in the corresponding variable gadget. This, in turn, determines the squares to be chosen from the clause chains. Since every clause α is satisfied, for each clause, at least one literal gets evaluated to True. Hence, the construction ensures that at least one central clause square of α is not selected, leaving enough room for placing a non-overlapping square that covers p_α. Thus, $k = c+m$ non-overlapping squares are selected to cover P_φ.

Now consider the reverse direction. Let the formula φ be a no-instance, i.e., it is not satisfiable. Suppose for a contradiction that $\mathcal{S}$ is a 1-ply unit square cover of P_φ of size at most $k = m + c$. Define the *budget* for a variable or clause as half the number of corresponding guiding squares placed during the reduction. By construction, for each variable x_i, only two distinct square patterns exist that can cover the *variable points* of x_i, while not exceeding the budget for x_i. The same is true for a clause α in φ. Moreover, covering the central clause points in P_φ requires at least m additional unit squares. Since φ is not satisfiable, for any truth assignment, there exists an unsatisfied clause, say, C_i. Since $\mathcal{S}$ must respect the budget for each variable/clause in C_i, to cover the points in C_i within the budget, it is necessary to choose all the three central clause squares of C_i. Hence, there is at least one clause for which the number of unit squares required will exceed the budget. Thus, the number of non-overlapping unit squares required to cover P_φ is strictly more than k. □

3 Minimum Ply Cover Using Tiling Objects

In this section, we characterize the minimum ply cover of objects that tile the plane. An object is called a *tiling object* if the entire plane can be tiled using translated copies of the object. In other words, the plane is an interior-disjoint union of translated copies of the object. Examples of tiling objects are a square and a regular hexagon. We give the following characterization:

Theorem 5. *Given a set P of n points in the plane and an object t, a 1-ply cover of P with translated copies of t exists if and only if t is a tiling object.*

Proof. Let t be a tiling object and consider a tiling of the plane with t. We obtain a 1-ply cover by ensuring that the translated copies of t are boundary disjoint as follows: Perform a uniform expansion of the boundary of t by a small amount $\delta, \delta > 0$ to obtain an expanded object t'. Consider the tiling T' of the plane using t' such that the points of P are at least δ distance away from the boundary of the tiling. Now, shrink the objects t' in tiling T' by an amount δ so that they become translated copies of t. As these translated copies are boundary-disjoint and they cover P, we get a 1-ply cover of P.

To prove the *only if* part, consider an object t that is not a tiling object. Consider a packing of the plane using translated copies of t that minimizes the size s of the smallest hole. The size s of a hole in a packing is defined as the diameter of the largest disk that can be inscribed within the hole. Let P be a grid of points with grid cell size $< s$. P cannot be covered using disjoint translated copies of t. □

We now make an observation on the size of the 1-ply cover of a tiling object t. Consider the 1-ply cover $\mathcal{C}$ of P consisting of non-empty objects of the tiling as constructed in the above proof. Let m_t be the maximum number of objects in $\mathcal{C}$ that can be intersected by a translated copy of t.

Lemma 4. *The size of the* 1*-ply cover* $\mathcal{C}$ *is an* m_t*-approximation to the minimum-sized* 1*-ply cover of* P*.*

Proof. Each object in the minimum-sized 1-ply cover O of P intersects at most m_t objects in $\mathcal{C}$. Also, since each object in $\mathcal{C}$ is non-empty, it is intersected by some object in O. Thus, the size of the cover $\mathcal{C}$ is at most $m_t|O|$. □

Remark: By the above lemma, we get a 4-approximation for squares and regular hexagons.

4 Unit Disk Cover

Given a set P of n points in the plane, the objective is to produce a set S of unit disks (disks with diameter 1) that cover all points in P while minimizing ply of P. In this section, we prove that a ply of 2 is necessary and sufficient.

4.1 2-Ply Unit Disk Cover

Lemma 5. *Given a set* P *of* n *points in the plane, a* 2*-ply unit disk cover can be constructed, which is at most* 7 *times the size of a minimum-sized* 2*-ply cover.*

Proof. Generating a 2-ply unit disk cover for a point set P consists of two steps. The first step is to apply the Separate(P) algorithm to distribute the points in P into boundary-disjoint vertical strips. This is achieved by using an interval length of $\frac{1}{\sqrt{2}}$ for the x-coordinates of the points. Next, the Separate(P) algorithm is again applied to distribute the points in P into boundary-disjoint horizontal strips based on their y-coordinates. This construction ensures that all points lie within the intersection of these vertical and horizontal strips, resulting in square regions of side length $1/\sqrt{2}$. Therefore, a 1-ply square cover is obtained.

The next step is to draw circumcircles over these squares, resulting in a unit disk cover, say $\mathcal{D}$. The vertical and horizontal alignment of the squares in the 1-ply cover ensures that the drawn circumcircles form a grid (Fig. 4). In this grid of circles, only vertical or horizontal neighbors can cross, but diagonally adjacent disks do not intersect or touch. In Fig. 4, $|AB|, |BC| > 1/\sqrt{2}$, hence $|AC| = \sqrt{|AB|^2 + |BC|^2} > 1$. This property guarantees that $\mathcal{D}$ is a 2-ply unit disk cover for the point set P.

We claim that any unit disk can intersect at most 7 unit disks of $\mathcal{D}$. Suppose not. Then, the 8 disks intersecting a unit disk, say d, must be from the 9 disks shown in Fig. 4. Hence, at least two diagonally opposite disks must be among them, which is a contradiction. Thus, by Lemma 4, we get a 7-approximation to the minimum-sized 2-ply unit disk cover of P. □

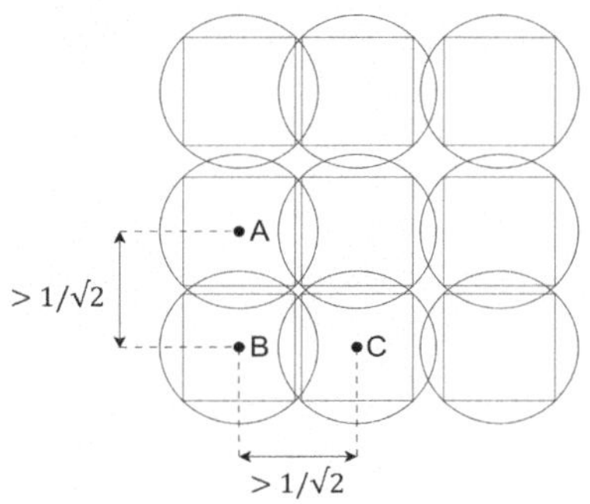

Fig. 4. Illustration of a 2-ply unit disk cover configuration.

It is known that there is a set of points for which a 1-ply unit disk cover cannot exist [1].

5 Convex Polygon Cover

Let P be a set of n points in the plane and let C be an arbitrary convex polygon with m vertices given as a sorted array. The goal is to find a set S of translations of C to cover all points in P while minimizing the ply.

We use the same terminology as in [4,7,18]. A pair of rectangles (r, R) is called *homothetic* if they are parallel and have the same aspect ratio (not necessarily axis-parallel). A homothetic pair (r, R) is an *approximating pair* for C if $r \subseteq C \subseteq R$. That is, r is enclosed in C, and C is enclosed in R, see Fig. 5(a). Let $\lambda(r, R)$ be the smallest ratio of the length of R to the length of r over all convex shapes. It was shown in [7,18] that $\lambda(r, R) \leq 2$ for any convex shape. Schwarzkopf et al. [7] also showed that if C is a convex polygon with m vertices given as a sorted array, then an approximating pair of rectangles with sides at most twice as long as each other can be computed in time $O(\log^2 m)$.

Theorem 6. *Given a set P of n points in the plane and an arbitrary convex polygon C with m vertices given as a sorted array, there exists an algorithm that can generate a set of translations of C to cover all points in P with a ply value of at most 4 in $O(n \log n + mn)$ time.*

Proof. We start by finding an approximating pair (r, R) for C where $\lambda(r, R) \leq 2$, and assume $\lambda(r, R) = 2$ for simplicity (this can be achieved by shrinking r). This step takes $O(\log^2 m)$ time. Let the dimensions of R be $l \times h$, with the corresponding dimensions for r being $\frac{l}{2} \times \frac{h}{2}$. Without loss of generality, we assume that R and r are axis-parallel. Using the results from Sect. 2.2, we generate a 1-ply cover for P using the inner rectangles (copies of r) in $O(n \log n)$ time. Finally, we replace the inner rectangles with the corresponding translations of C in $O(mn)$ time. Thus, the overall process takes $O(n \log n + nm)$ time. This process results in a valid cover since $r \subseteq C$. Furthermore, since $C \subseteq R$, the ply value resulting from C will be less than or equal to the ply value resulting after replacing the inner rectangles with the corresponding outer rectangles.

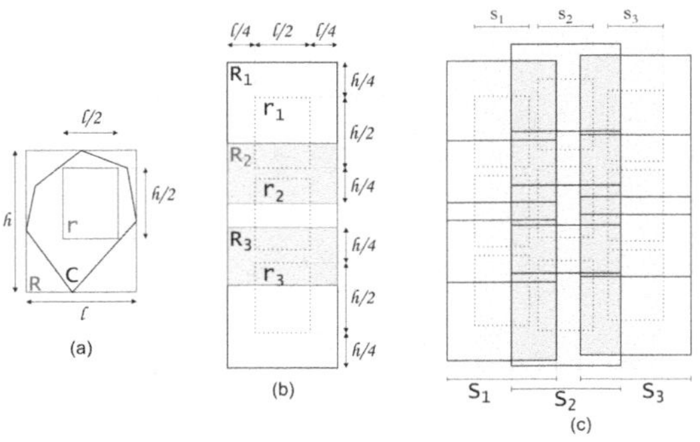

Fig. 5. Illustrations for Convex Polygon Cover: (a) Approximating pair (r, R) for polygon C. (b) Configuration of a single vertical strip. (c) Configuration of multiple vertical strips.

To simplify the analysis, we assume that r and R are concentric, which can be achieved by equally shifting all outer rectangles such that they become concentric with their respective inner rectangles while keeping the overall structure of all outer rectangles identical, thus keeping the ply unchanged.

We start by analyzing the ply for each vertical strip generated by the rectangle cover algorithm. Since the inner rectangles form a 1-ply cover, they are horizontally and vertically separated. As shown in Fig. 5(b), r_1 and r_2 are vertically separated, and r_2 and r_3 are also vertically separated. Since all (r, R) pairs are concentric, R_1 and R_3 will be vertically separated. Hence, $R_1 \cap R_3 = \emptyset$, but there can be intersections between R_1 and R_2, i.e., only between adjacent rectangles (blue regions in Fig. 5(b)). Thus, for each vertical strip, the ply value is at most 2.

Analogously, we can see that only adjacent vertical strips can intersect/touch. As shown in Fig. 5(c), s_1, s_2 and s_3 are vertical strips generated by the algorithm while computing the 1-ply inner rectangle cover. S_1, S_2, and S_3 are vertical strips obtained by replacing inner rectangles with outer rectangles. Since s_1, s_2 and s_3

are disjoint, $S_1 \cap S_3 = \emptyset$. Hence, the maximum ply region will result from the intersection of adjacent vertical strips (orange regions in Fig. 5(c)). Since each vertical strip has a maximum ply value of 2, the maximum ply value for the outer rectangle cover will be 4. Hence, after replacing the inner rectangles with the corresponding translations of C, we get a ply value at most 4. Hence, this is a 4-ply convex polygon cover.

References

1. Aloupis, G., Hearn, R.A., Iwasawa, H., Uehara, R.: Covering points with disjoint unit disks. In: Canadian Conference on Computational Geometry (2012). https://api.semanticscholar.org/CorpusID:16280099
2. Bandyapadhyay, S., Lochet, W., Saurabh, S., Xue, J.: Minimum-membership geometric set cover, revisited. In: International Symposium on Computational Geometry (2023)
3. Biedl, T., Biniaz, A., Lubiw, A.: Minimum ply covering of points with disks and squares. Comput. Geom. **94**, 101712 (2021)
4. Biniaz, A., Lin, Z.: Minimum ply covering of points with convex shapes. In: Proceedings of the 32nd Canadian Conference on Computational Geometry, pp. 2–5 (2020)
5. Biniaz, A., Liu, P., Maheshwari, A., Smid, M.: Approximation algorithms for the unit disk cover problem in 2d and 3d. Computational Geometry **60**, 8–18 (2017). the Twenty-Seventh Canadian Conference on Computational Geometry August 2015
6. Brönnimann, H., Goodrich, M.: Almost optimal set covers in finite vc-dimension. Discr. Comput. Geometry **14**(1), 463–479 (1995). cited by: 392. All Open Access, Bronze Open Access
7. Cheong, O., Fuchs, U., Rote, G., Welzl, E.: Approximation of convex figures by pairs of rectangles. Comput. Geom. **10**(2), 77–87 (1998)
8. Demaine, E.D., Feige, U., Hajiaghayi, M., Salavatipour, M.R.: Combination can be hard: approximability of the unique coverage problem. SIAM J. Comput. **38**(4), 1464–1483 (2008)
9. Durocher, S., Keil, J.M., Mondal, D.: Minimum ply covering of points with unit squares. In: WALCOM: Algorithms and Computation: 17th International Conference and Workshops. WALCOM 2023, Hsinchu, Taiwan, March 22–24, 2023, Proceedings, pp. 23–35. Springer-Verlag, Berlin, Heidelberg (2023)
10. Erlebach, T., van Leeuwen, E.J.: Approximating geometric coverage problems. In: Proceedings of the Nineteenth Annual ACM-SIAM Symposium on Discrete Algorithms. SODA '08. pp. 1267–1276. Society for Industrial and Applied Mathematics, USA (2008)
11. Fu, B., Chen, Z., Abdelguerfi, M.: An almost linear time 2.8334-approximation algorithm for the disc covering problem. Lecture Notes in Computer Science (including subseries Lecture Notes in Artificial Intelligence and Lecture Notes in Bioinformatics). LNCS, vol. 4508, pp. 317–326 (2007)
12. Goaoc, X., Kratochvíl, J., Okamoto, Y., Shin, C.S., Wolff, A.: Moving vertices to make drawings plane. In: Hong, S.H., Nishizeki, T., Quan, W. (eds.) Graph Drawing, pp. 101–112. Springer, Heidelberg (2008)

13. Gonzalez, T.F.: Covering a set of points in multidimensional space. Inf. Process. Lett. **40**(4), 181–188 (1991)
14. Govindarajan, S., Sarkar, S.: Improved algorithms for minimum-membership geometric set cover. In: Kalyanasundaram, S., Maheshwari, A. (eds.) Algorithms and Discrete Applied Mathematics, pp. 103–116. Springer, Cham (2024)
15. Hochbaum, D.S., Maass, W.: Approximation schemes for covering and packing problems in image processing and vlsi. J. ACM **32**(1), 130–136 (1985)
16. Kleinberg, J., Tardos, E.: Algorithm Design. Addison-Wesley Longman Publishing Co., Inc, USA (2005)
17. Knuth, D.E., Raghunathan, A.: The problem of compatible representatives. SIAM J. Discret. Math. **5**(3), 422–427 (1992)
18. Lassak, M.: Approximation of convex bodies by rectangles. Geom. Dedicata. **47**, 111–117 (1993). https://doi.org/10.1007/BF01263495
19. van Leeuwen, E.J.: Optimization and approximation on systems of geometric objects. Ph.D. thesis, University of Amsterdam (2009)
20. Lichtenstein, D.: Planar formulae and their uses. SIAM J. Comput. **11**(2), 329–343 (1982)

Economics and Computation

On the Distortion of Multi-winner Election Using Single-Candidate Ballots

Gennaro Auricchio[1], Zeyu Ren[2], Zihe Wang[2(✉)], and Jie Zhang[3]

[1] University of Padua, Via 8 Febbraio, 2, 35122 Padova, Veneto, Italy
gennaro.auricchio@unipd.it
[2] Renmin University of China, 59 Zhongguancun Street, Beijing 100872, China
{zeyuren,wang.zihe}@ruc.edu.cn
[3] University of Bath, Claverton Down, Bath, Somerset BA2 7AY, UK
jz2558@bath.ac.uk

Abstract. paper, we study the distortion bounds for voting mechanisms in multi-winner elections in general metric spaces. Our study pertains to the case in which each voter only reports her favorite candidate amongst m possible choices. Given that candidates' locations are undisclosed to the mechanism, the mechanism has to form a $w-$winner committee based solely on the number of votes received by candidates. We establish distortion bounds for both truthful and non-truthful mechanisms. Our research highlights the significance of the σ parameter, which represents the ratio between maximum and minimum distances among all candidate pairs. We show that the distortion is linear in σ. First, we demonstrate that all mechanisms possess a distortion greater than $1+\frac{w-1}{w+1}(\sigma-1)$. To give an upper bound, we study the Single Non-Transferable Vote (SNTV) mechanism, whose distortion is at most $1+2\sigma$. Second, we retrieve the upper bounds for strategyproof mechanisms. In particular, we infer an upper bound by examining the Random Sequential Dictator mechanism that achieves a distortion less than $1+4\sigma$ when $w=2$.

Keywords: Distortion · Multi-winner · Voting mechanisms

1 Introduction

Algorithmic mechanism design operates at the intersection of computer science, game theory, and economics. The field is concerned with designing mechanisms that achieve desired objectives while possessing appealing social properties, such as strategyproofness, which ensures that agents reveal their preferences. An important research area within algorithmic mechanism design is determining

Z.Wang was supported by National Natural Science Foundation of China (Grant No. 62172422) and by fund for building world-class universities (disciplines) of Renmin University of China.

J. Zhang was partially supported by a Leverhulme Trust Research Project Grant (2021–2024) and the EPSRC grant (EP/W014912/1).

F. V. Fomin and M. Xiao (Eds.): COCOON 2025, LNCS 15983, pp. 279–292, 2026.
https://doi.org/10.1007/978-981-95-0215-8_21

how, given the voters' preferences, a mechanism can select a committee of winners from a pool of candidates. A standard metric for quantifying the quality of a mechanism is the distortion, proposed by [24], which is the worst-case ratio between the social objective attained by the mechanism and the optimum. Determining the trade-offs between attaining an optimal outcome and other desirable social properties is a key problem in algorithmic mechanism design. The bounds on the distortion achievable by an election mechanism depend on the information disclosed by voters to the mechanism. The study by [14] examines three types of mechanism in the facility location-based voting problem: (i) *voting mechanisms*, where agents vote for their preferred candidate; (ii) *ranking mechanisms*, where agents report their ordinal preferences for candidates; and (iii) *location mechanisms*, where agents reveal their position in a political spectrum, i.e. their positions or view on different topics. The more information agents share, the more resources mechanisms have to reduce the distortion.

In this paper, we focus on the first category of mechanisms, that is voting mechanisms. This class of mechanisms is appealing for many reasons. First, voters may find it impossible to provide rankings for different candidates, primarily for cognitive reasons. In addition, agents' willingness to disclose their location can vary due to factors such as privacy concerns, fear of discrimination, and mistrust in how their information is used. Within this framework, we explore the distortion of multiwinner problems in the metric space; thus the mechanism collects only the first preference of voters and returns a committee of w winners, where $w > 1$.

Unfortunately, a simple example shows that the distortion is unbounded in this case: Let y_1, y_2 and y_3 be three candidates. Let us assume that there are three voters, each one of them shares the position with one of the candidates, and that the mechanism is asked to return a committee of two winners. Without loss of generality, we assume that the mechanism assigns the committee $\{y_1, y_2\}$ a positive probability. However, if the candidate profile is such that y_1, y_2 are located at the same point and y_3 is far away from them. Then the expected social cost achieved by the mechanism is positive and arbitrarily large, while the optimal cost is 0. In particular, the distortion is unbounded in this case. This is due to the fact that the mechanism has only access to the number of votes each candidate gets, but not to the candidates' locations. To overcome this issue, we introduce a parameter that represents the ratio between the maximum and minimum distance between candidate pairs, namely $\sigma \triangleq \frac{d_{\max}}{d_{\min}}$. The parameter σ allows us to better characterize both the lower and upper distortion bounds of the problem, allowing us to address the voting problem for any generic metric space. We consider both strategyproof and non-strategyproof mechanisms in our study.

1.1 Our Contribution and Technique Overview

In this paper, we study the distortion bounds of voting mechanisms for multiwinner elections in a general metric space. In particular, we assume that each voter submits only their favorite candidate to the mechanism. Based on this information, the mechanism must form a committee of w winners. In this framework, we first study the lower and upper bounds for general voting mechanisms. We show that both distortion bounds are linear in σ, which is the ratio between the maximum and minimum distance between any pair of candidates. In particular, we show that the distortion of any mechanism is lower bounded by $1+\frac{w-1}{w+1}(\sigma-1)$ and study the Single Non-Transferable Vote (SNTV) mechanism, a greedy mechanism that returns the w candidates who receive the most votes and arbitrarily breaks ties. We show that the SNTV mechanism has a distortion less than $1+2\sigma$.

We then study the distortion bounds for strategyproof mechanisms. We first focus on randomized mechanisms and then consider deterministic mechanisms. We show that every truthful mechanism is *independent of irrelevant candidates*, which ensures that the probability of any committee C remains unaffected by the votes received by candidates outside the committee. We then introduce the Random Sequential Dictator mechanism, which achieves an upper bound of at most $1+4\sigma$ when $w=2$. Lastly, we study deterministic truthful mechanisms. We establish that no anonymous deterministic strategyproof mechanism can achieve finite distortion. We show, however, that if we relax the anonymity condition and denote with n the number of voters, the Sequential Dictator mechanism achieves a distortion bound of $2(n-w)\sigma+1$.

The most significant findings of our research are (i) the role played by the ratio between the maximum and minimum distances of any pair of candidates, that is σ. Introducing the σ ratio allows for a more fine-grained analysis of distortion in multiwinner voting problems. (ii) a property of truthful mechanisms. Indeed, we show that every truthful mechanism is independent of irrelevant candidates and use this property to retrieve the lower bound of strategyproof mechanisms.

1.2 Related Work

To evaluate the performance of voting rules, Procaccia and Rosenschein [24] introduced the notion of distortion into the normalized social choice setting. Later, Caragiannis and Procaccia [7] employed this notion and analyzed the Plurality Rule. Boutilier et al. [5] studied the distortion of randomized rules and presented a simple rule with lower distortion. Caragiannis et al. [6] extended the framework to select a subset of alternatives.

Anshelevich et al. [1] first initiated the study of this problem in a metric space. Following their study, several papers analysed the distortion of many different rules: Skowron and Elkind [26] studied the class of scoring rules, Goel et al. [17] and Kempe [20] studied the Ranked Pairs Rule, Munagala and Wang [22] introduced a weighted tournament rule and improved the distortion, while Gkatzelis et al. [15] designed the plurality matching voting rule and proved that the optimal deterministic algorithm has distortion 3. Kizilkaya and Kempe proposed an extremely simple voting rule which achieves the same optimal distortion of 3. As for randomized mechanisms, Anshelevich and Post [3] showed that Random Dictatorship has distortion $3-\frac{2}{n}$. Kempe [21] and Gkatzelis et al. [15] improved on this result by providing a mechanism whose distortion is $3 - \frac{2}{m}$. Charikar et al. [10] further improved the results by showing that randomization over simple rules can achieve distortion less than 2.753, making it closer to the known lower bound of 2.1126, found by Charikar and Ramakrishnan [9]. Pulyassary and Swamy [25] independently showed a lower bound of 2.0631. Lastly, it is important to highlight the findings of [18], who investigated the distortion of a randomized mechanism known as 2-Agree. This mechanism operates by sequentially querying random voters for their top choices until a consensus is reached among at least two voters. Building on this line of research, Fain et al. [12] examined mechanisms with constant sample complexity and introduced the Random Oligarchy mechanism. For other distortion results, please refer to the survey by Anshelevich et al. [2].

Multi-winner Setting. Goel et al. [16] characterized the distortion of selecting a committee by repeatedly applying single-winner voting rules. Chen et al. [11] shifted their focus to the *single-loser* setting, which is $w = m - 1$. In this setting, the committee is formed by eliminating the least popular candidate. When considering the social cost objective, they demonstrate a tight distortion bound of 3 for deterministic mechanisms and $3 - 2/m$ for randomized mechanisms where m is the number of candidates. While this paper mentioned is relevant to our work, a significant difference lies in our primary focus on general multi-winner elections. Caragiannis et al. [8] studied ranking mechanisms, where agents report their ordinal preferences for candidates. They showed that with all ordinal preferences for candidates, the distortion is asymptotically linear in the number of agents when $w = 2$, and the distortion is unbounded when $w > 2$. We studied voting mechanisms where agents vote for their preferred candidate. Our conclusion is different because we consider mechanisms using different information, and we represent the distortion in terms of σ. When $w = 2$, we show a lower bound that is linear in σ. Since σ can be arbitrarily large, then our result implies that the distortion is unbounded for mechanisms using single-candidate ballots. The result can be easily generalized for $w > 2$. Aziz et al. [4], Kalayci et al. [19], and Peters and Skowron [23] studied the proportional representation. Please refer to [13] for other interesting topics in the multi-winner setting.

2 Preliminaries

Let $\Omega = (S, d)$ be a metric space. An election in Ω consists of n voters, which we denote with $N = \{1, \ldots, n\}$, and m candidates, which we denote with $M = \{1, \ldots, m\}$. All candidates and voters are points in S, thus we denote with $x_i \in S$ the positions of voters and with $y_j \in S$ the positions of candidates. We denote in bold letters the vector containing all the positions of the agents and all the positions of the candidates, that is $\mathbf{x} = (x_1, \ldots, x_n)$ and $\mathbf{y} = (y_1, \ldots, y_m)$. For *single-candidate ballots*, each voter i votes for one candidate, which we denote with a_i. The vote a_i of voter i is also referred to as its *action*. The distance between a voter i and her action a_i is $d(x_i, a_i)$. We denote the action profile by $\mathbf{a} = (a_1, \ldots, a_n)$, a vector containing all the actions of voters. An action profile $\mathbf{a}$

is *consistent* with $\mathbf{x}$ if $a_i \in \arg\min_{y\in M} d(x_i, y)$ for every $i \in \{1, \ldots, n\}$. Given a voter at x_i and a committee $C \subseteq M$, we define the cost of agent i as her distance from C, that is

$$\mathrm{cost}(x_i, C) = d(x_i, C) \triangleq \min_{y_k \in C} d(x_i, y_k).$$

Given $w \in \mathbb{N}$, we denote with $\mathcal{C}_w$ the set of all committees of w different candidates. The problem of *multi-winner voting* for single-candidate ballots consists in electing a committee $C^* \subseteq M$ of w winners that minimizes the sum of the voters' costs i.e.

$$C^* \in \arg\min_{C\in\mathcal{C}_w} \sum_{i\in[n]} \mathrm{cost}(x_i, C).$$

The sum of the voters' costs is also known as the Social Cost (SC) of the committee C, thus, henceforth, we set $\mathrm{SC}(\mathbf{x}, C) \triangleq \sum_{i\in[n]} \mathrm{cost}(x_i, C)$.

An election in the social choice problem under consideration is a tuple $\Gamma = (\Omega, M, \mathbf{a}, w)$. For simplicity, we may refer to an election Γ as its action profile $\mathbf{a}$.

A mechanism f takes an election Γ as input and outputs a committee $C \subseteq M$ of w winners. A mechanism is *deterministic* if, for each election, it outputs one committee. A mechanism is *randomized* if it outputs a probability distribution over committees in $\mathcal{C}_w$. The probability of the committee C being elected is $p_C(\mathbf{a})$.

Moreover, a mechanism f is *anonymous* if the output of the mechanism does not depend on the identities of the voters but only on the voters' aggregated information, that is the number of votes each candidate receives. For the sake of simplicity, given an anonymous mechanism f and an action profile $\mathbf{a}$, we denote the probability of a committee C being elected with $p_C(\mathbf{a}) = p_C(n_1, \ldots, n_m)$, where $(n_1, \cdots, n_m)$ is the m-tuple containing the number of votes of each candidate given a voters' action profile $\mathbf{a}$. When voters are self-interested, we assume that the position of every agent is their own private information. In this case, voters may act strategically if this lowers the cost. A mechanism f is *truthful* (or *strategyproof*) if no agent is able to lower its cost by misreporting their action, that is, for every $i \in N$ and every action a'_i, we have that $\mathrm{cost}(x_i, f(a_i, \mathbf{a}_{-i})) \leq \mathrm{cost}(x_i, f(a'_i, \mathbf{a}_{-i}))$ where a_i is an action consistent with the position of voter i and $\mathbf{a}_{-i}$ denotes the actions of the other voters. To evaluate the performance of a mechanism f, we consider its *distortion*. For a fixed election Γ and the action profile $\mathbf{a}$, the distortion of f over the election Γ is defined as the worst-case ratio between the expected SC returned by f and the optimal SC over all the agents' positions that are consistent with the action profile $\mathbf{a}$, that is

$$\mathrm{dist}(f, \Gamma) = \sup_{\mathbf{x}\in\chi(\Gamma)} \frac{\mathbb{E}[\mathrm{SC}(\mathbf{x}, f(\mathbf{a}))]}{\mathrm{OPT}(\mathbf{x})},$$

where $\chi(\Gamma)$ is the set of location profiles consistent with Γ and $\mathrm{OPT}(\mathbf{x})$ is the SC of the optimal solution. Finally, the distortion of a mechanism f is then $\mathrm{Dist}(f) = \sup_\Gamma \mathrm{dist}(f, \Gamma)$, which is the worst case in all elections. For deterministic mechanisms f, the distortion is defined similarly. In our study, we show that the distortion is strongly related to the ratio between the maximum and minimum distances among any two candidates. We denote this ratio as σ. Formally, we express the ratio as $\sigma = d_{\max}/d_{\min}$, where $d_{\max} = \max_{(y_k,y_l)\in M^2} d(y_k, y_l)$ and $d_{\min} = \min_{(y_k,y_l)\in M^2} d(y_k, y_l)$.

3 Distortion Without Strategyproofness

In this section, we study the distortion bounds of mechanisms that are not necessarily strategyproof. We start our analysis from the lower bound and show that no randomized voting mechanism can achieve an approximation ratio that is lower than $1 + \frac{w-1}{w+1}(\sigma - 1)$.

Theorem 1. *It is impossible for any randomized mechanism to achieve a distortion smaller than* $1 + \frac{w-1}{w+1}(\sigma - 1)$.

In order to prove the lower bound, we first introduce a candidate profile.

Definition 1 (Candidate Profile I). *Let us consider m candidates in an m-dimensional Euclidean space. The coordinate of the locations of $y_i (1 \le i \le m-2)$ has only one dimension that is non-zero. Given $r \in \mathbb{R}$, we denote the locations of $y_i (1 \le i \le m-2)$ as $y_1 = (r, 0, \ldots, 0), y_2 = (0, r, \ldots, 0), \ldots, y_{m-2} = (0, \ldots, 0, r, 0, 0)$. The locations of y_{m-1} and y_m are $y_{m-1} = (0, \ldots, 0, \sqrt{r^2-1}, 1)$ and $y_m = (0, \ldots, 0, \sqrt{r^2-1}, -1)$. When r is large enough, the distance between y_{m-1} and y_m is $d_{\min}$. Any other distance between two candidates is $d_{\max}$.*

Using the candidate profile, we are ready to prove Theorem 1.

Proof. Suppose that n voters are co-located at $w+1$ candidates, that is, each candidate gets $\frac{n}{w+1}$ votes. The $w+1$ candidates are $y_{m-w}, \cdots, y_{m-1}$ and y_m.

First, the committee $\{y_{m-w}, \cdots, y_{m-1}\}$ or $\{y_{m-w}, \cdots, y_{m-2}, y_m\}$ is the optimal. Thus, the optimal social cost is $\frac{n}{w+1} \cdot d_{\min}$.

Then, we consider the social cost achieved by the mechanism. Let p_t denote the probability of selecting w candidates from $w+1$ candidates that receive votes. Due to the mechanism designer does not know the exact locations of candidates, she should treat $w+1$ potential committees equally. It indicates that the probability of any committee is $\frac{p_t}{w+1}$. Otherwise, we can do a permutation over all the candidates. If the mechanism selects a candidate which does not receive votes, the social cost is at least $\frac{n}{w+1} \cdot d_{\max} + \frac{n}{w+1} \cdot d_{\min}$. Therefore, the social cost achieved by the mechanism is lower bounded by

$$
\begin{aligned}
&\frac{p_t}{w+1} \cdot ((w-1)d_{\max} + 2d_{min}) \cdot \frac{n}{w+1} + (1-p_t) \cdot \frac{n}{w+1} \cdot (d_{\max} + d_{\min}) \\
&= \frac{p_t \cdot n}{w+1} \left(-\frac{2}{w+1} d_{\max} - \frac{w-1}{w+1} d_{\min} \right) + \frac{n}{w+1} (d_{\max} + d_{\min}) \\
&\ge \frac{n}{w+1} \left(-\frac{2}{w+1} d_{\max} - \frac{w-1}{w+1} d_{\min} \right) + \frac{n}{w+1} (d_{\max} + d_{\min}) \\
&= \frac{n}{w+1} \left(\frac{w-1}{w+1} d_{\max} + \frac{2}{w+1} d_{\min} \right).
\end{aligned}
$$

We combine it with the optimal social cost. Thus, the distortion is lower bounded by

$$\begin{aligned}&\frac{\frac{n}{w+1}\left(\frac{w-1}{w+1}d_{\max}+\frac{2}{w+1}d_{\min}\right)}{\frac{n}{w+1}\cdot d_{\min}}\\=&\frac{(w-1)\sigma+2}{w+1}\\=&1+\frac{w-1}{w+1}(\sigma-1).\end{aligned}$$

□

We now study the upper bound on the distortion. We do this by analysing the SNTV mechanism, which achieves a distortion of at most $1+2\sigma$.

Mechanism 1 (SNTV Mechanism). *Given an election $\Gamma=(\Omega,M,\mathbf{a},w)$, then SNTV outputs a committee C such that $C=\arg\max_{C'\in\mathcal{C}_w}n_{C'}$ where $n_{C'}=\sum_{y_j\in C'}n_j$. If there are multiple committees getting the most votes, then the mechanism breaks ties arbitrarily.*

To study the distortion of SNTV mechanism, we introduce the following lemma by assuming that C^* is a committee in $\arg\min_{C\in\mathcal{C}_w}SC(\mathbf{x},C)$.

Lemma 1. *The distortion of any anonymous mechanism is at most*

$$1+\frac{2\sigma}{n-n_{C^*}}\sum_{C\neq C^*}p_C(n_1,...,n_m)(n-n_C).$$

Our main proof tool is the triangle inequality. We defer proof details to the appendix.

Then, we notice that, by definition, $C=\arg\max_{C'\in\mathcal{C}_w}n_{C'}$, thus $n-n_C\le n-n_{C^*}$. Lastly, owing to Lemma 1, we have

$$1+\frac{2\sigma}{n-n_{C^*}}\sum_{C\neq C^*}p_C(n_1,...,n_m)(n-n_C)\le 1+2\sigma.$$

Theorem 2. *The distortion of SNTV mechanism is at most $1+2\sigma$.*

4 Distortion With Strategyproofness

In this section, we focus our attention to strategyproof mechanisms. Our study hinges upon the fact that on suitable elections, any strategyproof mechanism is *independent of irrelevant candidates* (IIC).

Definition 2. *A mechanism is independent of irrelevant candidates (IIC) if the probability that the mechanism outputs a committee C is independent of the distribution of a fixed number of votes among candidates in $M \backslash C$ receive. Formally, for any committee $C = \{y_{l_1}, \ldots, y_{l_w}\} \subset M$, a mechanism is IIC if $p_C(\mathbf{a}) = p_C(n_{l_1}, \ldots, n_{l_w})$.*

In order to prove that every strategyproof mechanism is IIC, we need to introduce the following candidate profile.

Definition 3 (Candidate Profile II) *Let us consider m candidates in an $(m-1)$-dimensional Euclidean space. Given $r \in \mathbb{R}$, we denote their locations with $y_1 = (1, 0, \ldots, 0), \ldots, y_{m-1} = (0, 0, \ldots, 1)$, and $y_m = (r, r, \ldots, r)$. When r is large enough, $d_{\min}$ is the distance between any two of the first $m-1$ candidates, and $d_{\max}$ is the distance between y_m and any of the first $m-1$ candidates.*

We are then ready to state the following theorem.

Theorem 3. *In any election instance where candidates' positions are represented by Candidate Profile II, if $w < m-1$, any strategyproof mechanism must be IIC.*

The proof idea is that we construct a subspace to analyze the distances between the voter and the candidates, and the expected cost of the voter is expressed in terms of these distances. By leveraging the strategyproofness of the mechanism, we show that two probabilities must be equal regardless of whether the voter chooses y_i or y_j, thereby ensuring the mechanism satisfies IIC.

Proof. For any voter k and fixed actions of other voters $\mathbf{a}_{-k}$, we consider voter k's two different actions and the corresponding action profiles $\mathbf{a}^1 = (a_k = y_i, \mathbf{a}_{-k})$ and $\mathbf{a}^2 = (a_k = y_j, \mathbf{a}_{-k})$. Denote $L = \{l_1, ..., l_w\} \subset M$ a subset of candidates. We will show that for any committee $C = \{y_{l_1}, ..., y_{l_w}\}$ such that $y_i, y_j \notin C$, a strategyproof mechanism must have $p_C(\mathbf{a}^1) = p_C(\mathbf{a}^2)$. We then generalize this result to prove the theorem. There are two cases based on whether candidate y_m is one of the candidates y_i and y_j.

Case 1, if $y_i = y_m$ or $y_j = y_m$. Without loss of generality, we assume $y_j = y_m$. Let α_1 and α_2 be real numbers such that $r/2 \leq \alpha_1 \leq \alpha_2 \leq (r+1)/2$. We define a subspace $U_L(\alpha_1, \alpha_2)$ as follows.

$$\Big\{(t_1, t_2, ..., t_{m-1}) \in \mathbb{R}^{m-1} | t_{l_1} = \frac{(m-2)r}{2} - (w-1)\alpha_1 - (m-w-2)\alpha_2,$$
$$t_{l_2} = ... = t_{l_w} = \alpha_1, t_i = \frac{r+1}{2}, t_h = \alpha_2, \forall h \notin L \cup \{i\}\Big\}.$$

The construction of the subspace $U_L(\alpha_1, \alpha_2)$ has a twofold effect. First, the distances between voter $x_k \in U_L(\alpha_1, \alpha_2)$ to candidates y_i and y_j are the same. Second, the cost $cost(x_i, C)$ falls into three categories for all committees. This effect will facilitate us to represent voter x_k's expected cost.

In particular, for any $x_k \in U_L(\alpha_1, \alpha_2)$, we have $d(x_k, y_i) = d(x_k, y_j)$. The two distances can be written as

$$\left(\left(\frac{r-1}{2}\right)^2 + t_{l_1}^2 + (w-1)\alpha_1^2 + (m-w-2)\alpha_2^2\right)^{\frac{1}{2}}.$$

Let $\eta := d(x_k, y_i)^2 + r + 1$. For simplicity, we express the distances between x_k and other candidates in terms of η. By simple calculation, we have that

$$d(x_k, y_h) = \sqrt{\eta - 2\alpha_2}, \quad d(x_k, y_{l_1}) = \sqrt{\eta - 2t_{l_1}}$$
$$d(x_k, y_{l_2}) = ... = d(x_k, y_{l_w}) = \sqrt{\eta - 2\alpha_1}.$$

Since $r/2 \leq \alpha_1 \leq \alpha_2 \leq (r+1)/2$ and $h \notin L \cup \{i\}$, it is easy to check that $d(x_k, y_i) = d(x_k, y_j) \leq d(x_k, y_h) \leq d(x_k, y_{l_2}) = ... = d(x_k, y_{l_w}) \leq d(x_k, y_{l_1})$. Next, we consider voter k's cost. The distance from voter k to the nearest candidate has three possibilities: $d(x_k, y_i)$, $d(x_k, y_{l_2})$ and $d(x_k, y_h)$ for different committees.

Let C' be a committee different from C such that $y_i, y_j \notin C'$. Then, the summation $\sum_{C': y_i, y_j \notin C'} p_{C'}(\mathbf{a}^1)$ is the probability that the mechanism outputs one of these committees C' when the action of voter x_k is $\mathbf{a}^1$. Let $e_1(\mathbf{a}^1) = \sum_{C': y_i, y_j \notin C'} p_{C'}(\mathbf{a}^1) - p_C(\mathbf{a}^1)$ and $e_2(\mathbf{a}^1) = 1 - \sum_{C': y_i, y_j \notin C'} p_{C'}(\mathbf{a}^1)$. Thus, when $a_k = y_i$, we can write the expected cost of voter k as $\mathbb{E}[\text{cost}(x_k)] = p_C(\mathbf{a}^1) d(x_k, y_{l_2}) + e_1(\mathbf{a}^1) d(x_k, y_h) + e_2(\mathbf{a}^1) d(x_k, y_i)$.

Similarly, when $a_k = y_j$, we have that $\mathbb{E}[\text{cost}(x_k)] = p_C(\mathbf{a}^2) d(x_k, y_{l_2}) + e_1(\mathbf{a}^2) d(x_k, y_h) + e_2(\mathbf{a}^2) d(x_k, y_i)$.

Since the mechanism is strategyproof, voter x_k should derive the same cost when she votes for y_i or y_j. We then have that

$$\begin{aligned}&\left(p_C(\mathbf{a}^1) - p_C(\mathbf{a}^2)\right) d(x_k, y_{l_2}) + \left(e_1(\mathbf{a}^1) - e_1(\mathbf{a}^2)\right) d(x_k, y_h)\\ &= \left(e_2(\mathbf{a}^2) - e_2(\mathbf{a}^1)\right) d(x_k, y_i).\end{aligned} \tag{1}$$

We square both sides of the equation (1). The terms $d^2(x_k, y_{l_2})$, $d^2(x_k, y_h)$ and $d^2(x_k, y_i)$ do not involve a square root. However, the term $d(x_k, y_{l_2}) \cdot d(x_k, y_h)$ still contains a square root. Since we are considering the Euclidean distance, and the equation should hold for any α_1 and α_2 such that $r/2 \leq \alpha_1 \leq \alpha_2 \leq (r+1)/2$. Therefore, the coefficient of the term $d(x_k, y_{l_2}) \cdot d(x_k, y_h)$ must be 0. That is,

$$2 \cdot \left(p_C(\mathbf{a}^1) - p_C(\mathbf{a}^2)\right) \cdot \left(e_1(\mathbf{a}^1) - e_2(\mathbf{a}^2)\right) = 0.$$

Consequently, we conclude that $p_C(\mathbf{a}^1) = p_C(\mathbf{a}^2), e_1(\mathbf{a}^1) = e_1(\mathbf{a}^2)$ and $e_2(\mathbf{a}^1) = e_2(\mathbf{a}^2)$.

Case 2, if $y_i \neq y_m$ and $y_j \neq y_m$. The proof follows a similar approach as in Case 1, but the process of constructing the subspace to establish these equations is more intricate. To keep our primary discussion on track, we defer the detailed construction of the subspace for this case to the Appendix.

These two cases illustrate that, irrespective of the committee C, when a single voter's decision leads to the election of a candidate from $M \backslash C$, it does not impact the probability of the mechanism producing committee C. By recursively applying this proof, we can establish that the same outcome holds true even if multiple voters alter their choices, resulting in the election of candidates from $M \backslash C$. In other words, the mechanism remains unaffected by candidates who are irrelevant to the final result. $\square$

For any committee $C = \{y_{l_1}, \ldots, y_{l_w}\} \subset M$, the probability that an IIC mechanism outputs C depends only on the number of votes the candidates $y_{l_1}, \ldots, y_{l_w}$ receive. Note that given any strategyproof mechanism f', there always exists a randomized, anonymous strategyproof mechanism f whose distortion is not worse than f'. Specifically, f can be obtained by applying a uniformly chosen random permutation to the set of voters before applying f'. So, without loss of generality, we consider randomized, anonymous strategyproof mechanisms. Because of Theorem 3, these mechanisms are IIC.

Recall that the lower bound of distortion in Theorem 1 can also be applied to the strategyproof setting. Moving forward, we present an upper bound for randomized strategyproof mechanisms. En route to this result, we retrieve a sufficient condition for strategyproofness and define a mechanism that meets this sufficient condition.

Definition 4 *(Monotonicity). An IIC mechanism is monotone if the probability of a committee $C = \{y_{l_1}, \ldots, y_{l_w}\}$ being elected weakly increases when the number of votes received by a candidate in C weakly increases and others are unchanged. In particular, for a w-winner election, for any committee C and any $y_{l_i} \in C$, it holds that $p_C(n_{l_i} + 1, \mathbf{n}_{-\mathbf{l_i}}) \geq p_C(n_{l_i}, \mathbf{n}_{-\mathbf{l_i}})$, where $\mathbf{n}_{-\mathbf{l_i}}$ denotes the number of votes the candidates in C excepts y_{l_i} received.*

Theorem 4. *Any monotone mechanism is strategyproof.*

Next, we present an upper bound of distortion. Therefore, we need to study a strategyproof mechanism. Recalling the SNTV mechanism, it is not strategyproof.

Proposition 1. *The SNTV is not a strategyproof mechanism.*

Proof. Suppose there are four candidates: y_1, y_2, y_3 and y_4. The goal is to elect 2 winners. There are 7 voters. We have y_1, y_2 and y_3 receive 2 votes respectively. We consider the last voter. She locates at y_4. As for the distances, we assume that $d(y_1, y_4) > d(y_2, y_4) > d(y_3, y_4)$. If the last voter reports honestly, that is y_4, the candidate pair $\{y_1, y_2\}$, $\{y_1, y_3\}$ and $\{y_2, y_3\}$ should be elected arbitrarily by SNTV mechanism. If she misreports to y_3, then y_3 gets 3 votes. Thus, either $\{y_1, y_3\}$ or $\{y_2, y_3\}$ is elected by SNTV mechanism. The cost of the voter decreases. $\square$

To conclude, we introduce and study the Random Sequential Dictator mechanism.

Mechanism 2 *(Random Sequential Dictator). For a given election $\Gamma = (\Omega, M, \mathbf{a}, w)$, the voters' actions are arranged in a sequence randomly. The mechanism always outputs the first w different candidates as the committee.*

The strategyproofness of Random Sequential Dictator is apparent from its dictator nature. In the sequence of indexing the candidates, when it comes to voter x_i's action, if the first w candidates are not entirely determined yet, then voter x_i should vote for the nearest candidate to her to minimize her cost; if the first w candidates are already determined, then voter x_i's action will not change the output committee of the mechanism. Then, in order to derive the distortion of the mechanism, we need the probability function of each committee so that we can use Lemma 1.

Proposition 2. *When $w = 2$, the probability function of Random Sequential Dictator for committee $C = \{y_k, y_l\}$ is*

$$p_C(n_k, n_l) = \frac{n_k}{n - n_l} + \frac{n_l}{n - n_k} - \frac{n_k + n_l}{n}.$$

Proof. We consider the probability of y_k, y_l being selected. For the case where y_k is the first chosen candidate and y_l is the second, the probability is $\frac{n_k}{n} \cdot \frac{n_l}{n-n_k}$. For the case where y_l is the first chosen candidate and y_k is the second, the probability is $\frac{n_l}{n} \cdot \frac{n_k}{n-n_l}$. Thus, we have

$$\begin{aligned} p_C(n_k, n_l) &= \frac{n_k}{n} \cdot \frac{n_l}{n - n_k} + \frac{n_l}{n} \cdot \frac{n_k}{n - n_l} \\ &= n_l \cdot \frac{n_k}{n(n - n_k)} + n_k \cdot \frac{n_l}{n(n - n_l)} \\ &= \frac{n_k}{n - n_l} + \frac{n_l}{n - n_k} - \frac{n_k + n_l}{n}. \end{aligned}$$

□

Lastly, we compute the distortion upper bound of the mechanism.

Theorem 5. *The distortion of Random Sequential Dictator is upper bounded by $1 + 4\sigma$ when $w = 2$.*

When $w > 2$, it is difficult to analyze the performance of Random Sequential Dictator mechanism. Thus, we apply Lemma 1 to get an upper bound.

Proposition 3. *The distortion of Random Sequential Dictator is upper bounded by $1 + 2(n - w)\sigma$ when $w > 2$.*

Random Dictator is a special case of Random Sequential Dictator when $w = 1$. For every candidate $y_i \in M$, the winning probability is n_i/n. [14] showed that Random Dictator mechanism achieves a distortion of exactly 3. By leveraging on Theorem 3, we show that Random Dictator is the only anonymous strategyproof mechanism that has finite distortion.

Theorem 6. *In single-winner elections, the unique anonymous strategyproof mechanism using single-candidate ballots with finite distortion is Random Dictator.*

To conclude, we consider deterministic mechanisms. In particular, we show that no anonymous deterministic strategyproof mechanism can achieve finite distortion. Lastly, we show that anonymousness plays a crucial role into showing the unboundness of the lower bound by considering the Sequential Dictator mechanism and investigating its distortion.

Theorem 7. *For a w-winner election, no anonymous deterministic strategyproof mechanism can achieve finite distortion.*

Mechanism 3 *(Sequential Dictator). Given an election $\Gamma = (\Omega, M, \mathbf{a}, w)$, the voters cast their actions by a predetermined order. The mechanism always outputs the first w different candidates as the committee.*

Theorem 8. *Sequential Dictator is a deterministic strategyproof mechanism and its distortion is at most $2(n-w)\sigma + 1$.*

5 Conclusion and Future Work

In this paper, we studied the problem of multi-winner voting using single-candidate ballots. Our study examined mechanisms in non-strategyproof and strategyproof settings, where we identified the lower and upper bounds of distortion. Overall, our research contributes to understanding multi-winner voting problems, offering insights into distortion bounds and the significance of the parameter σ.

There are still several unresolved questions that merit further exploration. First, an intriguing alternative is to explore mechanisms where each voter is asked to provide their top-k choices or even a rank of all candidates rather than simply indicating their favorite. Secondly, a compelling question arises: can our approaches be adapted to address the max cost objective? If so, to what extent would this alteration impact the distortion bounds? These inquiries present formidable challenges and warrant in-depth investigation.

References

1. Anshelevich, E., Bhardwaj, O., Postl, J.: Approximating optimal social choice under metric preferences. In: 29th AAAI Conference on Artificial Intelligence (2015)
2. Anshelevich, E., Filos-Ratsikas, A., Shah, N., Voudouris, A.A.: Distortion in social choice problems: The first 15 years and beyond. In: Proceedings of the Thirtieth International Joint Conference on Artificial Intelligence Survey Track (2021)
3. Anshelevich, E., Postl, J.: Randomized social choice functions under metric preferences. J. Artif. Intell. Res. **58**, 797–827 (2017)
4. Aziz, H., Brill, M., Conitzer, V., Elkind, E., Freeman, R., Walsh, T.: Justified representation in approval-based committee voting. Soc. Choice Welfare **48**(2), 461–485 (2017). https://doi.org/10.1007/s00355-016-1019-3
5. Boutilier, C., Caragiannis, I., Haber, S., Lu, T., Procaccia, A.D., Sheffet, O.: Optimal social choice functions: a utilitarian view. Artif. Intell. **227**, 190–213 (2015)

6. Caragiannis, I., Nath, S., Procaccia, A.D., Shah, N.: Subset selection via implicit utilitarian voting. J. Artif. Intell. Res. **58**, 123–152 (2017)
7. Caragiannis, I., Procaccia, A.D.: Voting almost maximizes social welfare despite limited communication. Artif. Intell. **175**(9–10), 1655–1671 (2011)
8. Caragiannis, I., Shah, N., Voudouris, A.A.: The metric distortion of multiwinner voting. In: 36th AAAI Conference on Artificial Intelligence, p. 4 (2022)
9. Charikar, M., Ramakrishnan, P.: Metric distortion bounds for randomized social choice. In: Proceedings of the 2022 Annual ACM-SIAM Symposium on Discrete Algorithms (SODA), pp. 2986–3004. SIAM (2022)
10. Charikar, M., Ramakrishnan, P., Wang, K., Wu, H.: Breaking the metric voting distortion barrier. In: Proceedings of the 2024 Annual ACM-SIAM Symposium on Discrete Algorithms (SODA), pp. 1621–1640. SIAM (2024)
11. Chen, X., Li, M., Wang, C.: Favorite-candidate voting for eliminating the least popular candidate in a metric space. In: Proceedings of the AAAI Conference on Artificial Intelligence, vol. 34, pp. 1894–1901 (2020)
12. Fain, B., Goel, A., Munagala, K., Prabhu, N.: Random dictators with a random referee: Constant sample complexity mechanisms for social choice. In: Proceedings of the AAAI Conference on Artificial Intelligence, vol. 33, pp. 1893–1900 (2019)
13. Faliszewski, P., Skowron, P., Slinko, A., Talmon, N.: Multiwinner voting: a new challenge for social choice theory. Trends Comput. Soc. Choice **74**(2017), 27–47 (2017)
14. Feldman, M., Fiat, A., Golomb, I.: On voting and facility location. In: Conitzer, V., Bergemann, D., Chen, Y. (eds.) Proceedings of the 2016 ACM Conference on Economics and Computation, EC '16, Maastricht, The Netherlands, July 24–28, 2016. pp. 269–286. ACM (2016)
15. Gkatzelis, V., Halpern, D., Shah, N.: Resolving the optimal metric distortion conjecture. In: 2020 IEEE 61st Annual Symposium on Foundations of Computer Science (FOCS), pp. 1427–1438. IEEE (2020)
16. Goel, A., Hulett, R., Krishnaswamy, A.K.: Relating metric distortion and fairness of social choice rules. In: Proceedings of the 13th Workshop on Economics of Networks, Systems and Computation, pp. 1–1 (2018)
17. Goel, A., Krishnaswamy, A.K., Munagala, K.: Metric distortion of social choice rules: lower bounds and fairness properties. In: Proceedings of the 2017 ACM Conference on Economics and Computation, pp. 287–304 (2017)
18. Gross, S., Anshelevich, E., Xia, L.: Vote until two of you agree: Mechanisms with small distortion and sample complexity. In: Proceedings of the AAAI Conference on Artificial Intelligence, vol. 31 (2017)
19. Kalayci, Y., Kempe, D., Kher, V.: Proportional representation in metric spaces and low-distortion committee selection. In: Proceedings of the AAAI Conference on Artificial Intelligence, vol. 38, pp. 9815–9823 (2024)
20. Kempe, D.: An analysis framework for metric voting based on LP duality. In: Proceedings of the AAAI Conference on Artificial Intelligence, vol. 34, pp. 2079–2086 (2020)
21. Kempe, D.: Communication, distortion, and randomness in metric voting. In: Proceedings of the AAAI Conference on Artificial Intelligence, vol. 34, pp. 2087–2094 (2020)
22. Munagala, K., Wang, K.: Improved metric distortion for deterministic social choice rules. In: Proceedings of the 2019 ACM Conference on Economics and Computation, pp. 245–262 (2019)
23. Peters, D., Skowron, P.: Proportionality and the limits of welfarism. In: Proceedings of the 21st ACM Conference on Economics and Computation, pp. 793–794 (2020)

24. Procaccia, A.D., Rosenschein, J.S.: The distortion of cardinal preferences in voting. In: International Workshop on Cooperative Information Agents, pp. 317–331 (2006)
25. Pulyassary, H., Swamy, C.: On the randomized metric distortion conjecture. arXiv preprint arXiv:2111.08698 (2021)
26. Skowron, P.K., Elkind, E.: Social choice under metric preferences: scoring rules and STV. In: Thirty-First AAAI Conference on Artificial Intelligence (2017)

Fair and Efficient Graphical Resource Allocation with Matching-Induced Utilities

Zheng Chen[1(✉)], Bin Deng[2], Bo Li[3], Minming Li[4], Weidong Li[2], and Guochuan Zhang[1]

[1] College of Computer Science and Technology, Zhejiang University, Hangzhou, China
{21721122,zgc}@zju.edu.cn
[2] School of Mathematics and Statistics, Yunnan University, Yunnan, China
dengbin@mail.ynu.edu.cn
[3] Department of Computing, The Hong Kong Polytechnic University, Hong Kong, China
comp-bo.li@polyu.edu.hk
[4] Department of Computer Science, City University of Hong Kong, Hong Kong, China
minming.li@cityu.edu.hk

Abstract. We study the fair allocation of graphical resources, where the resources are the vertices in a graph. Upon receiving a set of resources, an agent's utility equals the weight of a maximum matching in the induced subgraph. We care about maximin share (MMS) fairness and envy-freeness up to one item (EF1). Regarding MMS fairness, the problem does not admit a finite approximation ratio for heterogeneous agents. For homogeneous agents, we design polynomial-time constant-approximation algorithms, and also note that significant amount of social welfare is sacrificed inevitably in order to ensure (approximate) MMS fairness. We then consider EF1 allocations whose existence is guaranteed. However, the social welfare guarantee of EF1 allocations cannot be better than $1/n$ for the general case, where n is the number of agents. Fortunately, for three special cases, two-agent, binary-weight and homogeneous-agent, we are able to design polynomial-time algorithms that also ensure a constant fraction of the maximum social welfare.

1 Introduction

Resource allocation has been actively studied due to its practical applications [19,23,34]. Traditionally, the utilities are assumed to be additive which means an agent's value for a bundle of resources equals the sum of each single item's utility. But in many real-world problems, the resources have graph structures and thus the agents' utilities are not additive but depend on the structural properties of the received resources. To illustrate our model, consider the following real-world

A full version of the paper is available at https://arxiv.org/abs/2212.01031.

F. V. Fomin and M. Xiao (Eds.): COCOON 2025, LNCS 15983, pp. 293–306, 2026.
https://doi.org/10.1007/978-981-95-0215-8_22

scenario. When a department in some company gets assigned a number of engineers who need to work in pairs, the utility of the department may depend on the way to pair these engineers. Due to the engineers' diverse expertise and personalities, an engineer can make different contribution by pairing with different people. We can use a graph to describe the above problem where the engineers are the vertices and the weight on the edge between two vertices represents the collective contribution by pairing these two engineers. Thus the department's utility on a set of vertices is decided by the maximum matching in the induced subgraph of these vertices. To create a good working environment, we want to fairly allocate the engineers to different departments. Besides fairness, the company also wants to maximize the overall efficiency of the assignment. Another example involves radio networks, where each node is assigned a neighbor to back up its data in case of eventual node failures [15]. Similar joint work also appears as long-trip coach driver vs. co-driver and accountant vs. cashier, optimizing the matching of whom can increase efficiency and bring more revenue to the transportation and finance companies [32].

The graphical nature of resources has been considered in the literature (see, e.g., [7,10,25,36]). In this line of research, the graph is used to characterize feasible allocations, such as the resources allocated to each agent should be connected, but the agents still have additive utilities over allocated items. We refer the readers to [37] for a comprehensive survey of constrained fair division. As shown by the previous examples, with graphical resources, the value of a set of resources does not solely depend on the vertex or the edge weights, but is decided by the combinatorial structure of the subgraph, such as the maximum matching in our problem.

Our problem also aligns with the research of balanced graph partition [33]. Although there are heuristic algorithms in the literature [4,28] that partition a graph when the subgraphs are evaluated by maximum matchings, these algorithms do not have theoretical guarantees. Our first fairness criterion is the *maximin share* (MMS) fairness proposed by [11], which generalizes the max-min objective in Santa Claus problem [3]. Informally, the MMS value of an agent is her best guarantee if she is to partition the graph into several subgraphs but receives the worst one. We aim at designing efficient algorithms with provable approximation guarantees. As will be clear later, to achieve (approximate) MMS fairness, a significant amount of social welfare has to be inevitably sacrificed. Our second fairness notion is *envy-freeness* (EF) [20]. In an EF allocation, no agent prefers the bundle of another agent to her own. Since the resources are indivisible, such an allocation rarely exists, and recent research in fair division focuses on achieving its relaxations instead. One of the most widely accepted and studied relaxations is *envy-freeness up to one item* (EF1) [11], which requires the envy to be eliminated after removing one item. [31] proved that an EF1 allocation always exists even with combinatorial valuations.[1] It is noted that an arbitrary EF1 allocation may have low social welfare, and our goal is to compute an EF1

[1] The algorithm in [31] was originally published in 2004 with a different targeting property. In 2011, [11] formally proposed the notion of EF1 fairness.

allocation which preserves a large fraction of the maximum social welfare without fairness constraints. The social welfare loss by enforcing the allocations to be EF1 is quantified by *price of EF1* [6].

1.1 A Summary of Results

We study the fair allocation of graphical resources when the resources are indivisible and correspond to the *vertices* in a graph, and agents' valuations are measured by the weight of the maximum matchings in the induced subgraphs. The fairness of an allocation is measured by maximin share (MMS) and envy-free up to one item (EF1). Our model strictly generalizes the additive setting: it degenerates to the additive setting when the graph consists of a set of independent edges by regarding each edge as an item whose value is the weight of the edge. This is because the removal of a vertex also removes the incident edge. We aim at designing efficient algorithms that compute fair allocations with large social welfare. Our main results are summarized as follows.

We first consider MMS fairness and find that no algorithm has bounded approximation ratio even if there are only two agents with binary weights. We thus focus on the homogeneous case when the agents have identical valuations. Then our problem degenerates to the max-min objective, i.e., partitioning the vertices so that the minimum weight of the maximum matchings in the subgraphs is maximized. It is easy to see that an MMS fair allocation always exists but finding it is NP-hard. Accordingly, we design a polynomial-time 1/2-approximate algorithm for arbitrary number of agents, and show that when the problem only involves two agents, the approximation ratio can be improved to 2/3.

It is noted that, to ensure any finite approximation of MMS fairness, significant amount of social welfare is inevitably sacrificed, which motivates the study of EF1 allocation whose existence is guaranteed [31]. We prove that there exist instances for which none of the EF1 allocations can ensure better than $1/n$ fraction of the maximum social welfare. But this result does not exclude the possibility of constant approximations for special cases. In particular, we consider three cases: (1) two-agent, (2)binary-weight functions, (3) homogeneous-agent. For each setting, as shown in Table 1, we design polynomial-time algorithms that compute EF1 allocations whose social welfare is at least a constant fraction of the maximum social welfare that can be achieved without fairness constraints.

Table 1. Main Results on the Social Welfare Guarantee of EF1 Allocations.

setting	two-agent	binary-weight	homogeneous-agent
approximation ratio	$\frac{1}{3}$ (Theorem 3)	$\frac{1}{3}$ (Theorem 5)	$\frac{2}{3}$ (Theorem 6)

1.2 Related Works

Partitioning graphs into balanced subgraphs has been extensively studied in operations research [8,33] and computer science [12]. There are several popular objectives for evaluating whether a partition is balanced. Among the most prominent ones are the max-min (or min-max) objectives, where the goal is to maximize (or minimize) the total weight of the minimum (or maximum) part. Particularly, the vehicle routing problem (VRP) [27,35,40], which generalizes the travelling salesperson problem (TSP), is closely related to our work. It asks for an optimal set of routes for a number of vehicles to visit a set of customers. Another related problem is the Equitable Connected Partition [9], which aims to find a balanced partition of a graph, ensuring that the size of each connected subgraph is approximately equal. There are many other combinatorial structures studied in graph partitioning problems. For example, in the min-max tree cover (a.k.a. nurse station location) problem, the task is to use trees to cover an edge-weighted graph such that the largest weight of the trees is minimized [26]. This problem also falls under the umbrella of the graph covering problem, where a set of pairwise disjoint subgraphs is used to cover a given graph, such as paths [18], vertex covers [17,39], cycles [38], and matchings [28].

Allocating a set of indivisible items among multiple agents is a fundamental problem in multi-agent systems and computational social choice, and we refer the readers to a recent survey [1] for more detailed discussion. Envy-freeness (EF) and maximin share fairness (MMS) are two well accepted and extensively studied solution concepts. However, with indivisible items, these requirements are demanding and thus the state-of-the-art research mostly studies their relaxations and approximations. For example, EF1 allocation is studied as a relaxation of EF which always exists [31]. Various constant approximation algorithms for MMS allocations are proposed in [21,29] for additive valuations and in [5,22] for subadditive valuations. Some recent works consider graph structures in fair division. For example, there is a line of research where the items are vertices in a graph and each agent is required to receive a connected piece [13,24]. Other works at the intersection of fair division and graphs can be seen, for example, in [2,14,16,30]. Our work is different from previous studies, where the graph structure is incorporated in the valuations instead of constraints.

2 Preliminaries

Denote by $G = (V, E)$ an undirected graph with no reflexive edges, where V contains all vertices and E contains all edges. The vertices are the resources, also called items, that are to be allocated to n heterogeneous agents, denoted by N. Each agent i has an edge weight function $w_i : E \to \mathbb{R}^+ \cup \{0\}$. If $w_i(e) \in \{0, 1\}$ for all $e \in E$, then the weight function is called binary. Let $\mathbf{w} = (w_1, \ldots, w_n)$. A matching $M \subseteq E$ is a set of vertex-disjoint edges, and let $w_i(M) = \sum_{e \in M} w_i(e)$. Let $M(V)$ be the maximum (weighted) matching within the induced subgraph $G[V]$. For any subgraph G', let $V(G')$ and $E(G')$ be the sets of vertices and edges in G', respectively. An allocation $\mathbf{X} = (X_1, \ldots, X_n)$ is a partition of V such that

$\bigcup_{i \in N} X_i = V$ and $X_i \cap X_j = \emptyset$ for $i \neq j$. If $\bigcup_{i \in N} X_i \subsetneq V$, the allocation is called partial. Each agent i has a utility function $u_i : 2^V \to \mathbb{R}^+ \cup \{0\}$, where $u_i(X_i)$ equals the weight of a maximum (weighted) matching in $G[X_i]$. When the agents have identical valuations (i.e., homogeneous agents), we omit the subscript and use $w(\cdot)$ and $u(\cdot)$ to denote all agents' weight and utility functions. A problem instance is denoted by $\mathcal{I} = (G, N)$. When we want to highlight the weight function, w is also included as a parameter, i.e., $\mathcal{I} = (G, N, w)$. Next, we introduce the solution concepts. Our first fairness notion is *maximin share* (MMS) [11]. Denote by $\Pi_n(V)$ the set of all n-partitions of V. The maximin share of agent i is

$$\mathsf{MMS}_i(\mathcal{I}) = \max_{\mathbf{X} \in \Pi_n(V)} \min_{j \in N} u_i(X_j).$$

We may write MMS_i for short if $\mathcal{I}$ is clear from the context. Therefore agent i is satisfied regarding MMS fairness if her utility is no smaller than MMS_i.

Definition 1 (α-MMS). *For any $\alpha \geq 0$, an allocation $\mathbf{X} = (X_1, \ldots, X_n)$ is called* α-approximate maximin share *(α-MMS) fair if for all agents $i \in N$,*

$$u_i(X_i) \geq \alpha \cdot \mathsf{MMS}_i.$$

The allocation is called MMS fair if $\alpha = 1$.

The second fairness notion is *envy-freeness* (EF). An allocation $\mathbf{X}$ is called EF if no agent envies any other agent's bundle, i.e.,

$$u_i(X_i) \geq u_i(X_j) \text{ for all agents } i, j \in N.$$

EF is very hard to satisfy; consider a simple example, where the graph is a triangle and two agents have weight 1 for all edges. Then in every allocation, there is one agent who gets at most one vertex (with utility 0) and the other agent gets at least two vertices (which contains an edge and thus has utility 1). Accordingly, we focus on *envy-free up to one item* [11] instead.

Definition 2 (EF1). *An allocation $\mathbf{X} = (X_1, \ldots, X_n)$ is called* envy-free up to one item *(EF1) if for any i and j, there exists $g \in X_j$ such that $u_i(X_i) \geq u_i(X_j \setminus \{g\})$.*

Besides fairness, we also want the allocation to be efficient. Given an allocation $\mathbf{X} = (X_1, \ldots, X_n)$, the *social welfare* of $\mathbf{X}$ is $\mathsf{sw}(\mathbf{X}) = \sum_{i \in N} u_i(X_i)$. Note that given any instance $\mathcal{I}$, the optimal social welfare of any allocation is the weight of a maximum matching in the graph G by setting the weight of each edge to $\max_{i \in N} w_i(e)$, which is denoted by $\mathsf{sw}^*(\mathcal{I})$. If the instance $\mathcal{I}$ is clear from the context, we also denote $\mathsf{sw}^*(\mathcal{I})$ as sw^* for short.

3 MMS Fair Allocations

3.1 Heterogeneous Agents

Theorem 1. *There exists an instance such that no α-MMS allocation exists for any $\alpha > 0$, even if there are only two agents with non-identical binary weight functions on the graph.*

Proof. Consider the example as shown in Fig. 1. The graph containing four nodes $\{v_1, v_2, v_3, v_4\}$ is allocated to two agents whose valuations (i.e., edge weights) are shown in Fig. 1(a) and 1(b) respectively. It can be verified that $\mathsf{MMS}_i = 1$ for both $i = 1, 2$. However, no matter how we allocate the vertices to the agents, one of them receives utility of 0.

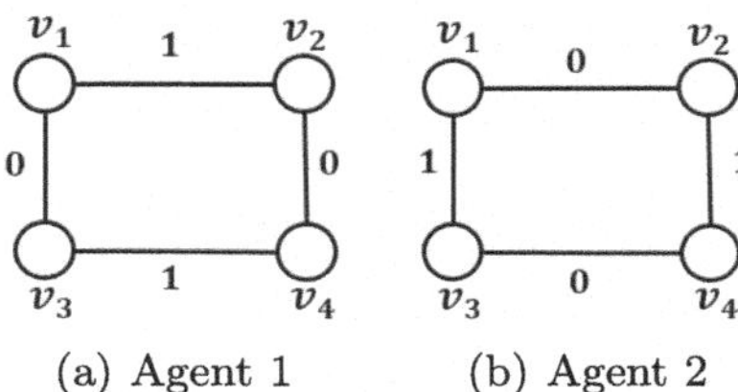

Fig. 1. A bad example for which no allocation has bounded approximation.

Theorem 1 is very strong in the sense that it excludes the possibility of designing algorithms with bounded approximation ratio for MMS even for the special cases of two agents and binary weight functions.

3.2 Homogeneous Agents

Due to the strong impossibility, we study the case of identical valuations, where MMS fairness degenerates to the max-min objective, where the problem is to partition a graph into n subgraphs so that the smallest weight of the maximum matchings in these subgraphs is maximized. It is easy to see that finding such an allocation is NP-hard even when there are two agents and the graph only consists of independent edges, which is essentially a Partition problem. Thus, our target is polynomial-time approximation algorithms. Without loss of generality, in this section, we assume $w(e) \geq 1$ for all $e \in E$. We emphasize that this can be regarded as a scaled operation, which does not affect the computation of the approximate MMS allocations. Specifically, we identify an edge $e \in E$ with the smallest weight $w(e)$, and then for each edge $e' \in E \setminus \{e\}$, we reset its weight to $w(e')/w(e)$. Since the agents are identical, the subscript in MMS_i is omitted. Our main result in this section is the following theorem.

Theorem 2. *For homogeneous agents, we can compute a* $1/2$*-MMS allocation in polynomial time.*

Given an instance $\mathcal{I} = (G, N)$, to design such an algorithm with guaranteed approximation of MMS, the first natural idea is to allocate a maximum matching in G. That is, we compute a maximum matching $M^* \subseteq E$, and then partition M^* into n bundles $(M_1, \ldots, M_n)$ where $w(M_1) \geq \cdots \geq w(M_n)$ such that $w(M_n)$ is as large as possible. However, such an allocation can be arbitrarily bad. Consider an example with two agents shown in Fig. 2 where $\Delta > 1$ is arbitrarily large. Any allocation with bounded approximation ratio of MMS fairness ensures that every agent has value 1, but by partitioning the maximum matching (which contains a single edge with weight Δ) the smaller bundle has value 0. Before describing our algorithm, we first define greedy partition of the maximum matching.

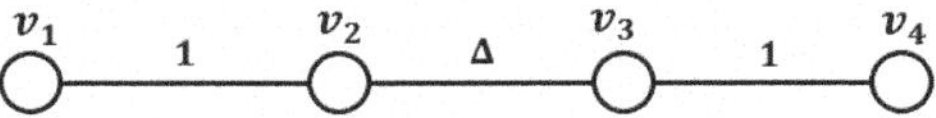

Fig. 2. A bad example when partitioning the maximum matching does not have bounded approximation of MMS.

Greedy Partition. Given a matching M, partition M into $\Gamma(M) = (M_1, \ldots, M_n)$ as follows.

- Sort and rename edges in M such that $w(e_1) \geq \cdots \geq w(e_k)$ where $k = |M|$.
- Initially set $M_1 = \cdots = M_n = \emptyset$.
- For $i = 1, \ldots, k$, select j such that $w(M_j) \leq w(M_{j'})$ for all j' and set $M_j = M_j \cup \{e_i\}$.
- Sort and rename $M_1, \ldots, M_n$ so that $w(M_1) \geq \cdots \geq w(M_n)$.

The greedy partition $(M_1, \ldots, M_n)$ of the maximum matching M^* corresponds to an allocation where unmatched vertices $V' = V \setminus \cup_{i \in N} V(M_i)$ can be allocated arbitrarily. Although this allocation may not be good in general, it ensures a good approximation when the graph is unweighted ($w(e) = w(e')$ for all $e, e' \in E$) or $|M_1| \geq 2$. We have the following lemmas.

Lemma 1. *If G is unweighted, the greedy partition $(M_1, \ldots, M_n)$ of M^* is an MMS allocation.*

Proof. Without loss of generality, assume all edges have weight 1. In the greedy partition $(M_1, \ldots, M_n)$ of M^*, for any $i \in N$,

$$|M_i| \geq |M_n| = \lfloor \frac{|M^*|}{n} \rfloor.$$

Let $(O_1, \ldots, O_n)$ $(u(O_1) \geq \cdots \geq u(O_n))$ be an optimal max-min allocation. If $\mathsf{opt} = u(O_n) > |M_n|$, then for all $i \in N$,

$$u(O_i) \geq \lfloor \frac{|M^*|}{n} \rfloor + 1.$$

Thus

$$\sum_{i \in N} u(O_i) \geq n \cdot \lfloor \frac{|M^*|}{n} \rfloor + n > |M^*|,$$

which is a contradiction with M^* being a maximum matching.

Lemma 2. *Given a greedy partition* $(M_1, \ldots, M_n)$ *of* M^*, *if* $|M_1| \geq 2$, $\Gamma(M^*)$ *corresponds to an allocation that is* 1/2*-MMS fair.*

Proof. Denote by $O = (O_1, O_2, \ldots, O_n)$ the optimal max-min allocation, where $u(O_1) \geq u(O_2) \geq \cdots \geq u(O_n)$ and $\mathsf{opt}(\mathcal{I}) = u(O_n)$. Under the maximum matching M^*, consider the greedy partition $(M_1, M_2, \ldots, M_n)$, where $w(M_1) \geq w(M_2) \geq \cdots \geq w(M_n)$. In the greedy partition procedure, all edges are sorted in descending order of their weights and each time we select the edge with the largest weight in the remaining edge set and allocate it to the bundle with the least total utility. If $|M_1| \geq 2$, consider the last edge e added to M_1. We have $w(M_n) \geq w(e)$, since there exists at least one edge added to M_n before edge e is added to M_1. Since in the greedy procedure, edges are added to the bundle with least utility, we have $w(M_n) \geq w(M_1 \setminus \{e\})$. Furthermore, we have

$$\begin{aligned} w(M_n) &\geq \frac{1}{2}(w(e) + w(M_1 \setminus \{e\})) \geq \frac{1}{2} w(M_1) \\ &\geq \frac{1}{2n} \sum_{i=1}^{n} w(M_i) \geq \frac{1}{2n} \sum_{i=1}^{n} u(O_i) \\ &\geq \frac{1}{2} u(O_n), \end{aligned}$$

and the lemma holds accordingly.

Now we are ready to describe our algorithm. Although the value of $\mathsf{MMS}(\mathcal{I})$ is unknown, it must lie within the interval $[1, w(M^*)]$, or $\mathsf{MMS}(\mathcal{I}) = 0$ since $w(e) \geq 1$ for each edge $e \in E$. Note that the case $\mathsf{MMS}(I) = 0$ is trivial, as any allocation is MMS fair. Consequently, we focus on the case where $\mathsf{MMS}(\mathcal{I}) \in [1, w(M^*)]$. Let $\epsilon = 1/(3(n-1))$. To guess the possible value of $\mathsf{MMS}(\mathcal{I})$, we divide $[1, w(M^*)]$ into several disjoint sub-intervals $[(1+\epsilon)^t, (1+\epsilon)^{t+1}), t = 0, 1, \ldots, \lceil \log_{1+\epsilon} w(M^*) \rceil - 1$. For each sub-interval $[(1+\epsilon)^t, (1+\epsilon)^{t+1})$, we update the edge weights. Let H be the set of edges with the weights larger than $(1+\epsilon)^{t+1}$, and decrease all their weights to $(1+\epsilon)^{t+1}$, then we find the greedy partition $\Gamma(M^*) = (M_1, \ldots, M_n)$ $(w(M_1) \geq \cdots \geq w(M_n))$ of the maximum matching M^*. For the sub-interval that the $\mathsf{MMS}(\mathcal{I})$ value belongs to, the greedy partition can ensure $w(M_n) \geq 1/2 \cdot \mathsf{MMS}(\mathcal{I})$. Since we consider all the possible sub-intervals and output the best value $w(M_n)$, the final allocation can ensure 1/2-MMS.

Before proving Theorem 2, we present the following lemma.

Lemma 3. *Let* $\mathcal{I}' = (G', N), G' = (V, E; \mathbf{w}')$ *be the instance obtained from* $\mathcal{I} = (G, N), G = (V, E; \mathbf{w})$ *by rounding all edge weights down to a constant* $k \geq \mathsf{MMS}(\mathcal{I})$. *Then* $\mathsf{MMS}(\mathcal{I}) = \mathsf{MMS}(\mathcal{I}')$.

Proof. For any O_i, if $w(e) \leq k$ for all $e \in M(O_i)$, then no weights of the edges in O_i are rounded down and thus $u(O_i)$ does not decrease. If $w(e) > k$ for some $e \in M(O_i)$, then $u(O_i) \geq w(e) \geq k \geq \mathsf{MMS}(\mathcal{I})$ and thus after decreasing the weights to k, $u'(O_i) \geq \mathsf{MMS}(\mathcal{I})$, implying the existence of an allocation with the minimum utility no smaller than $\mathsf{MMS}(\mathcal{I})$, which means $\mathsf{MMS}(\mathcal{I}') = \mathsf{MMS}(\mathcal{I})$.

Algorithm 1 Approximately MMS Fair Algorithm

Input: Instance $\mathcal{I} = (G, N)$ with $G = (V, E; w)$.
Output: Allocation $\mathbf{X} = (X_1, \ldots, X_n)$.

Initialize $M_i^* \leftarrow \emptyset, i \in N$, $k \leftarrow 1$, $t \leftarrow \lceil \log_{1+\epsilon} K \rceil - 1$.
Find a maximum matching M^* in G.
Initialize $K \leftarrow w(M^*)$.
while $t \geq 0$ **do**
  $k \leftarrow (1+\epsilon)^t$.
  Let $H = \{e \in E \mid w(e) > k(1+\epsilon)\}$.
  Set $w(e) \leftarrow k(1+\epsilon)$ for all $e \in H$.
  Find a maximum matching M^* in G.
  Find greedy partition $\Gamma(M^*) = (M_1, \ldots, M_n)$ of M^* s.t. $w(M_1) \geq \cdots \geq w(M_n)$.
  if $w(M_n) \geq w(M_n^*)$ **then**
    $M_i^* \leftarrow M_i, i \in N$.
  end if
  $t \leftarrow t - 1$.
end while
$X_i \leftarrow V(M_i^*), i \in N$.
Return the allocation $(X_1, \ldots, X_n)$.

Now, we are ready to prove Theorem 2.

Proof of Theorem 2. Note that $\mathsf{MMS}(\mathcal{I}) = 0$ is a trivial case, since any allocation returned by our algorithm is certainly an MMS allocation. It suffices to consider the case that $\mathsf{MMS}(\mathcal{I}) \geq 1$. Given a constant k, denote by $O^k = (O_1^k, \ldots, O_n^k)$ the optimal max-min allocation and $\Gamma(M^k) = (M_1^k, \ldots, M_n^k)$ (assume $w(M_1^k) \geq \cdots \geq w(M_n^k)$) the greedy partition of the maximum matching M^k after the weights of all edges in H are decreased to $k(1+\epsilon)$. $\mathbf{X} = (X_1, \ldots, X_n)$ is the allocation returned by Algorithm 1. We first discuss the case when $k \leq \mathsf{MMS}(\mathcal{I}) \leq k(1+\epsilon)$. There are two subcases.

- Subcase 1. $w(M_n^k) \geq \frac{1}{2} w(M_1^k)$;
- Subcase 2. $w(M_n^k) < \frac{1}{2} w(M_1^k)$.

First, consider Subcase 1. By Lemma 3, for any edge e such that $w(e) > k(1+\epsilon)$, decreasing the weight to $k(1+\epsilon) \geq \mathsf{MMS}(\mathcal{I})$ does not change the optimal solution, i.e., $u(O_n^k) = \mathsf{MMS}(\mathcal{I})$. It holds that

$$w(M_n^k) \geq \frac{1}{2n} \sum_{i=1}^{n} w(M_i^k) \geq \frac{1}{2n} \sum_{i=1}^{n} u(O_i^k) \geq \frac{1}{2} u(O_n^k) = \frac{1}{2} \mathsf{MMS}(\mathcal{I}).$$

If all edges have the same weight, the allocation is optimal by Lemma 1. Next, consider Subcase 2 and the edge weights are not identical. From the proof of Lemma 2, when $|M_1^k| \geq 2$, we know that $w(M_1^k) \leq 2w(M_n^k)$, which contradicts the condition of Subcase 2. Therefore, $|M_1^k| = 1$, implying that $w(M_1^k) \leq k(1+\epsilon)$. Thus,

$$w(M_n^k) \leq \cdots \leq w(M_1^k) \leq k(1+\epsilon) \leq (1+\epsilon)\mathsf{MMS}(\mathcal{I}).$$

Since

$$\sum_{i=1}^{n} w(M_i^k) \geq \sum_{i=1}^{n} u(O_i^k) \geq n \cdot u(O_n^k) = n \cdot \mathsf{MMS}(\mathcal{I}),$$

we have

$$\begin{aligned} w(M_n^k) &\geq n \cdot \mathsf{MMS}(\mathcal{I}) - \sum_{i=1}^{n-1} w(M_i^k) \geq n \cdot \mathsf{MMS}(\mathcal{I}) - (n-1)(1+\epsilon)\mathsf{MMS}(\mathcal{I}) \\ &= (1 - \epsilon(n-1))\mathsf{MMS}(\mathcal{I}) > \frac{1}{2}\mathsf{MMS}(\mathcal{I}). \end{aligned}$$

Therefore, $w(M_n^k) \geq \frac{1}{2}\mathsf{MMS}(\mathcal{I})$ when $k \leq \mathsf{MMS}(\mathcal{I}) \leq k(1+\epsilon)$. Note that Algorithm 1 traverses all the values $k = (1+\epsilon)^t, t = 0, 1, \ldots, \lceil \log_{1+\epsilon} w(M^*) \rceil - 1$ and returns the greedy partition $(M_1^*, \ldots, M_n^*)$ such that $w(M_n^*)$ is maximized. Therefore,

$$u(X_n) \geq \max_k w(M_n^k) \geq \frac{1}{2}\mathsf{MMS}(\mathcal{I}),$$

as desired. It remains to analyze the running time of Algorithm 1. Observe that the outer **while** loop will execute for $O(\log_{1+\epsilon} w(M^*))$ rounds. For each loop the weights of some edges are rounded down to a power of $1+\epsilon$, which is completed in $O(|V|^2)$ time. A maximum matching can be found in $O(|V|^3)$ time and a greedy partition can be computed in $O(|V|^2)$ time. Therefore, the time complexity of Algorithm 1 is $O(|V|^3 \log_{1+\epsilon} w(M^*))$. Note that

$$\log_{1+\epsilon} w(M^*) = \frac{\ln(w(M^*))}{\ln(1+\epsilon)} = \frac{\ln(w(M^*))}{\ln(1+\frac{1}{3(n-1)})} \leq \frac{\ln(w(M^*))}{\frac{1}{6(n-1)}} = 6(n-1)\ln(w(M^*)),$$

where the inequality follows from $\ln(1+x) \geq x/2, \forall x \in (0,1)$. Therefore, Algorithm 1 completes in $O(n|V|^3 \ln(w(M^*)))$ time and we complete the proof. The corresponding example showing that the analysis in Theorem 2 is asymptotically tight is presented as follows. Consider an instance $\mathcal{I}$ with $4n$ vertices and $4n-1$ edges such that

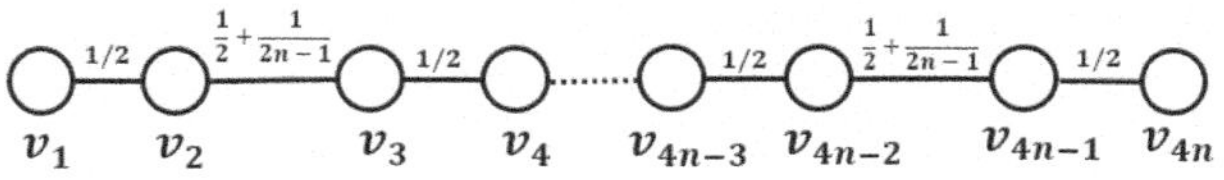

Fig. 3. An example showing that the analysis in Theorem 2 is asymptotically tight.

$$w((v_j, v_{j+1})) = \begin{cases} \dfrac{1}{2}, & \text{if } j \text{ is odd,} \\ \dfrac{1}{2} + \dfrac{1}{2n-1}, & \text{if } j \text{ is even,} \end{cases}$$

shown in Fig. 3. It can be easily verified that $\mathsf{MMS}(\mathcal{I}) = 1$ by assigning agent $i \in N$ a set $X_i = \{v_{4n-3}, v_{4n-2}, v_{4n-1}, v_{4n}\}$ of vertices. The maximum matching is $M^* = \{(v_j, v_{j+1}) \mid j \text{ is even}\}$ and

$$w(M^*) = 2(\frac{1}{2} + \frac{1}{2n-1})(n-1) + \frac{1}{2} + \frac{1}{2n-1} = n + \frac{1}{2}.$$

By Algorithm 1, $M_i = 1 + \frac{2}{2n-1}, i \in [n-1]$ and $M_n = \frac{1}{2} + \frac{1}{2n-1}$. Thus

$$\lim_{n \to +\infty} \frac{w(M_n)}{\mathsf{MMS}(\mathcal{I})} = \frac{1}{2}.$$

Remark. When $n = 2$, we improve Algorithm 1 and obtain a better approximation ratio of 2/3 in the full version.

4 EF1 Allocations

Recall the example in Fig. 2. The maximum social welfare is $\mathsf{sw}^* = \Delta$, but any bounded-approximate MMS allocation has social welfare $2 \ll \Delta$, which means to ensure (approximate) MMS, we have to sacrifice a significant amount of social welfare. Thus in this section, we turn to study EF1 allocations, whose existence is guaranteed [31]. However, an arbitrary EF1 allocation does not have any social welfare guarantee, and our goal in this section is to compute an EF1 allocation that also preserves relatively good social welfare. We begin with the two-agent case and present a polynomial-time algorithm which computes an EF1 allocation with at least 1/3 fraction of the optimal social welfare. Moreover, we present a counterexample showing that the approximate ratio 1/3 is tight, i.e., any EF1 allocation guarantees at most 1/3 fraction of the optimal social welfare. Then we have the following theorem, proved in the full version.

Theorem 3. *For any instance $\mathcal{I}$ with two heterogeneous agents, there exists an EF1 allocation with social welfare at least $1/3 \cdot \mathsf{sw}^*(\mathcal{I})$, which is optimal.*

Unfortunately, for the general case, we found that EF1 allocations cannot guarantee good social welfare either.

Theorem 4. *For heterogeneous agents, there exists an instance such that no EF1 allocation can guarantee better than $1/n$ fraction of the optimal social welfare without fairness constraints.*

We prove Theorem 4 in the full version, and complement it by a polynomial-time EF1 algorithm that achieves $\Omega(1/n^2)$ approximation of the social welfare for the general case. Given the hardness result for the general case, we now impose some restrictions on the valuation functions and present two special cases: binary weight functions and homogeneous agents. In these cases, EF1 allocations can guarantee a constant fraction of the optimal social welfare. Due to space limitations, we only present the main idea for the binary-weight-function case, while the analysis of another case is shown in the full version. Now, we show that if the agents have binary weight functions, we can compute an EF1 allocation whose social welfare is at least 1/3 of the optimal social welfare. Recall that the *envy-cycle elimination algorithm* proposed by [31] always returns an EF1 allocation. However, it does not have any social welfare guarantee, which is illustrated in the full version. Consequently, to increase the social welfare, in each round of our algorithm, we try to allocate an edge (i.e., two items) to the agent i with the smallest value so that the social welfare can increase by 1. However, we need to be very careful when allocating two items, since it may break the EF1 requirement even if i is not envied by the others. If allocating an edge e to i makes some agent j envy i for more than one item, we check whether i can maintain her utility by selecting a bundle from unallocated items. If so, we execute *exchange* procedure by asking j to (properly) select a bundle from X_i and asking i to (properly) select a bundle from unallocated items so that the social welfare is increased by 1. All the items in X_j and the items in X_i that are not selected by j are returned to the pool. If not, we try to allocate an edge to the agent with the second smallest value by executing the above procedures, and so on. Then we have the following theorem.

Theorem 5. *For any instance $\mathcal{I} = (G, N)$ with binary weights, there exists an EF1 allocation with social welfare at least $1/3 \cdot \mathsf{sw}^*(\mathcal{I})$, which can be computed in polynomial time*

Furthermore, regarding the homogeneous-agent case, we have Theorem 6.

Theorem 6. *For homogeneous instance $\mathcal{I}$, there exists an EF1 allocation with social welfare at least $(2/3) \cdot \mathsf{sw}^*(\mathcal{I})$, which can be computed in polynomial time.*

5 Conclusion and Future Directions

In this work, we study the fair (and efficient) allocation of graphical resources when the agents' utilities are determined by the weights of the maximum matchings in the obtained subgraphs. We provide a string of algorithmic results regarding MMS and EF1, but also leave some problems open. For example, regarding MMS, we can further improve the approximation ratio when the agents are homogeneous, and prove inapproximability results; regarding EF1, the social welfare guarantee for binary weight functions and homogeneous agents can be potentially improved. Our work also uncovers some other interesting future directions. First, regarding MMS, although we show that there is no bounded multiplicative approximation, it may admit good additive or bi-factor approximations. Second, in addition to matching, it is also intriguing to consider other combinatorial structures such as independent set, network flow and more. Third, we can extend the framework to the fair allocation of graphical chores when the agents have costs to complete the items, and the asymmetric situation when the agents have possibly different weights (like entitlements) to the system.

Acknowledgments. This work is supported in part by NSFC No. 12271477, HK RGC No. PolyU 25211321, and GDSTC No. 2024A1515011524.

References

1. Amanatidis, G., et al.: Fair division of indivisible goods: recent progress and open questions. Artif. Intell. **322**, 103965 (2023)
2. Aziz, H., Bouveret, S., Caragiannis, I., Giagkousi, I., Lang, J.: Knowledge, fairness, and social constraints. In: Proceedings of the Thirty-Second AAAI Conference on Artificial Intelligence, AAAI, pp. 4638–4645. AAAI Press (2018)
3. Bansal, N., Sviridenko, M.: The Santa Claus problem. In: STOC, pp. 31–40 (2006)
4. Barketau, M., Pesch, E., Shafransky, Y.M.: Minimizing maximum weight of subsets of a maximum matching in a bipartite graph. Discret. Appl. Math. **196**, 4–19 (2015)
5. Barman, S., Krishnamurthy, S.K.: Approximation algorithms for maximin fair division. ACM Trans. Economics Comput. **8**(1), 5:1–5:28 (2020)
6. Bei, X., Lu, X., Manurangsi, P., Suksompong, W.: The price of fairness for indivisible goods. Theory Comput. Syst. **65**(7), 1069–1093 (2021)
7. Bilò, V., et al.: Almost envy-free allocations with connected bundles. In: ITCS, vol. 124, pp. 14:1–14:21 (2019)
8. Blazej, V., Ganian, R., Knop, D., Pokorný, J., Schierreich, S., Simonov, K.: The parameterized complexity of network microaggregation. In: Thirty-Seventh AAAI Conference on Artificial Intelligence, AAAI, pp. 6262–6270. AAAI Press (2023)
9. Blazej, V., Knop, D., Pokorný, J., Schierreich, S.: Equitable connected partition and structural parameters revisited: N-fold beats lenstra. In: 49th International Symposium on Mathematical Foundations of Computer Science, MFCS. LIPIcs, vol. 306, pp. 29:1–29:16. Schloss Dagstuhl - Leibniz-Zentrum für Informatik (2024)
10. Bouveret, S., Cechlárová, K., Elkind, E., Igarashi, A., Peters, D.: Fair division of a graph. In: IJCAI, pp. 135–141. ijcai.org (2017)
11. Budish, E.: The combinatorial assignment problem: approximate competitive equilibrium from equal incomes. J. Polit. Econ. **119**(6), 1061–1103 (2011)
12. Buluç, A., Meyerhenke, H., Safro, I., Sanders, P., Schulz, C.: Recent advances in graph partitioning. In: Algorithm Engineering, Lecture Notes in Computer Science, vol. 9220, pp. 117–158 (2016)
13. Caragiannis, I., Micha, E., Shah, N.: A little charity guarantees fair connected graph partitioning. In: Thirty-Sixth AAAI Conference on Artificial Intelligence, AAAI, pp. 4908–4916. AAAI Press (2022)
14. Chevaleyre, Y., Endriss, U., Maudet, N.: Distributed fair allocation of indivisible goods. Artif. Intell. **242**, 1–22 (2017)
15. Dani, V., Gupta, A., Hayes, T.P., Pettie, S.: Wake up and join me! an energy-efficient algorithm for maximal matching in radio networks. Distributed Comput. **36**(3), 373–384 (2023)
16. Eiben, E., Ganian, R., Hamm, T., Ordyniak, S.: Parameterized complexity of envy-free resource allocation in social networks. Artif. Intell. **315**, 103826 (2023)
17. Epstein, L., Levin, A., Woeginger, G.J.: Vertex cover meets scheduling. Algorithmica **74**(3), 1148–1173 (2016)
18. Farbstein, B., Levin, A.: Min-max cover of a graph with a small number of parts. Discret. Optim. **16**, 51–61 (2015)
19. Flanigan, B., Gölz, P., Gupta, A., Hennig, B., Procaccia, A.D.: Fair algorithms for selecting citizens' assemblies. Nature 1–5 (2021)
20. Foley, D.K.: Resource allocation and the public sector. Yale Econ. Essays **7** (1967)
21. Garg, J., Taki, S.: An improved approximation algorithm for maximin shares. Artif. Intell. **300** (2021)

22. Ghodsi, M., Hajiaghayi, M.T., Seddighin, M., Seddighin, S., Yami, H.: Fair allocation of indivisible goods: improvements and generalizations. In: Proceedings of the 2018 ACM Conference on Economics and Computation, EC, pp. 539–556. ACM (2018)
23. Goldman, J.R., Procaccia, A.D.: Spliddit: unleashing fair division algorithms. SIGecom Exch. **13**(2), 41–46 (2014)
24. Igarashi, A.: How to cut a discrete cake fairly. In: Thirty-Seventh AAAI Conference on Artificial Intelligence, AAAI. pp. 5681–5688. AAAI Press (2023)
25. Igarashi, A., Peters, D.: Pareto-optimal allocation of indivisible goods with connectivity constraints. In: Thirty-Third AAAI Conference on Artificial Intelligence, AAAI, pp. 2045–2052. AAAI Press (2019)
26. Khani, M.R., Salavatipour, M.R.: Improved approximation algorithms for the min-max tree cover and bounded tree cover problems. Algorithmica **69**(2), 443–460 (2014)
27. Koç, Ç., Bektas, T., Jabali, O., Laporte, G.: Thirty years of heterogeneous vehicle routing. Eur. J. Oper. Res. **249**(1), 1–21 (2016)
28. Kress, D., Meiswinkel, S., Pesch, E.: The partitioning min-max weighted matching problem. Eur. J. Oper. Res. **247**(3), 745–754 (2015)
29. Kurokawa, D., Procaccia, A., Wang, J.: Fair enough: guaranteeing approximate maximin shares. J. ACM **65**(2), 8 (2018)
30. Li, L., Micha, E., Nikolov, A., Shah, N.: Partitioning friends fairly. In: Thirty-Seventh AAAI Conference on Artificial Intelligence, AAAI, pp. 5747–5754. AAAI Press (2023)
31. Lipton, R.J., Markakis, E., Mossel, E., Saberi, A.: On approximately fair allocations of indivisible goods. In: EC, pp. 125–131. ACM (2004)
32. Lovász, L., Plummer, M.D.: Matching theory, vol. 367. American Mathematical Soc. (2009)
33. Miyazawa, F.K., Moura, P., Ota, M.J., Wakabayashi, Y.: Partitioning a graph into balanced connected classes: formulations, separation and experiments. Eur. J. Oper. Res. **293**(3), 826–836 (2021)
34. Moulin, H.: Fair Division and Collective Welfare. MIT Press (2003)
35. Rathinam, S., Ravi, R., Bae, J., Sundar, K.: Primal-dual 2-approximation algorithm for the monotonic multiple depot heterogeneous traveling salesman problem. In: SWAT. LIPIcs, vol. 162, pp. 33:1–33:13 (2020)
36. Suksompong, W.: Fairly allocating contiguous blocks of indivisible items. Discret. Appl. Math. **260**, 227–236 (2019)
37. Suksompong, W.: Constraints in fair division. SIGecom Exch. **19**(2), 46–61 (2021)
38. Traub, V., Tröbst, T.: A fast $(2 + 2/7)$-approximation algorithm for capacitated cycle covering. In: IPCO, pp. 391–404. Springer (2020)
39. Wang, F., Li, B.: Fair surveillance assignment problem. In: WWW, pp. 178–186. ACM (2024)
40. Yaman, H.: Formulations and valid inequalities for the heterogeneous vehicle routing problem. Math. Program. **106**(2), 365–390 (2006)

Equivalence of Connected and Peak-Pit Maximal Condorcet Domains

Guanhao Li(✉)

Department of Mathematics, University of Auckland, Auckland, New Zealand
gli103@aucklanduni.ac.nz

Abstract. This paper provides a combinatorial proof to show that, in the study of maximal Condorcet domains, the class of peak-pit Condorcet domains, the class of connected Condorcet domains, and the class of directly connected Condorcet domains are all equivalent.

Keywords: Combinatorics · Condorcet Domains · Social Choice Theory

1 Introduction

Condorcet domains are sets of linear orders on a given set of alternatives such that any majority relation induced by these orders is acyclic; see [15] for seven different yet equivalent definitions. These domains hold a crucial place in combinatorics and are especially useful in social choice theory, where they help address voting paradoxes and aim for more rational ways to aggregate preferences [1]. Structurally, Condorcet domains can be represented as graphs, and research often focuses on their connectedness properties, particularly in peak-pit Condorcet domains. From a computational perspective, these domains present significant challenges, which have led to the development of efficient algorithms from combinatorial optimization to handle real-world voting scenarios [3]. Additionally, counting and listing preference orderings within these domains helps classify different types of domains [9]. One of the major research questions in the literature focuses on finding the largest size of Condorcet domains for a given number of alternatives [15]. A central research question involves characterising maximal Condorcet domains by combinatorical objects.

The class of peak-pit maximal Condorcet domains has been a primary object of investigation in the search for large Condorcet domains. Using combinatorial tools and computational methods, the largest sizes of this class have been proven to match the maximal cardinality of Condorcet domains for $n \leq 8$ alternatives [6–8,10]. Consequently, it has been shown that all such peak-pit maximal Condorcet domains are connected, where any two linear orders in the domain can be transformed into one another through a sequence of adjacent swaps within the domain. [4] demonstrated that connected Condorcet domains can support strategyproof social decision schemes. This means that within such connected

F. V. Fomin and M. Xiao (Eds.): COCOON 2025, LNCS 15983, pp. 307–319, 2026.
https://doi.org/10.1007/978-981-95-0215-8_23

domains, voters have no incentive to misrepresent their preferences, leading to more honest and reliable outcomes. Thus, understanding connected Condorcet domains is crucial for designing voting systems that are resistant to paradoxes and manipulation, ensuring fairer decision-making processes.

Previous research has shown some connections between these two subclasses of Condorcet domain, yet the question of their equivalence remains open. [5] proved that all peak-pit maximal Condorcet domains of maximal width are connected. [11] first used a mathematical method and proved that even without the restriction of maximal width, connectedness and peak-pittedness are equivalent on four alternatives. Later, [12] attempted to show this equivalence is valid for any arbitrary number of alternatives. However, [17] provided a counter-example showing that the work of [12] is incomplete. Since their counter-example does not correspond to a maximal Condorcet domain, the problem remains open. [16] conjectured this equivalence in their survey. In this paper, we aim to work from a different perspective and show the equivalence of connected and peak-pit maximal Condorcet domains.

Furthermore, within the class of connected Condorcet domains, the subclass of directly connected Condorcet domains consists of those domains where any two orders can be connected by a shortest path in the permutahedron, with all intermediate orders lying within the domain. This is equivalent to the non-restoration property identified by [18], according to which any two linear orders can be linked by a path along which no pair of alternatives is swapped more than once. [18] showed that direct connectedness is exactly the feature underpinning the equivalence between local and global strategy-proofness. Moreover, [18] proved that directly connected single-peaked domains admit concise characterisations of strategy-proof rules. Hence, it is practically important to characterise directly connected domains in the context of Condorcet domains. [14] showed that all peak-pit maximal Condorcet domains of maximal width are not just connected but in fact directly connected. In this paper, we will show that the connectedness property is equivalent to the directly connectedness property within the class of maximal Condorcet domains.

Finally, our findings establish that the class of *maximal peak-pit Condorcet domains* is the same as the class of *peak-pit maximal Condorcet domains*. While there is a belief in the literature that these two concepts are equivalent, and the terms are often used interchangeably. However, no formal mathematical definitions have been provided to clarify their relationship, which leads to confusion. This paper resolves this ambiguity by providing a formal definition and proving their equivalence.

Due to space constraints, the full proof of this paper is available in [13].

2 Condorcet Domains

Suppose A is a finite set of alternatives. Any set of linear orders on A is called **a domain of linear orders**, or shortly **a domain**. Let $\mathcal{L}(A)$ denote the set of all linear orders on A. Let $\mathcal{D} \subseteq \mathcal{L}(A)$ be a domain of linear orders on A. We denote the **restriction** of $\mathcal{D}$ to any subset S of A by $\mathcal{D}_S$, where $\mathcal{D}_S$ is also called the **restricted domain** of $\mathcal{D}$ to S. The restriction of $\mathcal{D}$ to $A \setminus \{a\}$ is denoted by $\mathcal{D}_{-a}$, where $a \in A$. The restriction of $\mathcal{D}$ to $A \setminus X$ is denoted by $\mathcal{D}_{-X}$, where $X \subset A$. For every linear order $R \in \mathcal{D}$, the **restriction** of R to $A \setminus \{a\}$ is denoted by R_{-a}, where R_{-a} is also called the **restricted linear order** of R to $A \setminus \{a\}$. The restriction of R to a subset S of A is denoted by R_S.

Definition 1. *Let A be a finite set of alternatives and $\mathcal{D} \subseteq \mathcal{L}(A)$ a domain of linear orders. Let $a, b, c \in A$ be a triple of distinct alternatives. Let $x \in \{a, b, c\}$ and $k \in \{1, 2, 3\}$. Then we say that the restricted domain $\mathcal{D}_{\{a,b,c\}}$ satisfies the* **never-condition** *$xN_{\{a,b,c\}}k$ if for all $R = x_1x_2x_3 \in \mathcal{D}_{\{a,b,c\}}$, one has $x \neq x_k$. We say that $\mathcal{D}_{\{a,b,c\}}$ satisfies a* **never-top condition** *if there exists an $x \in \{a, b, c\}$ such that $\mathcal{D}_{\{a,b,c\}}$ satisfies the never-condition $xN_{\{a,b,c\}}1$. Similarly, we say that $\mathcal{D}_{\{a,b,c\}}$ satisfies a* **never-bottom condition** *if there exists an $x \in \{a, b, c\}$ such that $\mathcal{D}_{\{a,b,c\}}$ satisfies the never-condition $xN_{\{a,b,c\}}3$. The definition for $\mathcal{D}_{\{a,b,c\}}$ satisfying a* **never-middle condition** *is similar.*

We say that $\mathcal{D}$ is a **never-top domain** *if for every distinct triple $a, b, c \in A$, the restricted domain $\mathcal{D}_{\{a,b,c\}}$ satisfies a never-top condition. Similarly, we say that $\mathcal{D}$ is a* **never-bottom domain** *if for every distinct triple $a, b, c \in A$, the restricted domain $\mathcal{D}_{\{a,b,c\}}$ satisfies a never-bottom condition. The definition for a* **never-middle domain** *is similar. We say that $\mathcal{D}$ is a* **Condorcet domain** *if for every distinct triple $a, b, c \in A$, the restricted domain $\mathcal{D}_{\{a,b,c\}}$ is a never-top domain or never-bottom domain or never-middle domain. A domain $\mathcal{D}$ is called a* **peak-pit Condorcet domain** *if for every distinct triple $a, b, c \in A$, the restricted domain $\mathcal{D}_{\{a,b,c\}}$ is a never-top domain or never-bottom domain. A never-condition that is a never-top or never-bottom condition is called a* **peak-pit condition**.

Remark 1. A domain $\mathcal{D}$ is a Condorcet domain if and only if for every distinct $a, b, c \in A$, the restricted domain $\mathcal{D}_{\{a,b,c\}}$ is a Condorcet domain.

Given a Condorcet domain $\mathcal{D}$, the **set of never-conditions satisfied by** $\mathcal{D}$ is denoted by $N(\mathcal{D})$, and the **set of peak-pit-conditions satisfied by** $\mathcal{D}$ is denoted by $N_p(\mathcal{D})$.

One can order Condorcet domains by set inclusion. A Condorcet domain $\mathcal{D} \subseteq \mathcal{L}(A)$ is called **maximal** if for every Condorcet domain $\mathcal{D}'$ from $\mathcal{L}(A)$ with $\mathcal{D} \subseteq \mathcal{D}'$, it follows that $\mathcal{D} = \mathcal{D}'$. Equivalently, it means that a Condorcet domain $\mathcal{D}$ is maximal if for every linear order $R \notin \mathcal{D}$, the union $\mathcal{D} \cup \{R\}$ is not a Condorcet domain. We are interested in maximal Condorcet domains which are peak-pit Condorcet domains. The maximal Condorcet domains for three alternatives are well-known (see, e.g., [6]). It will still be useful for us to have the following lemma.

Lemma 1. *Let a, b, c be three distinct alternatives. Define*

$$\begin{aligned} \mathcal{D}_{3,t} &= \{abc, acb, cab, cba\}, \\ \mathcal{D}_{3,m} &= \{abc, bca, acb, cba\} \text{ and} \\ \mathcal{D}_{3,b} &= \{abc, bac, bca, cba\}. \end{aligned}$$

Let $\mathcal{D} \subseteq \mathcal{L}(\{a, b, c\})$ be a Condorcet domain. Then the following are equivalent:

(i) *$\mathcal{D}$ is a maximal Condorcet domain;*
(ii) $|\mathcal{D}| = 4$.
(iii) *$\mathcal{D}$ is isomorphic to $\mathcal{D}_{3,t}$ or $\mathcal{D}_{3,m}$ or $\mathcal{D}_{3,b}$.*

Proof. Note that since $\mathcal{D}$ is a Condorcet domain, there are $x \in \{a, b, c\}$ and $i \in \{1, 2, 3\}$ such that $\mathcal{D}$ satisfies a never-condition $xN_{\{a,b,c\}}i$. Write $\{y, z\} = \{a, b, c\} \setminus \{x\}$. There are three cases for $\mathcal{D}$. Suppose $\mathcal{D}$ satisfies the never-top condition $xN_{\{a,b,c\}}1$. Then $\mathcal{D} \subseteq \{yxz, yzx, zxy, zyx\}$. Suppose $\mathcal{D}$ satisfies the never-middle condition $xN_{\{a,b,c\}}2$. Then $\mathcal{D} \subseteq \{xyz, xzy, yzx, zyx\}$. Suppose $\mathcal{D}$ satisfies the never-bottom condition $xN_{\{a,b,c\}}3$. Then $\mathcal{D} \subseteq \{xyz, xzy, yxz, zxy\}$. Note that $\{yxz, yzx, zxy, zyx\}$, $\{xyz, xzy, yzx, zyx\}$, and $\{xyz, xzy, yxz, zxy\}$ are Condorcet domains.

(i) $\Leftrightarrow$ (ii): Suppose $\mathcal{D}$ is a maximal Condorcet domain. Then maximality implies that $\mathcal{D}$ equals $\{yxz,\ yzx,\ zxy,\ zyx\}$ or $\{xyz,\ xzy,\ yzx,\ zyx\}$ or $\{xyz,\ xzy,\ yzx,\ zyx\}$. Hence, $|\mathcal{D}| = 4$.

On the other hand, suppose $|\mathcal{D}| = 4$. Then $\mathcal{D}$ equals $\{yxz,\ yzx,\ zxy,\ zyx\}$ or $\{xyz,\ xzy,\ yzx,\ zyx\}$ or $\{xyz, xzy, yzx, zyx\}$. Since each of these three possibilities is a maximal Condorcet domain, $\mathcal{D}$ is a maximal Condorcet domain.

(ii) $\Leftrightarrow$ (iii): Suppose $|\mathcal{D}| = 4$. Then $\mathcal{D}$ equals $\{yxz, yzx, zxy, zyx\}$ or $\{xyz,\ xzy,\ yzx,\ zyx\}$ or $\{xyz, xzy, yzx, zyx\}$. Now $\{yxz, yzx, zxy, zyx\}$ is isomorphic to $\mathcal{D}_{3,t} = \{abc, acb, cab, cba\}$, and $\{xyz, xzy, yzx, zyx\}$ is isomorphic to $\mathcal{D}_{3,m} = \{abc, bca, acb, cba\}$, and $\{xyz, xzy, yzx, zyx\}$ is isomorphic to $\mathcal{D}_{3,b} = \{abc, bac, bca, cba\}$. Hence, $\mathcal{D}$ is isomorphic to $\mathcal{D}_{3,t}$ or $\mathcal{D}_{3,m}$ or $\mathcal{D}_{3,b}$.

On the other hand, suppose $\mathcal{D}$ is isomorphic to $\mathcal{D}_{3,t}$ or $\mathcal{D}_{3,m}$ or $\mathcal{D}_{3,b}$. Since $\mathcal{D}_{3,t}$, $\mathcal{D}_{3,m}$ and $\mathcal{D}_{3,b}$ all have four linear orders, $|\mathcal{D}| = 4$.

We note that there are only three maximal Condorcet domains on the set $A = \{a, b, c\}$ up to an isomorphism, as listed in Lemma 1.

While the only never-condition that $\mathcal{D}_{3,t}$ satisfies is the never-top condition $bN_{\{a,b,c\}}1$, the domain $\mathcal{D}_{3,m}$ only satisfies the never-middle condition $aN_{\{a,b,c\}}2$, and $\mathcal{D}_{3,b}$ only satisfies the never-bottom condition $bN_{\{a,b,c\}}3$. Both $\mathcal{D}_{3,t}$ and $\mathcal{D}_{3,b}$ are peak-pit maximal Condorcet domains.

There are various definitions of connected Condorcet domains defined over the permutahedron. We will use the concept of paths of alike linear orders, which will be defined in next section.

3 Paths of Alike Linear Orders

Suppose $R = r_1 \cdots r_n$ and $T = t_1 \cdots t_n$ are linear orders of $\mathcal{L}(A)$. We say that R and T are **alike** if they differ by a swap of adjacent alternatives in them. Precisely, there is an $i \in [n-1]$ such that $r_i = t_{i+1}$, $r_{i+1} = t_i$, and $r_j = t_j$ for all $j \in [n] \setminus \{i, i+1\}$. So, $T = r_1 \cdots r_{i-1} r_{i+1} r_i r_{i+2} \cdots r_n$ in this case. We call (r_i, r_{i+1}) the **switching pair of alternatives** between R and T, or, shortly, switching pair between R and T. We do not specify the order of the alternatives in a switching pair, so the switching pair between R and T can also be written as (r_{i+1}, r_i).

Definition 2. *Let R and T be two linear order in $\mathcal{L}(A)$. We say that $\mathcal{A} = (R_1, \ldots, R_k)$ is a* **path of alike linear orders** *connecting R and T, or shortly a* **path connecting** *R and T, if $R_1 = R, R_k = T$, and R_i and R_{i+1} are alike for all $i \in [k-1]$. The number $k \in \mathbb{N}$ is called the* **length** *of the path $\mathcal{A}$ connecting R and T. We say that $\mathcal{A} = (R_1, \ldots, R_k)$ is a* **geodesic of alike linear orders** *connecting R and T, or shortly a* **geodesic connecting** *R and T, if*

$$k = \min\{\, m \in \mathbb{N} : (S_1, \ldots, S_m) \text{ is a path of alike linear orders connecting } R \text{ and } T\}.$$

Note that the length of a geodesic connecting linear orders R and T is one more than the length of the reduced decomposition of a permutation sending R to T (see, e.g., [2], Sec. 6.4). The Kendall tau distance between R and T is the number of pairs of alternatives that are ranked differently by R and T. The length of a geodesic connecting R and T is one more than the Kendall tau distance between R and T.

Example 1. Note that there are only six linear orders in $\mathcal{L}(\{a, b, c\})$, namely

$$\mathcal{L}(\{a, b, c\}) = \{abc, bac, bca, cba, cab, acb\}.$$

If an edge is used to connect any two alike linear orders in $\mathcal{L}(\{a, b, c\})$, we obtain a diagram that shows all paths connecting any two alike linear orders in $\mathcal{L}(\{a, b, c\})$, as shown in Fig. 1, where the switching pair between two adjacent linear orders is labeled on the edge that connects them.

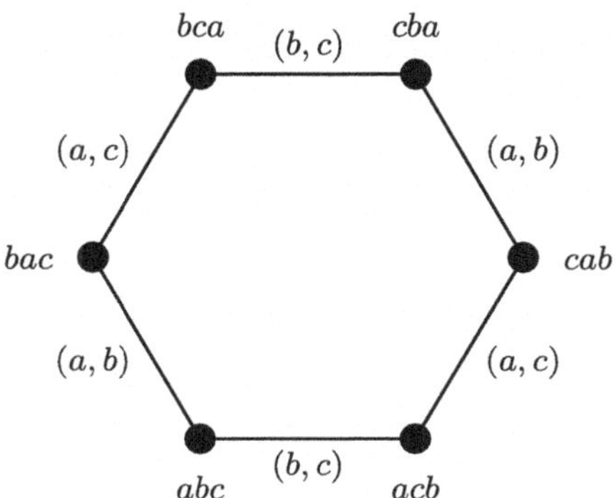

Fig. 1. All paths connecting any two linear orders in $\mathcal{L}(\{a, b, c\})$.

While $\mathcal{A} = (abc, bac, bca)$ and $\mathcal{B} = (abc, acb, cab, cba, bca)$ are both paths connecting the same linear orders, abc and bca, the path $\mathcal{A}$ is a geodesic but $\mathcal{B}$

is not since the number of linear orders in the latter one is not minimal. Note that abc and bca differ by three swaps of adjacent alternatives, and and this is also the number of swaps in the shortest path (geodesic) connecting abc and bca. This tells us some obvious facts about paths of alike linear orders.

Let $\mathcal{A} = (R_1, \ldots, R_k)$ be a path of alike linear orders, we use $K(\mathcal{A}) = \{R_1, \ldots, R_k\}$ to represent the **set of linear orders from** $\mathcal{A}$. We denote $S(\mathcal{A})$ to be its **sequence of switching pairs of alternatives**. Observe that each path of alike linear orders is uniquely determined by its sequence of switching pairs of alternatives. The number of switching pairs in $S(\mathcal{A})$ is called the **length** of $S(\mathcal{A})$. Note that the length of $S(\mathcal{A})$ is always one less than the length of $\mathcal{A}$.

Let $\mathcal{A} = (R_1, \ldots, R_k)$ be a path of alike linear orders on $\mathcal{L}(A)$. Let $a \in A$ and $B \subseteq A$. Then the **restricted path** $\mathcal{A}_B$ is the path of alike linear orders on $\mathcal{L}(B)$ obtained from $(R_1)_B, \ldots, (R_k)_B$ and removing duplicate equal linear orders which are next to each other. Similarly, the **restricted path** $\mathcal{A}_{-a}$ is the path of alike linear orders on $\mathcal{L}(A_{-a})$ obtained from $(R_1)_{-a}, \ldots, (R_k)_{-a}$ and removing duplicate equal linear orders which are next to each other. Note that the length of $\mathcal{A}_B$ can be less than the length of $\mathcal{A}$. We also note that switching pairs of alternatives in a path are defined over a pair of alike linear orders in the path. Hence, if (a, b) is a switching pair in a path, then it must occur at least once in it.

Let (a, b) and (c, d) be two switching pairs in a geodesic $\mathcal{A}$. If (a, b) occurs before (c, d) in $S(\mathcal{A})$, we denote that by

$$(a, b) \lhd (c, d).$$

We similarly define $\unlhd$.

We will now investigate the restriction of any geodesic to a triple of alternatives that differ by three swaps of adjacent alternatives. Let $R = abc$ and $T = cba$. Then $\mathcal{A}_1 = (abc, bac, bca, cba)$ and $\mathcal{A}_2 = (abc, acb, cab, cba)$ are the only possible geodesics connecting them, as shown in Fig. 1. Note that both geodesics rise to peak-pit Condorcet domains, These two geodesics can be represented by the wiring diagrams shown in Fig. 2.

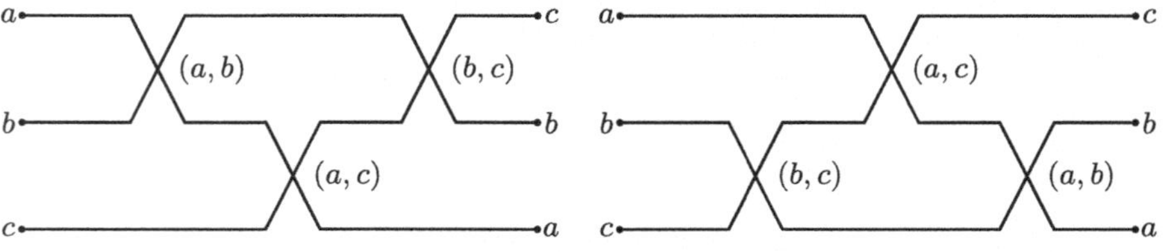

(a) Sequence of switching pairs for $\mathcal{A}_1$. (b) Sequence of switching pairs for $\mathcal{A}_2$.

Fig. 2. Sequences of switching pairs for two geodesics connecting abc and cba.

Note that the sequences of switching pairs for $\mathcal{A}_1$ and $\mathcal{A}_2$ are in reversed order. While the set of linear orders from $\mathcal{A}_1$ only satisfies one never-condition, which is the never-bottom condition $bN_{\{a,b,c\}}3$, the set from $\mathcal{A}_2$ also only satisfies one never-condition, which is the never-top condition $bN_{\{a,b,c\}}1$. As a result, we refer to the first geodesic $\mathcal{A}_1$ as a **never-bottom geodesic**, while the second geodesic $\mathcal{A}_2$ is referred to as a **never-top geodesic**. It is important to note that the switching pairs occurring in $\mathcal{A}_1$ involve first switching the top two alternatives, i.e., (a, b), while those in $\mathcal{A}_2$ involve first switching the bottom two ranked alternatives, i.e., (b, c). This dichotomy can be generalised into the following lemma.

Lemma 2. *Let a, b, c be distinct alternatives. Let $T \in \mathcal{L}(\{a, b, c\})$. Let $\mathcal{A} = (R_1, \ldots, R_k)$ be a geodesic on $\mathcal{L}(\{a, b, c\})$ connecting abc and T. Then the following are equivalent:*

(I) $T = cba$;
(II) *(a, b), (a, c) and (b, c) are switching pairs in $S(\mathcal{A})$.*

Now suppose $T = cba$. Then $\{R_1, \ldots, R_k\}$ satisfies a unique never-condition which is either $bN_{\{a,b,c\}}3$ or $bN_{\{a,b,c\}}1$. Moreover,

(a) *The following are equivalent:*
  (i) $N(\{R_1, \ldots, R_k\}) = \{bN_{\{a,b,c\}}3\}$;
  (ii) *$\mathcal{A} = (abc, bac, bca, cba)$, that is it is a never-bottom geodesic;*
  (iii) $S(\mathcal{A}) : (a, b) \lhd (a, c) \lhd (b, c)$.
(b) *The following are equivalent:*
  (i) $N(\{R_1, \ldots, R_k\}) = \{bN_{\{a,b,c\}}1\}$;
  (ii) *$\mathcal{A} = (abc, acb, cab, cba)$, that is it is a never-top geodesic;*
  (iii) $S(\mathcal{A}) : (b, c) \lhd (a, c) \lhd (a, b)$.

Proof. For (II) $\Rightarrow$ (I), since (a, b), (a, c) and (b, c) are switching pairs in $S(\mathcal{A})$, we have that *abc* and T differ by three swaps of adjacent alternatives. Hence, $T = cba$. For (I) $\Rightarrow$ (II), suppose $T = cba$. Then $\mathcal{A}$ be a geodesic connecting *abc* and *cba*. Inspecting Fig. 1, we note that there are exactly 2 geodesics on $\mathcal{L}(\{a, b, c\})$ connecting *abc* and *cba*, which are $\mathcal{A}_1 = (abc, bac, bca, cba)$ and $\mathcal{A}_2 = (abc, acb, cab, cba)$. In particular, (II) is valid. Note that $\mathcal{A}_1$ gives the domain $\{abc,$ $bac,$ $bca,$ $cba\}$ satisfying a unique never-condition $bN_{\{a,b,c\}}3$, while $\mathcal{A}_2$ gives the domain $\{abc, acb, cab, cba\}$ satisfying a unique never-condition $bN_{\{a,b,c\}}1$.

(a) Suppose $\{R_1, \ldots, R_k\}$ satisfies the never-top condition $bN_{\{a,b,c\}}3$. Since $\mathcal{A}$ connecting *abc* and *cba*, the linear order *bac* or *bca* that ranks b first needs to be in $\{R_1, \ldots, R_k\}$. Hence, $\mathcal{A} = \mathcal{A}_1 = (abc, bac, bca, cba)$. In particular, the sequence of switching pairs $S(\mathcal{A})$ is $(a, b), (a, c), (b, c)$, as shown in Fig. 2(a). On the other hand, suppose $S(\mathcal{A})$ is $(a, b), (a, c), (b, c)$. Since $\mathcal{A}$ connecting *abc* and *cba*, the geodesic $\mathcal{A} = \mathcal{A}_1 = (abc, bac, bca, cba)$. Thus, the domain $\{R_1, \ldots, R_k\} = \{abc, bac, bca, cba\}$ satisfies the never-top condition $bN_{\{a,b,c\}}3$.

(b) Suppose $\{R_1, \ldots, R_k\}$ satisfies the never-top condition $bN_{\{a,b,c\}}1$. Since $\mathcal{A}$ connecting abc and cba, the linear order acb or cab that ranks b last needs to be in $\{R_1, \ldots, R_k\}$. Hence, $\mathcal{A} = \mathcal{A}_2 = (abc, acb, cab, cba)$. In particular, the sequence of switching pairs $S(\mathcal{A})$ is $(b, c), (a, c), (a, b)$, as shown in Fig. 2(b). On the other hand, suppose $S(\mathcal{A})$ is $(b, c), (a, c), (a, b)$. Since $\mathcal{A}$ connecting abc and cba, the geodesic $\mathcal{A} = \mathcal{A}_2 = (abc, acb, cab, cba)$. Thus, the domain $\{R_1, \ldots, R_k\} = \{abc, acb, cab, cba\}$ satisfies the never-top condition $bN_{\{a,b,c\}}1$.

We will use paths of alike linear orders to define connectedness for a domain.

Definition 3. *Let $\mathcal{D} \subseteq \mathcal{L}(A)$ be a domain and $R, T \in \mathcal{D}$. We say that R and T are* **connected** *in $\mathcal{D}$ if there exists a path of alike linear orders $\mathcal{A} = (R_1, \ldots, R_k)$ connecting R and T such that $R_i \in \mathcal{D}$ for all $i \in [k]$. We say that R and T are* **directly connected** *in $\mathcal{D}$ if there exists a geodesic of alike linear orders $\mathcal{A} = (R_1, \ldots, R_k)$ connecting R and T such that $R_i \in \mathcal{D}$ for all $i \in [k]$. We then call $\mathcal{A}$ a* **geodesic on $\mathcal{D}$ connecting** *R and T. A domain $\mathcal{D}$ is called* **connected** *if for all $R, T \in \mathcal{D}$, the linear orders R and T are connected in $\mathcal{D}$. A domain $\mathcal{D}$ is called* **directly connected** *if for all $R, T \in \mathcal{D}$, the linear orders R and T are directly connected in $\mathcal{D}$.*

Obviously, if a domain $\mathcal{D}$ is directly connected, then it is connected by definition.

Example 2. Note that there are only two peak-pit maximal Condorcet domains for three alternatives a, b, c, and which contain abc, namely,

$$\mathcal{D}_{3,b} = \{abc, bac, bca, cba\} \text{ and } \mathcal{D}_{3,t} = \{abc, acb, cab, cba\}.$$

While the only never-condition that $\mathcal{D}_{3,b}$ satisfies is the never-bottom condition $bN_{\{a,b,c\}}3$, the domain $\mathcal{D}_{3,t}$ only satisfies the never-top condition $bN_{\{a,b,c\}}1$. However, both domains are connected and directly connected as shown in Fig. 1.

Remark 2. Note that our definition of connectedness is equivalent to that given by [18]. Moreover, the directly connected domains defined here are equivalent to those satisfying the no-restoration property from [18].

4 Equivalence of Connectedness and Peak-Pittedness

The following proposition is central to establishing the equivalence between connectedness and peak-pittedness. Due to space constraints, the proof of Proposition 1 is omitted and can be found in the extended version of this paper in [13].

Proposition 1. *Let $\mathcal{D} \subseteq \mathcal{L}(A)$ be a peak-pit Condorcet domain. Let $R, T \in \mathcal{D}$ be linear orders and $B \subseteq A$ be a subset such that $|B| \geq 3$. Then there exists a geodesic $\mathcal{A}$ connecting R_B and T_B such that $\mathcal{D}_B \cup K(\mathcal{A})$ is a peak-pit Condorcet domain.*

We need one more lemma before showing our main theorems.

Lemma 3. *Let A be a set of at least four alternatives. Let $\mathcal{D} \subseteq \mathcal{L}(A)$ be a domain, and $a \in A$. Then*

(a) *If $\mathcal{D}$ is connected, then $\mathcal{D}_{-a}$ is connected.*
(b) *If $\mathcal{D}$ is directly connected, then $\mathcal{D}_{-a}$ is directly connected.*

Proof. (a) Suppose $\mathcal{D} \subseteq \mathcal{L}(A)$ is connected, and $a \in A$. Let $R, T \in \mathcal{D}_{-a}$ be any two linear orders, then there are linear orders $R^*, T^* \in \mathcal{D}$ such that $R^*_{-a} = R$ and $T^*_{-a} = T$. Since $\mathcal{D}$ is connected, there is a path $\mathcal{A} = (R^1, \ldots, R^k)$ on $\mathcal{D}$ connecting R^* and T^*. Hence, the restricted path $\mathcal{A}_{-a} = (R^1_{-a}, \ldots, R^{k'}_{-a})$ is a path connecting R and T. Since $R^i \in \mathcal{D}$, we have $R^i_{-a} \in \mathcal{D}_{-a}$, for all $i \in [k]$. Hence, $\mathcal{A}_{-a}$ is a path on $\mathcal{D}_{-a}$ and so $\mathcal{D}_{-a}$ is connected.

(b) suppose $\mathcal{D} \subseteq \mathcal{L}(A)$ is directly connected, and $a \in A$. Let $R, T \in \mathcal{D}_{-a}$ be any two linear orders, then there are linear orders $R^*, T^* \in \mathcal{D}$ such that $R^*_{-a} = R$ and $T^*_{-a} = T$. Since $\mathcal{D}$ is directly connected, there is a geodesic $\mathcal{A} = (R^1, \ldots, R^k)$ on $\mathcal{D}$ connecting R^* and T^*. Hence, the restricted path $\mathcal{A}_{-a} = (R^1_{-a}, \ldots, R^{k'}_{-a})$ is a path connecting R and T. Moreover, since $\mathcal{A}$ is a geodesic and $S(\mathcal{A}_{-a})$ is obtained by removing all switching pairs that contain a from $S(\mathcal{A})$, we have that $\mathcal{A}_{-a}$ is also a geodesic. Since $R^i \in \mathcal{D}$, we have $R^i_{-a} \in \mathcal{D}_{-a}$, for all $i \in [k]$. Hence, $\mathcal{A}_{-a}$ is a geodesic on $\mathcal{D}_{-a}$ and so $\mathcal{D}_{-a}$ is directly connected.

Although our primary interest lies in maximal Condorcet domains that satisfy the peak-pit condition, it is still useful to understand the structure of the peak-pit condition itself, rather than focusing solely on the maximality of the domain in terms of size.

Definition 4. *Let $\mathcal{D} \subseteq \mathcal{L}(A)$ be a peak-pit Condorcet domain. If for every Condorcet domain $\mathcal{D}' \subseteq \mathcal{L}(A)$ with $\mathcal{D} \subseteq \mathcal{D}'$, one has $\mathcal{D} = \mathcal{D}'$, then $\mathcal{D}$ is called a* **peak-pit maximal Condorcet domain**. *If for every peak-pit Condorcet domain $\mathcal{D}' \subseteq \mathcal{L}(A)$ with $\mathcal{D} \subseteq \mathcal{D}'$, one has $\mathcal{D} = \mathcal{D}'$, then $\mathcal{D}$ is called a* **maximal peak-pit Condorcet domain**.

Using the concept of connectedness, we establish that the class of maximal peak-pit Condorcet domains is equivalent to the class of peak-pit maximal Condorcet domains. Before showing this, let us first see an important property of maximal peak-pit Condorcet domains.

Theorem 1. *Every maximal peak-pit Condorcet domain is directly connected.*

Proof. Let $\mathcal{D} \subseteq \mathcal{L}(A)$ be a maximal peak-pit Condorcet domain. If we consider $B = A$ as in Proposition 1, then for every pair of linear orders $R, T \in \mathcal{D}$, there exists a geodesic $\mathcal{A}$ connecting R and T such that $\mathcal{D} \cup K(\mathcal{A})$ is a peak-pit Condorcet domain. Since $\mathcal{D}$ is a maximal peak-pit Condorcet domain, we have $K(\mathcal{A}) \subseteq \mathcal{D}$. Thus, $\mathcal{A}$ is a geodesic on $\mathcal{D}$ and so $\mathcal{D}$ is directly connected.

Consequently, we have the following theorem.

Theorem 2. *Let $\mathcal{D}$ be a Condorcet domain. Then $\mathcal{D}$ is a peak-pit maximal Condorcet domain if and only if $\mathcal{D}$ is a maximal peak-pit Condorcet domain.*

Proof. Suppose $\mathcal{D}$ is a peak-pit maximal Condorcet domain. Since every peak-pit Condorcet domain is a Condorcet domain, $\mathcal{D}$ is a maximal peak-pit Condorcet domain by definition.

Conversely, suppose $\mathcal{D} \subseteq \mathcal{L}(A)$ is a maximal peak-pit Condorcet domain. By Theorem 1, the domain $\mathcal{D}$ is directly connected. Now we will show that $\mathcal{D}$ is a maximal Condorcet domain. Suppose not. Then there is a linear order $R \in \mathcal{L}(A)$ such that $\mathcal{D} \cup \{R\}$ is a Condorcet domain but not a peak-pit Condorcet domain. Hence, there is a triple $a, b, c \in A$ such that $(\mathcal{D} \cup \{R\})_{\{a,b,c\}}$ only satisfies a never-middle condition. Without loss of generality, suppose $(\mathcal{D} \cup \{R\})_{\{a,b,c\}}$ satisfies the never-middle condition $cN_{\{a,b,c\}}2$, namely, $(\mathcal{D} \cup \{R\})_{\{a,b,c\}} \subseteq \{abc, bac, cab, cba\}$. Because $(\mathcal{D} \cup \{R\})_{\{a,b,c\}}$ does not satisfy any never-top or never-bottom condition and any proper subset of $\{abc, bac, cab, cba\}$ satisfies a never-top or never-bottom condition, we have $(\mathcal{D} \cup \{R\})_{\{a,b,c\}} = \{abc, bac, cab, cba\}$, as shown in Fig. 3. Moreover, since $\mathcal{D}$ is a peak-pit Condorcet domain, we have $R_{\{a,b,c\}} \in \{abc, bac, cab, cba\}$. Hence, $\mathcal{D}_{\{a,b,c\}} = \{abc, bac, cab, cba\} \setminus \{R_{\{a,b,c\}}\}$. Note that removing any one element from $\{abc, bac, cab, cba\}$ will cause the set of remaining elements to be disconnected. In particular, if $R_{\{a,b,c\}}$ is abc or cab, then bac and cba is not directly connected in $\mathcal{D}_{\{a,b,c\}}$. Similarly, if $R_{\{a,b,c\}}$ is bac or cba, then abc and cab is not directly connected in $\mathcal{D}_{\{a,b,c\}}$. Since $\mathcal{D}$ is directly connected, that contradicts Lemma 3(b). Hence, $\mathcal{D}$ is a maximal Condorcet domain.

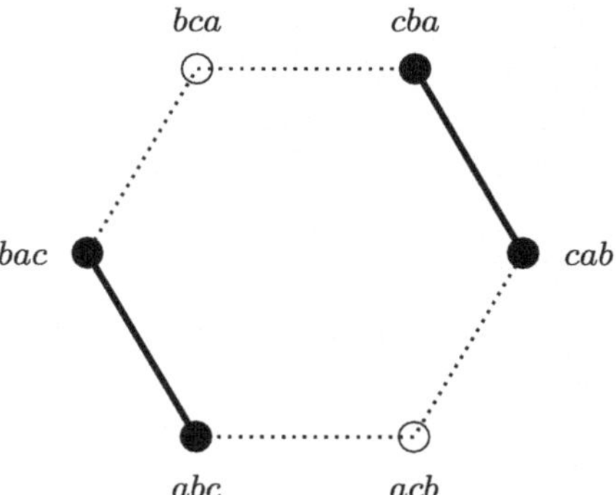

Fig. 3. A never-middle domain that is not connected in $\mathcal{L}(\{a, b, c\})$.

Next we will see that, in the class of maximal Condorcet domains, peak-pittedness, direct connectedness, and connectedness are actually equivalent.

Theorem 3. *Let $\mathcal{D}$ be a maximal Condorcet domain. Then the following are equivalent:*

(i) *$\mathcal{D}$ is a peak-pit Condorcet domain;*
(ii) *$\mathcal{D}$ is directly connected;*
(iii) *$\mathcal{D}$ is connected.*

Proof. **(i) $\Rightarrow$ (ii):** Suppose $\mathcal{D} \subseteq \mathcal{L}(A)$ is a peak-pit maximal Condorcet domain. By Theorem 2, $\mathcal{D}$ is a maximal peak-pit Condorcet domain. Hence by Theorem 1, it is directly connected.

(ii) $\Rightarrow$ (iii): Suppose $\mathcal{D}$ is directly connected, then every two linear orders in it are connected by a geodesic on $\mathcal{D}$. Since all geodesics are paths, $\mathcal{D}$ is connected.

(iii) $\Rightarrow$ (i): Suppose $\mathcal{D}$ is a connected maximal Condorcet domain. Suppose $\mathcal{D}$ is not a peak-pit Condorcet domain for a contradiction. Then there is a triple of distinct alternatives a, b, c such that $\mathcal{D}_{\{a,b,c\}}$ does not satisfy a never-top or never-bottom condition. Since $\mathcal{D}$ is a Condorcet domain, $\mathcal{D}_{\{a,b,c\}}$ satisfies a never-middle condition. Without loss of generality, suppose $\mathcal{D}_{\{a,b,c\}}$ satisfies the never-middle condition $cN_{\{a,b,c\}}2$, namely, $\mathcal{D}_{\{a,b,c\}} \subseteq \{abc, bac, cab, cba\}$. However, since $\mathcal{D}_{\{a,b,c\}}$ does not satisfy any never-top or never-bottom condition and any proper subset of $\{abc, bac, cab, cba\}$ would satisfy a never-top or never-bottom condition, we have $\mathcal{D}_{\{a,b,c\}} = \{abc, bac, cab, cba\}$, as shown in Fig. 3. Note that $\mathcal{D}_{\{a,b,c\}}$ is not connected, because no path on $\mathcal{D}_{\{a,b,c\}}$ connects abc and cba. However, since $\mathcal{D}$ is connected, $\mathcal{D}_{\{a,b,c\}}$ is also connected by Lemma 3(a). This is a contradiction and so $\mathcal{D}$ is a peak-pit Condorcet domain.

5 Conclusion

We have established the equivalence of three fundamental classes of maximal Condorcet domains: the class of connected Condorcet domains, the class of peak-pit Condorcet domains, and the class of directly connected Condorcet domains. By examining the structure of geodesics of alike linear orders, we resolved a long-standing open question and unified these previously distinct perspectives under the maximality condition. Our results not only advance the theoretical understanding of Condorcet domains but also clarify the relationship between maximal peak-pit Condorcet domains and peak-pit maximal Condorcet domains, providing formal definitions to resolve ambiguity in the literature.

With these new findings, the work of [12] can be carried forward. They introduced combinatorial tools, such as weakly separated ideals and generalised arrangements of pseudolines, to explore the combinatorial structure of peak-pit maximal Condorcet domains. Our results provide a solid foundation for future research, opening new avenues to investigate the deeper combinatorial properties of these domains.

Acknowledgments. I thank Tom ter Elst and Dominik Peters for their thorough review and invaluable feedback on the manuscript. I also thank the anonymous reviewers for their insightful comments. I gratefully acknowledge financial support from the Department of Mathematics at the University of Auckland.

Disclosure of Interests. The author has no competing interests to declare that are relevant to the content of this article.

References

1. Arrow, K.: Social Choice and Individual Values. Wiley, New York (1951)
2. Björner, A., Las Vergnas, M., Sturmfels, B., White, N., Ziegler, G.: Oriented matroids. Cambridge University Press (1999), https://doi.org/10.1017/CBO9780511586507
3. Brandt, F., Conitzer, V., Endriss, U., Lang, J., Procaccia, A. (eds.): Handbook of Computational Social Choice. Cambridge University Press (2016), https://doi.org/10.1017/CBO9781107446984
4. Brandt, F., Lederer, P., Tausch, S.: Strategyproof social decision schemes on super Condorcet domains. In: Appears at the 4th Games, Agents, and Incentives Workshop (GAIW 2022). Held as part of the Workshops at the 21st International Conference on Autonomous Agents and Multiagent Systems (2022), https://doi.org/10.48550/arXiv.2302.12140
5. Danilov, V.I., Karzanov, A.V., Koshevoy, G.A.: Condorcet domains of tiling type. Discrete Appl. Math. **160**(7-8), 933–940 (2012), https://doi.org/10.1016/j.dam.2011.08.001
6. Fishburn, P.C.: Acyclic sets of linear orders. Soc. Choice Welfare **14**(1), 113–124 (1996)
7. Fishburn, P.C.: Acyclic sets of linear orders: a progress report. Soc. Choice Welfare **19**(2), 431–447 (2002). https://doi.org/10.1007/s003550100120
8. Galambos, A., Reiner, V.: Acyclic sets of linear orders via the Bruhat orders. Soc. Choice Welfare **30**(2), 245–264 (2008), https://doi.org/10.1007/s00355-007-0228-1
9. Gehrlein, W.: Condorcet's Paradox, Theory and Decision Library C, vol. 40. Springer Berlin, Heidelberg (2006), https://doi.org/10.1007/3-540-33799-7
10. Leedham-Green, C., Markström, K., Riis, S.: The largest condorcet domain on 8 alternatives. Soc. Choice Welfare **62**, 109–116 (2024), https://doi.org/10.1007/s00355-023-01481-3
11. Li, G.: Maximal peak-pit domains on four alternatives. In: Student Research Conference. Department of Mathematics, University of Auckland (2020)
12. Li, G.: A classification of peak-pit maximal Condorcet domains. Math. Soc. Sci. **125**, 42–57 (2023), https://doi.org/10.1016/j.mathsocsci.2023.06.004
13. Li, G.: Equivalence of connected and peak-pit maximal Condorcet domains. arXiv preprint (2025), https://doi.org/10.48550/arXiv.2505.19520
14. Li, G., Puppe, C., Slinko, A.: Towards a classification of maximal peak-pit Condorcet domains. Math. Soc. Sci. **113**, 191–202 (2021), https://doi.org/10.1016/j.mathsocsci.2021.07.005
15. Monjardet, B.: Acyclic domains of linear orders: a survey. In: Brams, S., Gehrlein, W., Roberts, F. (eds.) The Mathematics of Preference, Choice and Order, pp. 139–160. Studies in Choice and Welfare, Springer, Berlin, Heidelberg (2009), https://doi.org/10.1007/978-3-540-79128-7_8
16. Puppe, C., Slinko, A.: Maximal Condorcet domains. A further progress report. Games Econ. Behav. **145**, 426–450 (2024), https://doi.org/10.1016/j.geb.2024.04.001

17. Puppe, C., Slinko, A.: Note on "A classification of peak-pit maximal Condorcet domains" by Guanhao Li, Mathematical Social Sciences 125 (2023), 42–57. Mathematical Social Sciences **128**, 16–17 (2024), https://doi.org/10.1016/j.mathsocsci.2024.01.006
18. Sato, S.: A sufficient condition for the equivalence of strategy-proofness and non-manipulability by preferences adjacent to the sincere one. J. Econ. Theor. **148**(1), 259–278 (2013), https://doi.org/10.1016/j.jet.2012.12.001

On the Oscillations in Cournot Games with Best Response Strategies

Zhengyang Liu[1], Haolin Lu[1], Liang Shan[2](✉), and Zihe Wang[2]

[1] Beijing Institute of Technology, Beijing, China
{zhengyang,haolin}@bit.edu.cn
[2] Renmin University of China, Beijing, China
{Shanliang,zihe.wang}@ruc.edu.cn

Abstract. In this paper, we consider the dynamic oscillation in the Cournot oligopoly model, which involves multiple firms producing homogeneous products. To explore the oscillation under the updates of best response strategies, we focus on the linear price functions. In this setting, we establish the existence of oscillations. In particular, we show that for the scenario of different costs among firms, the best response converges to either a unique equilibrium or a two-period oscillation. We further characterize the oscillations and propose linear-time algorithms for finding all types of two-period oscillations. To the best of our knowledge, our work is the first step toward fully analyzing the periodic oscillation in the Cournot oligopoly model.

Keywords: Cournot Oligopoly Model · Best Response · Oscillation

1 Introduction

The Cournot competition [14] is one of the most fundamental economic models in the oligopoly theory. In the classic Cournot game, n firms produce homogeneous products. Each firm has market power, as its output affects the market price, and aims to maximize its profits. All the firms are assumed to utilize the production quantities as strategic variables, which they decide simultaneously and independently of each other. The existence of equilibrium has been demonstrated in a broad range of models [16,40], while the question of whether firms can automatically converge to equilibrium through best response strategies remains a challenging problem.

When the firms update their strategies asynchronously, they ultimately reach the Nash equilibrium since the Cournot game can be considered as a potential game [32]. The scenario becomes more intriguing when the firms *synchronize* their strategy updates. That is, each firm chooses the best response strategy, while others keep their strategies (production quantities) unchanged. Palander and Theocharis [35,41] showed that convergence depends on the number of firms in the game. They demonstrated that equilibrium is attained with two firms,

F. V. Fomin and M. Xiao (Eds.): COCOON 2025, LNCS 15983, pp. 320–332, 2026.
https://doi.org/10.1007/978-981-95-0215-8_24

but with three firms, a stationary oscillation ensues. Beyond three firms, significant instability arises. Puu [36] conducted numerous simulations of the firms' actions. This line of research [13,21,31] considered the properties of local dynamics, assuming the firms' quantities are always non-negative. Consequently, the dynamics of their strategies can be described by the system of linear equations. Cánovas et al. [12] pioneered the examination of global dynamics, considering the constraint that supply quantities must be non-negative. Thus, the dynamics of their strategies are delineated by a set of non-linear equations. They identified specific scenarios where periodic orbits of period two exist within the dynamics. Furthermore, Cánovas et al. [10,11] delved into circumstances where the number of firms diminishes to a monopoly within the dynamics. To the end, all the previous works typically first assume a particular type of oscillation with a period of two, followed by establishing conditions under which such oscillation is feasible based on the costs incurred by the firms. Then a natural question arises:

Does an oscillation of a longer period exist in a general setting with multiple firms?

In this work, we address this issue and fully characterize the oscillations in the Cournot oligopoly model. Our model builds upon the foundation laid by Cánovas et al. [12]. We examine a Cournot oligopoly model with a population of n firms producing homogeneous products. All firms make decisions simultaneously, considering a linear price function. Production costs may vary among firms, and we impose the constraint that production quantities cannot be negative. The contributions of this paper are summarized as follows:

- Under linear inverse demand with non-negative quantity constraints, we prove that simultaneous best-response dynamics in an n-firm Cournot game must converge exclusively to either a Nash equilibrium or a two-period cycle.
- We further establish that any two-period cycle must conform to one of three specific structural patterns. The first type arises when each firm produces zero output in one of the two periods. The second type occurs when fewer than three firms produce a positive quantity in the first period, with additional firms producing positive quantities in the second period. The third type arises in the case of exactly three firms.
- We propose linear-time $O(n)$ algorithms that simultaneously detect oscillation patterns and classify their structural type. This computational contribution equips practitioners with a diagnostic toolkit for empirically analyzing strategic cycling behavior in real-world oligopolistic markets.

1.1 Related Work

In his seminal work, Cournot [14] introduced a mathematical model to characterize the features of duopoly markets. Since then, there has been considerable interest in whether firms could reach equilibrium through best response strategies, leading to numerous studies on equilibrium in the Cournot oligopoly model. Based on Cournot's model, Palander [35] discovered that convergence fails when the number of competitors exceeds three. When there are three competitors, the Cournot equilibrium tends towards endless but stationary oscillation. Theocharis [41] independently studied the Cournot model and obtained results similar to Palander's. Numerous of studies [4,6,36,37] investigated the impact of increasing the number of firms in the Cournot model on equilibrium outcomes. Cánovas et al. [10,11] explored conditions leading to the reduction in the number of competitors in the Cournot oligopoly model to duopoly or monopoly. Our work extends this line by thoroughly examining the periodic oscillations in Cournot competition with multiple firms, addressing the conditions under which these oscillations occur and characterizing their nature.

The significant influence of adjustment rules on the convergence to equilibrium within the Cournot oligopoly model has been extensively studied. Fisher [21] highlighted the pivotal role of adjustment speed in facilitating convergence in Cournot games involving multiple firms. Building upon the aforementioned insights, Nowaihi and Levine [33] explored scenarios involving both discrete and continuous adjustments. Okuguchi [34] delved into models where firms employ different adjustment rules, assuming that the i-th firm's expectation of the k-th firm's output is not necessarily the same as the j-th firm's. Matsumoto [30] investigated production differentiation oligopoly, focusing on adjustments in pricing and quantities. Furthermore, subsequent studies have conducted experiments examining adaptation rules within the Cournot game [15,25,26,38]. The best response strategy is not only widely applied in the field of Cournot games, but also extensively studied in various other related areas, such as lottery contests [23], continuous zero-sum games [24] and additive aggregation finite games [27]. Our contribution builds upon these studies by providing a detailed classification of oscillation types and introducing linear-time algorithms to identify these oscillations, thereby offering new insights into the dynamic behavior of firms in Cournot competition.

In another research line, Ahmed et al. [7] began to relax the assumption in Puu's work [36] that each firm knows the output of its competitors and is aware of the market demand function. These studies focused on bounded rationality, where firms lack complete information about the market demand function. Bischi and Naimzada [9] presented a dynamic Cournot model with bounded rationality. Following this, a series of works by Agiza [1–3,5] explored the bounded rationality with firms' strategy updates by using the marginal profit approach. The complex dynamics, bifurcations, and chaos are observed in the numerical simulations [19,43]. Even-Dar [18] showed that using gradient-based no-external regret procedures can guarantee convergence to equilibrium. Our research further differentiates itself by examining the constraints on non-negative production quantities.

There are also discussions regarding how the firms' adjustments, whether simultaneous or sequential, affect equilibrium outcomes. Tullock [42] constructed a rent-seeking model and proposed the possibility of an "Intellectual mire" situation that could lead to the non-existence of equilibrium. Leininger [28] later addressed the issues raised by Tullock using sequential play methods. Subsequently, a series of articles studied the dynamics of the first or second move in the duopoly model [22,39]. Building upon this foundation, these cases where more than two firms sequentially take action were further investigated [36,37]. Our work specifically focuses on simultaneous updates in a Cournot setting with multiple firms, exploring the resulting oscillations.

2 Preliminaries

We consider a Cournot oligopoly model, where a population of n firms in the market produces homogeneous products. All firms make simultaneous decisions about production quantities in each round. At round t, we denote by q_i^t the output decision of firm i, and the market price p^t is determined by the inverse demand function. In this work, we consider the *linear* price function as following [2,9,17,20,41]:

$$p^t := A - \sum_{i=1}^{n} q_i^t,$$

where A is the market capacity which is a positive constant.

Suppose each firm can have different production technology, firm i producing a unit item with the cost $c_i \geq 0$. Therefore, the production cost for firm i with producing q_i items is $c_i q_i$.

The utility for firm i at round t with production quantity q_i^t is:

$$u_i^t = (A - \sum_{j=1}^{n} q_j^t - c_i) q_i^t.$$

Assuming the update rules for all firms use the *best response* strategy, for $t \geq 1$, we have

$$q_i^{t+1} = \arg\max_{q_i \geq 0} \left\{ q_i (A - \sum_{j \neq i} q_j^t - q_i - c_i) \right\}. \tag{1}$$

The R.H.S. of Eq. (1) is a singleton set since the objective is quadratic w.r.t. q_i, and we abuse the equality sign here. Consider the production quantity strategy q_i^{t+1} at round $t+1$ is maximized at $q_i^{t+1} = \frac{A - \sum_{j \neq i} q_j^t - c_i}{2}$, may lead to a negative quantity. However, a firm cannot produce a negative number of products. Hence, we further limit the lower bound of the production quantity [11,12]. That is, for any $i \in [n]$,

$$q_i^{t+1} := \max\left\{ 0, \frac{A - \sum_{j \neq i} q_j^t - c_i}{2} \right\}. \tag{2}$$

Next, we give the definition of the Nash equilibrium in our setting.

Definition 1 (Nash equilibrium). *Given $\boldsymbol{q}^t = (q_1^t, q_2^t, \ldots, q_n^t)$ as a production quantity vector, we say it is a Nash equilibrium if it is unchanged under the update rule as Eq. (2). That is,*

$$(q_1^{t+1}, q_2^{t+1}, \ldots, q_n^{t+1}) = (q_1^t, q_2^t, \ldots, q_n^t).$$

3 Periodic Oscillation

In this section, we start to analyze the periodic oscillations. W.l.o.g., we assume that the firms are sorted according to the cost such that $0 \leq c_1 \leq c_2 \leq \cdots \leq c_n$. We first show that firms' production quantities are in an inverse relationship with the costs after a finite round of best response.

Lemma 1. *For any $i > j$ and any t, we let $Q^t = q_i^t - q_j^t$ and $\Delta c = c_i - c_j \geq 0$. There exists an integer $T > 0$ such that $Q^t \leq 0$ for any $t > T$. Moreover, if $\Delta c = 0$, then $\lim_{t\to\infty} Q^t = 0$.*

Proof. It follows from Eq. (2) that

$$\begin{aligned} Q^{t+1} &= \max\left\{0, \frac{A - \sum_{k\neq i} q_k^t - c_i}{2}\right\} - \max\left\{0, \frac{A - \sum_{k\neq j} q_k^t - c_j}{2}\right\} \\ &= \max\left\{0, \frac{A - \sum_{k\neq j} q_k^t - c_j}{2} + \frac{1}{2}Q^t - \frac{1}{2}\Delta c\right\} \\ &\quad - \max\left\{0, \frac{A - \sum_{k\neq j} q_k^t - c_j}{2}\right\}. \end{aligned}$$

Noting $\max\{0, x\} = \frac{1}{2}(x + |x|)$ and $\min\{0, x\} = \frac{1}{2}(x - |x|)$, we obtain

$$\min\{0, y\} \leq \max\{0, x + y\} - \max\{0, x\} \leq \max\{0, y\}.$$

Then we conclude

$$\min\left\{0, f(Q^t)\right\} \leq Q^{t+1} \leq \max\left\{0, f(Q^t)\right\}, \tag{3}$$

where $f(Q^t) = \frac{1}{2}Q^t - \frac{1}{2}\Delta c$. In Fig. 1, we demonstrate the relationship between Q^t and Q^{t+1}. Since $\Delta c \geq 0$, the shadowed part of the figure is the feasible domain for (Q^t, Q^{t+1}).

For the case $Q^t \leq \Delta c$, it follows from Eq. (3) that $Q^{t+1} \leq 0$, and then $Q^{t+k} \leq 0$ for any integer $k \geq 1$ by induction. Otherwise, that is, $Q^t > \Delta c$, it follows from Eq. (3) that

$$0 \leq Q^{t+1} \leq \frac{1}{2}Q^t - \frac{1}{2}\Delta c.$$

We can always find a finite t_0 such that $0 \leq Q^{t+t_0} \leq \Delta c$. Then from the above case, we get $Q^{t+t_0+k} \leq 0$ for any integer $k \geq 1$. To sum up, there exists a positive integer T such that $Q^t \leq 0$ for any $t > T$.

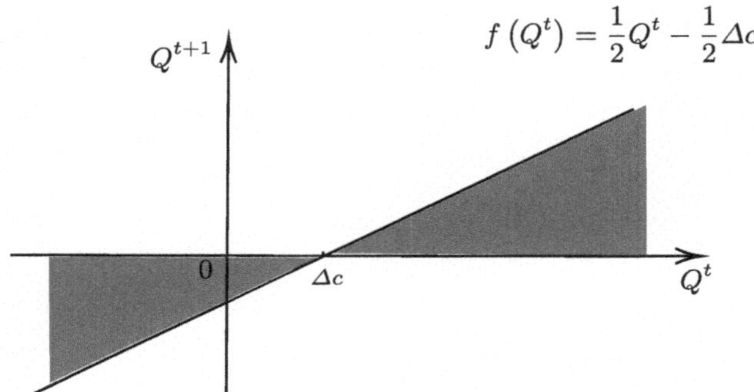

Fig. 1. The shadowed area is the feasible domain for values of (Q^t, Q^{t-1}).

When $\Delta c = 0$, it follows from Eq. (3) that

$$\min\left\{0, \frac{1}{2}Q^t\right\} \leq Q^{t+1} \leq \max\left\{0, \frac{1}{2}Q^t\right\} \Rightarrow |Q^{t+1}| \leq \frac{1}{2}|Q^t|.$$

It holds that $\lim_{t\to\infty} Q^t = 0$. □

Suppose the dynamic of the best response does not converge to a Nash equilibrium, then the product quantity vector keeps changing. To study the changing pattern of product quantity, we start by focusing on the number of survival firms, which are firms that produce a positive output. We use m^t to represent the number of survival firms in round t. We define

$$\underline{n} = \lim_{t\to\infty} \inf\{m^k, k > t\}.$$

Hence, there exists $T_1 > T$ such that $\inf\{m^k, k > T_1\} = \underline{n}$. Therefore, after round T_1, the first $\underline{n}$ firms would always survive by Lemma 1.

We focus on the dynamics after round T_1. To make it simpler, we design a new game and a new instance of dynamics within it where we temporarily ignore the difference between $\underline{n}$ firms. In particular, we construct a new game where we use $\underline{n}$ identical firms to replace the first $\underline{n}$ possibly heterogeneous original firms. For $i \leq \underline{n}$, firm i has an identical cost $\underline{c}_i = \frac{1}{\underline{n}}\sum_{j=1}^{\underline{n}} c_j$. For $i > \underline{n}$, firm i's cost does not change, i.e., $\underline{c}_i = c_i$. Then we design the following product quantity sequence $\underline{q}_i^t$ such that for any $t > 0$, $\underline{q}_i^t = \frac{1}{\underline{n}}\sum_{j=1}^{\underline{n}} q_j^{T_1+t}$ for $i \leq \underline{n}$ and $\underline{q}_i^t = q_i^{T_1+t}$ for $i > \underline{n}$. The following lemma shows that $\{\underline{\boldsymbol{q}}^t\}$ is still a valid product quantity sequence.

Lemma 2. *Sequence $\{\underline{\boldsymbol{q}}^t\}$ is a product quantity sequence in best response dynamics in the new game.*

We need to show the product quantity $\{\underline{\boldsymbol{q}}^{t+1}\}$ is still a best response to $\{\underline{\boldsymbol{q}}^t\}$, the idea is to divide the proof into two parts: one part addresses the case where $i > \underline{n}$, and the other part deals with the case where $i \leq \underline{n}$.

Proof. We first show that for firm $i > \underline{n}$, it still makes the best response decision in the designed product quantity sequence. Since it chooses the best strategy in the sequence $\{\boldsymbol{q}^t\}$, we have

$$q_i^{t+T_1} = \frac{1}{2}\max\left\{A - \sum_{j\neq i} q_j^{t+T_1} - c_i, 0\right\}.$$

According to the definition of $\underline{\boldsymbol{q}}^t$, we have

$$\begin{aligned}
\underline{q}_i^t &= \frac{1}{2}\max\left\{A - \sum_{j\leq \underline{n}} q_j^{t+T_1} - \sum_{j>\underline{n}, j\neq i} q_j^{t+T_1} - c_i, 0\right\} \\
&= \frac{1}{2}\max\left\{A - \sum_{j\leq \underline{n}} \underline{q}_j^t - \sum_{j>\underline{n}, j\neq i} \underline{q}_j^t - \underline{c}_i, 0\right\} \\
&= \frac{1}{2}\max\left\{A - \sum_{j\neq i} \underline{q}_j^t - \underline{c}_i, 0\right\}.
\end{aligned}$$

Hence, firm i indeed uses the optimal strategy.

Then, we consider the first $\underline{n}$ firms. For $i \leq \underline{n}$, since it uses the best response strategy in the sequence $\{\boldsymbol{q}^t\}$ and it always has a positive product quantity, we have

$$q_i^{t+T_1} = \frac{1}{2}\left(A - \sum_{j\neq i} q_j^{t+T_1} - c_i\right).$$

Sum up for all $i \leq \underline{n}$, we have

$$\begin{aligned}
\sum_{i\leq \underline{n}} q_i^{t+T_1} &= \frac{1}{2}\left(\underline{n}A - \sum_{i\leq \underline{n}} c_i - \underline{n}\sum_{j>\underline{n}} q_j^{t+T_1} - (\underline{n}-1)\sum_{j\leq \underline{n}} q_j^{t+T_1}\right), \\
\underline{n}\underline{q}_i^t &= \frac{1}{2}\left(\underline{n}A - \underline{n}\underline{c}_i - \underline{n}\sum_{j>\underline{n}} \underline{q}_j^t - \underline{n}\sum_{j\neq i, j\leq \underline{n}} \underline{q}_j^t\right), \\
\underline{q}_i^t &= \frac{1}{2}\left(A - \sum_{j\neq i} \underline{q}_j^t - \underline{c}_i\right).
\end{aligned}$$

Therefore, all firms are using the best response strategy in the new game. □

When we focus on the number of survival firms after round T_1, the above lemma tells us that, w.l.o.g., we can assume the first $\underline{n}$ firms share the same cost and the same strategies.

Next, we show the firms' behavior will eventually converge to either a Nash equilibrium or an oscillation. We first consider the best response dynamics in the designed game $\{\underline{q}^t\}$ and show there is a cyclical oscillation. The result is based on the cyclical variation in the production quantity of the first $\underline{n}$ firms. Then we extend the result to the best response dynamics in the original game.

Lemma 3. *Both the original and new dynamics converge to either periodic oscillation or Nash equilibrium, and share the same length of oscillation period.*

Proof. (1) We first consider the best response dynamics $\{\underline{q}^t\}_{t\in\mathbb{N}}$ in the new game, by definition, there exist infinite t's such that we have $m^t = \underline{n}$. Among them, we can always choose $t_1 < t_2$ such that $t_2 - t_1$ is an even number. Otherwise, there must exist $t_1 < t' < t_2$ such that $t_2 - t'$ and $t' - t_1$ are both odd, we can choose t_1 and t_2 instead.

According to the new dynamic, the first $\underline{n}$ entries of $\underline{q}^t$ are identical, and the remaining entries are zero, so it must be that either $\underline{q}^{t_1} \leq \underline{q}^{t_2}$, or $\underline{q}^{t_1} > \underline{q}^{t_2}$ element-wisely. We will only consider the first case. The result for the case $\underline{q}^{t_1} \geq \underline{q}^{t_2}$ holds similarly. Next we show that if $\underline{q}^{t_1} \leq \underline{q}^{t_2}$, that is, $\underline{q}_i^{t_1} \leq \underline{q}_i^{t_2}$ for each $i \in [n]$, then for any even number r, we have $\underline{q}^{t_1+r} \leq \underline{q}^{t_2+r}$. By our definition, we have for each $i \in [n]$,

$$\underline{q}_i^{t_1+1} = \max\left\{\frac{1}{2}(A - \sum_{j\neq i} \underline{q}_j^{t_1} - c_i), 0\right\},$$

and

$$\underline{q}_i^{t_2+1} = \max\left\{\frac{1}{2}(A - \sum_{j\neq i} \underline{q}_j^{t_2} - c_i), 0\right\}.$$

Note that in this case, we have $\sum_{j\neq i} \underline{q}_j^{t_1} \leq \sum_{j\neq i} \underline{q}_j^{t_2}$, so $\underline{q}_i^{t_1+1} \geq \underline{q}_i^{t_2+1}$ for any $i \in [n]$. By repeating the above process, we have that $\underline{q}_i^{t_1+2} \leq \underline{q}_i^{t_2+2}$ for each $i \in [n]$.

Since $\Delta := t_2 - t_1$ is even, we have that the sequence $\underline{q}^{t_1} \leq \underline{q}^{t_1+\Delta} \leq \underline{q}^{t_1+2\Delta} \leq \underline{q}^{t_1+3\Delta} \leq \cdots$. Since $\underline{q}_i^t$ is bounded by A for any i and t, we must have the sequence $\{\underline{q}^{t_1+k\Delta}\}_{k\in\mathbb{N}}$ converges to a limit.

Since the sequence $\{\underline{q}^{t_1+k\Delta}\}_{k\in\mathbb{N}}$ converges, we have the sequence $\{\underline{q}^{t_1+k\Delta+1}\}_{k\in\mathbb{N}}$ converges. It implies the sequence $\{\underline{q}^{t_1+k\Delta+2}\}_{k\in\mathbb{N}}$ converges and so on. Therefore, for any $\ell < \Delta$, we have the sequence $\{\underline{q}^{t_1+k\Delta+\ell}\}_{k\in\mathbb{N}}$ converges. We have now proven that, for the new dynamic $\{\underline{q}^t\}_{t\in\mathbb{N}}$, if it does not converge to a Nash equilibrium[1], it will result in periodic oscillations.

(2) Since the new dynamics $\{\underline{q}^t\}_{t\in\mathbb{N}}$ can converge to a periodic oscillation, it implies the total production of the first $\underline{n}$ firms and the production of the remaining firms can also converge to a periodic oscillation. The length of the oscillation period is the same as the one in the new dynamics, denoted by Δ. Hence, we have already shown that $\sum_{i=1}^{\underline{n}} q_i^t = \sum_{i=1}^{\underline{n}} q_i^{t+\Delta}$ and $q_j^t = q_j^{t+\Delta}$, for

[1] Nash equilibrium can also be regarded as an oscillation whose period equals to one.

any index $j \in [\underline{n}+1, n]$. It is left to prove that the production of each of the first $\underline{n}$ firms converges to a periodic oscillation with length Δ.

We consider the first firm. For the production of the first firm at time $t > T_1$ and $t+\Delta$, if q_1^t and $q_1^{t+\Delta}$ are not equal, there are two cases to consider. We only provide the proof for the case $q_1^t < q_1^{t+\Delta}$. The case where $q_1^t > q_1^{t+\Delta}$ follows similarly. According to the best response dynamics, we have:

$$q_1^{t+1} = \frac{1}{2}(A - \sum_{i=1}^{\underline{n}} q_i^t - \sum_{j=\underline{n}+1}^{n} q_j^t - c_1 + q_1^t),$$

and

$$q_1^{t+\Delta+1} = \frac{1}{2}(A - \sum_{i=1}^{\underline{n}} q_i^{t+\Delta} - \sum_{j=\underline{n}+1}^{n} q_j^{t+\Delta} - c_1 + q_1^{t+\Delta}).$$

Subtract these two equations, we get

$$q_1^{t+1} - q_1^{t+\Delta+1} = \frac{1}{2}(q_1^t - q_1^{t+\Delta}).$$

Thus we have $q_1^{t+1} < q_1^{t+\Delta+1}$. For any number r, it follows that $q_1^{t+r} < q_1^{t+\Delta+r}$. Furthermore, we obtain the inequality chain $q_1^t < q_1^{t+\Delta} < q_1^{t+2\Delta} < q_1^{t+3\Delta} < \cdots$. As production can not be arbitrarily large, the sequence $\{q_1^{t+k\Delta}\}_{k\in\mathbb{N}}$ converges to a limit.

Since the sequence $\{q_1^{t+k\Delta}\}_{k\in\mathbb{N}}$ converges, the sequence $\{q_1^{t+k\Delta+1}\}_{k\in\mathbb{N}}$ must also converge. This implies that the sequence $\{q_1^{t+k\Delta+2}\}_{k\in\mathbb{N}}$ converges, and so on. Thus, for any $l < \Delta$, the sequence $\{q_1^{t+k\Delta+l}\}_{k\in\mathbb{N}}$ converges as well. Then the sequence $\{q_1^t\}_{t\in\mathbb{N}}$ converges to an oscillation and the length of the oscillation is a factor of Δ. By the same reasoning, the production levels of the other $\underline{n}-1$ firms will also converge. The length of each cyclical oscillation is always a factor of Δ. Combining the remaining firms except the first $\underline{n}$ firms, we can conclude that the original dynamics $\{\boldsymbol{q}^t\}_{t\in\mathbb{N}}$ converge to a periodic oscillation with length Δ or a Nash equilibrium. □

Oscillations of period two have been observed in the literature. However, no oscillation of longer periods is observed or studied. Our main theorem shows that there is no oscillation of period greater than two.

Theorem 1. *The period of oscillation cannot be greater than two.*

According to Lemma 3, the original and new dynamics share the same period. Therefore, we will apply techniques similar to those in Lemma 2 to construct the new dynamic, with particular attention to analyzing this periodicity. For a better illustration of the dynamics, we introduce a matrix to represent the quantity of each firm in each round. Given $n, t > 0$, we say a matrix $M \in \mathbb{R}_+^{t\times n}$ is a *t-period quantity matrix* for n firms, if $M_{i^{+1},j}$ is the updated quantity of $(M_{i,1}, \ldots, M_{i,n})$ for firm j, where $i^{+1} := (i+1) \pmod{t} + 1$. That is, all firms update their production quantities by using the previous row of the matrix. Here is an example of the 2-period quantity matrix. There are four firms with $A = 20$ and $c_1 = c_2 = c_3 = c_4 = 0$. There is a 2-period oscillation in the dynamics described by the following 2-period quantity matrix.

$$\begin{bmatrix} 10 & 10 & 10 & 10 \\ 0 & 0 & 0 & 0 \end{bmatrix}.$$

Due to space constraints, the complete proof of Theorem 1 can be found in the full version [29].

Next, we study the forms for the two-period oscillation. We first classify by whether the number of companies producing products in both periods, k_1, is zero or not. For $k_1 > 0$, we further classify the forms based on the number of companies producing products only in one of the two periods. We investigate the time complexity of verifying whether a given form of oscillation exists and the time complexity of finding an oscillation in that form if it exists.[2]

Theorem 2. *Let k_1 denote the number of firms with consistently positive production during a two-period stable oscillation. Depending on the value of k_1, the system exhibits three distinct equilibrium types:*

1. ***Case*** $k_1 = 0$*: In one of the two periods, all firms have zero production.*
   - *Equilibrium existence can be verified in $O(n)$ time.*
   - *The unique equilibrium can be computed in $O(n)$ time.*
2. ***Case*** $0 < k_1 < 3$*: Exactly k_1 firms maintain positive production in both periods, while all others have zero production in one period.*
   - *Oscillation existence can be verified in $O(n)$ time.*
   - *Infinitely many oscillations may exist.*
   - *A solution can be find in $O(n)$ time.*
3. ***Case*** $k_1 \geq 3$ *: Stable oscillations emerge among three or more firms.*
   - *Oscillation existence can be verified in $O(1)$ time.*
   - *Infinitely many oscillations may exist.*
   - *A solution can be find in $O(1)$ time.*

Proof Sketch of Theorem 2. The complete argument involves (1) systematic classification of oscillation patterns through production quantity matrices, (2) equilibrium analysis using modified Cournot update rules, and (3) complexity-aware verification procedures for each case. Key technical steps include dimensional reduction via equivalent firm aggregation (for $0 < k_1 < 3$ cases) and parametric characterization of periodic quantity differences Δa (for $k_1 = 3$). The appendix further details the mathematical derivation of threshold conditions, solution space construction for infinite oscillations, and tight complexity bounds through algorithmic analysis of cost comparison operations.

[2] Due to space constraints, the complete proof of Theorem 2 can be found in the full version [29], while providing a high-level sketch below.

4 Conclusion

In this paper, we characterized specific patterns in the oscillations of production quantities, where firms' production decisions alternate between two distinct periods. This behavior arises due to the complexity of firms' best response strategies, especially when ensuring they produce non-negative quantities. We consider all three specific types of two-period oscillations. Each type can be found efficiently, taking linear time in the number of firms involved. This classification enhances our understanding of strategic interactions among firms producing homogeneous products, providing insights into the intricate dynamics of markets with multiple competitors. Furthermore, our techniques are robust and can be extended to other settings. In particular, Theorem 1 still holds when firms use the weighted average between the best response and the strategy in the previous round.

For the future work, it would be interesting to examine the convergence rate to reach an oscillation and the stability of different forms of oscillations. Besides, how to extend our results to Bertrand model [8] could be a challenging problem. In Bertrand model, each firm competes by setting prices instead of quantities.

Acknowledgments. This work is supported by the National Natural Science Foundation of China (Nos. 62472029 and 62172422) and the Key Laboratory of Interdisciplinary Research of Computation and Economics (Shanghai University of Finance and Economics), Ministry of Education.

References

1. Agiza, H.N., Elsadany, A.A.: Nonlinear dynamics in the cournot duopoly game with heterogeneous players. XXPhys. A **320**, 512–524 (2003)
2. Agiza, H.N., Elsadany, A.A.: Chaotic dynamics in nonlinear duopoly game with heterogeneous players. Appl. Math. Comput. **149**(3) (2004)
3. Agiza, H.N., Hegazi, A.S., Elsadany, A.A.: Complex dynamics and synchronization of a duopoly game with bounded rationality. Math. Comput. Simul. **58**(2), 133–146 (2002)
4. Agiza, H.: Explicit stability zones for cournot game with 3 and 4 competitors. Chaos, Solitons Fractals **9**(12), 1955–1966 (1998)
5. Agiza, H., Hegazi, A., Elsadany, A.: The dynamics of bowley's model with bounded rationality. Chaos, Solitons Fractals **12**(9), 1705–1717 (2001)
6. Ahmed, E., Agiza, H.: Dynamics of a cournot game with n-competitors. Chaos, Solitons Fractals **9**(9), 1513–1517 (1998)
7. Ahmed, E., Agiza, H., Hassan, S.: On modifications of puu's dynamical duopoly. Chaos, Solitons Fractals **11**(7), 1025–1028 (2000)
8. Bertrand, J.: Review of "theorie mathematique de la richesse sociale" and of "recherches sur les principles mathematiques de la theorie des richesses.". J. de Savants **67**, 499 (1883)
9. Bischi, G.I., Naimzada, A.: Global analysis of a dynamic duopoly game with bounded rationality. In: Advances in Dynamic Games and Applications, pp. 361–385. Springer (2000)

10. Cánovas, J.S.: Reducing competitors in a cournot-theocharis oligopoly model. J. Differ. Equations Appl. **15**(2), 153–165 (2009)
11. Cánovas, J.S., Muñoz-Guillermo, M.: Monopoly conditions in a cournot-theocharis oligopoly model under adaptive expectations. Discrete Continuous Dynamical Syst.-B **27**(5), 2817–2831 (2022)
12. Cánovas, J.S., Puu, T., Ruíz, M.: The cournot-theocharis problem reconsidered. Chaos, Solitons Fractals **37**(4), 1025–1039 (2008)
13. Chrysanthopoulos, N., Papavassilopoulos, G.P.: Adaptive rules for discrete-time cournot games of high competition level markets. Oper. Res. **21** (2021)
14. Cournot, A.A.: Researches into the Mathematical Principles of the Theory of Wealth. Hachette (1838)
15. Cox, J.C., Walker, M.: Learning to play cournot duopoly strategies. J. Econ. Behav. Organ. **36**(2), 141–161 (1998)
16. Debreu, G.: A social equilibrium existence theorem. Proc. Natl. Acad. Sci. **38**(10), 886–893 (1952)
17. Ding, Z., Wang, Q., Jiang, S.: Analysis on the dynamics of a cournot investment game with bounded rationality. Econ. Model. **39**, 204–212 (2014)
18. Even-Dar, E., Mansour, Y., Nadav, U.: On the convergence of regret minimization dynamics in concave games. In: Proceedings of the Forty-First Annual ACM Symposium on Theory of Computing, pp. 523–532 (2009)
19. Fan, Y., Xie, T., Du, J.: Complex dynamics of duopoly game with heterogeneous players: a further analysis of the output model. Appl. Math. Comput. **218**(15), 7829–7838 (2012)
20. Fiat, A., Koutsoupias, E., Ligett, K., Mansour, Y., Olonetsky, S.: Beyond myopic best response (in cournot competition). Games Econ. Behav. **113** (2019)
21. Fisher, F.M.: The stability of the cournot oligopoly solution: the effects of speeds of adjustment and increasing marginal costs. Rev. Econ. Stud. **28**(2), 125–135 (1961)
22. Gal-Or, E.: First mover and second mover advantages. Int. Econ. Rev. 649–653 (1985)
23. Ghosh, A., Goldberg, P.W.: Best-response dynamics in lottery contests. In: The Twenty-Fourth ACM Conference on Economics and Computation (EC 2023) (2023)
24. Hofbauer, J., Sorin, S.: Best response dynamics for continuous zero-sum games. Discrete Continuous Dyn. Syst. Ser. B **6**(1), 215 (2006)
25. Huck, S., Normann, H.T., Oechssler, J.: Learning in cournot oligopoly-an experiment. Econ. J. **109**(454), 80–95 (1999)
26. Huck, S., Normann, H.T., Oechssler, J.: Stability of the cournot process-experimental evidence. Internat. J. Game Theory **31**, 123–136 (2002)
27. Kukushkin, N.S.: Best response dynamics in finite games with additive aggregation. Games Econom. Behav. **48**(1), 94–110 (2004)
28. Leininger, W.: More efficient rent-seeking—a münchhausen solution. In: Efficient Rent-Seeking: Chronicle of an Intellectual Quagmire, pp. 187–206. Springer (1993)
29. Liu, Z., Lu, H., Shan, L., Wang, Z.: On the oscillations in cournot games with best response strategies. arXiv preprint arXiv:2410.09435 (2024)
30. Matsumoto, A., Szidarovszky, F.: Theocharis problem reconsidered in differentiated oligopoly. Econ. Res. Int. **2014** (2014)
31. McManus, M., Quandt, R.E.: Comments on the stability of the cournot oligipoly model. Rev. Econ. Stud. **28**(2), 136–139 (1961)
32. Monderer, D., Shapley, L.S.: Potential games. Games Econ. Behav. **14**(1), 124–143 (1996)

33. al Nowaihi, A., Levine, P.L.: The stability of the cournot oligopoly model: a reassessment. J. Econ. Theor. **35**(2), 307–321 (1985)
34. Okuguchi, K.: Adaptive expectations in an oligopoly model. Rev. Econ. Stud. **37**(2), 233–237 (1970)
35. Palander, T.: Konkurrens och marknadsjämvikt vid duopol och oligopol. Ekonomisk Tidskrift (r 3), 222–250 (1939)
36. Puu, T.: Complex dynamics with three oligopolists. Chaos, Solitons Fractals **7**(12), 2075–2081 (1996)
37. Puu, T.: The chaotic duopolists revisited. J. Econ. Behav. Organ. **33**(3–4), 385–394 (1998)
38. Rassenti, S., Reynolds, S.S., Smith, V.L., Szidarovszky, F.: Adaptation and convergence of behavior in repeated experimental cournot games. J. Econ. Behav. Organ. **41**(2), 117–146 (2000)
39. Reinganum, J.F.: A two-stage model of research and development with endogenous second-mover advantages. Int. J. Ind. Organ. **3**(3), 275–292 (1985)
40. Szidarovszky, F., Yakowitz, S.: A new proof of the existence and uniqueness of the cournot equilibrium. Int. Econ. Rev. 787–789 (1977)
41. Theocharis, R.D.: On the stability of the cournot solution on the oligopoly problem. Rev. Econ. Stud. **27**(2), 133–134 (1960)
42. Tullock, G.: Efficient rent seeking. In: Efficient Rent-Seeking: Chronicle of an Intellectual Quagmire, pp. 3–16. Springer (2001)
43. Zhang, J., Da, Q., Wang, Y.: Analysis of nonlinear duopoly game with heterogeneous players. Econ. Model. **24**(1), 138–148 (2007)

Simultaneous All-Pay Auctions with Budget Constraints

Yan Liu, Ying Qin, and Zihe Wang(✉)

Gaoling School of Artificial Intelligence, Renmin University of China, Beijing, China
{liuyan5816,2021201494,wang.zihe}@ruc.edu.cn

Abstract. The all-pay auction, a classic competitive model, is widely applied in scenarios such as political elections, sports competitions, and research and development, where all participants pay their bids regardless of winning or losing. However, in the traditional all-pay auction, players have no budget constraints, whereas in real-world scenarios, players typically face budget constraints. This paper studies the Nash equilibrium of two players with budget constraints across multiple heterogeneous items in a complete-information framework. The main contributions are as follows: (1) a comprehensive characterization of the Nash equilibrium in single-item auctions with asymmetric budgets and valuations; (2) the construction of a joint distribution Nash equilibrium for the two-item scenario; and (3) the construction of a joint distribution Nash equilibrium for the three-item scenario. Unlike the unconstrained all-pay auction, which always has a Nash equilibrium, a Nash equilibrium may not exist when players have budget constraints. Our findings highlight the intricate effects of budget constraints on bidding strategies, providing new perspectives and methodologies for theoretical analysis and practical applications of all-pay auctions.

Keywords: All-pay auction · Budget constraints · Nash equilibrium

1 Introduction

Competition is ubiquitous in life. For example, in political elections, two presidential candidates compete across multiple states, allocating their limited resources to different states in the hope of securing more votes. In the development of large language models, IT companies operate under budget constraints and must allocate their limited resources among computing power, data acquisition, and talent recruitment to develop the most advanced models. A common characteristic of these competitive scenarios is that ultimately, only one player wins and receives the reward, while every participant incurs costs, which may include money, time, or effort. The all-pay auction, as a classic competition model, effectively captures the nature of such competitive situations [11,14].

In an all-pay auction, all players compete against each other to obtain an item or prize. Specifically, each player submits a bid, and the player with the

F. V. Fomin and M. Xiao (Eds.): COCOON 2025, LNCS 15983, pp. 333–345, 2026.
https://doi.org/10.1007/978-981-95-0215-8_25

highest bid wins and receives the item. The key distinction from other auction formats, such as first-price or second-price auctions, is that in an all-pay auction, each player must pay their bid regardless of whether they win or lose. Because the all-pay auction provides a compelling model for simulating competitive scenarios, it has attracted significant interest from researchers in fields such as computer science, sociology, and economics [1–3,6,7,10]. In the classic all-pay auction, players do not have budget constraints. However, in real-world scenarios, players are often subject to budget constraints. Therefore, we investigate the all-pay auction with budget constraints under complete information. Specifically, the budgets of the two players are asymmetric, and the items are heterogeneous. Moreover, the value of each item is also asymmetric between the players. This general setting allows us to more realistically describe the competitive relationships between players.

In the study of all-pay auctions with budget constraints, Roberson's work is the most relevant to ours. However, in their setting, the items are homogeneous [16]. Another closely related work is that of Dulleck, but in their setting, the resources invested by players only satisfy the budget constraint in expectation, and players aim to maximize their winning probability rather than their utility [9]. In the above two studies, their conclusions only provide the strategy for players on each individual item but do not offer the joint strategy for players across multiple items. In our work, the budget constraint is a hard constraint rather than a soft constraint in expectation. Moreover, the Nash equilibrium we provide will be the joint distribution across multiple items, rather than merely the marginal distribution on a single item.

However, analyzing the Nash equilibrium in all-pay auctions with multiple items and budget constraints is a highly challenging problem, primarily due to the following reasons: when players do not have budget constraints, they can bid freely on each item, making their bidding strategies more independent. In contrast, when players face budget constraints, their bids become interdependent, introducing a correlation between their bidding decisions across items. In summary, when players lack budget constraints, their bids on each item follow an independent distribution, meaning that bids on different items do not directly influence each other [3]. However, once players have budget constraints, their bid distributions typically form joint distributions, where bids on different items may be correlated [16].

1.1 Our Contribution

In this paper, we are the first to study the joint distributions of two budget-constrained players bidding on multiple items with heterogeneous values. Our contributions are as follows:

- For a single item, under the conditions of budget asymmetry and asymmetric valuations, we fully characterize the Nash equilibrium.
- We extend our analysis from a single item to two items and construct a Nash equilibrium, which takes the form of a two-dimensional joint distribution.

- For three items, under the assumption that item values are symmetric between players, we construct a Nash equilibrium, which is a three-dimensional joint distribution.

1.2 Related Work

There is a rich body of literature on all-pay auctions. We provide an overview of all-pay auctions with budget constraints under complete information.

Dulleck et al. consider the Nash equilibrium strategies of two players across multiple items. However, they only require that the amount of resources allocated does not exceed the budget in expectation, and each player focuses solely on maximizing their probability of winning rather than their utility [9]. Dekel et al. analyze how allowing two players to alternate bids, with the possibility of jumping bids, affects auction outcomes [8]. Roberson and Kvasov provide the marginal distributions of the Nash equilibrium strategies for two players with asymmetric budgets across multiple homogeneous items [16]. Hwang et al. analyze the equilibrium allocation strategies of two players with symmetric budget constraints in continuous items, which approximate an environment with an arbitrarily large (but finite) number of items [12].

Another competitive model related to the all-pay auction is the Colonel Blotto game [15]. In this model, players also have budget constraints and place bids across multiple items. For each item, the player with the highest bid wins and obtains the item, but they do not have to pay their respective bids. As a result, players tend to exhaust their budgets in competition. In contrast, in the all-pay auction, since players must pay their bids regardless of winning or losing, they may not necessarily spend their entire budgets. Researchers have conducted extensive studies on the Nash equilibrium of the Colonel Blotto game and its variants [5,13]. However, under the conditions of multiple heterogeneous items and asymmetric budgets among players, solving for the Nash equilibrium remains an open problem [15].

2 Preliminaries

We consider a model with two players and n items. Each player $i \in \{1,2\}$ has a budget constraint $B_i \geq 0$ and values item $j \in \{1, 2, \ldots, n\}$ at $v_{ij} > 0$. Player i's pure strategy is an n-dimensional bid vector $\boldsymbol{x}_i = (x_{i1}, x_{i2}, \ldots, x_{in})$, where $x_{ij} \geq 0$ represents their bid on item j, subject to the budget constraint $\sum_{j=1}^{n} x_{ij} \leq B_i$. Denote $\boldsymbol{X}_i$ as player i's set of pure strategies:

$$\boldsymbol{X}_i = \left\{(x_{ij})_{j=1}^{n} : \sum_{j=1}^{n} x_{ij} \leq B_i, \text{ and } x_{ij} \geq 0\right\}.$$

We consider a setting where players have full information about each other's values and budgets and choose their strategies simultaneously[1]. In the game, the player who places the higher bid on item j wins it. If both players bid the same amount for an item j, that is, $x_{1j} = x_{2j}$, a tie-breaking rule is applied. Denote $-i$ as the opponent of player i, where $i \in \{1, 2\}$.

- When $x_{1j} = x_{2j} = \min\{B_1, B_2, v_{1j}, v_{2j}\}$, if $\min\{B_i, v_{ij}\} > \min\{B_{-i}, v_{-ij}\}$ for some $i \in \{1, 2\}$, then player i wins item j.
- In the case of all other ties, each player wins with probability $\frac{1}{2}$.

The utility of player i for item j is given by:

$$u_{ij}(x_{ij}, x_{-ij}) = \begin{cases} v_{ij} - x_{ij}, & \text{if player } i \text{ wins item } j; \\ -x_{ij}, & \text{if player } i \text{ loses.} \end{cases}$$

The total utility of player i is the sum of the utilities from all items:

$$u_i(\boldsymbol{x}_i, \boldsymbol{x}_{-i}) = \sum_{j=1}^{n} u_{ij}(x_{ij}, x_{-ij}).$$

A game is represented as follows:

$$\mathcal{G} = \{\{1, 2\}, \{1, 2, \ldots, n\}, B_1, B_2, (v_{1j})_{j=1}^{n}, (v_{2j})_{j=1}^{n}, u_1, u_2\}.$$

A mixed strategy of player i is an n-dimensional probability distribution over their pure strategy set $\boldsymbol{X}_i$, characterized by the cumulative distribution function $F_i(\boldsymbol{x}_i)$ and the probability density function $f_i(\boldsymbol{x}_i)$. Given that player $-i$ follows a strategy with cumulative distribution function $F_{-i}(\boldsymbol{x}_{-i})$, when player i bids a pure strategy $\boldsymbol{x}_i$, their expected utility is

$$u_i(\boldsymbol{x}_i, F_{-i}) = \mathbb{E}_{\boldsymbol{X}_{-i} \sim F_{-i}} \left[u_i(\boldsymbol{x}_i, \boldsymbol{X}_{-i}) \right].$$

Accordingly, the overall expected utility of player i under the strategy F_i is given by:

$$\mathbb{E}[u_i(F_i, F_{-i})] = \mathbb{E}_{\boldsymbol{x}_i \sim F_i} \left[u_i(\boldsymbol{x}_i, F_{-i}) \right].$$

A strategy profile (F_i^*, F_{-i}^*) is a Nash Equilibrium if and only if

$$\mathbb{E}[u_i(F_i^*, F_{-i}^*)] = \max_{F_i} \mathbb{E}[u_i(F_i, F_{-i}^*)], \quad \forall i \in \{1, 2\}.$$

Now, we provide an example to illustrate the reasoning behind the tie-breaking rule we have proposed.

[1] This full-information setting serves as a baseline for analysis before introducing uncertainty or incomplete information. It is also practical in certain industries due to regulatory transparency, structural disclosure requirements, and the availability of historical data and valuation frameworks.

Example: Suppose that player 1 and player 2 have budgets of $B_1 = 0$ and $B_2 = 1$, respectively, and their valuations for the item are $v_1 = 0$ and $v_2 = 1$, respectively. Clearly, we have $\min\{B_2, v_2\} > \min\{B_1, v_1\}$, and player 1 only bids 0. If ties are resolved by assigning each player a 0.5 probability of winning, player 2's utility is 0.5 when bidding 0. However, player 2 can bid $\varepsilon \in (0, 0.5)$, increasing his utility to $1 - \varepsilon > 0.5$. Since ε can be infinitely close to 0, player 2 has no best response to player 1's strategy, leading to no Nash equilibrium under this tie-breaking rule. By contrast, if we adopt a tie-breaking rule where lets player 2 win the item, both players bidding 0 would form a Nash equilibrium, with player 1's utility being 0 and player 2's utility being 1.

3 Nash Equilibrium with Single Item

In this section, we analyze the Nash equilibrium for two players in a single-item scenario. For simplicity, we omit the footnote j. Let $F_i(x_i)$ represent player i's strategy, and let $Supp(F_i)$ denote the support of player i's strategy. Define $\overline{x}_i = \sup Supp(F_i)$ and $\underline{x}_i = \inf Supp(F_i)$ as the supremum and infimum of the support of player i's strategy, respectively. We begin by discussing the pure strategy Nash equilibrium. Then we analyze the mixed strategy Nash equilibria, focusing on the structure of the players' strategy supports and their utilities. Finally, we derive the Nash equilibria for two players in the single-item case.

The following Lemma provides the necessary and sufficient condition for the existence of a pure strategy Nash equilibrium.

Lemma 1. *Given a profile (B_1, B_2, v_1, v_2), if $B_1 = B_2$ and $B_2 \leq \frac{1}{2}\min\{v_1, v_2\}$, then (B_1, B_2) is the unique pure strategy Nash equilibrium; otherwise, pure strategy Nash equilibrium does not exist.*

When a pure strategy Nash equilibrium does not exist, we analyze the mixed strategy Nash equilibrium. First, we characterize the supremum of the supports of the two players' strategies. Lemma 2 shows that the suprema of the players' strategies are identical and are determined by the minimum value among (B_1, B_2, v_1, v_2). In this paper, we denote this supremum by L.

Lemma 2. *Let $L = \min\{B_1, B_2, v_1, v_2\}$, we have $\overline{x}_1 = \overline{x}_2 = L$ in the Nash equilibrium.*

According to Lemma 2, we know that $\overline{x}_1 = \overline{x}_2 = L = \min\{B_1, B_2, v_1, v_2\}$. Next, we analyze the infimum of the support of each player's strategy. There are three possible cases: the first case occurs when the infima and suprema of both players' strategy supports are equal. The case corresponds to Lemma 1. The second case arises when the infimum of the support of only one player's strategy is smaller than the supremum. The case corresponds to Lemma 3. The third case involves the infima of both players' strategy supports are smaller than their respective suprema. The case corresponds to Lemma 4.

Lemma 3. *In Nash equilibrium, if for some $i \in \{1, 2\}$, we have $\underline{x}_i < \overline{x}_i$ and $\underline{x}_{-i} = \overline{x}_{-i}$, then $Supp(F_i) = \{0, 1\}$ and $F_i(0) \leq 1 - \frac{2L}{v_{-i}}$. Additionally, player i's utility is 0, and $-i$'s utility is at most $v_{-i} - 2L$.*

Lemma 4. *In Nash equilibrium, if* $\underline{x}_1 < \overline{x}_1$ *and* $\underline{x}_2 < \overline{x}_2$*, then* $\underline{x}_1 = \underline{x}_2 = 0$.

When $\underline{x}_1 = \underline{x}_2 = 0$, we next show that when the bids lie within the interval $(0, L)$, both players' strategies are continuously randomized, and there are no mass points within this interval. To establish this result, we employ a simplified version of Theorem 2 from literature [4] as a lemma to support our analysis. This simplified version characterizes the support properties of a player's strategy in a Nash equilibrium without budget constraints.

Lemma 5. *(Simplified version of Theorem 2 from [Baye et al., 1996]) Two budget-unconstrained players have valuations as* v_1 *and* v_2 *for one item,* $v_1 \geq v_2$*, respectively. In Nash equilibrium, each player randomizes continuously, and there is no mass point in the interval* $(0, v_2)$ *for either player.*

Using Lemma 5, we derive Lemma 6, which states that the strategy of each player with a budget constraint is continuous on the interval $(0, L)$ and contains no mass points within this interval. Specifically, by constructing an auction model $\hat{\mathcal{G}}$ without budget constraints, we can transform the problem with budget constraints into one without them. By defining new strategies $\hat{F}_1$ and $\hat{F}_2$, we map the strategies from the original auction model to the budget-free model. This mapping preserves the properties of the strategies, ensuring that the equilibrium condition still holds in the new model. By applying the simplified version of Lemma 5, we conclude that the strategies of players in an auction with budget constraints retain the same properties as in the budget-free setting.

Lemma 6. *In the Nash equilibrium, if* $\underline{x}_1 = \underline{x}_2 = 0$*, then the strategy* $F_i(x_i)$*,* $\forall i \in \{1, 2\}$*, is continuously randomized over* $(0, L)$*, and there is no mass point within the interval* $(0, L)$.

Now, we have identified all possible structures of the support of the players' strategies, namely:

- $Supp(F_i) = \{L\}$, $\forall i \in \{1, 2\}$;
- $Supp(F_i) = \{0, L\}$ and $Supp(F_{-i}) = \{L\}$, $\exists i \in \{1, 2\}$;
- $\overline{Supp(F_i)} = \{x | 0 \leq x \leq L\}$, $\forall i \in \{1, 2\}$, where $\overline{Supp(F_i)}$ is the closed set of $Supp(F_i)$.

We provide an instance to show that Nash equilibrium does not exist when $B_1 = B_2$. Specifically, when $B_1 = B_2$ and $B_1 > \frac{1}{2}\min\{v_1, v_2\}$, it is possible that the Nash equilibrium does not exist.

Proposition 1. *When* $B_1 = B_2$*, there exists an instance in which the Nash equilibrium does not exist.*

Let's analyze the situation where $B_1 \neq B_2$ and $\overline{Supp(F_i)} = \{x | 0 \leq x \leq L\}$, $\forall i \in \{1, 2\}$. In this situation, we already know that each player's strategy is continuous on the interval $(0, L)$ and contains no mass points. Now, we focus on analyzing the behavior of players at the endpoints. Lemma 7 characterizes the endpoint measurements and the utilities of both players.

Lemma 7. *When $B_1 \neq B_2$, and $\underline{x}_1 = \underline{x}_2 = 0$ in Nash equilibrium, define $s = \arg\max_{i \in \{1,2\}} B_i$ and w to be the other player. There are two cases in the Nash equilibrium, namely*

- ***Case 1:*** *if $v_s > L$, then the expected utility of player s is $v_s - L$, and the expected utility of player w is 0. The probability that player w bids at 0 is $\frac{v_s - L}{v_s}$ and the probability of bidding at L is 0. The probability that player s bids at L is $1 - \frac{L}{v_w}$ and the probability of bidding at 0 is 0.*
- ***Case 2:*** *if $v_s = L$, then the expected utility of player s is 0, and the expected utility of player w is $v_w - L$. The probability that player s bids at 0 is $\frac{v_w - L}{v_w}$ and the probability of bidding at L is 0. The probability that player w bids at 0 is 0 and the probability of bidding at L is 0.*

Based on Lemmas 1 to 7, we deduce Lemma 8, which describes the possible structure of $Supp(F_i)$ in Nash equilibrium.

Lemma 8. *Given the profile (B_1, B_2, v_1, v_2), if Nash equilibrium exists, we have:*

- *Case (1): if $B_1 = B_2$ and $B_2 < \frac{1}{2}\min\{v_1, v_2\}$, then $\underline{x}_i = \overline{x}_i$ for any $i \in \{1, 2\}$;*
- *Case (2): if $B_1 = B_2$ and $B_1 = \frac{1}{2}\min\{v_1, v_2\}$, let $i = \arg\min_{i' \in \{1,2\}} v_{i'}$, then $\underline{x}_i = \overline{x}_i$, $\underline{x}_{-i} = \overline{x}_{-i}$, or $\underline{x}_i = 0$ and $\underline{x}_{-i} = \overline{x}_{-i}$;*
- *Case (3): if $B_1 \neq B_2$, then $\underline{x}_1 = \underline{x}_2 = 0$.*

Based on Lemmas 1 to 8, we derive the main theorem of this section, which characterizes the Nash equilibrium for two players with budget constraints in a single-item auction.

Theorem 1. *Given the profile (B_1, B_2, v_1, v_2), the Nash equilibrium is as follows:*

Case (1): *When $B_1 = B_2$ and $B_2 < \frac{1}{2}\min\{v_1, v_2\}$, then the unique Nash equilibrium is given by the pure strategy profile (B_1, B_2).*

Case (2): *When $B_1 = B_2$ and $B_1 = \frac{1}{2}\min\{v_1, v_2\}$, let $i = \arg\min_{i' \in \{1,2\}} v_{i'}$, then Nash equilibrium is as follows: $Supp(F_i) = \{0, L\}$ where $F_i(0) \leq 1 - \frac{2L}{v_{-i}}$, and $Supp(F_{-i}) = \{L\}$.*

Case (3): *When $B_1 \neq B_2$, define $s = \arg\max_{i \in \{1,2\}} B_i$ and w to be the another player.*

- ***When*** $v_s > L$:

$$F_s(x) = \begin{cases} \frac{x}{v_w}, & x \in [0, L), \\ 1, & x = L, \end{cases} \qquad F_w(x) = \begin{cases} \frac{v_s - L}{v_s}, & x = 0, \\ \frac{v_s - L + x}{v_s}, & x \in (0, L]. \end{cases}$$

- ***When*** $v_s = L$:

$$F_s(x) = \begin{cases} \frac{v_w - L}{v_w}, & x = 0, \\ \frac{v_w - L + x}{v_w}, & x \in (0, L], \end{cases} \qquad F_w(x) = \frac{x}{v_s}, \quad x \in [0, L].$$

4 Nash Equilibrium with Two Items

In this section, we study the Nash equilibrium for two players competing over two items. Each player's strategy is a two-dimensional joint distribution that satisfies their budget constraints. Given the profile $(B_1, B_2, v_{11}, v_{12}, v_{21}, v_{22})$, we construct a strategy profile for both players and then verify that the constructed profile forms a Nash equilibrium. Let $Supp(F_i)$ denote the support of player i's strategy, and let $Supp(F_{ij})$ denote the support of player i's strategy on item j. When there are multiple items, we use the joint density function f_i to represent player i's strategy.

First, we focus on the case of complete symmetry.

Theorem 2. *Given the profile $(B_1, B_2, v_{11}, v_{12}, v_{21}, v_{22})$, if $B_1 = B_2$ and $v_{11} = v_{12} = v_{21} = v_{22}$, then*

$$f_i(x_{i1}, x_{i2}) = \frac{1}{\sqrt{2}c}, \quad x_{ij} \in [0, c], x_{i2} = -x_{i1} + c, i \in \{1, 2\}, j \in \{1, 2\},$$

where $c = \min\{v_{11}, B_1\}$, can form a Nash equilibrium.

According to Theorem 2, we consider the following instance: $B_1 = B_2 = 1$ and $v_{11} = v_{12} = v_{21} = v_{22} = 3$. Clearly, $\forall i \in \{1, 2\}$ and $\forall j \in \{1, 2\}$, $x_{ij} \in [0, 1]$, $x_{i2} = -x_{i1} + 1$, $f_i(x_{i1}, x_{i2}) = \frac{1}{\sqrt{2}}$ can form a Nash equilibrium. However, the marginal distribution of player i on item j is $F_{ij} = x$, $x \in [0, 1]$. By Theorem 1, the Nash equilibrium of two players on a single item should be $(1, 1)$. This implies that under the Nash equilibrium for two items, the marginal distribution of a player's two-dimensional joint strategy on a single item does not necessarily form a Nash equilibrium on that item. Thus, we obtain the following corollary.

Corollary 1. *For each item j, $\forall j \in \{1, 2\}$, the strategies induced on each item j, F_{ij}, $\forall i \in \{1, 2\}$, do not necessarily constitute a Nash equilibrium.*

Next, we analyze the case where $B_1 \neq B_2$. Define $s = \arg\max_{i \in \{1,2\}} B_i$ as the player with the larger budget, and let w denote the other player. For each item $j \in \{1, 2\}$, define $L_j = \min\{B_1, B_2, v_{1j}, v_{2j}\}$. Without loss of generality, we assume $L_1 \geq L_2$.

According to Theorem 1, we need to analyze four cases. For each case, we construct a Nash equilibrium $(f_1(x_{11}, x_{12}), f_2(x_{21}, x_{22}))$. The cases are as follows:

- Case 1: $v_{s1} > L_1$ and $v_{s2} > L_2$, see Theorem 3;
- Case 2: $v_{s1} = L_1$ and $v_{s2} = L_2$, see Theorem 4.
- Case 3: $v_{s1} > L_1$ and $v_{s2} = L_2$, see Theorem 5;
- Case 4: $v_{s1} = L_1$ and $v_{s2} > L_2$, see Theorem 6.

Theorem 3. *Given the profile $(B_1, B_2, v_{11}, v_{12}, v_{21}, v_{22})$ with $B_1 \neq B_2$, if $v_{s1} > L_1$ and $v_{s2} > L_2$, then the following strategies can form a Nash equilibrium:*

$$f_s(x_{s1}, x_{s2}) = \begin{cases} \frac{1}{v_{w1}}, & x_{s1} \in [0, T_1], x_{s2} = L_2; \\ \frac{1}{v_{w2}}, & x_{s1} = L_1, x_{s2} \in [0, T_2]; \\ \frac{\frac{L_2}{v_{w2}} + \frac{L_1}{v_{w1}} - 1}{\sqrt{(L_1 - T_1)^2 + (L_2 - T_2)^2}}, & x_{s1} \in (T_1, L_1), x_{s2} \in (T_2, L_2), \\ & \frac{T_2 - L_2}{L_1 - T_1} x_{s1} + T_2 - \frac{T_2 - L_2}{L_1 - T_1} L_1 = x_{s2}, \end{cases}$$

where $T_1 = v_{w1} - \frac{v_{w1}}{v_{w2}} L_2$, $T_2 = v_{w2} - \frac{v_{w2}}{v_{w1}} L_1$.

$$f_w(x_{w1}, x_{w2}) = \begin{cases} \frac{1}{v_{s1}}, & x_{w1} \in [T_3, L_1],\ x_{w2} = 0; \\ \frac{1}{v_{s2}}, & x_{w1} = 0,\ x_{w2} \in [T_4, L_2]; \\ \frac{\frac{L_2}{v_{s2}} + \frac{L_1}{v_{s1}} - 1}{\sqrt{T_3^2 + T_4^2}}, & x_{w1} \in (0, T_3),\ x_{w2} \in (0, T_4),\ x_{w2} = -\frac{T_4}{T_3} x_{w1} + T_4, \end{cases}$$

where $T_3 = L_1 - \frac{v_{s2} - L_2}{v_{s2}} v_{s1}$, $T_4 = L_2 - \frac{v_{s1} - L_1}{v_{s1}} v_{s2}$.

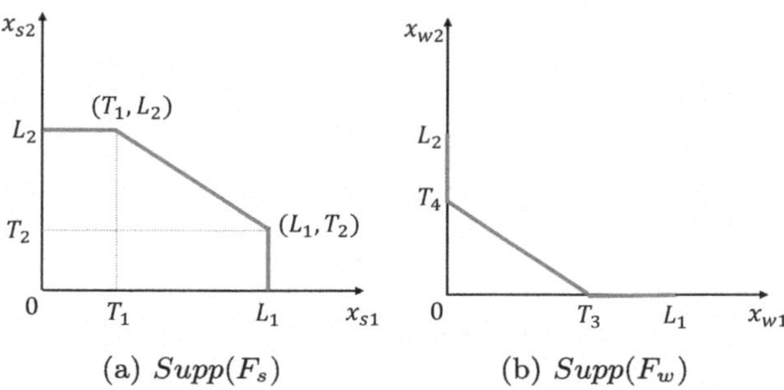

(a) *Supp*(F_s) (b) *Supp*(F_w)

Fig. 1. The blue line represents the support of players' strategies for Case 1.

Figure 1 illustrates the structure of the support of both players' strategies in Case 1. We observe that both players' strategies still form a Nash equilibrium within a single item. Player s forms a mass point at both $x_{s1} = L_1$ and $x_{s2} = L_2$, respectively. Meanwhile, player w forms a mass point at both $x_{w1} = 0$ and $x_{w2} = 0$, respectively.

Theorem 4. *Given the profile* $(B_1, B_2, v_{11}, v_{12}, v_{21}, v_{22})$ *with* $B_1 \neq B_2$, *if* $v_{s1} = L_1$ *and* $v_{s2} = L_2$, *then the following strategies can form a Nash equilibrium:*

$$f_s(x_{s1}, x_{s2}) = \begin{cases} \frac{\frac{L_1}{v_{w1}} + \frac{L_2}{v_{w2}} - 1}{\sqrt{T_2^2 + T_1^2}}, & x_{s1} \in [0, T_1),\ x_{s2} \in [0, T_2),\ x_{s2} = -\frac{T_2}{T_1} x_{s1} + T_2; \\ \frac{1}{v_{w1}}, & x_{s1} \in [T_1, L_1],\ x_{s2} = 0; \\ \frac{1}{v_{w2}}, & x_{s1} = 0,\ x_{s2} \in [T_2, L_2], \end{cases}$$

where $T_1 = L_1 - v_{w1} + \frac{L_2}{v_{w2}} v_{w1}$, $T_2 = L_2 - v_{w2} + \frac{L_1}{v_{w1}} v_{w2}$.

$$f_w(x_{w1}, x_{w2}) = \frac{1}{\sqrt{L_1^2 + L_2^2}}, \quad x_{w1} \in [0, L_1],\ x_{w2} \in [0, L_2],\ x_{w2} = -\frac{L_2}{L_1} x_{w1} + L_2.$$

Figure 2 illustrates the structure of the support of both players' strategies in Case 2. We observe that both players' strategies also form a Nash equilibrium within a single item. Since player s has a low valuation for every item, he forms a mass point at both $x_{s1} = 0$ and $x_{s2} = 0$, respectively. In contrast, player w has no mass points in his support.

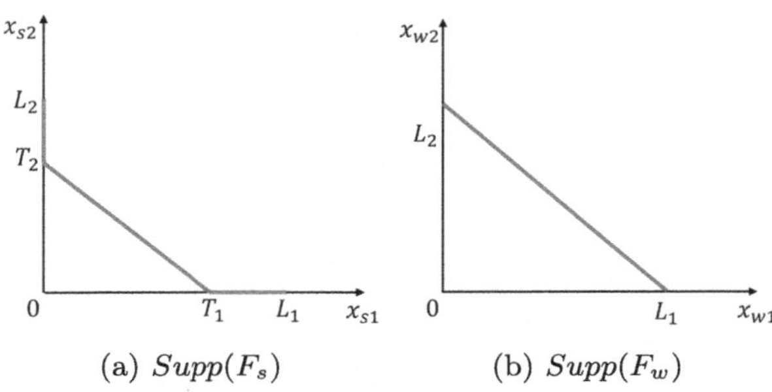

(a) $Supp(F_s)$ (b) $Supp(F_w)$

Fig. 2. The blue line represents the support of players' strategies for Case 2.

Theorem 5. *Given the profile* $(B_1, B_2, v_{11}, v_{12}, v_{21}, v_{22})$ *with* $B_1 \neq B_2$*, if* $v_{s1} > L_1$ *and* $v_{s2} = L_2$*, then the following strategies form a Nash equilibrium:*

$$f_s(x_{s1}, x_{s2}) = \begin{cases} \frac{1}{v_{w2}}, & x_{s1} = L_1,\ x_{s2} \in (0, T_2]; \\ \frac{\frac{L_1}{v_{w1}}}{\sqrt{L_1^2 + (L_2 - T_2)^2}}, & x_{s1} \in [0, L_1),\ x_{s2} \in (T_2, L_2], \\ & x_{s2} = -\frac{L_2 - T_2}{L_1} x_{s1} + L_2, \end{cases}$$

where $T_2 = -\frac{L_1}{v_{w1}} v_{w2} + L_2$*. When* $x_{s1} = L_1$*,* $x_{s2} = 0$*, there is a mass point, the probability value is* $1 - \frac{L_2}{v_{w2}}$*.*

$$f_w(x_{w1}, x_{w2}) = \begin{cases} \frac{1}{v_{s2}}, & x_{w1} = 0,\ x_{w2} \in [T_4, L_2]; \\ \frac{\frac{L_1}{v_{s1}}}{\sqrt{L_1^2 + T_4^2}}, & x_{w1} \in (0, L_1],\ x_{w2} \in [0, T_4),\ x_{w2} = -\frac{T_4}{L_1} x_{w1} + T_4, \end{cases}$$

where $T_4 = L_2 - \frac{v_{s1} - L_1}{v_{s1}} v_{s2}$*.*

Theorem 6. *Given the profile* $(B_1, B_2, v_{11}, v_{12}, v_{21}, v_{22})$ *with* $B_1 \neq B_2$*, if* $v_{s1} = L_1$ *and* $v_{s2} > L_2$*, then the following strategies form a Nash equilibrium:*

$$f_s(x_{s1}, x_{s2}) = \begin{cases} \frac{1}{v_{w1}}, & x_{s1} = (0, T_1],\ x_{s2} = L_2; \\ \frac{\frac{L_2}{v_{w2}}}{\sqrt{L_2^2 + (L_1 - T_1)^2}}, & x_{s1} \in (T_1, L_1],\ x_{s2} \in [0, L_2), \\ & x_{s2} = \frac{L_2}{T_1 - L_1} x_{s1} - \frac{L_2}{T_1 - L_1} L_1, \end{cases}$$

where $T_1 = -\frac{L_2}{v_{w2}} v_{w1} + L_1$*. When* $x_{s1} = 0$*,* $x_{s2} = L_2$*, there is a mass point, the probability value is* $1 - \frac{L_1}{v_{w1}}$*.*

$$f_w(x_{w1}, x_{w2}) = \begin{cases} \frac{1}{v_{s1}}, & x_{w1} \in [T_3, L_1],\ x_{w2} = 0; \\ \frac{\frac{L_2}{v_{s2}}}{\sqrt{L_2^2 + T_3^2}}, & x_{w1} \in [0, T_3),\ x_{w2} \in (0, L_2], x_{w2} = -\frac{L_2}{T_3} x_{w1} + L_2, \end{cases}$$

where $T_3 = L_1 - v_{s1} + \frac{L_2}{v_{s2}} v_{s1}$*.*

5 Nash Equilibrium with Three Items

In this section, we study the Nash equilibrium for three items. When analyzing the two-item case, the values of the items are asymmetric between players. However, when there are three items, analyzing the Nash equilibrium becomes highly complex. To simplify the analysis, we consider the symmetric valuation case, where the value of each item is the same for both players, namely $v_{1j} = v_{2j}$ for $j \in \{1, 2, 3\}$. We construct a strategy for each player and, in the end, verify that the constructed strategies form a Nash equilibrium.

For three items, let v_j denote the valuation of item j, without loss of generality, we assume $v_1 \geq v_2 \geq v_3$ and $B_1 \geq B_2$. The following theorem provides a Nash equilibrium when $B_i \geq \max\{\frac{v_1+v_2+v_3}{2}, v_1\}$, $\forall i \in \{1, 2\}$.

Theorem 7. *Given the profile* $(B_1, B_2, v_1, v_2, v_3)$*, if* $B_i \geq \max\{\frac{v_1+v_2+v_3}{2}, v_1\}$*,* $\forall i \in \{1, 2\}$*, then we have two cases:*

- *Case 1:* $\frac{v_1+v_2+v_3}{2} > v_1$. *Let* $z = \frac{v_1+v_2+v_3}{2}$, $A = (v_1, 0, z - v_1)$, $B = (z - v_2, v_2, 0)$, $C = (0, z - v_3, v_3)$. *Let* L_{AB}, L_{BC}, *and* L_{CA} *denote the line segments with endpoints A and B, B and C, and C and A, respectively. Let* $|AB|$, $|BC|$, *and* $|CA|$ *denote the lengths of the line segments* L_{AB}, L_{BC}, *and* L_{CA}, *respectively. Then for player* $i \in \{1, 2\}$,

$$f_i(x_{i1}, x_{i2}, x_{i3}) = \begin{cases} \frac{P_{AB}}{|AB|}, & if\ (x_{i1}, x_{i2}, x_{i3})\ is\ in\ L_{AB}; \\ \frac{P_{BC}}{|BC|}, & if\ (x_{i1}, x_{i2}, x_{i3})\ is\ in\ L_{BC}; \\ \frac{P_{CA}}{|CA|}, & if\ (x_{i1}, x_{i2}, x_{i3})\ is\ in\ L_{CA}, \end{cases}$$

where $P_{AB} = \frac{1}{\frac{z-v_3}{z-v_2}+1+\frac{z-v_3}{z-v_1}} \cdot \frac{z-v_3}{z-v_2}$, $P_{BC} = \frac{1}{\frac{z-v_3}{z-v_2}+1+\frac{z-v_3}{z-v_1}}$, $P_{CA} = \frac{1}{\frac{z-v_3}{z-v_2}+1+\frac{z-v_3}{z-v_1}} \cdot \frac{z-v_3}{z-v_1}$, *can form a Nash equilibrium.*
- *Case 2:* $\frac{v_1+v_2+v_3}{2} \leq v_1$. *Define* $A = (v_1, 0, 0)$ *and* $B = (0, v_2. v_3)$. *Let* $f_i(x_{i1}, x_{i2}, x_{i3})$ *be the uniform joint density function on the line segment AB, i.e.,* $f_i(x_{i1}, x_{i2}, x_{i3}) = \frac{1}{|AB|}$. *Then,* $f_i(x_{i1}, x_{i2}, x_{i3})$ *for player* $i \in \{1, 2\}$ *can form a Nash equilibrium.*

Figure 3 illustrates the support of both players' strategies. It can be verified that the three-dimensional joint density function we constructed has uniform marginal density functions in each dimension. From the perspective of a single item, both players still form a Nash equilibrium on each individual item.

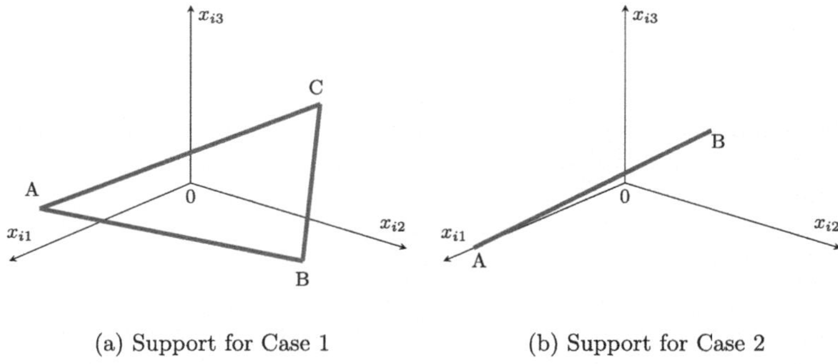

(a) Support for Case 1 (b) Support for Case 2

Fig. 3. The support of players' strategies for three items

6 Conclusion and Future Works

We investigate the Nash equilibrium in an all-pay auction where two players with budget constraints compete for multiple items. We find that when players have budget constraints, a Nash equilibrium does not always exist, and for a single item, the Nash equilibrium may not be unique. For multiple items, we construct Nash equilibrium strategies for the two players. However, we have only discussed cases with up to three items. For a larger number of items, the existence of a Nash equilibrium remains an open question. Additionally, for the three-item case, we have provided Nash equilibrium only for certain special cases. In the future, analyzing Nash equilibrium in more general settings for the three-item case will be our research focus.

Acknowledgements. We thank the fnancial support from National Natural Science Foundation of China (Grant No. 62172422), and Key Laboratory of Interdisciplinary Research of Computation and Economics (Shanghai University of Finance and Economics), Ministry of Education.

References

1. Avni, G., Ibsen-Jensen, R., Tkadlec, J.: All-pay bidding games on graphs. In: Proceedings of the AAAI Conference on Artificial Intelligence, vol. 34, pp. 1798–1805 (2020)
2. Avni, G., Jecker, I., Žikelić, Đ.: Infinite-duration all-pay bidding games. In: Proceedings of the 2021 ACM-SIAM Symposium on Discrete Algorithms (SODA), pp. 617–636. SIAM (2021)
3. Barut, Y., Kovenock, D.: The symmetric multiple prize all-pay auction with complete information. Eur. J. Polit. Econ. **14**(4), 627–644 (1998)
4. Baye, M.R., Kovenock, D., De Vries, C.G.: The all-pay auction with complete information. Econ. Theor. **8**(2), 291–305 (1996)

5. Boix-Adserà, E., Edelman, B.L., Jayanti, S.: The multiplayer colonel blotto game. In: Proceedings of the 21st ACM Conference on Economics and Computation, pp. 47–48 (2020)
6. Che, Y.K., Gale, I.: Expected revenue of all-pay auctions and first-price sealed-bid auctions with budget constraints. Econ. Lett. **50**(3), 373–379 (1996)
7. Dechenaux, E., Kovenock, D., Sheremeta, R.M.: A survey of experimental research on contests, all-pay auctions and tournaments. Exp. Econ. **18**, 609–669 (2015)
8. Dekel, E., Jackson, M.O., Wolinsky, A.: Jump bidding and budget constraints in all-pay auctions and wars of attrition. Technical Report, Discussion Paper (2007)
9. Dulleck, U., Frijters, P., Podczeck, K.: All-pay auctions with budget constraints and fair insurance. Technical Report, Working Paper (2006)
10. Dziubiński, M., Jahn, K.: Discrete two player all-pay auction with complete information. In: Proceedings of the Thirty-Second International Joint Conference on Artificial Intelligence, pp. 2659–2666 (2023)
11. Hillman, A.L., Riley, J.G.: Politically contestable rents and transfers. Econ. Polit. **1**(1), 17–39 (1989)
12. Hwang, S.H., Koh, Y., Lu, J.: Constrained contests with a continuum of battles. Games Econom. Behav. **142**, 992–1011 (2023)
13. Kovenock, D., Roberson, B.: Generalizations of the general lotto and colonel blotto games. Econ. Theor. **71**, 997–1032 (2021)
14. Nalebuff, B.J., Stiglitz, J.E.: Prizes and incentives: towards a general theory of compensation and competition. Bell J. Econ. 21–43 (1983)
15. Roberson, B.: The colonel blotto game. Econ. Theor. **29**(1), 1–24 (2006)
16. Roberson, B., Kvasov, D.: The non-constant-sum colonel blotto game. Econ. Theor. **51**, 397–433 (2012)

Online Budget Allocation Maximization Problem on Two Uniform Machines with a Common Due Date

Yaru Yang, Honglin Ding(✉), and Wentao He

School of Mathematics and Statistics, Yunnan University, Kunming 650504, China
Ding-HL@outlook.com

Abstract. In this paper, we delve into the online budget allocation maximization (BAM) problem on two uniform machines. Given two uniform machines, M_1 and M_2, with speeds $s \geq 1$ and 1, respectively, all jobs $\mathcal{J} = \{J_1, \cdots, J_n\}$ are ordered in a list and arrive one by one. Each job J_j has a size p_j and a common due date d for $j = 1, \cdots, n$. The objective is to determine a schedule that maximizes the total size of jobs processed by these two machines before the due date. For this problem, we propose an online algorithm with a competitive ratio of $(\sqrt{s^2 + 10s + 9} - s - 1)/2$, which is bounded by a constant value of 2 and reaches $\sqrt{5} - 1$ when $s = 1$. For a special case, known as the $\mathrm{BAM}_{\leq 1}$ problem, we devise an online algorithm with a competitive ratio of $(\sqrt{9s^2 + 10s + 1} - s - 1)/2s$, which is no greater than $\sqrt{5} - 1$.

Keywords: Budget allocation · Uniform machines · Common due date · Online algorithm · Competitive ratio

1 Introduction

In the literature, research on the parallel machine scheduling problem, which involves assigning n jobs $\{J_1, J_2, \ldots, J_n\}$ to m parallel machines with the aim of minimizing the makespan, i.e., the maximum completion time, is typically divided into three types: offline, online, and semi-online. In the case of offline scheduling, all job sizes, i.e., processing times, are known before starting computation. Conversely, in online scheduling, the jobs $J_1, J_2, \ldots, J_n$ arrive one by one, and when each job J_i is received, no information is available about the number or size of later jobs J_j with $j > i$. Semi-online scheduling occupies a middle ground, where some preliminary information about all jobs can be gleaned in advance, in contrast to the online scheduling, such as the total processing time, the largest processing time, and so on.

Algorithms designed to solve (semi-) online problems are known as (semi-) online algorithms. The performance of such (semi-) online algorithms is commonly evaluated using the competitive ratio. Suppose there is an (semi-) online algorithm A to solve some (semi-) online scheduling problem. For any given

F. V. Fomin and M. Xiao (Eds.): COCOON 2025, LNCS 15983, pp. 346–357, 2026.
https://doi.org/10.1007/978-981-95-0215-8_26

instance I of this problem, once no further jobs are incoming, we consider that all jobs of I are fully determined. For convenience, we typically refer to the offline scheduling for the same set of jobs as the offline version of I. Let $C^*(I)$ denote the optimal value for this offline version, and $C^A(I)$ represent the value produced by (semi-) online algorithm A for intance I. The competitive ratio of the algorithm A is defined as the infimum of ρ such that for any instances I, $C^*(I) \leq \rho \cdot C^A(I)$ always holds. If no (semi-) online algorithm can achieve a competitive ratio strictly smaller than ρ for this problem, then ρ is called its lower bound. Furthermore, the algorithm A is referred to as the optimal (semi-) online algorithm for this (semi-) online scheduling problem.

In recent years, scheduling problems with various constraints have been extensively studied, including the early work maximization problem on identical machines and some scheduling problems on uniform machines. We mainly investigate these two types of scheduling problems, with a particular focus on the scenario involving two parallel machines.

For the (semi-) online early work maximization problem on identical machines with a common due date, Chen et al. [1] first studied it in 2016. They developed an algorithm with a competitive ratio of $\frac{\sqrt{2m^2-2m+1}-1}{m-1}$ for the online version of the problem. When the number of machines $m = 2$, this algorithm achieves a competitive ratio of $\sqrt{5} - 1$ and was proven to be the optimal online algorithm. In 2021, Chen et al. [2] studied the semi-online vesion of this problem. When the total processing time of all jobs is provided in advance, they devised an optimal semi-online algorithm with a competitive ratio of 6/5. In the scenario where both the total and the largest processing times are known in advance, they presented an optimal semi-online algorithm with competitive ratio 10/9. When only the largest processing time is known, they established a lower bound 1.1231 and developed a semi-online algorithm with a competitive ratio of 1.1375.

For the offline early work maximization problem on two identical machines with a common due date, Chen et al. [1] proved that it is binary NP-hard in 2016. Subsequently, Sterna et al. [3] designed a PTAS for it in 2017. Chen et al. [4] demonstrated that the longest processing time (LPT) algorithm attains an approximation ratio of 10/9 for the problem in 2020. Jiang et al. [5] further confirmed that the LPT algorithm achieves an approximation ratio of 12/11 in 2021. Moreover, Chen et al. [6] proposed an fully polynomial-time approximation scheme (FPTAS) to address this problem in 2020, when the number of machines is a constant m. In 2022, Li [7] enhanced Chen et al.'s FPTAS and developed an efficient PTAS for the scenario where the number of machines m is not fixed.

For the (semi-) online early work maximization problem on two hierarchical machines, Xiao et al. [8] devised an optimal online algorithm with a competitive ratio of $\sqrt{2}$, and an optimal semi-online algorithm with a competitive ratio of 4/3 for the scenario where the total processing time is known in advance. In addition, when the largest processing time is known, they presented two optimal algorithms with a competitive ratio of $\frac{6}{5}$ if the largest job is a lower hierarchy one, and of $\sqrt{5} - 1$ if the largest job is a higher hierarchy one, respectively. Xiao et al. [9] explored two semi-online models for this problem, where they considered the use of a buffer or rearrangements. With a buffer size of K, they designed an optimal online algorithm with a competitive ratio of 4/3 and when the ability to reassign at most K jobs is permitted, they developed an optimal online algorithm with the same competitive ratio. Wei et al. [10] considered the hierarchical scheduling on m identical machines in shared manufacturing to minimize the total completion time, where each job has a unit-size processing time. It is worth noting that Li et al. [11] studied the max-min allocation problem under a grade of service provision, they designed three PTASes and one FPTAS for it.

In recent years, various generalizations of early work maximization problem have garnered significant research interest. In 2025, Li et al. [12] studied a parallel machine scheduling problem with a common due date to maximize total weighted early work and designed an EPTAS for solving. In addition, they presented an FPTAS for a special case of it. That same year, Li et al. [13] introduced a novel scheduling criterion named as a discounted profit, which could be considered as a generalization of early work. They proved that the competitive ratio of the classical List Scheduling algorithm for the online case. Moreover, they also proved that the Longest Processing Time algorithm has an approximation ratio for the offline case.

We also explored the research results of scheduling problems on two uniform machines with speeds s and 1. For the minimum makespan scheduling problem on two uniform machines, in 2001, Epstein et al. [14] proved that the list scheduling (LS) is an optimal online algorithm with a competitive ratio of $\min\{\frac{1+2s}{1+s}, \frac{1+s}{s}\}$. In recent years, Ng et al. [15], Dosa et al. [16,17] have successively studied two semi-online versions of this problem, one with known optimal value and the other with known total sum, and provided some semi-online algorithms and lower bounds. For the hierarchical scheduling on two uniform machines with the objective to minimize makespan, Lu et al. [18] studied three semi-online vesions of this problem, where the optimal makespan, the total size of jobs, and the largest job size are known in advance respectively, and provided optimal algorithms for all of them. In 2010, Tan et al. [19] considered the online vesion of this problem and devised two optimal online algorithms. The first algorithm has a competitive ratio of $\min\{1+s, 1+\frac{1+s}{1+s+s^2}\}$ for $0 < s < 1$, while the second has a competitive ratio of $\min\{\frac{1+s}{s}, 1+\frac{2s}{1+s+s^2}\}$ for $s > 1$. For the early work maximization problem on two hierarchical uniform machines, in 2023, Xiao et al. [20] studied three semi-online vesions of this problem, where the total sizes of all jobs, high-hierarchy jobs, low-hierarchy jobs are known in advance respectively, and designed optimal algorithms for each scenario. More related results and can be found in the recent surveys [21,22].

Consider a real-world problem: two audit firms have joined forces to audit a batch of budget plans. These budget plans are gradually submitted to the auditing firm as the audit work progresses. The audit speeds of the two companies may be different, and all buget plans share the same deadline. How should these budget plans be allocated to ensure that the amount audited by both companies reaches its maximum before the deadline? Since the budget plans are submitted incrementally during the audit work of the two audit firms, when allocating the current budget plan, we do not have the specific information about future budget plans that may be submitted. Therefore, this is an online optimization problem. For convenience, we assume that once the current budget plan is allocated, the next budget plan is submitted immediately, without any waiting time. We treat the two audit firms as two machines of scheduling problem, with their auditing speeds corresponding to the processing speeds of machines. Each budget plan is regarded as a job, and its budget amount is considered the size of the job.

Motivated by the above application and corresponding assumptions, we propose the online budget allocation maximization problem on two uniform machines. For convenience, it is referred to as the BAM problem and is defined as follows: Given two uniform machines $\mathcal{M} = \{M_1, M_2\}$ with respective processing speeds s_1 and s_2, all jobs $\mathcal{J} = \{J_1, \cdots, J_n\}$ are ordered in a list and arrive one by one. Each job J_j has a size p_j and a common due date d, $j = 1, \cdots, n$. By the due date, the total size of jobs processed by each machine M_i can't exceed $s_i d$. For any schedule, denoted by $\mathcal{S} = (S_1, S_2)$, the valid load of each machine M_i is $X_i = \min\{s_i d, \sum_{J_j \in S_i} p_j\}$ for $i = 1, 2$. The objective is to determine a schedule that maximizes the total valid load $X = X_1 + X_2$.

The reminder of this paper is organized as follows. In Sect. 2, we introduce some key notations and assumtions. In Sect. 3, we establish a lower bound for the BAM problem. In Sect. 4, we devise an online algorithm with a competitive ratio of $(\sqrt{s^2 + 10s + 9} - s - 1)/2$ for the BAM problem. In Sect. 5, we design an online algorithm with a competitive ratio of $(\sqrt{9s^2 + 10s + 1} - s - 1)/2s$ for the $\text{BAM}_{\leq 1}$ problem. In Sect. 6, we made a conlusion.

2 Preliminaries

For the BAM problem, the machine with the slower processing speed is referred to as the slow machine, while the machine with the faster processing speed is referred to as the fast machine. Without loss of generality, we assume that the machine M_2 is the slow machine. For simplicity, the processing speed s_2 of the slow machine M_2 is normalized to 1, i.e., $s_2 = 1$. The processing speed s_1 of the fast machine is denoted as s, i.e., $s_1 = s \geq 1$. Moreover, the common due date can be set to 1, i.e., $d = 1$. Consequently, the maximum total size of jobs processed by the fast machine is $s \cdot 1 = s$, which is also referred to as the capacity of the fast machine. Similarly, the capacity of the slow machine is defined as $1 \cdot 1 = 1$.

A schedule of the BAM problem is a partition $\mathcal{S} = (S_1, S_2)$ of the job set $\mathcal{J}$, such that $S_1 \cup S_2 = \mathcal{J}$, $S_1 \cap S_2 = \emptyset$. Let L_i and X_i denote the load and the valid load of each machine M_i, respectivly. This implies

$$L_i = \sum_{J_j \in S_i} p_j \text{ and } X_i = \min\{L_i, s_i\} \text{ for } i = 1, 2.$$

Therefore, the total valid load is

$$X = X_1 + X_2 = \min\{L_1, s_1\} + \min\{L_2, s_2\} = \min\{L_1, s\} + \min\{L_2, 1\}. \quad (1)$$

Furthermore, each machine M_i is considered fully loaded if its load L_i is no less than the capacity s_i, i.e., $L_i \geq s_i$; otherwise, it is referred to as underloaded.

Base on the above assumptions, let $I = (\mathcal{P}, s_1, s_2)$ denote an instance of the BAM problem, where $\mathcal{P} = (p_1, p_2, \dots, p_n)$ represents the list of sizes for the job set $\mathcal{J} = \{J_1, \cdots, J_n\}$, s_1 and s_2 are the processing speeds of the fast machine M_1 and the slow machine M_2, respectively. Noticing that $s_1 = s$ and $s_2 = 1$, $I = (\mathcal{P}, s_1, s_2)$ can be simplified to $I = (\mathcal{P}, s)$. The objective is to determine a schedule $\mathcal{S} = (S_1, S_2)$ that maximizes the total valid load $X = X_1 + X_2$.

3 A Lower Bound

In this section, we will demonstrate that when $s_1 = s_2$, the BAM problem is equivalent to the early work maximization (EWM) problem on two identical machines with a common due date. This problem can be formulated as follows. Given two identical machines $\mathcal{M} = \{M_1, M_2\}$, all jobs $\mathcal{J} = \{J_1, \cdots, J_n\}$ are ordered in a list and arrive sequentially. Each job J_j has a processing time p_j and a common due date d, $j = 1, \cdots, n$. For any schedule $\mathcal{S} = (S_1, S_2)$, the early work on M_i is defined as $X_i = \min\{d, \sum_{J_j \in S_i} p_j\}$ for $i = 1, 2$. The objective is to determine a schedule that maximizes the total early work $X = X_1 + X_2$. For convenience, such a problem is referred to as the EWM problem. We denote an instance of this problem by $\Pi = (\mathcal{P}, d)$, where $\mathcal{P} = (p_1, p_2, \dots, p_n)$ represents the list of processing times for the job set $\mathcal{J}$, and d is the common due date.

Clearly, for any instance $I = (\mathcal{P}, s_1, s_2)$ of the BAM problem with $s_1 = s_2$, the instance $\Pi = (\mathcal{P}, d = s_1)$ is an instance of the EWM problem. Both instances have the same schedules and the same corresponding objective function values. The reverse is also true. Therefore, we can deduce the following Lemma 1.

Lemma 1. *When both machines have the same processing speed, the BAM problem is equivalent to the EWM problem.*

Since Chen et al. [1] developed an optimal online algorithm for the EWM problem with a competitive ratio of $\sqrt{5} - 1$, which implies that $\sqrt{5} - 1$ serves as a lower bound for the EWM problem. By Lemma 1, it follows that Theorem 1 holds, and its proof is therefore omitted here.

Theorem 1. *For the BAM problem, there is no online algorithm with a competitive ratio strictly less than $\sqrt{5} - 1$.*

4 An Online Algorithm for the BAM Problem

For the instancce $I = (\mathcal{P}, s)$ of the BAM problem, let L_i^j denote the load on machine M_i after the job J_j has been assigned, for $i = 1, 2$ and $j = 1, 2, \dots, n$. Consequently, $L_i^n = L_i$ for $i = 1, 2$. Meanwhile, we use C_i^j to represent the remaining capacity of machine M_i after the job J_j has been assigned, i.e.,

$$C_1^j = \max\{0, s - L_1^j\} \text{ and } C_2^j = \max\{0, 1 - L_2^j\},$$

for $j = 1, 2, \ldots, n$. Additionally, for the instance I, let X^* denote the offline optimal value, and X^A denote the output value produced by an online algorithm A. The following two lower bounds on the offline optimal value X^* is apparent, and the proof is omitted here.

Lemma 2. *The offline optimal value X^* satisfies that*

$$X^* \leq \min\{s+1, L_1 + L_2\}.$$

In 2016, Chen et al. [1] devised a clever algorithm, denoted as EFF_m, for solving the early work maximization problem on m identical machines with a common due date. This algorithm assigns a new job to the first suitable machine or to the machine with the minimum load, if no suitable machine is available. A machine is considered suitable if, after the assignment of a new job, its load will not exceed the assumed ratio $r_m d$. Here, $r_m \geq 1$ is a pending ratio, which is the competitive ratio of the EFF_m algorithm in fact. Specifically, when $m = 2$, EFF_2 is an optimal online algorithm with a competitive ratio of $\sqrt{5} - 1$ for the early work maximization problem on two identical machines with a common due date. In the EFF_2 algorithm, it essentially only requires the use of the ratio r_m as a load limit on a single machine. Based on this observation and Chen et al.'s strategy, we can employ the following approach to solve our problem. Similarly, we introduce a pending ratio $r \geq 1$, which is also the competitive ratio of our algorithm. When assigning each job J_j, if the load on the slow machine M_2 increases by p_j without exceeding r, then assign the job J_j to M_2; otherwise, assign it to the machine with the largest remaining capacity. The specifics of this approach are detailed in Algorithm A.

Algorithm 1: A

```
1  Initially, let j = 1, L_1^0 = L_2^0 = 0.
2  When a new job J_j comes,
3  if L_2^{j-1} + p_j <= r then
4      Assign the job J_j to the slow machine M_2.
5  else
6      if C_1^{j-1} + C_2^{j-1} > 0 then
7          Assign the job J_j to the machine with the largest remaining capacity.
8      else
9          Assign the job J_j to the machine with the smallest current load.
10 If there is another job, let j = j + 1, go to Line 2. Otherwise, stop.
```

In Theorem 2 below, we determine the competitive ratio of Algorithm A, which is a function $r(s) = (\sqrt{s^2 + 10s + 9} - s - 1)/2$ dependent on the speed s of the fast machine.

Theorem 2. *Algorithm A has a competitive ratio of* $r(s) = \frac{\sqrt{s^2+10s+9}-s-1}{2}$*, which is bounded by a constant value of 2. Moreover, when* $s = 1$*,* $r(s) = \sqrt{5}-1$*.*

Proof. Assume that Algorithm A terminates after n jobs have been assigned, then $L1 = L_1^n$ and $L_2 = L_2^n$. By Lemma 2, if $L_1 \geq s$ and $L_2 \geq 1$, we have $X^A = s+1 \geq X^*$. Conversely, if $L_1 \leq s$ and $L_2 \leq 1$, we have $X^A = L_1+L_2 \geq X^*$. This indicates we can obtain the optimal solution for these two case. Therefore, we only need to proceed with our proof for two cases.

Case 1 If $L_1 > s$ and $L_2 < 1$, then $X^A = s + L_2$ follows from (1). We consider the following two cases.

Case 1.1 If $s < L_1 \leq r \cdot s$, by Lemma 2, we can derive

$$\frac{X^*}{X^A} \leq \frac{L_1 + L_2}{s + L_2} \leq \frac{L_1}{s} \leq \frac{rs}{s} \leq r. \tag{2}$$

Case 1.2 If $L_1 > r \cdot s$, let J_l be the last job assigned to M_1. Then, the jobs assigned to M_1 may be classified into two cases.

Case 1.2 (A) There are no jobs assigned to M_1 before the job J_l is assigned to M_1, which implies $L_1^{l-1} = 0$. Noticing that $L_2 < 1, r \geq 1$, and based on Lines 6 and 7 of Algorithm A, we can deduce that $p_l = L_1 > rs \geq s$, indicating that J_l is the only job assigned to M_1 and has a "big" size $p_l > s$. Furthermore, the total size of the other jobs, excluding J_l, is less than 1, i.e., $\sum_{J_j \in \mathcal{J} \setminus \{J_l\}} p_j < 1$. As a result, we have $X^A = s + L_2 = s + \sum_{J_j \in \mathcal{J} \setminus \{J_l\}} p_j = X^*$, which indicates an optimal solution has been achieved.

Case 1.2 (B) There is at least one job assigned to M_1 before the job J_l is assigned to M_2, which implies $L_1^{l-1} > 0$. According to Lines 3–4 of Algorithm A, we can deduce

$$L_1^{l-1} + L_2^{l-1} > r. \tag{3}$$

Otherwise, the other jobs on M_1 (excluding J_l) should be assigned to M_2, which leads to a contradiction. Clearly, the assignment of job J_l to M_1 is executed in Line 7, which implies that the machine M_1 currently has the largest remaining capacity. Hence, we have

$$s - L_1^{l-1} \geq 1 - L_2^{l-1}. \tag{4}$$

By the inequalities (3) and (4), we can derive a lower bound for L_2^{l-1}, which is $L_2^{l-1} \geq (r + 1 - s)/2$. Since $L_2 \geq L_2^{l-1}$, we have

$$\frac{X^*}{X^A} \leq \frac{s+1}{s+L_2} \leq \frac{s+1}{s+\frac{r+1-s}{2}} = \frac{2s+2}{s+r+1}. \tag{5}$$

Case 2 If $L_1 < s$ and $L_2 > 1$, then $X^A = 1 + L_1$ similarly. We also consider two cases.

Case 2.1 If $1 < L_2 \leq r$, by Lemma 2, we can obtain

$$\frac{X^*}{X^A} \leq \frac{L_1 + L_2}{1 + L_1} \leq L_2 \leq r. \tag{6}$$

Case 2.2 If $L_2 > r$, let J_l be the last job assigned to M_2. Clearly, the assignment of job J_l to M_2 is executed in Line 7, which implies that the machine M_2 currently has the largest remaining capacity. Hence, we have

$$1 - L_2^{l-1} \geq s - L_1^{l-1}. \tag{7}$$

At this moment, the jobs assigned to M_1 can be classified into two cases.
Case 2.2 (A) There are no jobs assigned to M_1 before the job J_l is assigned to M_2, which means $L_1^{l-1} = 0$. Then, we can obtain $L_2^{l-1} = 0$ and $s = 1$ according to the inequality (7). It indicates that there are no jobs assigned to M_2 at this time. This indicates that J_l is the first job J_1. From the given assumptions, we know that $L_1 < s = 1$ and $L_2 > r \geq 1$. Therefore, we can deduce that J_l is the only job assigned to M_2 and has a "big" size $p_l > 1$, while the total size of the other jobs (excluding J_l) is less than $s = 1$, i.e., $\sum_{J_j \in \mathcal{J} \setminus \{J_l\}} p_j < 1$. As a result, we have $X^A = 1 + L_1 = 1 + \sum_{J_j \in \mathcal{J} \setminus \{J_l\}} p_j = X^*$, indicating that an optimal solution has been achieved.
Case 2.2 (B) There is at least one job assigned to M_1 before the job J_l is assigned to M_2, which means $L_1^{l-1} > 0$. Similarly, we can obtain

$$L_1^{l-1} + L_2^{l-1} > r. \tag{8}$$

By the inequalities (7) and (8), we can derive a lower bound for L_1^{l-1}, which is $L_1^{l-1} \geq (s + r - 1)/2$. Since $L_1 \geq L_1^{l-1}$, we can obtain

$$\frac{X^*}{X^A} \leq \frac{s+1}{1+L_1} \leq \frac{s+1}{1+\frac{s+r-1}{2}} = \frac{2\,s+2}{s+r+1}. \tag{9}$$

Finally, we compare the sizes of $(2s+2)/(s+r+1)$ and r within the region of $s \geq 1, r \geq 1$. For each fixed $s \geq 1$, $(2s+2)/(s+r+1)$ decreases as r increases. When $r = 1$, we have $(2s+2)/(s+r+1) = (2s+2)/(s+2) > 1 = r$. Therefore, for each $s \geq 1$, $\max\{(2s+2)/(s+r+1), r\}$ reaches its minimum value when $(2s+2)/(s+r+1) = r$. Let $(2s+2)/(s+r+1) = r$ and solve it to obtain

$$r = \frac{\sqrt{s^2 + 10\,s + 9} - s - 1}{2} \geq 1.$$

Accroding to the inequalities (2),(5),(6) and (9), we can conclude that the competitive ratio of Algorithm A is $r(s) = (\sqrt{s^2 + 10s + 9} - s - 1)/2$. □

5 An Online Algorithm for the BAM$_{\leq 1}$ Problem

In reality, if a client submits a budget plan to an audit firm, it indicates that the audit firm possesses the capability to handle the budget plan before the deadline, meaning that the time required for processing the budget plan will not exceed the deadline. On the other hand, an audit firm may also refuse to accept a budget plan that is beyond its capability in order to protect its reputation. As a result, for an instance $I = (\mathcal{P}, s_1, s_2)$ of the BAM problem, the processing time required for each M_i to process each J_j should not exceed the due date d, i.e., $p_j/s_i \leq d$. Observing that $s_1 \geq 1, s_2 = 1$ and $d = 1$, we can thus assume that each size $p_j \leq 1$ for all $j = 1, 2, \ldots, n$. For convenience, this special case is referred to as the $\text{BAM}_{\leq 1}$ problem.

For the $\text{BAM}_{\leq 1}$ problem, we utilize a similar approach to solve it. The only difference is that the pending ratio r will be used for the fast machine M_1, and the predetermined load limit has increased from r to $r \cdot s$. That is, we prioritize considering whether each new job can be allocated to M_1 such that the current load not exceeds rs. The details of this approach are given in Algorithm B.

Algorithm 2: B

1 Initially, let $j = 1$, $L_1^0 = L_2^0 = 0$.
2 When a new job J_j comes,
3 **if** $L_1^{j-1} + p_j \leq r \cdot s$ **then**
4   Assign the job J_j to the fast machine M_1.
5 **else**
6   **if** $C_1^{j-1} + C_2^{j-1} > 0$ **then**
7   Assign the job J_j to the machine with the largest remaining capacity.
8   **else**
9   Assign the job J_j to the machine with the smallest current load.
10 If there is another job, let $j = j + 1$, go to **Line 2**. Otherwise, stop.

In Theorem 3 below, we determine the competitive ratio of Algorithm B, which is a function $r(s) = (\sqrt{9s^2 + 10s + 1} - s - 1)/2s$ dependent on the speed s of the fast machine.

Theorem 3. *Algorithm B has a competitive ratio of* $r(s) = \frac{\sqrt{9s^2+10s+1}-s-1}{2s}$, *which is bounded by a constant value of* $\sqrt{5} - 1$.

Proof. Similarly, we only need to prove for two cases.
Case 1 If $L_1 > s$ and $L_2 < 1$, then $X^A = s + L_2$. We consider two cases.
Case 1.1 If $s < L_1 \leq r \cdot s$, we can deduce that $X^*/X^B \leq r$ in the same way as in Theorem 2.
Case 1.2 If $L_1 > r \cdot s$, let J_l be the last job assigned to M_1. Clearly, the assignment of job J_l to M_1 is executed in Line 7, which implies that the machine M_1 currently has the largest remaining capacity. Therefore, we can deduce that

$$s - L_1^{l-1} \geq 1 - L_2^{l-1}. \tag{10}$$

Since we have assumed that the size of each J_j satisfies $p_j \leq 1$ in the beginning of this section, we can infer that both M_1 and M_2 have been assigned some jobs before the job J_l is assigned. Similarly, we can derive

$$L_1^{l-1} + L_2^{l-1} > rs. \tag{11}$$

by (10) and (11), we derive a lower bound for L_2^{l-1}, which is $L_2^{l-1} \geq (rs-s+1)/2$. Since $L_2 \geq L_2^{l-1}$, we have

$$\frac{X^*}{X^B} \leq \frac{s+1}{s+L_2} \leq \frac{s+1}{s+\frac{rs-s+1}{2}} = \frac{2s+2}{rs+s+1}. \tag{12}$$

Case 2 If $L_1 < s$ and $L_2 > 1$, then $X^A = 1 + L_1$. We also consider two cases. **Case 2.1** If $1 < L_2 \leq r$, we can deduce that $X^*/X^B \leq r$ in a similar manner. **Case 2.2** If $L_2 > r$, let J_l be the last job assigned to the machine M_2. Similarly, this implies that the machine M_2 currently has the largest remaining capacity when assigning J_l to M_2. Therefore, we can deduce that

$$1 - L_2^{l-1} \geq s - L_1^{l-1}. \tag{13}$$

Similarly, we can infer that both M_1 and M_2 have been assigned some jobs before the job J_l is assigned, and we can obtain

$$L_1^{l-1} + L_2^{l-1} > rs. \tag{14}$$

Otherwise, the other jobs on M_2 (excluding J_l) should be assigned to M_1, which leads to a contradiction. By the inequalities (13) and (14), we can obtain a lower bound for L_1^{l-1}, which is $L_1^{l-1} \geq (rs+s-1)/2$. Since $L_1 \geq L_1^{l-1}$, we have

$$\frac{X^*}{X^B} \leq \frac{s+1}{1+L_1} \leq \frac{s+1}{s+\frac{rs+s-1}{2}} = \frac{2s+2}{rs+3s-1} \leq \frac{2s+2}{rs+s+1}. \tag{15}$$

Similarly, for each $s \geq 1$, $\max\{(2s+2)/(rs+s+1), r\}$ reaches its minimum value when $(2s+2)/(rs+s+1) = r$. Let $(2s+2)/(rs+s+1) = r$ and solve it to obtain

$$r = \frac{\sqrt{9s^2+10s+1}-s-1}{2s} \geq 1.$$

In summary, we can conclude that the competitive ratio of Algorithm B is $r(s) = (\sqrt{9s^2+10s+1} - s - 1)/2s$. □

6 Conclusion

This paper focuses on the online budget allocation maximization (BAM) problem on two uniform machines, M_1 and M_2, with speeds $s \geq 1$ and 1, repectively. We propose an online algorithm with a competitive ratio of $(\sqrt{s^2+10s+9}-s-1)/2$, which is bounded by a constant 2 and reaches $\sqrt{5}-1$ when $s=1$. For a special case, the $\text{BAM}_{\leq 1}$ problem, we devise an online algorithm with a competitive ratio of $(\sqrt{9s^2+10s+1}-s-1)/2s$, which is no greater than $\sqrt{5}-1$. In the future, we will explore the semi-online and offline versions of the BAM problem.

References

1. Chen, X., Sterna, M., Han, X., Blazewicz, J.: Scheduling on parallel identical machines with late work criterion: off-line and online cases. J. Sched. **19**, 729–736 (2016)
2. Chen, X., Kovalev, S., Liu, Y., Sterna, M., Chalamon, I., Blazewicz, J.: Semi-online scheduling on two identical machines with a common due date to maximize total early work. Discret. Appl. Math. **290**, 71–78 (2021)
3. Sterna, M., Czerniachowska, K.: Polynomial time approximation scheme for two parallel machines scheduling with a common due date to maximize early work. J. Optim. Theory Appl. **174**, 927–944 (2017)
4. Chen, X., Wang, W., Xie, P., Zhang, X., Sterna, M., Blazewicz, J.: Exact and heuristic algorithms for scheduling on two identical machines with early work maximization. Comput. Ind. Eng. **144**, 106449 (2020)
5. Jiang, Y., Guan, L., Zhang, K., Liu, C., Cheng, T., Ji, M.: A note on scheduling on two identical machines with early work maximization. Comput. Ind. Eng. **153**, 107091 (2021)
6. Chen, X., Liang, Y., Sterna, M., Wang, W., Blazewicz, J.: Fully polynomial time approximation scheme to maximize early work on parallel machines with common due date. Eur. J. Oper. Res. **284**(1), 67–74 (2020)
7. Li, W.: Improved approximation schemes for early work scheduling on identical parallel machines with a common due date. J. Oper. Res. Soc. China **12**, 341–350 (2024)
8. Xiao, M., Liu, X., Li, W., Chen, X., Sterna, M., Blazewicz, J.: Online and semi-online scheduling on two hierarchical machines with a common due date to maximize the total early work. Int. J. Appl. Math. Comput. Sci. **34**(2), (2024)
9. Xiao, M., Bai, X., Li, W.: Online early work maximization problem on two hierarchical machines with buffer or rearrangements. In: International Conference on Algorithmic Applications in Management, pp. 46–54 (2022)
10. Wei, Q., Wu, Y., Cheng, T. C. E., Sun, F., Jiang, Y.: Online hierarchical parallel-machine scheduling in shared manufacturing to minimize the total completion time. J. Oper. Res. Soc. 1–23 (2022)
11. Li, J., Li, W., Li, J.: Polynomial approximation schemes for the max-min allocation problem under a grade of service provision. Discrete Math. Algorithms Appl. **1**(3), 355–368 (2009)
12. Li, W., Ou, J.: Approximation schemes for parallel machine scheduling to maximize total weighted early work with a common due date. Nav. Res. Logist. **72**, 454–464 (2025)
13. Li, W., Yang, Y., Xiao, M., Chen, X., Sterna, M., Blazewicz, J.: Scheduling with a discounted profit criterion on identical machines. Discret. Appl. Math. **367**, 195–209 (2025)
14. Epstein, L., Noga, J., Seiden, S., Sgall, J., Woeginger, G.: Randomized on-line scheduling on two uniform machines. J. Sched. **4**(2), 71–92 (2001)
15. Ng, C.T., Tan, Z., He, Y., Cheng, T.E.: Two semi-online scheduling problems on two uniform machines. Theoret. Comput. Sci. **410**(8–10), 776–792 (2009)
16. Dosa, G., Speranza, M.G., Tuza, Z.: Two uniform machines with nearly equal speeds: unified approach to known sum and known optimum in semi-online scheduling. J. Comb. Optim. **21**, 458–480 (2011)
17. Dósa, G., Fügenschuh, A., Tan, Z., Tuza, Z., Węsek, K.: Tight lower bounds for semi-online scheduling on two uniform machines with known optimum. CEJOR **27**, 1107–1130 (2019)

18. Lu, X., Liu, Z.: Semi-online scheduling problems on two uniform machines under a grade of service provision. Theoret. Comput. Sci. **489**, 58–66 (2013)
19. Tan, Z., Zhang, A.: A note on hierarchical scheduling on two uniform machines. J. Comb. Optim. **20**(1), 85–95 (2010)
20. Xiao, M., Liu, X., Li, W.: Semi-online early work maximization problems on two hierarchical uniform machines with partial information of processing time. J. Comb. Optim. **46**, 21 (2023)
21. Lin, L., Tan, Z.: Online scheduling on parallel machines: a survey. Scientia Sinica Mathematica **50**(9), 1183–1200 (2020)
22. Dwibedy, D., Mohanty, R.: Semi-online scheduling: a survey. Comput. Oper. Res. **139**, 105646 (2022)

Author Index

F. V. Fomin and M. Xiao (Eds.): COCOON 2025, LNCS 15983, pp. 359–361, 2026.
https://doi.org/10.1007/978-981-95-0215-8

The manufacturer's authorised representative in the EU is Springer Nature Customer Service Centre GmbH, Europaplatz 3, 69115 Heidelberg, Germany. If you have any concerns regarding our products, please contact ProductSafety@springernature.com

Printed and bound by CPI Group (UK) Ltd, Croydon, CR0 4YY
01/05/2026
02101055-0007